Residual Media

Residual Media

CHARLES R. ACLAND, *Editor*

 University of Minnesota Press

Minneapolis

London

Copyright 2007 by the Regents of the University of Minnesota

"Vinyl Junkies and the Soul Sonic Force of the 12-Inch Dance Single"
copyright 2007 by Hillegonda C. Rietveld

All rights reserved. No part of this publication may be reproduced, stored in a retrieval system, or transmitted, in any form or by any means, electronic, mechanical, photocopying, recording, or otherwise, without the prior written permission of the publisher.

Published by the University of Minnesota Press
111 Third Avenue South, Suite 290
Minneapolis, MN 55401-2520
http://www.upress.umn.edu

Library of Congress Cataloging-in-Publication Data

Residual media / Charles R. Acland, editor.
  p. cm.
 Includes bibliographical references and index.
 ISBN-13: 978-0-8166-4471-1 (hc : alk. paper)
 ISBN-10: 0-8166-4471-3 (hc : alk. paper)
 ISBN-13: 978-0-8166-4472-8 (pb : alk. paper)
 ISBN-10: 0-8166-4472-1 (pb : alk. paper)
1. Mass media and technology. 2. Technology—Social aspects. 3. Mass media and culture. I. Acland, Charles R., 1963–
P96.T42R47 2007
302.23—dc22   2006026092

Printed in the United States of America on acid-free paper

The University of Minnesota is an equal-opportunity educator and employer.

14 13 12 11 10 09 08 07   10 9 8 7 6 5 4 3 2 1

To Lillian Ava

Contents

	Acknowledgments	xi
	Introduction: Residual Media *Charles R. Acland*	xiii

PART I Mechanics of Obsolescence

1	Embedded Memories *Will Straw*	3
2	Out with the Trash: On the Future of New Media *Jonathan Sterne*	16
3	Falling Apart: Electronics Salvaging and the Global Media Economy *Lisa Parks*	32
4	New Lamps for Old: Photography, Obsolescence, and Social Change *Michelle Henning*	48

PART II Residual Uses

5	"Automatic Cinema" and Illustrated Radio: Multimedia in the Museum *Alison Griffiths*	69
6	Vinyl Junkies and the Soul Sonic Force of the 12-Inch Dance Single *Hillegonda C. Rietveld*	97
7	Reporting by Phone *Collette Snowden*	115
8	Vaudeville: The Incarnation, Transformation, and Resilience of an Entertainment Form *JoAnne Stober*	133

PART III	Collecting and Circulating Material	
9	Every Home an Art Museum: Mediating and Merchandising the Metropolitan *Haidee Wasson*	159
10	Recovering a Trashed Communication Genre: Letters as Memory, Art, and Collectible *Jennifer Adams*	185
11	The Celebration of a "Proper Product": Exploring the Residual Collectible through the "Video Nasty" *Kate Egan*	200
12	Going Analog: Vinylphiles and the Consumption of the "Obsolete" Vinyl Record *John Davis*	222

PART IV	Media, Mediation, and Historiography	
13	Neglected News: Women and Print Media, 1890–1928 *Maria DiCenzo and Leila Ryan*	239
14	The New Techno-Communitarianism and the Residual Logic of Mediation *James Hay*	257
15	Unearthing Broadcasting in the Anglophone World *James Hamilton*	283

PART V	Training, Technology, and Modern Subjectivity	
16	The Musicking Machine *Jody Berland*	303
17	Mississippi MSS: Twain, Typing, and the Moving Panorama of Literary Production *Lisa Gitelman*	329
18	Streamlining the Eye: Speed Reading and the Revolution of Words, 1870–1940 *Sue Currell*	344

| 19 | The Swift View: Tachistoscopes and the Residual Modern
Charles R. Acland | 361 |

Contributors 385
Index 389

Acknowledgments

A hearty cheer of gratitude goes out to the contributors, all of whom made the task of editing an agreeable and rewarding experience. Many others helped me with critiques and suggestions as this project developed. Marty Allor, Anne Balsamo, Tim Brennan, Monika Kin Gagnon, Keya Ganguly, Ron Greene, Larry Grossberg, Keir Keightley, Gil Rodman, Matt Soar, Kitty Scott, Johanne Sloan, and Peter van Wyck were especially influential, and I thank them all.

In the Department of Communication Studies, Concordia University, Montreal, I have a uniquely energizing environment made up of fine colleagues and students. I enjoy a razor-sharp testing ground for new ideas there, and I thank all concerned. The Conjunctures Working Group acts as an equally exacting arena for exchange, one from which this project benefited greatly. Presentations I gave as part of the Harold Innis Commemorative Panel (University of Toronto, 2002) and at the Visual Knowledges Conference (University of Edinburgh, 2003), along with the subsequent discussion at each event, helped shape and direct this research.

The Social Science and Humanities Research Council of Canada and Concordia University provided research funding for this project.

The people at the University of Minnesota Press offered an ideal combination of encouragement and professional guidance. At different stages of production, Doug Armato, Jason Weidemann, Kathy Glidden, and Andrea Kleinhuber provided expert and welcome advice. I thank the people who read and reviewed this project as a proposal and manuscript; their labor made this a better book.

Peter Lester supplied excellent research assistance, for which I cannot thank him often enough.

I would like to issue special appreciation to my parents, Joan and Derek. Over the years they have shown how serious intellectual pursuit and debate can be part of the mortar that sustains a close family. I am indebted to my brothers, Stephen and Bruce, and my sisters-in-law, Meghan Lauber and Leah Acland, who create a warm environment from which I am able to draw support and inspiration. I also send my thanks to Anne Acland in whom I have a model of independent and sustained engagement with history and culture.

The most resounding note of gratitude goes to Haidee Wasson, who offers her generous intellect and her profound understanding to all things scholarly, emotional, and quotidian. More than anyone else, she provides me with an essential blend of insight and humor, for which I am infinitely thankful and adoring.

INTRODUCTION
Residual Media

Charles R. Acland

In Honoré de Balzac's *Colonel Chabert,* a Napoleonic soldier, long thought to be deceased, returns to upset the world that had settled in his absence. The old man relates a wild tale of how he survived the battlefield, arousing suspicion about the authenticity of his identity and claims to "his" wife's fortune. The veteran tells of being mistakenly buried in a mass grave, of digging his way through the corpses with a severed arm as an improvised shovel, and of the blood of soldiers, or perhaps a piece of horse skin, that helped close his head wound, allowing him to recover. Upon returning to Paris, he encounters disbelief and bureaucratic insensitivity. Although Colonel Chabert is a legendary figure, the man who claims that identity is merely a reminder of past heroics and atrocities, a memory that troubles a now peaceful social order. And with this disruption, he finds, "I've been buried beneath the dead, but now I'm buried beneath the living."[1]

Among many things, Balzac's narrative captures how surprising and unsettling figures from the past can be. They creep up to remind us of their existence and of the influence they wield in the present. For an era such as ours that puts a premium on advancement and change above all else, declarations of the presence of the past can be confusing or alarming. There is nothing like that old party pooper "historical consciousness" to dull the gleeful celebrations of progress and the new. Yet, regardless of the efforts to keep the fete alive, the surfacing of the past is inevitable. Indeed, it is essential. Accordingly, and parallel to *Colonel Chabert*'s themes, Michel de Certeau proposes that historiography is partly a way for the young to encounter and become acquainted with the dead.[2]

Confronting the historical traces that reside in "the new" need not be as troubling as the colonel's reappearance. Consider, for instance, the Trachtenburg Family Slideshow Players, a band originally from Seattle whose performances of calculatedly naive music include a slide show of family photographs found at flea markets and estate sales. With a couple and their daughter as members of the trio, they construct a sweet atmosphere of familial bliss. Deploying the immediately recognizable amateur snapshot accentuates this atmosphere. With its arrangement of furniture, seating, and darkened viewing conditions, the home slide show is mostly imagined or remembered now. By referencing it, the Trachtenburg Family Slideshow Players signal an unspecified past of the 1950s or 1970s, of one's own family or someone else's. In this harkening, the band conjures up an idea of youth. Their music and performances rest on collective ideas of a time past, that is, on an *impression of childhood* as expressed in simple sounds, occasions of family commune, and media that served as the alibi for purportedly happy gatherings. Whether or not anyone ever enjoyed this benign contentment, this act elicits a desire to imagine such comforts.

The Trachtenburg Family would be just another oddball irony-laden musical act were it not a part of the broader appeal of popular pleasures that revel in the rediscovery of vintage artifacts and styles.[3] Among the many examples of this secondhand retrospective sensibility is Toronto's Hidden Cameras, an ensemble playing a hybrid of Belle and Sebastian–style art rock music, inflected with a thrift-store, low-fi aesthetic. The retrieval and revalorization of discarded artifacts and fashions can also be seen in venues like *Found Magazine* and their Web site foundmagazine.com, to which people contribute Polaroids, utility bills, audio fragments, cryptic personal pleas, and a variety of other paper ephemera from shopping lists to threats scratched on notepaper. This venue builds on the lasting qualities and abundance of paper, sound recording, and photography, all of it repackaged for Web or magazine display. More than a virtual cabinet of curiosities, curated selections tour regularly as a mobile museum and party. *Found Magazine* captures the wonderment of having discovered expressive scraps, versions of which people ordinarily stumble across. Their public posting betrays a prideful appeal for acknowledgment of discovery—"Hey, look what I found." *Found Magazine* is not alone. In his chapter, Will Straw reads this site alongside another "rediscovery" service, longlostperfume.com, and Jonathan Sterne has discussed elsewhere National Public Radio's *Lost and Found*, which broadcasts unusual retrieved sound recordings.[4]

What are we to make of such gestures, these fond references to things and sentiments that won't stay lost, dead, and buried? Certainly, the ironic sensibility evident in these illustrations summons an innocent—but aware—view to the world, or what might be called a knowing unknowingness. Susan Stewart's *On Longing* elaborated on the desire to remember through the objects we elect to save and treasure long past their initial use-value has been expended. Her argument about souvenirs, collections, and scale asserts that nostalgia is "a sadness without an object," a "desire for desire," and a yearning for authentic presence.[5] Stuart Tannock challenges arguments that pit memory against nostalgia. Instead of treating nostalgia with suspicion, he sees in it a productive, contemporary structure of feeling, one that "invokes a positively evaluated past world in response to a deficient present world"[6] and one that does not necessarily express conservative longing. Expanding Fred Davis's argument in *Yearning for Yesterday*,[7] Tannock emphasizes the search for historical continuity lodged in nostalgic gestures, and he suggests that such expressions work to identify historical rupture.

The nostalgia for the vintage and the underappreciated cannot be seen apart from the very nature of material objects in consumer society. The proliferation of media artifacts—for instance, the vinyl records studied by Hillegonda C. Rietveld and John Davis in their respective chapters—is unambiguously a part of the everyday of modernity, their circulation and cheapness extending from market forces. Accumulated artifacts do more than gather dust and lie cluttered and neglected in storage facilities. They can become an essential part of secondary markets, from garage sales to antique collectibles, altering both commercial and semiotic value, as has been investigated by such disparate figures as Jean Baudrillard and Arjun Appadurai.[8] Things operate in circuits of value, which themselves can be spatially located and temporally varying. Dear treasures in one part of town may be garbage, and laughable, in another. Indeed, Pierre Bourdieu noted that there is perhaps no more telling indication of cultural capital than the ability to bestow value where there had previously been presumed to be none.[9] Art markets play an important role here, and artists have long reorchestrated their practices in relation to discarded materials and antiquarian techniques.[10] Similarly, connoisseurs of all types, cherishing everything from sheet music to snow domes, convert modes of appreciation and expertise into an organization of taste and value.

Cultural producers and connoisseurs are not alone here. We all confront the sheer abundance of artifacts that make up the visible layers of modern life. Noting this, Marc Augé forcefully argues that an essential quality of what he calls "supermodernity" is the excess of documentation and the rapid dissemination of events, lending heightened visibility to what are taken as instances of historical consequence and placing the most fleetingly mundane of incidents alongside revolutions. Augé writes, "History is on our heels, following us like our shadows, like death,"[11] adding to the morbid analogies just cited above. Events that are meaningful to vast populations escalate in number. The result, he reasons, is the acceleration of history, which poses special problems for historians for whom an "excess of time" makes the demand for general coherence of historical narrative and interpretation a near impossibility.

Evident in this overproduction, then, is an economic and cultural orientation toward novelty and innovation. This orientation helps sustain a gargantuan accumulation of materials, to be followed by increasingly intricate modes of accommodating the leftovers. These accumulating and accommodating markets contain views to cultural change itself. They are sensibilities that offer ways of encountering and experiencing freshness in cultural life via a relationship with its enduring elements and a retrieval of familiar signs and objects. In 2002, the Museum of Modern Art gift shop, in New York, carried a T-shirt for infants, emblazoned with a bright red and yellow starburst design, declaring in bold lettering, simply, NEW, jokingly presenting the bearer as a newly launched product (and everyone else "old" by default). Without this clear orientation toward newness, we appear to be lost. A character in *28 Days Later* (Danny Boyle, 2002) describes their apocalyptic situation by lamenting that they will never read a book that has not already been written, never hear music that has not already been recorded, never see a film that has not already been shot. So, although the new holds special appeal, it nevertheless betrays a concern about the past. Put differently, among the lasting and the novel we find an orientation between the past and present, rather like a compass. We hear of things described as "so five minutes ago." *Entertainment Weekly* offers an occasionally witty chart declaring the relative currency of cultural practices (e.g., "out—personal leave, five minutes ago—sabbatical, in—early retirement").[12] And think of the mobility of the fashion assessment "gray is the new black," variations of which have infiltrated all manners of metaphoric inversion ("Thursday is the new Saturday"; Jon Stewart declared, referring to the 2003–4 television season's excessive number of shows about the wealthy, "Rich is the new gay"; a Health Canada report on the rising seriousness of obesity pronounced, "Fat is the new tobacco").

Importantly, beyond an obsession with the "new" as it elbows its way alongside the ever-expanding "old," the material conditions within which we live with the past have changed. The fact that things and practices hang around long past their supposed use-by date confounds conventional ideas of historical progress, including how we make room for the new. Among the scholarly responses, Michael Thompson's "rubbish theory" opened up a sociological investigation of how waste figures in changing attributions of value in a culture of consumption.[13] The lasting presence of culture through various storage media has come to the foreground of media theory with the debates around the work of Friedrich Kittler.[14] Equally invested in the storage dimension of media, Greg Urban has attempted to construct an anthropological model for cultural change, seeing "newness" as one of the dominant metacultural discourses of modernity. He pays special attention to what he calls the "ceramicized" aspect of culture, alluding to the inscription of stories on pottery. This material, and hence exterior and potentially public element, corresponds to an interior and immaterial sense of affiliation, that is, the shared in culture. He sees the operation of cultural replication—carrying versions of stories forward through time—as being overtaken with dissemination—a metacultural valuation of the new, carrying versions through space.[15] Decades earlier, Harold Innis worried about the lasting time bias of media succumbing to the "present-mindedness" of space biased media, adding that a tilting toward spatial control is a characteristic of the growth of empires. In general, there has been a growing recognition of pathfinding work of figures like Harold Innis and James Carey,[16] of which Collette Snowden's chapter is one example. And in the past decade, the full force of Siegfried Kracauer's Weimar essays has been felt on the writing of history in English-language intellectual circles, directing us toward the fragments of texts, places, ornaments, and commodities that amass through time and that provide a critical vantage point for access to the characteristics of modernity.[17]

On this, Walter Benjamin has offered one of the most sustained investigations, and his influence is unmistakable in several chapters to follow, including the research by Michelle Henning, Jennifer Adams, and Kate Egan. He felt the modern dynamic of commodity circulation produced a unique experience of temporality, such that the "confrontation of the most recent past with the present is something historically new."[18] Susan Buck-Morss draws out Benjamin's special understanding of the role of modern childhood. In reference to *The Arcades Project,* she writes, "The world of the modern city appears in these writings as a mythic and magical one in which the child Benjamin 'discovers the new anew' and the adult Benjamin recognizes it as a rediscovery of the old."[19] The detritus of capital and commodity serve the dual purpose of announcing their own historicity and residing as a standing reserve, as Heidegger might have put it, for conversion into subsequent artifacts, memories, and stories. As commodities become a type of raw material once again, Benjamin alerts us to the transformative possibilities implied. As Buck-Morss indicates, Benjamin's generation "not only recognizes its own youth, the most recent one, but also an earlier childhood speaks to it. It is thus as a 'memory trace' that the discarded object possesses the potential for revolutionary motivation."[20]

That revolutionary kernel may be rarely realized, but still, the discarded object carries a semiotic richness ripe for appropriation. For instance, as schools dispose of outmoded instructional technology, an "AV-geek" culture has emerged, in which the low-tech clunkiness of vintage forms like filmstrips and educational films achieves an unexpected fan status.[21] In addition to the hardware of projectors and such, piles of discarded celluloid film prints are retrieved as found-footage films, as material for classroom editing assignments, and with a newfound historical significance. The Library of Congress acquired Rick Prelinger's collection of previously devalued industrial and classroom films in 2002. Anthology Film Archives began to present programs entirely made up of home movies.[22] Retrieval inspires a range of contemporary art practices. Rodney Graham's "Rheinmetall/Victoria 8" installation consists of a 35mm silent film that lovingly portrays a 1930s German typewriter, apparently acquired by the artist in a Vancouver junk shop, as fake snow falls and covers the keyboard. A 1950s Italian 35mm motion picture projector, situated in the middle of the gallery space, presents the film, with its noisy operation becoming the soundtrack, making the projector as much a part of the installation as the film. *Decasia* (Bill Morrison, 2002) presents found film footage along with whatever deterioration has occurred, so the splotches on the emulsion and the distortions of color and image seem almost lyrical. *Tribulation 99: Alien Anomalies under America* (Craig Baldwin, 1992) pieces together footage from disparate sources to tell a paranoid tale of contemporary U.S. history. Although versions of compilation films have been around at least since Soviet filmmaker Esfir Shub's reconstruction films of the 1920s, more recent found-footage films are notable for the way they explicitly announce their historicity. One cannot watch *Tribulation 99* or *Decasia*, see Rodney Graham's work, or attend a screening of amateur home movies without being alerted to the aged and aging qualities of the artifacts presented. The footage and hardware connote their status as "retrieved." The illustrations all point to the context of material accumulation and the related strategies of accommodation.

Alongside the material entwinement of the old and the new is a particular experience and understanding of the passing of time and historical change. The ineluctability of accumulation and selective retrieval means cultural analysis needs to focus on the unique and specific processes of transformation. A Hewlett-Packard advertising campaign, designed to announce the corporation's commitment to recycling, illustrates the relationship between the transformation of commodities and ordinary corporate operations.[23] This campaign is an effort to address the mounting landfill contributions stemming from the accelerating obsolescence of computer hardware, an issue taken up by Sterne and Lisa Parks in their respective chapters. As depicted in this print advertisement, we see a desktop monitor go through multiple blurry steps to become a bright red tricycle (Figure I.1). The process appears to yield an icon of childhood, with its mechanical organization of human energy, from the electronic totem of the information age. The visual representation of sequential transformation is curious. The graphics cascade down the page, visually constructing an ontological link between each endpoint through these intermediary hybrid computer-to-tricycle stages. The fact that only the final whole tricycle casts a shadow implies that the transformative process occurs in some immaterial realm, one that noticeably drags the corporate slogan along with it. Although HP's ad campaign is explicitly drawing attention to the newly produced children's toy, the conversion process is the graphically central item on the page. The advertisement is essentially about the very idea of transformation, here imagined as an illusionist's stunt to remake and reinvigorate the old. The formula, as the copy states, is "recycling + hp = everything is possible," intimating if we can make a tricycle out of a computer, imagine what else we could make.

The magical change represented here is like that of the digital morph. More than a pleasurable special effect ubiquitous in popular films, music videos, and television advertising, Vivian Sobchack argues that morphing is a resonant image of becoming.[24] As with the conventions of horror films, there is a fascination with the moments in between, as Dr. Jekyll becomes Mr. Hyde, as Larry Talbot becomes the Wolfman, as Irena Dubrovna becomes the panther. In the print advertising example, a magically transformative capitalism, lifted from the constraints of the material, assures a smooth movement from the old to the new, that is, a system arranged to deal with the main course and its leftovers. It may be a representation of becoming, yes, but it is also a visualization of the management of resources, brands, time, and aging (as well as a masking of the hand of labor).

This ideology of transformation and immateriality—that is, a language of capital's easy (re)production process—offers one explanation for the central place of "new media" in our social and critical context. For "new media" designates not only a motor and vehicle of historical change, but also a place at which we experience, in a direct fashion, the rapid obsolescing and remaking of things and skills, where typing becomes data entry, Windows Me becomes Windows XP, and movie outtakes, posters, and scripts become DVD extras. And yet, the casual and profuse usage of the term "new media" has left it vague and imprecise. At every turn—policy, budget, economic, pedagogic—we confront yet another triumphal proclamation of our age of freshly minted media trinkets. The mythologization of new media would not be surprising, but for the fact that it also appears among scholars and agenda setters who, well, should know better. There is often a shocking continuity between aesthetic avant-gardists and the glossy pages of free business inserts in daily newspapers. Reading the multitude of art and commerce manifestos for the variously identified *new* new media, you get the impression that mobility, miniaturization, decentralization, media self-referentiality, blurred

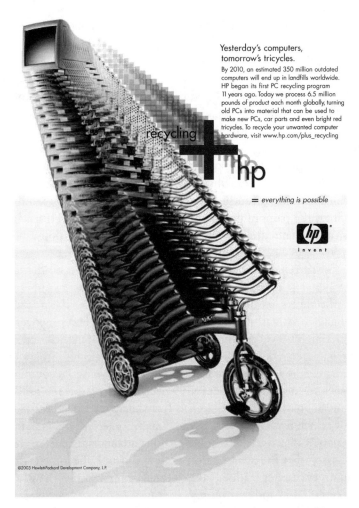

FIGURE I.1 *Hewlett-Packard print advertisement, "Everything Is Possible, Recycling," by Goodby, Silverstein & Partners, 2003. Courtesy of Hewlett-Packard.*

relations between copy and original, shifting boundaries between producer and consumer, and even the storage and transport of ideas apart from the singular human form appear only after 1994. In fact, I would go so far as to suggest that few phrases have been evacuated of meaning, and have outlived their critical usefulness, faster than "new media." If there is a reigning myth of media, it is that technological change necessarily involves the "new" and consists solely of rupture from the past. This preoccupation neglects the crucial role of continuity in historical processes as well as the accumulation and accommodation just described. It ignores the way the dynamics of culture bump along unevenly, dragging the familiar into novel contexts. This dynamic is precisely the focus of the present volume.

Residual Media addresses media and cultural history, bringing together studies of technology and cultural practice. It contributes a corrective to contemporary scholarship's fetishization of the "new." An inappropriate amount of energy has gone into the study of new media, new genres, new communities, and new bodies, that is, into the contemporary forms. Often, the methods of doing so have been at the expense of taking account of continuity, fixity, and dialectical relations with existing practices, systems, and artifacts. One response to this emphasis has been Bruce Sterling's, who productively and entertainingly assembled an online collection of "dead media." The research assembled here in *Residual Media,* however, might be better represented as studies of "living dead" culture.

This volume does not present a closed, singular, polemical approach to "residual media," and the contributions each investigate a variety of aspects of the concept. The residual can be artifacts that occupy space in storage houses, are shipped to other parts of the world, are converted for other uses, accumulate in landfills, and relate to increasingly arcane skills. How do some things—whether in archives or attics, minds or training manuals—become the background for the introduction of other forms? In what manner do the products of technological change reappear as environmental problems, as the "new" elsewhere, as collectables, as memories, and as art? What are the qualities of our everyday engagement with the half-life of media forms and practices and with the formerly state of the art? The interdisciplinary research included in this collection explores reconfigured, renewed, recycled, neglected, abandoned, and trashed media technologies and practices. The chapter authors combine theoretical challenges to media historiography with studies of specific residual ideas, activities, and materials. They ask, each from their own perspective, what the process is by which media, devices, texts, objects, spaces, and competencies fade away or persist. Collectively, the contributions advance a study of the aging of culture, asserting that the introduction of new cultural phenomena and materials rests on an encounter with existing forms and practices. The result is both material accumulation and ever more elaborate modes of accommodation.

Raymond Williams provides an often-cited conceptual apparatus for the examination of technology, ordinary cultural life, and the tone of a time and place. Especially inspirational for this volume, his attention to emergent, dominant, and residual forms has long alerted us to the varying forces of historical change and to the lacunae that result from examining any of the three in isolation of one another.[25] In appreciation of the system-in-process dimension of cultural materialism, Williams talked about temporal lags in cultural change, taking seriously "Marx's 1857 observation that the 'peaks' of art 'by no means correspond to the general development of society; nor do they therefore to the material substructure, the skeleton as it were of its organization.'"[26] Thus Williams is led to expand the base and superstructure model. Moreover, he was compelled to elaborate the categories of the residual and the emergent, which act as prefiguring formations of the not yet fully dominant. The emergent takes account of the fact "that new meanings and values, new practices, new significances and experiences, are continually being created."[27] Of the residual, Williams wrote, "I mean that some experiences, meanings and values, which cannot be expressed in terms of the dominant culture, are nevertheless lived and practised on the basis of the residue—cultural as well as social—of some previous social formation,"[28] using the persistence of religious values in a secular society as a case in point. He later rewrites the posthyphenated phrase as "some previous social and cultural institution or formation,"[29] and elaborates, "The residual, by definition, has been effectively formed in the past, but it is still active in the cultural process, not only and often not at all as an element of the past, but as an effective element of the present."[30] Lags in cultural change leave us with bundles of things, beliefs, and practices with varying temporalities, at different points in their own life cycle.

Williams guides us to this presence of the past. The residual formations' distance from the dominant, however razor thin at times, is the source of its oppositional potential, if only "because they represent areas of human experience, aspiration, and achievement which the dominant culture neglects, undervalues, opposes, represses, or even cannot recognize."[31] A clear, if complex, attempt to capture cultural change begins to emerge, in part, through his acknowledgment of the provisional nature of such designations. In keeping with the vital and unpredetermined nature of these categories, Williams provides a contrast with the more predictable *archaic,* which is "wholly recognized as an element of the past."[32] Note the wide net he casts in this discussion, one that is consistent with his expanded sense of the material; for Williams, the residual, emergent, or dominant can refer to experiences, practices, values, artifacts, institutions, and meanings. We are not just talking about computer hardware or wired communities!

The historical model here describes a materialist process, wherein ever-shifting social conditions arise from ever-present preexisting ones, as well as opportunities for struggle. In this relationship between the old and the new, the edges of class formation and class consciousness are made visible, and Williams saw this as part of the process of the reproduction of social relations. With the concepts of a living past and an unsettled future, we cannot expect to pinpoint which category is operating prior to analysis, and "there is indeed regular confusion between the locally residual (as a form of resistance to incorporation) and the generally emergent."[33] In this way, Williams illustrates how continuity and discontinuity both play roles in the structuring of a living culture, leaving behind any supposition about the simplistic, static, and well-defined boundaries of the province of the dominant, let alone the new. Accordingly, we might best wonder about those who speak with unqualified conviction of a uniformly new practice or form. In sum, Williams's dynamic model throws a wrench into the impression of the smooth dynastic sweep of social change, whatever the motivating force. And for the authors appearing in *Residual Media,* it highlights a suspicion about the overemphasis on media revolutions, as it is frequently an attention that drags along fixed equations of social formations, of which technological determinism is the most prevalent.

Others have begun to assess comparable approaches to cultural change. Jay Bolter and Richard Grusin have attempted to remedy some of the elisions in work on new media with their book on "remediation," providing a general theory about the interrelationship of the content and form of media, productively reprising the most familiar of McLuhanisms of the 1960s.[34] They present the two irreducible logics of remediation as the transparency of media (immediacy) and the exaggerated presence of media (hypermediacy). They assert when new media act as though they are transparent windows, they actually make other media more "present." Bolter and Grusin are part of a growing literature, and other works have critically explored the relations among media or the culture (or cult) of convergence.[35] This idea rings through the following chapters, most of which are not bound by a belief that media are singular and unified, conversely stressing intermedia connections.

Equally noteworthy is the burgeoning interest in the intersection between cultural and media history. Over the past few decades there has been an enlivening of cultural history with its attention to the everyday and the attendant relations of power.[36] Victoria Bonnell and Lynn Hunt typify this as part of a wider "cultural turn" taking place in the social sciences and humanities.[37] Clearly informed by cultural studies and social history, this turn valorizes "history from below" and the "thick description" of context, and it has been instrumental in extending the problematization of narrative and representation in historiography. Most assuredly, cultural history is less concerned with monolithic narratives or the history of events (*histoire événementielle*). Emphasis is instead on the connections between the micro- and macrohistorical—"a methodological swing . . . between close-ups and extreme long shots."[38] Yet any move between the detail and the panorama poses analytical quandaries. Bonnell and Hunt suggest that one of the lasting influences of the "cultural turn" has been a rethinking of causality, such that we "no longer assume . . . that causal explanation automatically traces everything cultural or mental back or down to its more fundamental components in the material world of economics and social relations."[39] As conceptually overwrought as the category of the everyday may be, such a project has made the history of everyday life come into focus, with special consideration given to the "view from below" that had been excluded from conventional historical narratives.

Having presented an enduring and interdisciplinary challenge to the priorities and methods of historiography, cultural history has been taken up by media scholars, despite what John Nerone describes as the "embarrassing" persistence of grand narrative tropes among media historians.[40] Maria DiCenzo and Leila Ryan trace some of the challenges that persist in this respect, and other chapters, including those by Haidee Wasson and Jody Berland, respond by drawing connections between media history and everyday life. Comparable cultural approaches to media history are evident in the examination of modernity in film and media studies, the scholarly ferment of what is loosely designated as the domain of visual culture, and the increasing critical attention to overlooked media and practices, such as home movies, panoramas, filing cabinets, wax museums, optical devices, stereoscopes, and so on.[41] A particularly influential example is *Cinema and the Invention of Modern Life,* through which students of cinema have reoriented themselves toward media history more broadly, expanding their study to include cities, newspapers, folk museums, and catalogs, to name but a few sites of analysis.[42] That collection also aided in the expansion of the historical purview of film historians, and in the process it took to task some core assumptions about periodization and the presumed newness of motion pictures in the late nineteenth century. In the present volume, Alison Griffiths, JoAnne Stober, and my own chapter are among the work that reflects this turn.

A top-notch collection that mirrors some of the present interests, *New Media, 1740–1915* challenges the myopic view that the "new" necessarily signifies "today," building on directions Carolyn Marvin advanced in her book, *When Old Technologies Were New*.[43] As with that earlier research, the essays in Gitelman and Pingree's collection zero in on the point of introduction of new media forms, where their uses and meanings have not yet been fixed and instead demonstrate both variability and possibility. The contributions by Parks, James Hamilton, Sue Currell, and Lisa Gitelman all develop comparable approaches.

These are just some of the theoretical and historical touchstones for the chapters in this volume. Given the rich conceptual legacy already in play, and in light of an expanding research investment in the "new" of past eras, it is striking how infrequently work has addressed the aging of media and cultural forms directly. It is generally acknowledged that the modern pursuit of technological progress has left us with an ever-intensifying valuation of the new, and a significant literature examines and qualifies this drive. And yet, conversely, a paucity of research has concentrated on the tenacity of existing technologies or on their related materials and practices that do not magically vanish with the appearance of each successive technology. For instance, PowerPoint software reinforces, rather than overturns, early twentieth-century approaches to visual instruction. Massive advertising for the "new media" of computers and cellular telephones has reinvigorated financially the "old media" of newspapers and broadcast television. Internet commerce could not operate without telephone and elaborate delivery services, themselves relying on organizations of transportation routes and container construction from another era.

The chapter lineup boasts research from, roughly, four overlapping interdisciplinary domains: media studies, film studies, cultural studies, and American studies. Together, the collection is defined not by a time period, or by specific media, but instead by an interest in the processes by which technological forms and related cultural practices age and are selectively revitalized. Some chapters examine neglected historical examples and forces. Others reevaluate familiar cultural forms. All explore the interrelationship of the old and the new in multiple media discourses, forms, and contexts. Obviously, this is not a comprehensive study of residual culture, but the contributions do represent an eclectic range of media forms: vinyl records, radio, desktop computers, television sets, telephones, sound cinema, antiquarian photography, discarded letters, video "nasties," newspapers, player pianos, typewriters, and mechanical reading devices. Further, they study a variety of practices: Internet shopping, computer disposal, planned obsolescence, recycling, vaudeville and musical performance, journalism, DJ mixing, collecting, writing, and speed reading.

Part I, "Mechanics of Obsolescence," includes research that contemplates the aging of media as well as the commercial, recycling, or artistic initiatives developed to manage material as it obsolesces. Will Straw examines the persistence of cultural value, as propelled by Internet and videotape markets, and shows that artifactual and semiotic abundance now typify a dominant cultural formation. Jonathan Sterne and Lisa Parks, focusing on computers and televisions, respectively, both dissect the culture of hardware obsolescence. Sterne takes computer "trash" as a vantage point from which to elucidate the very dynamics and economics of technological change; Parks sees "trashed" television as a meeting point for both devalued cultural forms and for the potentially disruptive reassembling of materials. Michelle Henning's chapter details shifts in image-making technologies and illuminates the role "old" techniques can play for artists. Doing so, she addresses the related political valence of symptoms of the past.

Part II, "Residual Uses," is made up of research on the practices of specific groups of professionals that arise through strategic deployments of old cultural forms. Alison Griffiths's chapter presents the long and confounding relationship between museums and multimedia, showing that the tensions between the two partly concern competing (and compromised) models of spectatorship. She demonstrates how museum practice, as a repository of the old and an assimilator of the new, can mold and be molded by media experiments. This leads to Hillegonda C. Rietveld's examination of the 12-inch single as a technological artifact, in which she concentrates on the inventiveness of DJs as they build musical styles and conventions on this supposedly out-of-date format. Similarly interested in professional practices built around "old" media, Collette Snowden elaborates on the century-plus entwinement of telephony and journalism. With this history, she challenges dominant presumptions about the revolutionary impact of new media forms in that location. JoAnne Stober winds up this part showing how some motion picture exhibitors, far from uniformly embracing sync-sound technology in the late 1920s, actually reintroduced the receding entertainment of live vaudeville. In so doing, they gave it one last spark of popularity before it left the public stage to find a new home on television.

Part III, "Collecting and Circulating Material," focuses on the everyday domestic assembly and archiving of objects. Haidee Wasson documents efforts to make the museum mobile through the mail-order sale of miniature reproductions of art museum collections and the development of museum gift shops. In this, she sees a rise in the material and discursive proximity between the home and the museum, giving old art a new life in the domestic environment. Jennifer Adams's chapter wonders about the life of saved love letters in the context of e-mail's disposability and its cheapening of intimate paper-bound written discourse. She reminds us, too, that arguments about virtual presence need not be solely the purview of cyberspace. Following this research are two chapters on collecting, both of which reveal the value that can accrue to illegitimate or "obsolete" cultural forms. Kate Egan studies collectors of rare horror videotape, and John Davis looks at record collectors, or "vinylphiles."

Part IV, "Media, Mediation, and Historiography," includes essays that investigate the writing of history and its role in providing access to previously neglected cultural forces or unrecognized influences, especially as they pertain to questions of community and the public sphere. The chapter by Maria DiCenzo and Leila Ryan draws our attention to significant blind spots in media history that have resulted from a reluctance to mine the feminist press of the suffragette movement. Their argument demonstrates how a more complete examination of the historical record shifts the very location and terms of public discourse and political agency. James Hay picks up on this theme, diving into some of the reigning and purportedly novel ideas about communities and politics, ideas whose longevity has effectively disguised their ideological import. Hay begins by addressing plans for a postoccupation Iraq, and he links this to taken-for-granted discourses of communitarianism that reside in work on new media. In a wider historical sweep, James Hamilton traces out a genealogy of broadcasting, from agricultural to vanguard to commercial formations, giving voice to its association with a particular set of ideas about individuals and publics. All three chapters make it clear that the historical connections they discuss have a residual presence in contemporary political life.

The echoes of historical continuity ring through the concluding grouping of research. Part V, "Training, Technology, and Modern Subjectivity," examines the shaping of bodies and skills in relation to technology. The chapters in this part address the technological interface as one link between new and old media, as well as one that establishes a bodily and institutional memory even as the source technology fades away. Jody Berland looks to the player piano and sees a precursor to more recent complications concerning performance and authenticity found in debates around digital music, and she provides a study of how media change is understood. On a parallel track, Lisa Gitelman explores changing concepts of authorship and the materiality of text evident in the early years of the typewriter. Sue Currell's chapter chronicles the changing place of reading in the context of the developing domain of experimental psychology and new educational technology. My own contribution complements Currell's research, focusing on an experimental instrument and accelerated learning machine, the tachistoscope. In this final chapter, I chart the tachistoscope's movement from the psychology laboratory to the classroom, arguing that it brought with it a configuration of priorities and assumptions about modern life and experience as it adapted to different institutional settings.

Although not an encyclopedic project, together the chapters in all of these parts mark out key issues facing cultural and media studies, offering insight into our understanding of the life cycle of the technologies, ideas, and practices in our midst, whether trashed or treasured. Ideally, this research will assist in a reinvigoration of how we approach cultural and media history. And I hope, in some fashion, this work will move us in the direction Benjamin set out for us: "Every epoch, in fact, not only dreams the one to follow but, in dreaming, precipitates its awakening. . . . With the destabilizing of the market economy, we begin to recognize the monuments of the bourgeoisie as ruins even before they have crumbled."[44]

Notes

Thanks goes to Haidee Wasson, who generously provided commentary on this chapter.

1. Honoré de Balzac, *Colonel Chabert*, trans. Carol Cosman (New York: New Directions, 1832/1997), 26.

2. Michel de Certeau, *The Writing of History* (New York: Columbia University Press, 1975/1988), 1.

3. Christopher Noxon, "I Don't Want to Grow Up," *New York Times*, August 31, 2003, sec. 9, 1.

4. Jonathan Sterne, *The Audible Past: Cultural Origins of Sound Reproduction* (Durham, N.C.: Duke University Press, 2002), 349.

5. Susan Stewart, *On Longing: Narratives of the Miniature, the Gigantic, the Souvenir, the Collection* (Durham, N.C.: Duke University Press, 1993), 23–24.
6. Stuart Tannock, "Nostalgia Critique," *Cultural Studies* 9, no. 3 (1995): 454.
7. Fred Davis, *Yearning for Yesterday: A Sociology of Nostalgia* (New York: Free Press, 1979).
8. Jean Baudrillard, *The System of Objects*, trans. James Benedict (New York: Verso, 1968/1996); Arjun Appadurai, ed., *The Social Life of Things: Commodities in Cultural Perspective* (Cambridge: University of Cambridge Press, 1986). For a more recent study, see Will Straw, "Exhausted Commodities: The Material Culture of Music," *Canadian Journal of Communication* 25 (2000): 175–85.
9. Pierre Bourdieu, *Distinctions: A Social Critique of the Judgement of Taste,* trans. Richard Nice (Cambridge: Harvard University Press, 1979/1984), 5.
10. See, for instance, *October* 100 (Spring 2000), a special issue on obsolescence.
11. Marc Augé, *Non-Places: Introduction to an Anthropology of Supermodernity,* trans. John Howe (New York: Verso, 1992/1995), 26–27.
12. Jessica Shaw, "The Shaw Report," *Entertainment Weekly,* April 23, 2004, 16.
13. Michael Thompson, *Rubbish Theory: The Creation and Destruction of Value* (Oxford: Oxford University Press, 1979).
14. Friedrich A. Kittler, *Discourse Networks, 1880/1900,* trans. Michael Metteer with Chris Cullens (Stanford: Stanford University Press, 1985/1990); and *Gramophone, Film, Typewriter,* trans. Geoffrey Winthrop-Young and Michael Wutz (Stanford: Stanford University Press, 1986/1999).
15. Greg Urban, *Metaculture: How Culture Moves through the World* (Minneapolis: University of Minnesota Press, 2001).
16. Discussions of this lasting impact can be found in Eve Stryker Munson and Catherine A. Warren, eds., *James Carey: A Critical Reader* (Minneapolis: University of Minnesota Press, 1997); and Charles R. Acland and William J. Buxton, eds., *Harold Innis in the New Century: Reflections and Refractions* (Montreal: McGill-Queen's University Press, 1999).
17. Siegfried Kracauer, *The Mass Ornament: Weimar Essays,* trans. Thomas Levin (Cambridge: Harvard University Press, 1995).
18. Walter Benjamin, *The Arcades Project,* trans. Howard Eiland and Kevin McLaughlin, ed. Rolf Tiedemann (Cambridge: Belknap Press of Harvard University Press, 1982/1999), 328.
19. Susan Buck-Morss, "Dream World of Mass Culture: Walter Benjamin's Theory of Modernity and the Dialectics of Seeing," in *Modernity and the Hegemony of Vision,* ed. David Michael Levin (Berkeley: University of California Press, 1993), 328.
20. Buck-Morss, "Dream World of Mass Culture," 327.
21. See, for instance, North Carolina–based http://www.avgeeks.com.
22. Lawrence Levi, "Lowly Home Movies Get a Day as High Art," *New York Times,* August 18, 2003, B1, B5.
23. Advertisement "HP + Recycling," *New York Times,* July 30, 2003, A7.
24. Vivian Sobchack, "'At the Still Point of the Turning World': Meta-Morphing and Meta-Stasis," in *Meta-Morphing: Visual Transformation and the Culture of Quick-Change* (Minneapolis: University of Minnesota Press, 2000), 131–58.
25. Raymond Williams, *Marxism and Literature* (New York: Oxford University Press, 1977); and "Base and Superstructure in Marxist Cultural Theory," in *Problems in Materialism and Culture* (New York: Verso, 1973/1980), 31–49.
26. Williams, *Marxism and Literature,* 79.
27. Williams, "Base and Superstructure in Marxist Cultural Theory," 41.
28. Ibid., 40.
29. Williams, *Marxism and Literature,* 122.
30. Ibid.
31. Ibid., 123–24.
32. Ibid., 122.
33. Ibid., 125.
34. J. D. Bolter and Richard Grusin, *Remediation: Understanding New Media* (Cambridge: MIT Press, 1999).
35. Anna Everett and John T. Caldwell, eds., *New Media: Theories and Practices of Digitextuality* (New York: Routledge, 2003); Henry Jenkins and David Thorburn, eds., *Rethinking Media Change: The Aesthetics of Transition* (Cambridge: MIT Press, 2003); and William Boddy, *New Media and the Popular Imagination: Launching Radio, Television, and Digital Media in the U.S.* (Oxford: Oxford University Press, 2004).

36. Peter Burke, ed., *New Perspectives on Historical Writing*, 2nd ed. (University Park: Pennsylvania State University Press, 2001); Roger Chartier, *Cultural History: Between Practices and Representations*, trans. Lydia G. Cochrane (Ithaca, N.Y.: Cornell University Press, 1988).

37. Victoria E. Bonnell and Lynn Hunt, eds., *Beyond the Cultural Turn: New Directions in the Study of Society and Culture* (Berkeley: University of California Press, 1999).

38. John Czaplicak, Andreas Huyssen, and Anson Rabinbach, *New German Critique* 65 (Spring/Summer 1995): 11; see Helmut Lethen, "Kracauer's Pendulum: Thoughts on German Cultural History," *New German Critique* 65 (Spring/Summer 1995): 37–45.

39. Bonnell and Hunt, *Beyond the Cultural Turn*, 26.

40. John Nerone, "Theory and History," *Communication Theory* 32 (May 1993): 153.

41. A sample of work that advances the intersection of cultural and media history includes Richard Cavell, *McLuhan in Space: A Cultural Geography* (Toronto: University of Toronto Press, 2002); Jonathan Crary, *Suspensions of Perception: Attention, Spectacle, and Modern Culture* (Cambridge: MIT Press, 2000); Alison Griffiths, *Wondrous Difference: Cinema, Anthropology, and Turn-of-the-Century Visual Culture* (New York: Columbia University Press, 2002); Tom Gunning, "An Aesthetic of Astonishment: Early Film and the (In)Credulous Spectator," in *Viewing Positions: Ways of Seeing Film*, ed. Linda Williams (New Brunswick, N.J.: Rutgers University Press, 1989/1997), 114–33; Erkki Huhtamo, "From Kaleidomaniac to Cybernerd: Notes Toward an Archaeology of Media," *Leonardo* 30, no. 3 (1997): 221–24; Keir Keightley, "You Keep Coming Back Like a Song: Adult Audiences, Taste Panics, and the Idea of the Standard," *Journal of Popular Music Studies* 13 (2001): 7–40; Michelle Martin, '*Hello Central?*': *Gender, Technology and Culture in the Formation of Telephone Systems* (Montreal: McGill-Queen's University Press, 1991); Anna McCarthy, *Ambient Television* (Durham, N.C.: Duke University Press, 2001); Lauren Rabinovich and Abraham Geil, eds., *Memory Bytes: History, Technology, and Digital Culture* (Durham, N.C.: Duke University Press, 2004); Sterne, *The Audible Past*.

42. Leo Charney and Vanessa Schwartz, eds., *Cinema and the Invention of Modern Life* (Berkeley: University of California Press, 1995).

43. Lisa Gitelman and Geoffrey B. Pingree, eds., *New Media, 1740–1915* (Cambridge: MIT Press, 2003); and Carolyn Marvin, *When Old Technologies Were New* (Oxford: Oxford University Press, 1988). See, too, Lisa Gitelman, *Always Already New: Media, History, and the Data of Culture* (Cambridge: MIT Press, 2006).

44. Walter Benjamin, "Paris, the Capital of the Nineteenth Century," in *The Arcades Project*, 13.

PART I

Mechanics of Obsolescence

1 Embedded Memories

Will Straw

In 2000, the *Wall Street Journal* reported on the success of the commercial Web site longlostperfume.com.[1] Longlostperfume.com specializes in selling fragrances from twenty or thirty years ago, fragrances that it replicates and advertises on the Internet. (The company's motto is "Perfume beyond the touch of time.") These fragrances are copies of perfumes—such as Afrodesia or Deneuve—that long ago disappeared from the shelves of retail stores. The scents of older perfumes, we are told in the *Wall Street Journal* article, were stronger and more distinct than those of perfumes popular today. As a result, they now appeal to a dwindling group of consumers. Longlostperfume.com successfully pulls together isolated customers for these perfumes, assembling a market sufficiently large to warrant their ongoing production.

This example invites us to explore the Internet's relationship to a cultural past that it reinvigorates and invests with value. The relationship of the Internet to the past is typically talked about in terms of remediation, a process by which new media come to enclose the old.[2] The emphasis in remediation theory is often on the persistence of the old amid the new, in analyses committed to demonstrating the stubborn survival of perceptual frames, affective attachments, or ideological pre-givens. The Internet has, in some fashion, enclosed within itself old and forgotten perfumes, but it obviously does not (or does not yet) transmit or otherwise carry their smells. It does, however, provide the preconditions for their perpetuation as material culture (as liquids in bottles), as sensory artifacts, and as marketable commodities. The Internet does this in two ways. One is that, binding together otherwise isolated interests, it reconstitutes viable markets from market fragments, an often-commented-on feature of Internet commerce. At the same time, the Internet, like other media with virtually unlimited storage capacity, provides the terrain on which sentimental attachments, vernacular knowledges, and a multitude of other relationships to the material culture of the past are magnified and given coherence. Longlostperfume.com has accomplished this in interesting ways. It has revitalized certain perfumes from the past, pulling them into a present marked by their simultaneous availability. At the same time, however, the agglomeration of older perfumes on its Web site allows us to imagine a "back catalog" of perfumes, to understand perfumes as historical phenomena in the way that we now understand reissued musical recordings or films.

A significant effect of the Internet, I would argue, is precisely this reinvigoration of early forms of material culture. It is not simply that the Internet, as a new medium, refashions the past within the languages of the present, so that vestiges of the past may be kept alive. Like most new media, in fact, the Internet has strengthened the cultural weight of the past, increasing its intelligibility and accessibility. On the Internet, the past is produced as a field of ever greater coherence, through the gathering together of disparate artifacts into sets or collections, and through the commentary and annotation that cluster around such agglomerations, made possible in part by high-capacity storage mechanisms. That we can think of lost perfumes as the objects of collective interest has much to do with the role of the Internet in creating spaces that magnify the significance of such phenomena, making even the most trivial objects the focus of a popular but highly ordered knowledge. Longlostperfume.com makes manifest the Internet's paradoxical relationship to physical artifactuality. It has pulled, from obscure places of domestic or industrial storage, artifacts whose value and intelligibility it manages to restore.

The Internet is clearly not a place in which old fragrances are stored, but neither is it sufficient to say that longlostperfume.com is merely "about" old fragrances or simply a place in which they are described or announced. The Internet is a container for old fragrances in the sense that, within it, these fragrances are given solidity as a category of artifacts, made to persist and to interact with other cultural phenomena.[3] With these other phenomena—with the retro fashion photographs that adorn the longlostperfume Web site, for example—the fragrances resonate to form clusters of cultural knowledge. The Internet carries and transports such clusters as the substance of its own cultural weight and authority. In one respect, the Internet is a distinctive apparatus, producing its own characteristic forms of knowing (particular balances of text, image, and sound). At the same time, by drawing together innumerable forms of scholarly and vernacular textuality, the Internet has become a repository for wide varieties of knowledge that have predated it: the rhetorics of old fandoms, folksy family genealogies, film buff checklists, and so on. Around something as minor as old perfumes, the Internet has gathered together the resources (old photographs, personal reminiscences, and the logos of now forgotten companies) that pull old objects into the limelight of cultural recognition and understanding. The Internet becomes a comfortable place, less because we are adept at handling its futuristic openness than because it has rendered the already familiar all the more coherent and solid. Paradoxically, that coherence and solidity serve to naturalize versions of the past, whose artifacts, we believe, the Internet has merely discovered.

We might think of all media forms as institutions marked by distinct ratios between their capacities for accumulation and the sorts of mobility they enable. In his book *Metaculture*, the anthropologist Greg Urban argues that culture is not something to which movement happens. Rather, he suggests, culture derives from movement, drawings its character from distinctive relationships between stasis and movement, inertia and acceleration.[4] *Metaculture* is one of several theoretical resources we might mobilize in thinking about media technologies and their role in perpetuating a cultural past. The inertial forces that interest Urban might be specified using concepts developed elsewhere, in writings on urban space and domestic collections. In a book of architectural theory entitled *Anytime*, the Turkish authors Mennan, Kutukcuoglu, and Yazgan offer productive new foundations for the study of cultural temporalilties. "Any culture produces a system of delays," they write. "This system secures the integrity of a culture, which by definition builds upon a set of procedural and behavioral codes that transform it into a machine for delay."[5] Mennan et al. go on to discuss cities as machines for delay, and there is much that is convincing here. The porousness of city life and the multiplicity of urban spaces within which practices may hide and persist—all of these slow or block the drives toward obsolescence and displacement that are more famously taken to characterize urban modernity. Any culture's system of delays may be tied, as well, to what Janelle Watson has called its "mode of accumulation": the manner in which, within that culture, artifacts are "stored, displayed and disposed of."[6] The relationship between a culture's system of delays and its mode of accumulation is easy to imagine. Both ideas suggest the presence of blockages and weighty buildup around which culture must travel in order to move forward.

Let us pursue these ideas through consideration of the video store, one of the most banal and familiar of media institutions. The video store has reshaped cultural time in at least two ways. Expanding the availability of films from the past, it has acted as a drag of sorts on the forward movement of cinematic culture, slowing the disappearance and commercial obsolescence of films as they pass out of their theatrical runs. In this, the video store is an inertial force. At the same time, however, it contributes to the acceleration of first-run film culture. It does so, in part, through the ways in which it has helped reduce the typical run of a film's theatrical release, and so increased the rate of turnover in commercial cinemas. More insidiously, though, the video store has nourished the first-run cinema by ensuring that the background knowledges that new films require as the basis of their intelligibility are perpetually accessible.

In 1988, Barry London, president of marketing and domestic distribution for Paramount Pictures, noted in an interview with the *New York Times* that older people were finally coming back to the movies.[7] Over the course of the late 1970s and 1980s, movie industry lore told us, people older than fifty had retreated from the practice of seeing films in theatrical release. This disengagement from first-run moviegoing was seen as a reaction to new, alienating moral standards and stylistic languages within Hollywood cinema. With time, the argument went, this estrangement had become self-perpetuating. Gradually, older audiences had lost familiarity with successive generations of new stars or with the cross-referencing and generic patterns that made new films intelligible. The longer older people stayed away from the cinema, of course, the more unintelligible the current cinema became. By the mid-1980s, the first-run film industry was seen as dominated by teenage moviegoers, whose tastes in films had come to dictate industry decision making.

At the end of the 1980s, however, this trend appeared to be reversing itself. Mike Medavoy, vice president of Orion Pictures, told the *New York Times* that moviegoing by people over fifty had recently risen by 50 percent (although older adults remained a minority within the larger moviegoing audience).[8] Invited to explain this shift, Barry London suggested that the film industry's reconciliation with an older audience was the result of two technologies that the industry had long accused of shrinking movie attendance. "Older people have been re-exposed to movies on video cassettes and cable," London claimed, "and now that they've seen most of those library movies, they're going out to see new movies."[9] Let us pass over London's ludicrously literal image of older people venturing out to the first-run cinema only in that desperate moment when the supply of older films on tape or cable had been completely exhausted. The more significant insight here is that the videocassette has played a role in reuniting audiences with a cinema that they had long believed they could no longer understand. The videocassette carried the aesthetic styles and moral economies of contemporary films into the homes of older audiences. There, older people might test and resuscitate their comfort with current cinema, in circumstances that involved little risk.

Accurate or not, London's analysis invites us to think of the videocassette as a technology through which cultural knowledges are stored and transported. By the late 1980s, the videocassette was disseminating films to larger numbers of viewers than were first-run cinemas. We are used to imagining this dissemination in geographical terms, but London's example points to the demographic distances that the videocassette had been able to traverse. It brought films into age groups from which the normal habits of moviegoing might otherwise have kept them. Something more than the discrete experience of any given film is disseminated in the renting of that film, however. The return of older people to cinemas in the late 1980s suggests the rented videocassette served more broadly as a pedagogical tool that accomplished a reduction of cultural distance. It disseminated those knowledges and protocols of reading through which older people might reacquaint themselves with the contemporary cinema.

In the age of the videocassette, new films in theatrical release displaced those that had come before them, as they always had, but older films piled up behind new releases to an extent unknown in the days before home video. In this accumulation, older films served as perpetually available tools of instruction for anyone wishing to renew a contact with the contemporary cinema. The cultural effects of the videocassette have been discussed primarily in terms of the repetition of unitary experiences that it permitted, for the manner in which it transformed our possible relationships to any given film. We might begin to see the videocassette as functioning more broadly as a tool of orientation, as an instrument for cultural way-finding. In the sequences of their release and consumption, videocassettes solidify the process by which older films enhance the readability and public resonance of those that come after them. They permit those whose relationship to current cinema has been disrupted or weakened to find their way back to that cinema.

Charles Acland has pointed to the multiple temporalities that had emerged in the film industry by the late 1980s as the release of first-run films often served to announce the video release that would follow. Audiences increasingly "see the current cinema as largely a rough catalogue of future cultural consumption," a means of orienting themselves to the sequence of future releases on video or to television.[10] More broadly, the passage of audiovisual materials across virtually all media over the last two decades may be seen to create multiple vectors of pathfinding across the cinematic field. In the analysis of media technologies, the challenge here is that of developing an account of such pathfinding that operates between two extremes, between, on the one hand, recognition of the specific promotional role played by distinct texts in relationship to each other and, on the other, the banal acknowledgment that films become intelligible within a broadly dispersed intertextual field. At an intermediary level, we may explore the variety of ways in which cultural knowledges are absorbed within particular texts, embedded within media forms and transported across sites of consumption.

The videocassette contains textual forms that are the expressive residues of more broadly based cultural knowledges and aesthetic languages. These residues allow the videocassette to serve as an instrument of instruction for those untrained in the protocols of the cinema around them. Like any container, the videocassette may serve to both transport and stockpile the cultural knowledges held within it. It will transport these across geographical and demographic boundaries, and, through such transportation, contribute to the mobility of contemporary cultural life. At the same time, in stockpiling those knowledges, the videocassette, like any medium of storage, allows them to pile up and to persist. If only because it provides a perpetual presence for the no longer new, this storage may block processes of innovation or commercial turnover within the cultural field. Some of that blockage is evident in the perpetuation of stale careers, to which we turn in a moment. It may be found in the persistence of genres such as the erotic thriller, which required the stabilization of star persona and generic conventions across multiple examples to acquire a generic identity and air of semilegitimacy.

For those older adults whose return to first-run filmgoing was noted in the late 1980s, the videocassette had served to smooth the transitions of a history that might otherwise have remained marked by significant, disruptive discontinuities of form or tone. The videocassette and the video store helped reorder that history in ways we might imagine in cartographic terms. The vast quantity of movies circulating on videocassette by the late 1980s offered multiple points of entry to the current cinema. Videocassettes provided numerous bridges or mediating series of texts through which a generation gap within the filmgoing audience might be traversed. These routes might be circuitous—taking viewers, perhaps, from *Driving Miss Daisy* (Bruce Beresford, 1989) to *Born on the Fourth of July* (Oliver Stone, 1989), and from there on to—who knows?—*Drugstore Cowboy* (Gus Van Sant, 1989) or *sex, lies, and videotape* (Steven Soderbergh, 1989). At the same time, the multiplicity of such routes and strings of films had made recent cinematic history sludgier, more porous, and, it might be said, more spatial. The sense of a unidirectional sequence of new releases, pulling further and further away from the sensibilities of an older audience, gave way to a sense of cinematic culture as much more densely populated, offering multiple points of contact and pathways through successive films.

Another example from film industry discourse is pertinent here. In the early 1990s, industry observers noted that the typical careers of actors and actresses seemed longer in the age of the videocassette than had previously been the case. This observation came amid industry complaints that actor salaries had become too high—that even the stars of seemingly endless series of flops could command higher salaries than executives felt they merited. Actors like Charlie Sheen, Harvey Keitel, or Mickey Rourke, whose careers were often marked by failures, seemed to always live to act another day.[11] They maintained the status of celebrity within the public mind through one failure after another because their names continued to resonate with the public. The sense that they still possessed an unchallenged celebrity persisted. In large measure, a film industry journalist suggested, this was because films that failed at the box office no longer passed from public view, rendering their performers invisible through that failure. The videocassette had become an institution of cultural memory. Like photos of missing persons, videocassette versions of films kept failing actors within the public eye. A pattern of ongoing disappearance had given way to a logic of accumulation, through which failed movies and their performers piled up on videotape and within the inventories of video stores. Rigorous displacement over time had given way to sedimentation and piling up across space. It is not simply that videotape and the video store might give a second commercial life to failed films, as often happened—famously, in some cases, like that of *The Shawshank Redemption* (Frank Darabont, 1994). More subtly, failed films and failing stars retained their visibility through the videocassette, which ensured that they were glanced over by customers at each visit to the video store. Home video, in this account, had slowed the disappearance of films and of their stars, maintaining them within a long-term if not permanent visibility and availability. At the very least, the failed theatrical film, now out on video, remained a carrier for the iconography of its lead actors and its own title. Even when no longer rented with any frequency, it sat in a relation of partial contiguity to more successful recent films, in the visual and artifactual landscape of the video store, and derived at least minor legitimacy from this.

As a technology, the videocassette was often imagined in terms of its lightness and velocity of travel. This fueled the fantasy that the rise of the videocassette would expand the pluralism of cinematic culture because marginal films could move more easily and with reduced investment around the world and across a variety of viewing contexts. Harold Innis, of course, would ask us to examine the variety of ways in which that same lightness contributed to the solidification of film industry structures and patterns of dominance.[12] It is not simply that the videocassette extended Hollywood's reach into demographic or geographical corners that might have resisted it. The piling up of videocassettes, within video store inventories and entertainment culture more broadly, allowed for the knitting together of multiple lines of association between films—lines of association that, most of the time, served to enhance the intelligibility and reduce the strangeness of any single one. The multiple mazelike sequences through which people rent sequences of films will almost certainly lead them, at various points, into the center of Hollywood film culture, and the knowledge learned there will lead them out across other lines of connection. If the themes and stars of failed Hollywood films still seem to resonate, however weakly, within the busy semiosis of the video store, this enhances the legitimacy and legibility of any new film that might employ those themes or stars. One result of this, since the mid-1980s, has been a weakening of processes of obsolescence as new films have resuscitated others backward in time or across the markers of shared personnel and genre.

The videocassette's capacity to reorder the continuities between films was first noted with the success of sequels like *Terminator 2*, whose first-run audiences were greater than those for their predecessors. The first *Terminator* film was a hit in its first-run release, of course, but its circulation on video helped more broadly to circulate the knowledges on which the appeal and intelligibility of its sequel depended. This back-and-forth reinvigoration of films old and new still left room on the margins for independent or international films with unknown stars or unfamiliar themes. Arguably, however, the weaving together of most of the others within compact relations of mutual reference and interconnection has reinforced the sense that cinematic culture has a center. Every videocassette or DVD transports the particular text that marks its distinctiveness, but each, as well, transports and stores sets of cultural knowledges that may be mobilized in the viewing of other texts. This is a commonplace of cultural analysis but might be profitably reworked within a theory of media storage and transportation. As Greg Urban notes, new texts reinvigorate elements of the surrounding cultural texture in ways we are often invited to see as novel, as initiating cultural movement. Each such mopping up, however, represents a storage and retransmission of these clusters of meaning, a reassertion of their cultural presence and authority.[13]

Roughly halfway into the film *Austin Powers: Goldmember* (Jay Roach, 2002), a long scene transpires within a mid-1970s New York disco. In this scene, the 1970s blaxploitation figure Foxy Cleopatra kibitzes with 1960s British spy figure Austin Powers. Their interplay is mediated by the Broadway actor Nathan Lane, whom, it is assumed, large numbers of audience members will recognize. The scene ends as Austin ventures into a private room at the club, to meet his father, played by Michael Caine in a send-up of his own image from *Alfie* (Lewis Gilbert, 1966) and from a number of mid-to-late 1960s British spy films. Like the film as a whole, this scene is chronotopic, in the sense that it enacts a particular joining of historical time to geographical place. The Austin Powers films participate in a widely dispersed labor of imagining, which, across dozens of films and musical revivals, has cast the period of the mid-to-late 1960s as essentially British, or at least most emblematically so. The mid-1970s, likewise, have come to take shape, within a broadly shared cultural sensibility, as essentially American and urban and funky.

The smooth intelligibility of the scene I have just described is easily explained in terms of an intertextual background that has settled into U.S. cinema over the last two or three decades. This background has been strengthened with each James Bond movie, which revives interest in the series as a whole and stimulates the recirculation of iconography from that series. It has been solidified in large measure through the regular parodies, remakes, and revivals of 1970s blaxploitation films we've seen over the past twenty years. There is an ever-increasing thickness to these renderings of the 1960s and 1970s, it seems to me—a sense that the overlaying of films and other texts that cast these periods in particular ways has made it difficult to move around them through alternate routes. Even Roman Coppola's attempt to reimagine the late 1960s as French, in his film *CQ* (2001), could not escape the temptation to flirt with Austin Powersisms.

A common way of talking about the *Goldmember* excerpt would invoke a theory of readability. The intelligibility of *Austin Powers: Goldmember* presumes a set of cultural knowledges that have taken shape elsewhere, which each viewing of *Goldmember* is intended to evoke. I would like to desubjectify this process, however—to move it away from a theory of reading and to look at the film itself as a container for cultural knowledges. *Austin Powers: Goldmember* serves as a storage device, through which a set of historical references is held, delivered to various places, and allowed to occupy cultural space. In particular, the film gathers up a flurry of recent ways of remembering the mid-1970s and warehouses them, then sets them down in the middle of a global popular cultural field. In so doing, it renews their cultural centrality and reinvigorates the value of the texts from which they are drawn. The film is not simply the recapitulation of these ways of remembering; it is a device for carrying them forward in time and, through the film's own monumental success, extending the cultural space that they occupy.

There are worse ways of imagining the mid-1970s, of course, and I will happily take that offered in *Goldmember* over one that installed Cat Stevens or Ryan O'Neal at its center. The movement of blaxploitation culture to the heart of a collectively imagined mid-1970s has involved time and cultural labor, and it is as welcome for being African American as it is suspicious for being so exclusively American. Nevertheless, I have watched similar films and sequences in Brazil, where colleagues noted the displacement of their own, quite different imaginings of the 1960s and 1970s—imaginings that have been given cultural form only in relatively isolated, disconnected texts, such as *City of God* (Katia Lund, Fernando Meirelles, 2002). This cultural logic operates all the more strongly when a film like *Goldmember* accumulates within the video store, no longer disappearing—to give way to new renderings of the 1960s and 1970s—but accumulating alongside other films that circle back on each other to render their own version of these historical periods weighty and durable. *Goldmember*, in this sense, is a kind of weighty delay, set up like a roadblock against transformative processes of cultural reimagining. We do not need the technology of the DVD or the videocassette for *Goldmember* to do this work—for it to set down a densely packed reading of the mid-1970s around which alternative readings have to tiptoe and circle. The *Austin Powers* series, nevertheless, is a group of films whose original points of reference—swinging London in the 1960s—were relatively obscure for the audiences at which they were directed (and the first film in the series performed weakly at the box office). Through the back-and-forth viewing of originals and sequels, however, it has drilled a set of historical and cultural reference points deep into the cultural field.

The inertia that marks these reference points is the result in part of two ways in which new media technologies, such as the videocassette or DVD, have reshaped cultural temporalities. In the first place, these technologies have set in place a spatialization of audiovisual culture. Within this spatialization, the principal relationships between texts are those of interreference and mutual support rather than succession and displacement. Secondly, the enormous storage capacity of the video store lets *Goldmember* take semantic nourishment, for years and years, from the *Pulp Fictions*, blaxploitation parodies, and different reiterations of itself that linger on and regularly resuscitate each other. This effect is different from that claimed in many analyses of "recombinant culture," which find, in processes of pastiche and juxtaposition, evidence of the random and fleeting character of cultural citation.[14] On the contrary, one result of the processes just described is that new media come to be characterized by a sluggishness, a lumpy weight. This sluggishness stems, in part, from the tendency for new media artifacts to come embedded within durable material forms that accumulate, rather than succeeding each other as events (as did the broadcasting of old films on television from the 1950s through the 1970s). This is true as much of the Web site as of the DVD or CD. It is not simply that, in their durability, new media artifacts pile up and, in so doing, increase the dense overlayering of all the artifacts available at any one time. It is also that, as I noted with respect to the video store, the circuits of reference that bind one artifact to another reverse chronologies or cross sequences of development in a way that muddies any sense of historical time. Processes of obsolescence, through which artifacts might regularly disappear, so that novelty within the cultural field might be perceptible, have been slowed.

At the same time, it is common for each new example of such media to possess higher bandwidth than that which preceded it: from vinyl album to CD, videocassette to DVD, and from all of these to the Internet. Of the many things said about this expansion of bandwidth, one of the least acknowledged is the extent to which it has altered the weight of the cultural past. New technologies do not simply take over from earlier technologies, but, increasingly, they build on a previous technology's work of collection or agglomeration. As Robert Cantwell notes, the vinyl LP gathered up dispersed examples of folk music or vocal jazz, helping produce the coherence and historical weight of these styles by allowing them to resonate with each other on the space of the long-playing record.[15] CD reissues continued this process of absorbing and reordering musical history within an abundant, simultaneous availability. The Internet perpetuates and magnifies this even further.

The effects of these processes warrant extended consideration. They have given us innumerable collections of music, films, or written textuality from the past, which, in their careful arrangement and elaborate annotation, have served to historicize these cultural forms in almost fetishistic ways. Each successive technology has enhanced the significance of older cultural artifacts by allowing them to be joined to others that clarify and embellish them, in packages more and more characterized by extensive textual—almost curatorial—documentation. At the same time, the backfilling of a weighty and finely differentiated past has diminished the ripples that any instance of present-day novelty is able to set off. As a result, historical sequence has been slowed and muddied. The most interesting effects of increased bandwidth, then, are these: the consolidation of historical archives that goes on within new media, the mopping up of dispersed texts from the past into collections of music or film, the expansive annotation that embeds such collections within complex knowledges, and so on. In all these processes, the weight of the past relative to the present seems to increase.

We might map this as a global process in which the gathering up and convergence of cultural artifacts in places of storage and annotation produces particular clusters of cultural authority and weight. In the global recording industry, the fate of national musical heritages is at play here. The CD, the retail superstore, the decline of tariffs, and other factors have all rendered national musical markets more open to an abundance of transnationally circulating musical forms. This openness threatens to diminish the weight of national musical traditions within abundant inventories of music from elsewhere. At the same time, however, in countries like Brazil and Mexico, dispersed catalogs of small or regional recording companies within these countries have been brought together within the reissue strategies of a few large companies that have bought them up. Boxed sets and annotated compilations of 1930s Brazilian samba or 1960s tropicalia now embed these musics within packages marked with scholarly prestige and characterized by a semiotic and physical weightiness. New storage media, like the CD or the Internet, are creating weighty, meaningful clusters of agglomerated cultural artifacts, clusters whose grounding in region or nation is often striking. The same processes have given weight to the packages that arrive from a cultural center—the box sets of *Sopranos* seasons arriving from the United States, for example—but they have also created various kinds of cultural ballast in these other countries. Regional or national cultural expression is gathered up in the boxed sets of Brazilian telenovelas or reissues of Shaw Brothers films from Hong Kong.

The CD, the DVD, and the Internet are technologies whose "newness" should not obscure their role in renewing the economic and semiotic life of older artifacts. Reinvigorating the past, and slowing down processes of obsolescence, new technologies have consistently rendered the past more richly variegated and dense. This has reduced the sense of relentless change that might otherwise have marked their introduction and dissemination. Indeed, as is sometimes claimed, the reliance of each new media technology on repertory from the past may be one of the features of new media that reduce their potentially intimidating novelty. At the same time, these pasts are put together with variable solidity and weight. New media, of course, will contain the music or films of a multitude of cultures. The solidity and perceived significance of given kinds of music or cinema, however, will have much to do with the lines of interconnection that flow between them across various media forms—the annotation, commentary, and packaging that accompany them as they circulate.

There is another dimension to these processes that merits comment, however. Longlostperfume.com reminds us that a key effect of storage media is the convergence of various sensory experiences within the realm of the visual. The Internet reassembles artifacts from the past within a space of simultaneous availability, but the locus of that simultaneity is the visual (and only very rarely the audiovisual). In the processes just described, the perfumes of the past become part of the visual culture of the present, inasmuch as their presence on the Internet has taken the form of graphics, photographs, and texts. Just as the Internet has renewed the economic value of innumerable cultural artifacts from the past (from the high school yearbooks sold on eBay through the Old-Time Radio programs available on Web sites), so it has hastened their convergence on the realm of the visible. In the realm of the visible, they come to be adorned with textual commentary and forms of graphic display that have become the basis of their public presence. On the Internet, aural, tactile, and other forms move toward the visual as the manner of their self-announcement. Textual artifacts from the past, similarly, migrate toward the gridlike graphic visuality of the Web page.

The visual has become the mode of public presence of cultural forms in their commodity or memorial state. This is true for music, written textuality, and so on. This was long true of the department store catalog, for example, as it was of the encyclopedia and of the paintings through which we acquaint ourselves with the material culture of the past. This is one of the conditions that allowed the surrealists, the situationists, foundmagazine.com, and innumerable other movements to fix on the sublimely disorientating character of the found object. The visual bias implicit in this idea of being found is what allows all these movements to imagine cultural artifacts, originally marked by distinctive forms of sensual expressivity, as occupying space in roughly equivalent ways. The Internet, with its graphic renderings or annotations of perfumes, Pez machines, 1960s soul music, or forms of ballroom dancing perpetually reinstalls the visual as the realm of historical cultural understanding. This is all the more paradoxical in the case of longlostperfume.com: an archive whose principle of selection is the intensity of odor cannot convey this odor in its public presentation.

Arguably, new media diminish the fetishistic properties of the old, through the simultaneous and ongoing availability that they permit and through the pacifying forms of annotation and commentary that have come to surround even the most lurid or passionate of cultural artifacts. At the same time, however, new media forms train us to make the connections through which the coherence of historical styles comes to be recognized. It is in the congealing of such styles that fetishism makes its return. In the second volume of his *Recherches*, Marcel Proust writes the following:

> All the products of one period resemble one another; the artists who illustrate the poetry of their generation are the same artists who are employed by the big financial houses. And nothing reminds me more strongly of the installments of *Notre-Dame de Paris* and of various works of Gèrard de Nerval, that used to hang outside the grocer's door at Combray, than does a registered share in the Water Company.[16]

The affinities that Proust notes, however, between stock certificates and illustrations to Nerval, are not inevitable. They would only be noted in a culture whose abundance of accumulated and discarded artifacts allowed the passage of time to be noted in deeply sedimented and richly resonating clusters of objects. We need large inventories of such objects in order that they may knit together within densely intertextual packages, so that the affinities among the hairstyles, record album jackets, movie posters, and childhood toys of, say, 1985, start to assume a historical solidity. New media, in the ordered ways by which they gather together historical artifacts and thus endow them with historical weight, are perpetually producing the past in various forms of coherence.

It is here, though, that global political-economic analysis of these processes seems warranted. In the journal *October*, Christian Thorne argued that the retro fascination with postwar lounge culture, ratpack style, and diner breakfasts was an "unabashedly nationalist project" that sought to create "a distinctively U.S. idiom, one redolent of Fordist prosperity, an American aesthetic culled from the American century, a version of Yankee high design able to compete, at last, with its vaunted European counterparts."[17] The solidification of these motifs and practices within a weighty and coherent cultural sensibility takes shape through the dialogue between cultural artifacts, which, I suggested earlier, is part of the spatializing mutual sustenance through which such artifacts speak to each other, in the video store or on the Internet. Retro, Christian Thorne suggests, "is the form that national tradition takes in a capitalist culture."[18] The Internet, like the video store, is about abundance, but it is also about the inertial movements that bring commodities and images together into clusters and networks whose solidity decides their cultural and symbolic weight. This weight is one of the stakes in local, regional, and national struggles over cultural authority.

Notes

1. Robert Johnson, "Fragrances of Yesteryear Are Back, in Supermarkets and on the Web," *Wall Street Journal*, November 2, 2000, B1.

2. See, for the major statement on remediation, David J. Bolter and Richard Grusin, *Remediation: Understanding New Media* (Cambridge, Mass.: MIT Press, 1999), esp. p. 19.

3. A paper by Jonathan Sterne on the MP3 helped focus my interest in the "container" character of media, an intermittent concern, as Sterne reminds us, within the theoretical work of Lewis Mumford and others, and one that has found further elaboration in the work by Kittler and others on the storage activity of media. I am grateful to Jonathan for this inspiration, and he bears no responsibility for any misuse. Jonathan Sterne, "Cultural Origins of Digital Audio: The Case of the MP3," public lecture, McGill University, Department of Art History and Communications Studies, January 23, 2004.

4. Greg Urban, *Metaculture: How Culture Moves Through the World* (Minneapolis: University of Minnesota Press, 2000), 6.

5. Zeynep Mennan, Mehmet Kutukcuoglu and Kerem Yazgan, "title," in *Anytime*, ed. Cynthia C. Davidson (New York and Cambridge, Mass.: Anyone Corporation and MIT Press, 1999), 71.

6. Janelle Watson, *Literature and Material Culture from Balzac to Proust: The Collection and Consumption of Curiosities* (Cambridge: Cambridge University Press, 1999), 40.

7. Aljean Harmetz, "Hollywood Catches a Wave," *New York Times*, January 20, 1988, C15.

8. Ibid., C15. See also Aljean Harmetz, "Hollywood Pays Court to the Young Adult," *New York Times*, May 13, 1990, C15.

9. Aljean Harmetz, "Hollywood Catches a Wave," *New York Times*, January 20, 1988, C11.

10. Charles Acland, *Screen Traffic: Movies, Multiplexes and Global Culture* (Durham, N.C.: Duke University Press, 2003), 65.

11. Doug Desjardins, "Video Stars," *Video Store*, October 30, 1994, 15.

12. See, for example, Daniel Drache, ed., *Harold Innis, Staples, Markets and Cultural Change: Selected Essays* (Montreal and Kingston: McGill-Queen's University Press, 1995), 366 and passim.

13. Urban, *Metaculture*, 61.

14. See Todd Gitlin, "Flat and Happy," in *Media in America*, ed. Douglas Gomery (Washington, D.C., and Baltimore: Woodrow Wilson Center Press, 1998), 222–32.

15. Robert Cantwell, *When We Were Good: The Folk Revival* (Cambridge, Mass.: Harvard University Press, 1996), 36.

16. Marcel Proust, *Within a Budding Grove*, trans. C. K. Scott Mongrieff and Terence Kilmartin (New York: Modern Library, 1992), 34.

17. Christian Thorne, "The Revolutionary Energy of the Outmoded," *October* 104 (Spring 2003): 103.

18. Ibid.

2 Out with the Trash: On the Future of New Media

Jonathan Sterne

Upon arrival at my new office in 1999, I found it waiting for me: a state-of-the-art ca. 1982 300-baud modem. The unit was filthy and connected to wires for interfaces that had long ago fallen into disuse. The power cable worked, and the red LEDs lit up to signal the unit's readiness when I plugged it in. So naturally, I unplugged it again and put it in a storage cabinet. There was no practical use for the device in 1999 and its only use today is as an example. The modem was a relic or an artifact but it was useless for its original purpose. It was obsolete. So too were the old Pentium I computer and monitor I wedged between a metal cabinet and the ceiling later that year, in anticipation that someone would come and take them away. They waited for the entire five years I occupied that office.

For academics like me, the acquisition of a new computer at work is, to use Pierre Bourdieu's phrase, one of those "rites of institution."[1] The new computer, or the budget for one, is a standard part of the hiring package for new faculty members in many North American colleges and universities. It is a sign of institutional health when the university replaces computers just to be up to date. But these are banal facts, apparent to anyone in academia. That's why it is surprising how little this knowledge impacts academic writing about new communication technologies. Perhaps this is because the shadows of Augustine and Descartes still loom darkly over new media studies. Emphasis on virtuality, the ethereal, ideational, immaterial, and experiential dimensions of new media leads many writers to accept the myriad strategies that states, institutions, and individuals use to move computer trash into the backspaces of modern life. If we are willing to look seriously at those backspaces, our garbage can tell us a lot about the relationship between our past and future.

Many thanks to Carrie Rentschler, Ken Hillis, Bill Fusfield, Charles Acland, Ariana Moscote Freire, Jeremy Morris, and audiences in Toronto, Montreal, and New Orleans for their comments on earlier versions of this chapter.

"New" media technologies as we know them, and all of their components, are defined by their own future decomposition. Obsolescence is a nice word for disposability and waste. Billions of pieces of computers, Internet hardware, cellphones, portable music devices, and countless other consumer electronics have already been trashed or await their turn. The entire edifice of new communication technology is a giant trash heap waiting to happen, a monument to the hubris of computing and the peculiar shape of digital capitalism.

In modern large-scale societies, every form of communication involves the physical disposition of bodies and, for lack of a more elegant way to put it, the physical disposition of stuff. If you can call something a medium, then it has a physical infrastructure. Take the Internet, for instance. A great deal of the literature on new media that discusses users' experience does so in terms of "disembodiment," as if the medium somehow removes the body from the mind. For Descartes, this was an exercise in abstraction—one body, one soul, was his equation, although he believed that he could indeed forget his body and his senses.[2] Modern Cartesians are less certain on that score—a body may lead to multiple souls (or subjects). Sandy Stone's widely cited *War of Desire and Technology at the Close of the Mechanical Age* contrasts two stories: a tale of the trial of a man accused of raping a woman with multiple personality disorder, and the case of a psychologist, Sanford Lewin, who passed as a woman in his online encounters via Compuserv chat groups. Stone's pairing suggests that one effect of computers and the Internet is to assist in the splitting of subject and body, and to open up a range of possible subject positions for a single body.[3] This perspective takes Cartesianism for granted and expands on the theme of a split between the body and the subject. Analytically, the reader is supposed to identify with the "subject" part of that dyad and leave aside the body.

Step back from that identification for a moment. Imagine yourself standing next to a person who is using a computer to connect to the Internet to become "disembodied." By imagining ourselves looking at him or her instead of being the disembodied surfer, we are forced to confront the body left behind in standard tales of online subjectivity. To put it another way: a little distance from the event forces our attention to move from subjects to objects. This is how it might look: our surfer friend more often than not sits in a less-than-optimally-ergonomic position in front of a keyboard, mouse, and monitor hooked to a computer, and in the neighborhood of a phone line, DSL, Ethernet, or other kind of connection. If we follow this connection, it will take us on a spiral of continuously increasing scale: we quickly find ourselves arriving at bits of infrastructure proper. First the body, then the interface, then the computer. Then routers, servers, T1 lines, backbones, switches, mirrors, telephone lines, local area networks, and networks of the networks—until the networks are so networked that we call them the Internet. In its very name, the Internet signals hardware and infrastructure. In this perspective, our new media subjects are not only embodied, but they are surrounded by piles and piles of humanmade stuff. Much of this stuff is going to be taken out of service long before it no longer works. It will sit in offices and warehouses. And then it will be trashed.

If computers and data lines are today's machines in the garden, then trash hardware is the giant modernist statue in the front yard that offends the neighbors. The disposability of computers may be one of the truly distinctive features of new media in our age. Or rather, it is the perception of their disposability that is so novel and interesting. Although advertisements and press releases suggest that every new machine is supposed to manifest a revolution, even our most casual understandings of digital technologies imply their own decomposition. Computers have become disposable consumer goods, and all the while the fact of their disposal is largely hidden from the front spaces of social life.

As a result, our understandings of what constitutes a "new" medium have shifted in a subtle but significant way. For the better part of the nineteenth and twentieth centuries, "new" media were primarily understood as "new" with respect to other media: "new" media forms replaced older media forms. When most people write the phrase "new media," they probably think that they are talking about the newness of computers and digital hardware in contrast to other, older analog media forms. Yet computers and other digital media actually embody a different model of newness: computers have reached a point where their "newness" references other computers and not other media. This is not a wholly novel turn of events—after all, the harmonic telegraph was meant to supplant the single-channel Morse "sounder" and color television sets supplanted black-and-white ones. But the dynamics of computers as "new" media differ significantly from the usual stories about innovation in media history. In short, there are really two models of "newness" to which scholars of media change need to attend: (1) the "newness" of a medium with respect to other media, and (2) the so-called state of the art in design and function *within* a given medium. Scholars, journalists, and many others who write about computers have tended to collapse the second sense of newness into the first. That is why a magazine like *Wired* can call a new operating system a "revolution" with a straight face, and that is why scholars are willing to call computers "new" media even though they have been around for decades. A short detour through U.S. media history will illustrate how central trash is to this shift from comparison across media to comparison within a single medium.

When U.S. reporters took notice of telegraphy in the 1840s, they understood it as a "new" medium in comparison with other communication media of the day: mainly, the post. In the 1880s, both engineers and public commentators elaborated an understanding of the "newness" of the telephone by comparing it with telegraphy. Indeed, telegraph wires eventually came down as phone companies established hegemony in the United States and elsewhere.[4] This pattern extends well into the twentieth century: when television exploded on the American scene, it was understood as a "new" medium with respect to radio, telephony, and film. Leo Bogart's classic *The Age of Television* is clear and unequivocal on this fact. As Bogart noted, the emergence of television required radio to recreate itself.[5] A more recent version of this story would have compact discs replacing long-playing records.[6]

One might imagine the same to be true of digital media today. Many of the most widely cited cultural commentaries on digital media express their newness in terms of their difference from "old" media such as television, print, or photography.[7] But if this were a sufficient explanation of the imagined "newness" of digital media, we could expect the moniker "new" to have declined in general usage by now. Strictly speaking, microcomputers—now simply called "computers"—are approximately forty years old. For the sake of argument, I am dating the history of microcomputers from the introduction of Digital Equipment's PDP-1 in 1960. It was the first commercial computer with a keyboard and a monitor. Compare that with the histories of other media at age forty. Telephony is usually dated from 1876. By 1916, commentators were no longer calling the telephone a "new" medium. Although telephony would not reach the majority of American homes until after World War II, it was a well-established feature of the cultural landscape. Radio is conventionally dated from 1899. By 1939, radio was not only no longer a "new" medium, it was essentially a consensus medium, reaching the vast majority of American households. Proportionally speaking, computer and Internet diffusion in America is greater than telephone diffusion was in 1916.[8] It is a well-established aspect of middle-class life: it receives extensive coverage in the news and appears regularly in fictional texts. Granted, the Digital Equipment PDP-1 was very different from the modern "personal computer," but Guglielmo Marconi's first radio (which he called a "wireless telegraph") was also quite different from the home radio set of 1939. On the scale of media history, computers have been around for quite some time, yet we persist in calling them "new media." The question is why.

In a weird, recursive way, new media are "new" primarily with reference to themselves. Computer culture has reached a truly bizarre equilibrium. Today, computers and other digital hardware displace their own counterparts more than anything else. "Newness" in computers is defined with primary reference to old computers. Along with cell phones, they are designed to become obsolete after a short period of use. They are designed to be trash, to make room for future profits, additional hardware sales, and performance upgrades. Certainly, computers have become a vehicle through which users can encounter "the new" in media technology. But even more, computers' apparently interminable status as a "new" medium speaks to the degree that we, who write about computer technology, have mistaken the "state of the art" in a single communications industry for the ongoing total transformation of the media environment. Journalists, scholars, and savants alike have collapsed the two meanings of "new" in our descriptions of media. Where other media industries certainly found ways to sell new hardware, the digital hardware industry has rationalized, accelerated, and made regular the process of equipment turnover.

One common available explanation of this tendency is that computers have not yet stabilized as a medium, since the technology is still evolving. One would expect, based on the history of other media, that innovation in computing would eventually settle down and a stable format and product would result. But within the occupational ideology of computer engineering, there is actually an opposite impulse: Moore's law. Intel founder Gordon Moore observed in 1965 a "doubling of transistor density on a manufactured die every year," which has been extrapolated to mean that computer power doubles every year. Moore's observation has been canonized as a "law" by computer engineers and others. In fact, it is less of a law of computer evolution than it is a fantasy the industry wishes to uphold. There is no better evidence of this than the varying reports of the length of time for the doubled density. Moore's original proposition was a single year. Other observers have suggested that a more accurate duration is eighteen months. The Intel Web site has itself resorted to fudging the time period as "every couple years." In other words, Moore's law is more of an imperative than a law. Consider Moore's own reflections many years after his initial observation:

> To be honest, I did not expect this law to still be true some 30 years later, but I am now confident that it will be true for another 20 years. By the year 2012, Intel should have the ability to integrate 1 billion transistors onto a production die that will be operating at 10 GHz. This could result in a performance of 100,000 MIPS, the same increase over the currently cutting edge Pentium II processor as the Pentium II processor was to the 386! We see no fundamental barriers in our path to Micro 2012, and it's not until the year 2017 that we see the physical limitations of wafer fabrication technology being reached.[9]

The important thing to note here is that the computing industry, at least according to Moore, not only does not expect to stabilize anytime soon, it *does not want to*. Innovation, so the logic goes, keeps a market fresh and guarantees computer sales. Although this seems like a reasonable enough proposition from the perspective of marketing, keep in mind that the *pace* of innovation is considerably faster than other, "old" media industries like television or sound recording. Apple Computer, for instance, which does not have a significant share of the business computing market, segments its market into three divisions: "creative professionals" who replace their computers every twenty-four to thirty-six months; consumers who replace their computers every five years; and educational institutions that replace their computers every five to six years.[10] Although these figures might seem overvalued, they actually reflect a *decreasing* rate of purchase because of the market slowdown in North America.

Susan Strasser writes that, by the 1920s, "Economic growth was fueled by what had once been understood as waste."[11] A high rate of machine turnover marks a condition of tremendous profitability for the computer hardware and software industry, and it plays on a long-established principle in American consumer economics. Anyone familiar with the history of marketing and consumer culture will recognize the patterns and attitudes I describe as a variation in a long history of obsolescence. Sociologists and marketers have often divided obsolescence into two types: stylistic and technological. Stylistic obsolescence was decried by social critics and celebrated by marketers: the idea that objects go out of fashion and need to be replaced was clearly wasteful from an environmental or social-critical point of view. But it was also clearly a good thing from a marketing perspective because it kept markets open. In fact, stylistic obsolescence was the basis of the first modern forays into *planned* obsolescence. In 1923, General

Motors, in an effort to increase its market share against Ford, introduced the "yearly model change" to its line of cars. Although even their CEO, Alfred Sloan, acknowledged the additional cost of research and design that accompanied the move to deliberately render their stock obsolete on an annual basis, the benefits were immense. In addition to generating annual publicity for GM cars, the scheduled redesign allowed GM to rationalize its own innovation process with stylistic changes every year and technological changes every three years (based on the life expectancy of the dies used to stamp the metal). The phrase "planned obsolescence" did not itself find general usage until a 1955 *Business Week* article noted that GM's model of industrial design, which had caught on in the automobile industry, was moving to other consumer industries as well.[12]

The idea of technological obsolescence extends even further back than stylistic obsolescence. In *Democracy in America*, Alexis de Toqueville writes of Americans' belief in the "indefinite perfectibility of man." Such a belief, he argued, flows throughout U.S. society: "continual changes then pass at each instant before the eyes of each man." Obsolescence was the technological manifestation of this never-ending change: "I meet an American sailor and I ask him why his country's vessels are built to last a short time, and he replies to me without hesitation that the art of navigation makes such rapid progress daily that the most beautiful ship would soon become almost useless if its existence were prolonged beyond a few years."[13] In this example, obsolescence is coupled with that equally loaded term "progress." Technological obsolescence was supposed to represent genuine innovation, utility, and, to some degree, necessity. Certainly, that is de Tocqueville's reading of the sailor's reasoning. Although it is specific to the United States, it would not be much of a stretch to suggest this reasoning has been carried forward by an international computing industry.

The similarity between the sailor's analysis of innovations in shipbuilding and Moore's analysis of innovations in microprocessor manufacturing should not be lost on us: they are two species of the same genus. Almost a century later, the same texts that decry stylistic obsolescence would still celebrate the obsolescence derived from "progress." Indeed, with the rise of consumer culture in the intermediate decades, the ideology and practice of technological progress were disseminated more fully into everyday life. Classics of consumer culture, like Christine Frederick's *Selling Mrs. Consumer*, argued that "obsolete" objects should be replaced with "modern," up-to-date ones.[14] In essence, those "when to upgrade your computer" columns in newspapers and magazines descend from Frederick's consumer advice of the 1920s. The most basic aesthetic dimensions of commercial and fictional representations of computers (assuming for a moment there is a difference) follow this line as well. Mainstream depictions of computers in films and television class them along with cars and other consumer electronics: they are clean, new, and generally work well unless sabotaged. "New technology" conjures up well-lit images of sleekly designed computers and monitors; bright colors, spotless, smooth surfaces, clear screens, and quick applications. This obtains even though the average condition of computers is closer to dust-covered CPUs and monitors, screens dotted with fingerprints, and keyboards darkened by use. In the extreme, these so-called new technologies run on operating systems so loaded down with applications and extensions that they crash on startup, with hard drives so full that they crawl along lazily seeking data. Of course it would be silly to expect the art of commercials to present audiences with computers as they are—advertisements are all about fantasies of novelty and change. But the ethos of commercial art extends into almost all public depictions of computing. In the vast range of available representations, computers are by and large clean, attractive, functional, and new. The exception would be the many online user group bulletin boards and listservs for particular brands of software and hardware. For most programs and platforms it is easy to find a litany of complaints and queries as to stability. Yet, even in these spaces, there is still often an unflappable belief—or at least a hope—that the latest new version will resolve all the existing problems. Today, the new computer is an object of industry and consumer fantasies alike.

So what makes computer obsolescence important, different, or new? The answer is that the computer industry has applied the logic of planned obsolescence to media hardware more thoroughly than any other media industry before it. Computers and digital media are no longer "new" with respect to other media. They are new primarily with respect to themselves. Designers know this. By the time a new IBM, Pentium, or Athlon processor rolls off the line, engineers are already working on a new model to render it obsolete. Marketers know this as well. Advertisements for "digital lifestyle" products from Dell, Apple, and Gateway (among other brands) all try to convince their viewers that new computers with faster processors, more RAM, and bigger hard drives will be necessary as a lifestyle accessory. Users know this too. People expect to replace their computers over time, and power users eagerly await the opportunity to replace their machines. Corporations and public institutions anticipate computer replacement in their budgets. Not only can U.S. citizens write off the purchase of a new computer under some circumstances, they can also write off the depreciation of the computer over time.

Obsolescence is not only planned but also forced or engineered. The boundary between a durable and obsolete has as much to do with social relations as it does with the decline or decay of the object. Groups of people choose to make an object obsolescent, or they choose to sustain an object long after it would have begun to fall apart on their own. As I have already suggested, this phenomenon is not limited to the computer industry. By way of comparison, consider Michael Thompson's analysis of how housing becomes "rubbish":

> The fact that buildings last for generations is dependent upon their receiving "reasonable maintenance." The amount of maintenance that is deemed reasonable is not a quantity deriving naturally from the intrinsic physical properties of the house and its environment. The level of maintenance that is deemed reasonable for a building is a function of its expected life-span and its expected life-span is a function of the cultural category to which that building is at any moment assigned, and, if its category membership changes, so will its expected life-span and its reasonable level of maintenance.[15]

Thompson's point is that buildings fall into disrepair and become obsolete because people decide not to maintain them anymore. "Obviously," writes Thompson, "it is much easier to impose durability on a solid granite-faced Edwardian bank than on a thatched wattle-and-daub cottage, yet we frequently choose the more difficult alternative."[16] His point is well taken: obsolescence and durability are first and foremost socially imposed categories, and they are only about the physical properties of things in a last instance that rarely, if ever, arrives.

As it is for buildings, so it is for computers. By limiting "backward compatibility" between older and newer computer systems—and this can be accomplished via changes to hardware or software—manufacturers make it increasingly difficult to interface between older and newer machines. Yet even this condition is not wholly made up and is not wholly a matter of an industry seeking the last drop of profit from a world of narrow margins and cutthroat competition. Most digital media are what Arnold Pacey calls "halfway" technologies. When we think of technology, we normally think of it as fully accomplished and reasonably functional—as in the sexy computers we see in magazine ads and on television. But computer technology is more like advanced medical procedures, missile defense, and other not-fully-accomplished technologies. It sort of works, but not in a flawless or entirely predictable fashion. Part of the problem, argues Pacey, is that technologies are often built to solve problems that are only half understood.[17] This approach actually fuels replacement: one reason people are so willing to replace infrastructure is that it doesn't work so well.

Combined with the "halfwayness" of most new media, planned obsolescence guarantees the continued recursive experience of digital media as "new." The "newness" of new media is sustained by people continually disposing of the equipment they have in anticipation of something better. The hope is always that the next generation will work better, be more stable, be more functional. Many computer users are aware that new versions of software and hardware are released well before all the bugs are ironed out. But by force of circumstance, they do not act on that knowledge. The halfwayness of computerized communication defines it. Users are ready for servers to be down and messages to bounce. They are ready to be infected with viruses, Trojan horses, and spyware. And they are ready to replace their machines. A certain faith in the promise of computing persists even if the reality of computing isn't all it's cracked up to be. One might be inclined to see this faith as tied to a cultural interpretation of computers' economic value. They are expensive to purchase and maintain, they must be imagined as better, more efficient or more flawless than they actually are.

Indeed, scholars interested in waste from a cultural perspective have framed the problem in terms of value.[18] But value alone is too blunt an analytical instrument here. Value is one particular kind of classification among many. A computer's social life might best be described as a kind of symbolic journey. It undergoes a series of symbolic transformations: it travels through categories from new, to useful, to obsolete, to unused, to trash. "Taxonomy is . . . not only an epistemological instrument (a means for organizing information) but it is also (as it comes to organize the organizers), an instrument for the construction of society."[19] The production of computer trash is thus a fundamentally taxonomic process. A computer passes through several classifications in its lifetime and only some of them have to do with "value." But every time a computer passes through the threshold from one classification to another, its meaning and function do change.

Assuming that a brand-new computer confronts its first users in the "new" category, the first threshold through which a computer passes is the passage from "new" to "useful." This passage is accompanied by a significant drop in cash value. Usually, depreciation first occurs within six months of the computer's purchase, and it is mainly the result of manufacturers routinely updating their lines to compete with one another. A side effect of this routine update is to devalue all machines the company sold within the past few months. Once in the "useful" category, a computer's cash value on the used market will steadily decrease every few months for the life of the machine. It will likely remain compatible with existing software and peripherals for several years, but within a single year, its value may have dropped 50 percent or more. Computers thus spend most of their working lives outside the much vaunted category of "the new" even as they remain perfectly functional for their users.[20] As should be clear from my narrative here, "newness" is a function of marketing practices. A six-month-old computer's passage over the threshold from "new" to "useful" is the result of practices by the machine's manufacturer. In changing their line of machines in production, manufacturers actually begin to produce the obsolescence of old machines as well.

For a device to become obsolete, it must be devalued again. Although the first devaluation of a computer has to do with its resale price, obsolescence requires an attack on or erosion of the machine's use-value. This can be accomplished through companies' refusal to update software; their decision to change software, hardware, peripherals, or networks to protocols or formats that are no longer compatible with a generation of older machines; or their development of software that won't run on older machines. Ivan Illich calls this kind of technological regulation a "radical monopoly": a monopoly of technological form, as opposed to the domination of a market by a single company.[21] The result of a radical monopoly is coercive participation. People who wish to use a technological system must do so on terms dictated to them by the people who control the radical monopoly. This is not meant as a conspiracy theory: Microsoft has to adjust its software to PC developers' operating equipment and vice versa, but together their interactions define the parameters of a radical monopoly in computing at any given time—at least for those who use Microsoft products on PCs.

Even when companies deliberately or accidentally "obsolete" old computers, they do not automatically become trash. Some users will persist for years with an old machine until they are forced to update or upgrade. The moment of coercion might come when their computer has a mechanical failure, or it might come when users want to integrate with some kind of network that has baseline requirements beyond their computers' capabilities. This is one of the important subtleties that other kinds of studies of waste often leave aside. The usual argument is that when an object loses its value, it becomes trash. But in the world of computing equipment, there is an important continuum between that kind of progression and a more insidious gap between obsolescence and trash. There is often a significant gap in time between the reclassification of a computer as obsolete and its fall into disuse. There is another gap between the time when a user ceases to use the machine and when it is finally thrown out or recycled.[22]

Computers' long decline from newness shapes their travels through the front and back spaces of social life. Computer obsolescence is a spatial problem. When computers exist in a marginal category—between "useful" and "garbage"—they often wind up in marginal spaces like warehouses, attics, and basements.[23] After they stop using a computer, users, whether individuals or companies, most often store them for a time after the cessation of use. The two most likely reasons for this are ignorance and denial. People may not know how to get rid of their computers. Or, more likely, the memory of dropping well over $1,000 (and probably considerably more) still lingers. As a result, there remains a belief that the machine might have some value—either economically or metaphorically. Chances are, at least among middle-class people in North America, that most people know someone with at least one or maybe more obsolete computers in their offices or homes. In fact, you all know of one such person thanks to the first-generation Pentium that lived on top of my supply cabinet and the modem that now sits on display in my new office.

This tendency toward storage creates a host of problems. Although a machine's condition is generally known before it is put in storage, computers that have been stored for some time have to be tested by someone. Organizations have to pay a technician to do it. This expense adds to the financial liability of computer recycling for organizations, and it becomes a structural incentive to throw them out. So, let's follow the logic: because computers are so expensive, people are less likely to get rid of their old machines—even after replacing them. Because they don't get rid of their computers right away, they are more likely eventually to throw them out. Yet computers are *so* valuable to many lines of work and leisure that they must be replaced. That is a twisted but true logic: computers are too valuable, so we eventually throw them out and buy new ones.[24]

As long as computers, software, and the Internet are conceivable as new technologies rather than as plain old vanilla technologies, we will be confronted by torrential rains of new machinery and applications. Those rains will mix together their own brand of toxic sludge as trash hardware piles up in basements, warehouses, and landfills. The U.S. Environmental Protection Agency projects that sometime this year (2005), for every new computer manufactured, another one will become obsolete. In the United States alone, more than 24 million computers became obsolete in 1999. Of those, only 4 million were properly recycled or donated. The remaining 20 million computers were dumped into landfills, incinerated, shipped as waste exports (and probably dumped or incinerated upon arriving at their destination), or stored. That was a single year in the United States alone. We are clearly talking about the disposal of hundreds of millions of computers in a very short span of time, all of which occupy the vague and fraught category of "obsolete" but may function perfectly well.[25]

Computer junkyards have sprung up across the United States. While a graduate student at the University of Illinois at Urbana-Champaign, I had the opportunity to observe the evolution of one such junkyard. It began as a repository of all the university's obsolete computers. It was a whole warehouse filled with piles of computers, monitors, disk drives, server racks, testing equipment. You name it, they had it. But most of this junk could not even be sold as junk. Now, the owner won't accept anything less than a Pentium II because he can't sell it. The rest of the materials, including the pile of dual-floppy systems and outmoded servers he inherited from the first owner, go into a landfill.

Some old technologies do make a comeback from the trash heap, but not computers. Old records, for instance, have surfaced from their status as once-dead commodities. Some can cross the threshold from "rubbish" back to "useful" through successive waves of nostalgic revivals. As Will Straw writes, global, "centrifugal" tendencies, "nourished by the scavenger-like record collecting tendencies of dance club disc jockeys, lounge music revivalists, curator-compilers like David Byrne, and by the activities of marginal reissue labels [are] dragging back, into the realms of hip credibility, musical currents long dismissed as false imitations or examples of debased exploitation."[26] In contrast, there is a tiny, barely existent vintage market in computer hardware. Old dual-floppy systems do not become fashionable again; they do not regain their value through discovery by hip members of the creative classes.[27] Until they are "obsoleted," many computers show no significant signs of wearing out. It is only when people stop caring for them that many of them begin to fall apart. So in an important way, computers' economic decay *hastens* the process of physical decay.[28]

Computers' physical decay raises other issues as well. The threshold of trash is an incredibly important one for computers because that is the moment when computers move from indoors to outdoors. Once it is reclassified as trash, the unit will be exposed to the elements and begin the long process of decomposition or decay. Of course, some computers will be recycled. But many more will be tossed out in the trash. When thrown into landfills or incinerated, computers and computer monitors can release hazardous materials and heavy metals into the environment such as lead, mercury, and hexavalent chromium. Each of these substances poses unique dangers to human beings and their environment. In landfills, these substances will eventually leak into the drinking water supply and the human food chain. Incinerating computers and parts releases toxic chemicals into the air, where people and animals breathe them in. It also creates ash and slag-containing toxic substances, which require specialized disposal. Some of the pollutants released through computer disposal, like lead, do not disappear over time.

As a result, many local, regional, and national governments are currently in the process of declaring computers to be hazardous waste. This means that they require special means of disposal and cannot be dumped into landfills or processed with other garbage. The disposal of computers has also become an issue worldwide. Some governments are exploring the idea of extended product responsibility (EPR). The idea behind EPR is to make companies responsible for the products they manufacture throughout the product's life cycle. Germany, the

Netherlands, Norway, Switzerland, and Denmark have all enacted EPR-related laws, and other European countries are following suit. Under EPR, theoretically, if you manufactured a computer monitor and five years later it surfaced in a landfill, it would be your problem rather than the state's problem. With EPR, companies are held responsible for the physical management of their products, the costs of the waste created by their products, the liability for environmental damage caused by their products, and for informing consumers about the possible environmental effects of a product at different times in its life cycle. In response to this kind of pressure, Hewlett-Packard and IBM will both, for a fee, dispose of your old computer for you.[29]

It is easy to see government involvement in computer disposal as a kind of natural advance of the liberal state (or what's left of it in some places). Prima facie, it seems reasonable that the state would step in to regulate computer waste, once the corpses of unloved computers reach a critical mass in storage lockers, warehouses, and landfills. But this very tendency to see state intervention as a natural outcome brings us back to the role of governments in all this. Dominique Laporte's provocation that "the state is the sewer" is apropos here.[30] The orderly management of computer waste is not simply an environmental problem but also a problem of legitimacy. It is the other side of innovation—as corporations manage or mismanage the introduction of new software and hardware into everyday life, so too must someone regulate the exit of computers from the social stage. EPR is government policy that acknowledges the future of all digital hardware is in the trash heap. It is a political response to an economic and cultural fact.

The state's managerial interest in waste is directly political. By managing waste products, by keeping them out of view and off its citizens' minds, the state maintains faith in infrastructure and the affirmative character of social life as mythically pure. Gay Hawkins calls knowledge of waste disposal a "public secret" because one of the state's most important symbolic roles is to help its citizens forget about their own excrement and other waste products.[31] The managed departure of computers from the social stage and into dumps follows a similar logic. To twist around Marx's famous handmaiden metaphor, the state is an administrative assistant to the computer industry when it comes to the disposal of computer trash. As with other kinds of refuse, computer trash works best as a public secret. If users can ignore their own computer trash once it leaves the home or office, it becomes that much easier to maintain an image of computers as new media. So even environmental regulations designed to restrict some of the damage done by computer disposal also help perpetuate the cycle of computer purchase, use, warehousing, and eventual disposal.

If that is not enough, in practice the regulations themselves do not so much reduce environmental harm as hide it from the middle classes of wealthy nations. A 2002 report coauthored by the Basel Action Network, the Silicon Valley Toxics Coalition, Toxic Link India, SCOPE (Pakistan), and Greenpeace China documents that "technotrash" is more often than not exported under the guise of recycling, only to be dumped in the villages and countrysides of Asian nations, especially China, India, and Pakistan. Computers thus become part of a global trade in toxic materials, in which "recycling" means hazardous materials are moved from richer to poorer nations, with traffickers turning a tidy profit.[32]

Of course it would be silly to oppose some environmental regulations applied to computer disposal on the grounds that they are not strong enough. The interest of states in computer trash is a good thing, but it is clearly not enough. A "successful" environmental program can be based on classification of computers as hazardous waste or it can be based on some version of EPR. But both approaches essentially manage and legitimate the continued onslaught of computer trash and the ongoing manufacture of obsolescence by the hardware industry; they also inadvertently support the global trade in toxic materials. Ultimately, both hazardous waste and EPR approaches are preferable to real reform from the industry's perspective. From an industry perspective, the real fear must lie in the manufacture of a computer that is finally "good enough." Then the computer industry will find itself in the same position as manufacturers of radio sets in the late 1920s. Having sold as many sets as practical, companies started to go out of business. The television industry learned from radio. When they reached market saturation, they moved to campaigns for families to purchase a second set, and they introduced color televisions—at first as a luxury good and then as a necessity.[33] All the computer industry has done is to rationalize and speed up this process of obsolescence in consumer electronics. They have done so through a faster pace of innovation, a willingness to release computers and components as halfway technologies, and a constant onslaught of advertising and punditry.

Pacey argues that although some halfway technologies are the result of attempts to solve half-understood problems, the other part of the problem is that there are some things "which professionals are almost trained to ignore."[34] Knowledge of sustainability or "green computing" is one area that is simply written out of computer design at the moment. In fact, some computer components are considerably less durable now than in earlier models. As a cutting-edge technology, computers are built not to last. Like the ships of the 1830s, computers are built with an eye toward their own replacement.

This need not be the case. Illich uses the term "conviviality" to connote the following characteristics of technologies: ease of use, flexibility in implementation, harmony with the environment, and ease of integration into truly democratic forms of social life.[35] Obviously, Illich's vision is a utopian one, but his measure of a technology's conviviality seems relevant to the question of computer trash. We need a "convivial" computer, or rather a whole convivial system of digital components, a convivial digital infrastructure. Imagine a company that took its time developing a computer that could last, could be easily updated, repaired, and upgraded, was easy to learn and use, worked well with other platforms, and that was less environmentally hazardous when it did finally decompose. The dream is not unrealistic: we expect our cars and consumer appliances to work for a decade or more. Major appliances are supposed to last even longer, and more specialized technologies like musical instruments can last decades or even centuries. The models are out there. But for a computer company to engage in such an undertaking would be viewed as commercial suicide.

Imagine if a company successfully designed a computer that would last more than a few years, could easily be repaired and upgraded, was "forward compatible" with as-yet-uninvented devices as well as backward compatible with older ones, and was made of less hazardous materials. Imagine if such a product took the consumer market by storm, wiped out the competition, and became the dominant unit and platform in homes and private and public institutions all over the world. Eventually, the successful company's profits would level out or even decline as the market saturated. In the current economic climate, such stability would be read as a sign of economic weakness on the company's part.

It is tempting to label this scenario a paradox, but it is not a paradox at all. It suggests that contemporary corporate culture, with its drive for growth, increase in market share, and larger profit margins is a fundamentally inhospitable environment for any form of convivial computer. The truly sad thing about it is that a convivial computer is not a revolutionary idea. It does not require a fundamentally different economic system. It simply requires a manufacturer that would be more interested in long-term stability than near-term growth. No such manufacturer exists in the current economic environment. For now, it is up to academics, designers, policymakers, and artists to come up with convivial models of computing, and we will have to do it on our own time, with our own resources. But it can be done. We need digital hardware that is more democratic, slower to change, easier to use, and less damaging to the environment.[36]

In the meantime, the anticipation of their own decomposition defines our new technologies. I could write with Georges Bataille that hardware trash is the accursed share of the digital economy—that bit of excess that must be disposed of "gloriously or catastrophically."[37] Or I could write with a more modulated John Frow, who argues that waste is not excess but "a generative dynamic in the destruction and formation of value."[38] Either way, it is computer trash that turns digital technologies into "new" media. Whether metaphorical or real, our trash heaps are public secrets. Computer trash is a catastrophic dimension of that middle space between fantasy and accomplishment occupied by so much digital halfway technology. A seemingly endless cycle of creation and disposal is driven by the dreams of users who seek that killer application and by manufacturers who stay above the bottom line only so long as they anticipate the underground burial of next year's new product.

Notes

1. Pierre Bourdieu, *Language and Symbolic Power*, trans. Gino Raymond and Matthew Adamson (Cambridge: Harvard University Press, 1991), 125.

2. René Descartes, *Discourse on Methods and Meditations on First Philosophy*, 4th ed., trans. Donald A. Cress (Indianapolis: Hackett Publishing, 1999).

3. Allucquère Rosanne Stone, *The War of Desire and Technology at the Close of the Mechanical Age* (Cambridge: MIT Press, 1995). See also Sherry Turkle, *Life on the Screen: Identity in the Age of the Internet* (New York: Simon & Schuster, 1995).

4. See Daniel Czitrom, *Media and the American Mind: From Morse to McLuhan* (Chapel Hill: University of North Carolina Press, 1982); Steven Lubar, *Infoculture* (Boston: Houghton Mifflin, 1993); Carolyn Marvin, *When Old Technologies Were New: Thinking about Electrical Communication in the Nineteenth Century* (New York: Oxford University Press, 1988).

5. Leo Bogart, *The Age of Television: A Study of Viewing Habits and the Impact of Television on American Life* (New York: Frederick Ungar Publishing, 1956), 106–23.

6. The shift from LP to CD was as much a matter of engineering a market as it was a feat of technological innovation: retailers switched over to CD as quickly as they did because record companies stopped accepting returns on vinyl records. See Negativland, *Shiny, Aluminum, Plastic and Digital* (August 2003), http://www.negativland.com/minidis.html.

7. Jay Bolter and Richard Grusin, *Remediation: Understanding New Media* (Cambridge: MIT Press, 2000); Lev Manovich, *The Language of New Media* (Cambridge: MIT Press, 2001); Nicholas Negroponte, *Being Digital* (New York: Knopf, 1995).

8. David Crowley, "Where Are We Now?: Contours of the Internet in Canada," *Canadian Journal of Communication* 27, no. 4 (2002); Susan Douglas, *Listening In: Radio and the American Imagination from Amos 'N Andy and Edward R. Murrow to Wolfman Jack and Howard Stern* (New York: Times Books/Random House, 1999); Claude S. Fischer, *America Calling: A Social History of the Telephone to 1940* (Berkeley: University of California Press, 1992); Tom Spooner, "Internet Use by Region in the United States," (Washington, D.C.: Pew Foundation Internet in American Life Project, 2003).

9. Gordon Moore, *The Continuing Silicon Technology Evolution inside the PC Platform* (October 2002), http://www.intel.com/update/archive/issue2/feature.htm.

10. Joan Hoover, Author Interview, 3 October, 2002.

11. Susan Strasser, *Waste and Want: A Social History of Trash* (New York: Henry Holt and Company, 1999), 204.

12. Jeffrey L. Meikle, *Twentieth Century Limited: Industrial Design in America, 1925–1939* (Philadelphia: Temple University Press, 1979), 12–13.; Strasser, *Waste and Want: A Social History of Trash*, 195, 274–75; Richard Tedlow, *New and Improved: The Story of Mass Marketing in America* (New York: Basic Books, 1990), 167–68.

13. Alexis de Tocqueville, *Democracy in America*, trans. Harvey C. Mansfield and Delba Winthrop (Chicago: University of Chicago Press, 2000), 428.

14. Christine Frederick, *Selling Mrs. Consumer* (New York: The Business Bourse, 1929).

15. Michael Thompson, *Rubbish Theory: The Creation and Destruction of Value* (New York: Oxford University Press, 1979), 37.

16. Ibid.

17. Arnold Pacey, *The Culture of Technology* (Cambridge: MIT Press, 1983), 35–37. See also Lewis Thomas, "Notes of a Biology-Watcher: The Technology of Medicine," *New England Journal of Medicine* 285 (1971). For a fascinating account of how these factors play out in missile technology, see Donald MacKenzie, *Inventing Accuracy: A Historical Sociology of Nuclear Missile Guidance* (Cambridge: MIT Press, 1990).

18. In fact, two of the key cultural analyses of waste signal this connection in their very titles: Stephen Muecke and Gay Hawkins, eds., *Culture and Waste: The Creation and Destruction of Value* (New York: Rowman & Littlefield, 2003); Thompson, *Rubbish Theory*.

19. Bruce Lincoln, *Discourse and the Construction of Society: Comparative Studies of Myth, Ritual, and Classification* (New York: Oxford University Press, 1989), 5–6.

20. This is slightly different from the scheme of devaluation elaborated by Michael Thompson. His studies of Victorian ephemera and old houses suggest that value and utility decrease together, and that so long as a thing is needed or desired, it would retain a certain amount of value. Although this makes sense in a study of houses, the case of computers is obviously a little different. Thompson, *Rubbish Theory*.

21. Ivan Illich, *Tools for Conviviality* (New York: Harper & Row, 1973), 51–57.

22. Jonathan Sterne, "Disposal of Computers," in *The Encyclopedia of New Media*, ed. Steve Jones (Thousand Oaks, Calif.: Sage, 2003).

23. On spatial sorting, see Strasser, *Waste and Want*, 6. See also Mary Douglas, *Purity and Danger: An Analysis of the Concepts of Pollution and Taboo* (London: Routledge and Kegan Paul, 1966).

24. Sterne, "Disposal of Computers."

25. Ibid.

26. Will Straw, "Exhausted Commodities: The Material Culture of Music," *Canadian Journal of Communication* 25, no. 1 (2000): 183. See also Thompson, *Rubbish Theory*, 10–11; McKenzie Wark, "Fashioning the Future: Fashion, Clothing, and the Manufacturing of Post-Fordist Culture," *Cultural Studies* 5, no. 1 (1991).

27. One striking exception to this generalization was the band Man or Astroman? which has composed music for old dot matrix printers and made use of old equipment as part of an elaborate

stage show. Computer trash artists have also made use of old machines, but their work belongs to the larger and more general genre of trash art.

28. Although there is a rising market for retro video games (and even video game systems, like the Atari 2600), these remain fairly inexpensive for the time being, and many of the ROMs are available for free on the Web.

29. Sterne, "Disposal of Computers."

30. Dominique Laporte, *History of Shit*, trans. Nadia Benabid and Rodolphe el-Khoury (Cambridge: MIT Press, 2000), 56.

31. Gay Hawkins, "Down the Drain: Shit and the Politics of Disturbance," in *Culture and Waste: The Creation and Destruction of Value*, ed. Stephen Muecke and Gay Hawkins (New York: Rowman & Littlefield, 2003), 47–52.

32. Jim Puckett et al., *Exporting Harm: The High Tech Trashing of Asia*. Silicon Valley Toxics Coalition (February 2002), http://www.svtc.org/cleancc/pubs/technotrash.htm.

33. Erik Barnouw, *Tube of Plenty: The Evolution of American Television* (New York: Oxford University Press, 1990); Lynn Spigel, *Make Room for TV: Television and the Family Ideal in Postwar America* (Chicago: University of Chicago Press, 1992).

34. Pacey, *The Culture of Technology*, 36.

35. Illich, *Tools for Conviviality*, 11, 13.

36. Linux has been lauded as an alterative, more democratic operating system—a step in the right direction—but it raises issues about labor and management because its maintenance requires programmers who have a lot of free time and are willing to work for free.

37. Georges Bataille, *The Accursed Share: An Essay on General Economy*, vol. 1, *Consumption*, trans. Robert Hurley (New York: Zone Books, 1988).

38. John Frow, "Invidious Distinction: Waste, Difference, and Classy Stuff," in *Culture and Waste: The Creation and Destruction of Value*, eds. Stephen Muecke and Gay Hawkins (New York: Rowman & Littlefield, 2003), 36.

3 Falling Apart: Electronics Salvaging and the Global Media Economy

Lisa Parks

In 1996, media artist Ivo Dekovic deposited an installation called *Monitors* 30 meters beneath the Adriatic Sea (Figure 3.1). The project consisted of nine cement television monitors molded from an old television set that once glowed in the living room of his family's home in Razanj, Croatia. When Dekovic poured cement into each mold, he placed small personal objects related to different people he admired inside and created a Braille-like system of bumps on their exteriors so that he would be able to later differentiate and identify them. Each autumn, Dekovic dives down to visit the monitors and documents the changes registered across their surfaces and environs—the sea mosses and crustaceans that accumulate on them, the fish that turn them into habitats, and the way they slowly fall apart. By treating television monitors as ruins, Dekovic stirs up a series of concerns about old, new, and residual media technologies that are also at the heart of this chapter. First, Dekovic metaphorically treats the television monitor as a technological object whose materiality and significance is allowed to linger in time, as opposed to a commodity with a limited life span that must necessarily be discarded and upgraded. Second, by placing personal mementos inside the monitors, Dekovic foregrounds television's ambivalent status as an object of mass production and personal psychic investment. Finally, by replicating the television set as a cement mold, Dekovic mimics the manufacturing process but instantly turns the "product" into rubble, pushing us to consider the problem of accumulation, specifically the electronic waste streams that have resulted from the past century of consumerism.

In television studies, the term "residuals" is typically understood as the financial royalties that are generated through the syndication and rebroadcasting of television series. In this chapter, however, I associate the term "residuals" with the accumulation of used media hardware that has emerged since the dawn of television and that greatly accelerated with the growth of the global digital economy in the 1990s. Sometimes referred to as "e-waste," residuals can be understood as

FIGURE 3.1 *Monitors*, an underwater art installation by Ivo Dekovic located in the Adriatic Sea near Razanj, Croatia. Courtesy of Frank Schroeter.

the old radio and television sets, computers, stereos, VCRs, telephones, and printers that have piled up in peoples' basements and garages, neighborhood repair shops and thrift stores, and electronics recycling and salvaging centers. Residuals are the waste products of a media and information society. By considering electronic hardware in such a way, that is, as material objects that linger or persist, I hope to complicate reductive bifurcations of "old" and "new" media. What lies at the core of this distinction, I want to suggest, is not just a formalist concern about the shift from analog to digital aesthetics. Often lurking within the differentiation of old and new media is also an idle acceptance of capitalist logics (such as structured obsolescence) used to regulate the life cycles of electronic and computer hardware. By continuing to use terms such as old and new media without reflection or analysis, critical media scholars risk inadvertently reinforcing the imperatives of electronics manufacturers and marketers who have everything to gain from such distinctions.

Some media scholars, of course, have already begun to address these issues. Carolyn Marvin, for instance, approached the problem of old and new quite directly by entitling her social history of electronic communication *When Old Technologies Were New*, stressing in her introduction that "New technologies is a historically relative term . . . new practices do not so much flow directly from technologies that inspire them as they are improvised out of old practices that no longer work in new settings?"[1] According to Marvin, the old/new distinction should be conceptualized as part of a variety of overlapping technologies and practices. She continues, "New media, broadly understood to include the use of new communications technology for old or new purposes, new ways of using old technologies, and in principle, all other possibilities of exchange of social meaning, are always introduced into a pattern of tension created by the coexistence of old and new, which is far richer than any single medium that becomes a focus of interest because it is novel."[2] In their ironically titled collection *New Media, 1740–1915*, Lisa Gitelman and Geoffrey Pingree assemble a range of early case studies to create a genealogy of new media suggesting, "When we forget or ignore the histories of each of these new media we lose a kind of understanding more substantive than either the commercially interested definitions spun by today's media corporations or the causal plots of technological innovation offered by some historians."[3] Finally, in *The Language of New Media*, Lev Manovich also adopts a genealogical approach, asking, "What are the ways in which new media relies on older cultural forms and languages, and what are the ways in which it breaks with them?"[4] Manovich delineates a language of "new media," but only by comparing and contrasting digital aesthetics with earlier photographic and cinematic forms. In other words, there is dialogic relation between old and new media that is fundamental to the way Manovich defines what is new about new media.

Although these scholars have framed old and new media technologies in insightful ways, there remains in humanistic discussions of new media a tendency to sidestep or deny the significance of both television and old hardware. As a way of confronting this issue in this chapter, I explore various practices of electronics salvaging and technological repurposing. In such practices we can begin to recognize the way old hardware persists and becomes ruins, like Dekovic's monitors, even as new media emerge. Historically, the word "salvage" referred to the "act of saving vessels or their cargoes from loss at sea, or the act of saving imperiled property from loss; the property so saved; something saved from destruction or waste and put to further use; to gain something beneficial from a failure."[5] Electronics salvaging involves saving, repurposing, and/or benefiting from old hardware. By considering different practices of electronics salvaging, I hope to shift away from designations of old and new, toward a model of residual media, exploring how hardware persists, lingers, and refuses to disappear even despite the dictates of a market economy. To develop this point, I begin with a description of the problem of electronic waste that emerges in the United States at the end of the twentieth century. After discussing various legislative, corporate, and activist practices that have formed in relation to this problem, I critically examine a cable television series called *Junkyard Wars* that is based on competitive practices of technological salvaging and repurposing. I evaluate whether its popularization of such practices marks an important intervention in a public culture so regulated by structured obsolescence—so geared toward the replacement of usable machines with "new" and "upgraded" ones. I also treat this show's ethos of repurposing as symptomatic of structures of cable TV programming more generally. It is highly symbolic that a cable network is extracting material from junkyards to generate TV content. With its emphasis on the competitive repurposing of already used materials *Junkyard Wars* gives metonymic expression to the extractive and derivative processes that give shape to cable programming in general. The chapter closes with a more theoretical discussion of the relationship between television and materialism, using this particular series as a site through which to consider "television" as an object of analysis, a set of historiographic concerns, and as part of the global economy.

Old Hardware

Residuals—accumulations of still functioning media hardware—emerge in part because the logic of structured obsolescence organizes the mode of production in capitalist societies. Structured obsolescence is an economic strategy whereby a consumer technology is manufactured with the assumption that it has a limited life span and will need replacement with a newer and upgraded model within a given number of years. This logic benefits manufacturers and attempts to build a company's financial future based on consumer band loyalty. The concept of structured obsolescence is hardwired into consumer technologies ranging from the refrigerator to the radio, from the computer to the car, and has been operational in the consumer products industry since the late nineteenth century. One of its effects has been to generate an excess of functional machines that are never exploited to their full potential. They are only partially used and then discarded when a new version, model, or upgrade becomes available on the market. Contemporary junkyards, thrift shops, and garages have become shrines to structured obsolescence. In these secondhand commerce zones lies an unwieldy accumulation of machines with low use-value precisely because they have already been used.[6]

When the personal computer industry boomed during the 1990s, the logic of structured obsolescence intensified as it intersected with a concept known as Moore's law. Developed by Gordon Moore, one of the founders of the chip maker Intel, this law predicted that the computer power available on a chip would approximately double every eighteen months. As John Seeley Brown and Paul Duguid explain, "It's this law that can make it hard to buy a computer. Whenever you buy, you always know that within eighteen months the same capabilities will be available at half the price."[7] Some consumers replaced their personal computers every two to three years just to keep current with software upgrades. The effect of Moore's law was to accelerate rates of computer consumption, which resulted in a corresponding accumulation of old hardware or "obsolete" computers. Once again, most of these "old" computers still functioned, but their use-value diminished because they no longer processed information as fast as new top-of-the-line models.

In the mid-1990s, digital enthusiasts ranging from Bill Gates to Al Gore celebrated the information economy as a green-friendly industry, but they failed to anticipate the waste accumulations that would result from such rapid rates of technological growth. In the United States alone, 20 million or more personal computers became obsolete each year, meaning more than 315 million computers were tossed by 2004.[8] In 2002, the Environmental Protection Agency (EPA) estimated that "250 million computers will be retired over the next five years." As one environmental outreach coordinator put it, "Never have so many purchased so much that's become obsolete so quickly. Computers are now the fastest-growing component of the waste stream in the industrialized world."[9]

During the 1990s, U.S. researchers began to study environmental problems associated with electronic hardware accumulation. A Carnegie Mellon study revealed that 2 million tons of scrap electronics goes into landfills nationwide each year, and it indicated that unless recycling catches on, more than 150 million computers would end up in U.S. landfills by 2005.[10] In 1998, the University of Florida released an influential report that estimated that there were 300 million television sets and computer monitors in the United States and that the average household has four TV sets.[11] The study was primarily concerned about the disposal of the cathode ray tubes (CRTs) inside television sets and computer monitors, and it determined that these components contained enough lead to be classified as "hazardous waste." The results of the study led local officials across the country to forbid the dumping of television and computer monitors into landfills because excessive lead jeopardizes groundwater and can damage landfill linings. In 2000, Massachusetts became the first state to ban CRTs from landfills, and most states have since followed suit. Researchers also claimed that e-waste accumulation was not yet at a crisis stage because most monitors were still held in private storage, but they urged municipal waste authorities to formulate policies for managing their disposal, especially because, as the study concluded, "Americans may be poised to throw away hundreds of millions of TV sets and monitors as digital television becomes popular and people continue to upgrade to ever-faster, ever-cheaper computers."[12]

As U.S. states and municipalities have prohibited the disposal of computer and television monitors in local landfills, environmental organizations have called for the passage of new legislation that would pressure manufacturers to design environment-friendly electronic and computer equipment that would not use so much lead and other toxic materials. In 2003, the state of California passed the Electronic Waste Recycling Act, which requires the reduction of hazardous substances used in certain electronic products sold in California, the collection of an electronic waste recycling fee at the point of sale of certain products, the distribution of recovery and recycling payments to qualified entities covering the cost of electronic waste collection and recycling, and a directive to establish environmentally preferred purchasing criteria for state agency purchases of electronic equipment.[13] Other states have filed similar legislation, and federal agencies like the EPA have issued public statements encouraging Americans to recycle computer equipment, television sets, and other electronics.[14]

Rather than support the passage of new state and federal regulations, manufacturers have perhaps not surprisingly adopted internal measures of self-regulation. Companies such as Dell, Epson, Hewlett-Packard,[15] Gateway, and IBM have all established programs to help consumers with recycling old computers as part of the process of buying new ones. Dell has perhaps been the most proactive with its national e-recycling campaign. In 2003, the company offered $120,000 in grants to cities and universities organizing collection day events that would help spread the motto "No Computer Should Go to Waste." Dell also recycles its customers' old computers for $7.50 per computer and has asset recovery programs for businesses. Dell's National Recycling Tour reached 40 million consumers and collected 2 million pounds of unwanted computer equipment.[16] Other manufacturers have helped organize electronic waste drop-off and collection days in cities such as Madison, Dallas, Portland, and Orlando.

Trade organizations such as the Electronics Industries Alliance (EIA), which includes manufacturers such as Canon, HP, JVC, Kodak, Nokia, Panasonic,

Philips Electronics, Sharp, Sony, and Thomson, have also initiated an electronics collection and recycling pilot project that funds state recycling initiatives and awards grants for electronics recycling events.[17] The organization sponsors consumer education initiatives and collection events. On national recycling day in 2003, for instance, the EIA arranged an electronic and computer drop-off event for government employees and residents of the Washington, D.C., metropolitan area. Companies in other technologically saturated communities such as Silicon Valley have organized similar collection events. On a single day in October 2003, a thousand people from the Bay Area flocked to Stanford University for free recycling of outmoded modems, monitors, processing units, mice, and printers, many of which were brand new only five years ago. By the end of the day, the pile of equipment weighed 45 tons and filled three large truck trailers. The drop-off event cost organizers $25,000 to $30,000 and was also used to train other local officials to organize their own events in other communities.[18]

Although these initiatives and events are no doubt helpful in diverting hazardous material away from landfills, they do not necessarily educate consumers about where this old hardware ends up. Most of the e-waste is handled by electronics salvaging firms that have been operating in the e-waste business for over a decade. Entrepreneurs have turned e-waste into a lucrative enterprise either by refurbishing old equipment and reselling it or by paring it down to basic elements such as glass, plastic, and heavy metals that can be resold as raw materials. In California, the firm Silicon Salvage has operated a junkyard since 1994 filled with old electronics that arrive in good condition from clients such as Lockheed Corporation, Southern California Gas Company, and UCLA. Rumarson Technologies formed in 1991 to specialize in "Nused" or "newly used" computers, and its owner insists, "Our company is environmentalist by heart and capitalist by trade."[19] Cerplex is another such company founded in 1990. By 1994, it generated an annual revenue of $130 million and had 1,500 employees worldwide.[20] Noranda, a Canadian company, specializes in the extraction of precious metals from salvaged Soviet electronics culled from Baltic states. Finally, Goodwill Computer Works refurbishes and resells old electronics and computers. The company has amassed such a stockpile of old or nused components that it functions like an auto parts wholesaler in the electronics field. As one of its spokespersons explained, "Nothing gets wasted. Whatever we can't sell goes to a recycler. Most of these systems people consider junk—they'd get dumped if we weren't around."[21] In four years, Goodwill claims to have kept more than a million PCs out of landfills by refurbishing and reselling used computers and parts. Thus as local municipalities have taken steps to implement ordinances to handle e-waste at solid waste management facilities and manufacturers have assisted with e-waste collection and recycling, a host of new companies have also formed to turn e-waste into a profitable commodity.

When I tried to dispose of my own old computer equipment in November 2003 at the Del Norte waste facility in Oxnard, California, which uses prison laborers to revamp old computers for use in local schools, I was turned away because I was not a city resident. I had driven 30 miles from Carpinteria because my hometown does not have yet have an e-waste facility, but I had to take it back

FIGURE 3.2 *The Del Norte e-waste facility in Oxnard, California, disposes of old electronic hardware for city residents. Photograph by the author.*

home and put it in my closet. While I was at Del Norte, however, I toured their facilities and learned about their e-waste disposal and recycling programs. I discovered that if I had disposed of my computer monitor there, it would have likely ended up at a salvaging center in Asia because the e-waste collected at this facility is trucked away to HMR, a multinational electronics salvager that operates distribution hubs in San Francisco and Los Angeles. This global conglomerate runs its corporate headquarters in Australia, operates collection and recovery centers in the United States, and distributes salvaged electronics to facilities in the Philippines, Malaysia, and Vietnam. HMR promotes itself as "providing a range of environmentally responsible surplus asset-chain management" practices. It first emerged in Australia in 1982 as Harrington Metal Recyclers and has since become a major player in the Asia/Pacific e-waste industry. HMR offices in Australia and the United States focus on the acquisition and distribution of old hardware, and centers in Asia handle electronics dismantling, recovery, and refurbishing. The HMR Web site indicates, "We believe the most effective way to manage obsolete equipment is to re-purpose it" and offers its clients "maximum return on unused assets." The company's Web site shows photos of cathode ray tubes piled up at its demanufacturing facility and assures customers, "Our continuing goal is to achieve the most efficient re-use of all materials, thus minimizing waste."[22]

HMR and other electronics salvagers boast about their environmentally friendly practices, but what remains unspoken and invisible is the way toxic e-waste flows from Western postindustrial to Asian developing countries.[23] Multinationals like HMR operate in global contexts to evade stringent environmental regulations in the United States, Australia, and Europe, dumping hazardous e-waste in parts of the world where there are little or no environmental regulations. It was precisely this situation that spurred the nonprofit organizations the Basel Action Network and the Silicon Valley Toxic Waste Coalition to conduct an onsite investigation in Guiyu, China, in 2001. BAN widely circulated its report entitled "Exporting Harm: The High Tech Trashing of Asia," which explained how e-waste from the United States, Western Europe, Australia, and Japan is ending up in Asian electronics salvaging centers like the one in Guiyu.[24] Members of the investigative team interviewed migrant workers in the region, visited workplaces, collected soil and water samples, and shot photographs and video to document living and working conditions. Along with the report, BAN released a twenty-minute documentary video that graphically displays where the postindustrial West's old hardware ends up.[25] The video serves as an activist polemic that exposes a dark side of the global digital economy so often eclipsed by bright, antiseptic visions of the clean room or advertisements that eulogize a technologized global village.

Given the growing problem of e-waste accumulation and the troubling flow of hazardous materials from postindustrial to developing societies, such educational and consciousness-raising efforts are crucial. What is arguably as important, however, is the publicization of technological reuse and repurposing in postindustrial societies. There are very few sites in our public culture—especially mass commercial culture—that emphasize practices of technological salvaging, repurposing,

FIGURE 3.3 *E-scrappers working at a facility in Guiyu, China. Courtesy of the Basel Action Network.*

and recycling. Instead we are bombarded with advertisements for technological products that valorize novelty, speed, and style over the virtues of longevity, tinkering, and making do. For more than fifty years, U.S. commercial television has been pitching technological products ranging from the car to the computer, fueling and reinforcing fantasies organized around technological novelty, but few programs cultivate an ethos of reuse or repurposing. Such technological practices are largely invisible in the sphere of mass culture. Even when U.S. consumers drop off their old hardware at collection day events or municipal facilities, they do not encounter the toppling mounds of salvage that must be sorted, compacted, and distributed. The piles are concealed by large containers or held behind closed doors. There is, in fact, a long history of concealing or hiding waste in urban and rural spaces that intersects with the history of class and racial politics in the United States.[26] Nuclear waste has been dumped onto the lands of American Indians in the Southwest, and city landfills often sit next to low-income mobile home parks.

As a way of exploring the (in)visibility of waste, salvaging, and repurposing in greater detail, I now shift to a discussion of the cable television series *Junkyard Wars*, which poses somewhat of a challenge to consumerist logics by celebrating technological resourcefulness and adaptive reuse of old machines and junkyard waste. While companies like HMR occupy space in Los Angeles gathering old electronics for redistribution, television producers have transformed an enormous junkyard in the city into the set of *Junkyard Wars*. Both the junkyard and commercial television series are spaces of accumulation. Where the junkyard is made up of old machines and spare parts that settle there, commercial television functions as what Todd Gitlin calls a "recombinant" form, made out of recycled genres, settings, characters, and plotlines.[27] Thus both the junkyard and the commercial television show are, in a sense, generated through practices of salvaging, whether in a literal or metaphoric manner. What is intriguing about *Junkyard Wars* is that the series' concept and format make this quite literal—the show is staged and uses materials from a junkyard, and, at the same time, it recombines elements of the game show, science documentary, and the how-to book. By enacting the collective salvaging and repurposing of machine waste week after week, *Junkyard Wars* publicizes practices that are typically suppressed and avoided in television culture, and, like Dekovic's "Monitors," structures a space for consumers/viewers to rethink their relationship to old hardware.

Wasteland TV

One of the reverberating refrains in U.S. television history came out of the mouth of FCC commissioner Newton Minow on May 9, 1961, when he addressed the National Association of Broadcasters imploring network executives to sit down and watch for themselves. "I can assure you," he told them, "that you will observe a vast wasteland."[28] Who could have predicted that forty years later, viewers of The Learning Channel would be able literally to plunk themselves in a vast wasteland, a wasteland also known as the set of *Junkyard Wars*? I invoke the term "wasteland" in this context certainly not to echo the moral outcries over sexual and violent television content that have since congealed into the term "trash TV," but rather to shift attention to an entirely different set of concerns—the technological waste streams that have resulted from decades worth of television-inspired consumerism discussed in the first section of this chapter.[29] Actual wastelands and junkyards are rarely ever represented on television because of the industry's reliance on the shiny and the new. To place junk in the small screen would require viewers to recognize the toppling mounds of residue that have formed in the wake of happy-go-lucky consumerism. A decade after Minow's speech, though, Norman Lear and Bud Yorkin created the sitcom *Sanford and Son*, adapted from a British series, featuring father-and-son salvagers who operated a junkyard at their home in South Central Los Angeles to make ends meet. Where this dramatic series used the junkyard setting to highlight working-class struggles, *Junkyard Wars* turns acts of technological salvaging and repurposing into elaborately staged competitions.

Produced by British company RDF Media, *Junkyard Wars* first emerged in 1998 on Great Britain's Channel 4 as *Scrapheap Challenge*. After three successful seasons in the UK, Discovery Communication bought the rights to air the show in the United States on The Learning Channel (TLC) as *Junkyard Wars*. The show's producers solicit self-selected teams of hobbyists and tinkerers, plop them into a 5-acre "monster junkyard," and give them ten hours to create assigned machines. Two teams of contestants must compete and work together (with the help of experts hired by producers) to assemble a contraption using only tools and parts found in the junkyard. Whether the challenge is to design Jet Trikes, Land Yachts, or Power Rafts, there is no cash prize; only a trophy is provided to the winning team. The show celebrates technical ingenuity, teamwork, and resourcefulness and has garnered a cult following in the UK and the United States. In 2001, *Junkyard Wars* was nominated for an Emmy Award, and *Time* magazine named it one of the year's top-ten TV shows.

Developed by the nonprofit group American Community Service Network, The Learning Channel (TLC) first emerged in 1980 and aired educational programming for a decade before being purchased by Discovery Communications in 1991. TLC is now one of a handful of other cable networks (including the Travel Channel, Animal Planet, and Science Channel) owned by the Discovery group, and it has undergone a series of transformations during the past decade. Recently in an effort to differentiate TLC from the Discovery channel, executives changed the network's motto to "life unscripted" and added a handful of reality-based series to its schedule. Shows such as *A Wedding Story*, *A Baby Story*, *A Makeover Story*, and *Trading Spaces* (to name a few) boosted the network's ratings 83 percent between 1998 and 2000, and by 2003, TLC ranked among the top-twenty cable networks at number sixteen with 84.7 million subscribers.[30] Commercial television has always mined everyday life for program concepts, but TLC has institutionalized this practice to new ends. Its life-unscripted strategy appropriates activities that people would do anyway—get married, have babies, get haircuts, put on makeup, remodel their homes, tinker in their garages—as fodder for an entire cable lineup. This strategy also draws attention to the extractive and derivative logics of television. Whereas extractive logics involve processes in which television appropriates aspects of lived social experience whether people's lives, homes, bodies, and/or junk, derivative logics involve the reuse, repurposing, and/or recombination of material already in use in the television industry whether formats, stars, settings, or story lines.

TLC's *Junkyard Wars* symbolically exemplifies these logics by transforming the leftovers of twentieth-century consumers into raw materials for an ongoing series.[31] Most of the show's participants are white men wearing brightly colored form-fitting jumpsuits with their team name—whether the Megalo Maniacs, Rusty Juveniles, Kinetic Kids—imprinted on the back side. Many of the contestants are former mechanics, aeronautics experts, engineers, military technicians, or repairmen. The series balances masculine bravado with collective labor, individual brilliance with common sense, and resourcefulness with spectacular innovation.

Each episode opens with the host's introduction of the teams, delivery of the assignment, and announcement that the participants have ten hours in which to create their devices. The teams scurry to their workshops, briefly discuss their strategies, and then salvagers from each side vanish into the junkyard searching for parts. Their ten hours of tinkering is compressed into a thirty-minute segment that features acts of welding, hammering, sawing, screwing, and so on, while the hosts and a judge comment on the teams' strategies, trying to anticipate who will build the best contraption. Their commentaries are interspersed with shots of the teams working on and discussing their machines, animated schematic drawings, and historical footage of the machine's earlier incarnations—whether a hovercraft, rocket, dragster, or walking machine.

The pleasures of the show are organized in two ways. First, there is the experience of witnessing a working machine being made out of junk.[32] In one episode, for instance, Texas Scrap Daddies and the Long Brothers used such parts as a jet engine casing, a shopping cart, an old tent, a boat motor, a car engine, ladders, a sewing machine, and wood planks to assemble their Hovercrafts. The program treats acts of salvaging and repurposing not as tactics of survival or profit-making activity but as creative collaborative practices. And the hosts and judges emphasize this creative salvaging by highlighting intriguing or surprising uses of old machines or parts. For instance, in "Sky Rockets," the Long Brothers use an old pair of pantyhose as a casing for their ostrich egg cargo. In "Amphibious Vehicle," the Navy Blues wrench the fiberglass roof off an old van and use it as a boat. In "Walking Machines," an old refrigerator part becomes a sitting platform. The hosts and judges use a technical vernacular designed to attract viewers with different knowledge and skill levels. As executive producer of program production for TLC Alexandra Middendorf explains, the show aspires to "deliver information and knowledge in a very nonacademic way."[33] The commentaries and animated illustrations inform viewers about various aspects of the design process while building anticipation of the final competition and featuring humorous workshop banter and mishaps along the way.

The show's pleasures are also organized around the final competition, an event that tests the durability of the machine, the viability of its design, and the operational competence of team members. These competitions are staged in remote locations where the contraptions can be deployed in a field, water body, desert floor, or other setting away from urban hazards and suburban interruptions. Set against these empty landscapes, the makeshift machines look all the more compelling and the activities surrounding them all the more bizarre. These segments, which typically last five to ten minutes, feature devices that sometimes perform wonderfully and sometimes fail miserably, and they are punctuated with the sporting gimmick of the instant replay, especially in the American version of the series. In "Hovercraft," for instance, the instant replay is used five times, extracting maximum return from a sequence in which one of the team members flies onto a moist salt flat and as the craft's fabric tears. Indeed, the machine failures are as impressive and thrilling as the successes: it is the unpredictability of cobbled together technologies that makes the show so compelling. As a former contestant puts it, "A dramatic crash can be just as spectacular as a win."

As it exposes a range of usable and unusable parts, functional and malfunctioning machines, *Junkyard Wars* popularizes the notion that technological objects have no fundamental coherence or essential uses and can take on a variety of appearances and be used for different purposes for which they were not invented. In this sense, the show establishes a provocative set of relations among technologies, knowledge, and viewers/consumers that valorizes creative reuse and repurposing of already existing machines rather than the invention and/or consumption of new ones. (This logic is, of course, undermined somewhat because the series is also part of TLC's flow where the promotion of myriad new gadgets interrupts and contradicts the program's ethos of techno-recycling.) Still, one might say the show functions as a metonym for cable television programming in general because the cable schedule itself is filled up with content and forms derived from preexisting elements that have already been used and tested elsewhere.

Although the treatment of creative salvaging as a *competitive standoff* may generate expressive technological practices—awkward machines with outrageous functionality or low-tech devices that accomplish high-tech ends—it is also symptomatic of the military discourses that underpin the series. For as much as *Junkyard Wars* works to popularize salvaging and repurposing as creative acts, it also promulgates the idea that technological innovation is ultimately derived through warring factions, many of whom are pulled directly from military-industrial institutions. This is, of course, hinted at in the series' title, but it is articulated more fully in the way military elements become part of the show's design. Not only do the hosts deliver assignments for the construction of military technologies such as assault vehicles, field artillery, amphibious vehicles, aerial bombers, rockets, torpedoes, and so on, military personnel are invited to serve as experts, judges, and contestants, and the show regularly integrates footage from military training videos and documentaries. This aspect of the show's design is significant because it excavates a repressed dimension in television technology's own history; that is, television emerged in the field of military experimentation before it became a domestic pleasure, and this aspect of television's history itself needs to be salvaged.

During fall 2001, as the ratings of U.S. cable news networks skyrocketed with citizens' concerns about the war on global terror, TLC and RDF Media decided to "retool" *Junkyard Wars*, renaming it *Junkyard Mega-Wars* and explaining, "mega comes from the Greek *megas*, which means 'large, greatly surpassing others of its kind.'" The show's Web site pronounced, "We've got new hosts, harder challenges, permanent team captains, a menacing new look, an engineering sage, a fistful of awesome locations and two dozen of the meanest, toughest and smartest engineers in North America."[34] TLC tested this new format in November 2001, airing a special two-hour episode featuring an international competition among the American Raptors, the British Bulldogs, and the Russian Bears, who faced the challenge of building a military transport device that could race across salt flats, climb a rock pile, and cross a lake. Team members were culled from U.S. defense contractor Lockheed Martin, the British Republican Guard, and science universities in Russia (aided by a German expert).[35]

In this episode the junkyard is figured as a post–Cold War playground crammed with old parts that once circulated in the global economy, a place where teams can find old engines from the UK, the United States, Germany, and Japan. In this zone of accumulation, multiple languages are spoken and different technological styles flourish.[36] While the Americans predictably used the largest Chevy engine they could find to build a monster amphibious vehicle, the Russians fabricated their more lightweight device out of an old rowboat and several motorcycle wheels.

While watching the episode I could not help but think about television's own history because the early tinkering that led to its development was conducted by Russians, Brits, Americans, and Germans who likely repurposed all kinds of scraps and materials in the process. Russian electrician Constantin Perskyi first coined the word "television" in a paper he delivered to the First International Congress of Electricity at the Paris World's Fair in 1900. In the 1910s, English

FIGURE 3.4 *Publicity still from* Junkyard Mega-Wars.

inventor A. A. Campbell-Swinton and Russian scientist Boris Rosing were working independently to develop the cathode ray tube, and by the 1920s, Russian émigré Vladimir Zworykin and Philo Farnsworth assembled different electronic television systems called the Electric Eye (1923) and the Image Dissector (1927). Not only was television technology's development an international affair, it was interwoven with militaristic imperatives as well. As William Urrichio reminds us, the Germans were the first to institutionalize TV broadcasting in 1935, and they continued to experiment with military applications throughout World War II with TV-guided missiles, bombs, and torpedoes until the fall of the Nazi regime. From 1942 to 1945, the United States also used television to guide missiles in the Pacific. Television itself, then, might be imagined as the product of a metaphoric and historical junkyard war that continues to this day.[37]

When Minow called television a wasteland, he intended to publicly condemn and shame the industry. But thinking about television in relation to the wasteland or the junkyard, I believe, can be quite constructive. *Junkyard Wars* pushes us to consider television's relationship to several areas such as global waste streams, structures of international technological innovation, military appropriations, and practices of historiography. The technological and military histories of television are often sidelined in favor of its more pleasurable forms. But with its emphasis on salvaging and repurposing of technological objects, *Junkyard Wars* reminds us that "television" itself, as a mode of production and object of study, can and should be constituted in different ways. Dekovic's *Monitors*, the salvaging of old hardware, and *Junkyard Wars* all suggest different ways of imagining television's materiality, whether through artistic practice, economic repurposing of the technology, or programming for cable television. When television is viewed as an underwater art installation, an object disassembled by migrant workers, or a series staged in the junkyard, it is decentered from its typical sites of analysis.

We have tended to develop research paradigms that treat television first and foremost as a system of commercial entertainment or public broadcasting to the exclusion of other modalities, whether artistic, militaristic, educational, or scientific. The formation of cable and satellite television systems and channels that brand practices of learning, discovery, history, and science (just to name a few) compels us to imagine television in ways that complement and extend beyond paradigms of commercial entertainment and popular pleasure to explore more fully the medium's relation to various epistemological systems and practices of knowledge production. In other words, perhaps we have not been as active and imaginative as we could be at reusing, reconfiguring, reimagining television itself as an object of analysis, as an object that has different material relations to practices that are both derived from and have relevance beyond the discipline of television studies.

The practice of salvaging may be a useful metaphor for television historiography. It is an apt way of describing the process by which television scholars land on and find use within particular mediated sites, plumbing them for scraps, pieces, and parts, and then remolding and refiguring them to satisfy or fulfill lines of critical and historical inquiry. But the term "salvaging," which refers to the act of saving imperiled property from loss, is also important because it implies a creative and exhaustive search for something yet to be rediscovered, and in this sense, it becomes a crucial description for the way television's history is often treated—or perhaps more accurately, negated—in relation to the emerging fields of digital or new media studies. It is as if television was not just a vast wasteland but an ongoing cultural nightmare best forgotten and substituted with more flexible digital forms. If we accept John Hartley's suggestion that television is irreducible,[38] then it may be the historian's compulsion to look for, salvage, and reassemble its parts in new ways that will help us insinuate and materialize it within debates and discussions about new media.

Notes

1. Carolyn Marvin, *When Old Technologies Were New* (Oxford: Oxford University Press, 1990), 5.
2. Ibid., 8.
3. Lisa Gitelman and Geoffrey Pingree, eds., *New Media 1740–1915* (Cambridge: MIT Press, 2003), xv.
4. Lev Manovich, *The Language of New Media* (Cambridge: MIT Press, 2001), 8.
5. It comes from the old French *salvaige*, which refers to right of salvage, and the Latin, *slavus*, which means "safe" as in saved. *The American Heritage Dictionary of the English Language: Fourth Edition* (2000), www.bartleby.com/61/8/S0050800.html. Also see word reference.com at www.wordreference.com/english/definition.asp?en=salvage.
6. Perry Hoberman's art installation *Faraday's Garden* is an intriguing work that appropriates and recontextualizes recycled home appliances. As he explains, "Participants walk through a landscape of innumerable household and office appliances, power tools, projectors, radios, phonographs, and various other personal comfort devices . . . collected from thrift stores, flea markets and garage sales." *Unexpected Obstacles: The Work of Perry Hoberman 1982–1997*, eds. Paivi Talasmaa and Erkki Huhtamo (Helsinki: Paino, 1997), 51.
7. John Seely Brown and Paul Duguid, *The Social Life of Information* (Boston: Harvard Business School Press, 2002), 14.
8. Monte Enbysk, "Don't Dump Your PCs in a Dump," MSN Business, SmallTech, http://www.bcentral.com/articles/enbysk/157.asp.
9. Eric Levin, "Donating MACs Is PC: Refurbished Computers Good for Community and Environment," *The Montclair Times*, December 3, 2003, http://www.montclairtimes.com/page.php?page=6600.
10. Cited in Johnathon E. Briggs and David Haldane, "Recycling Gives Old Computers New Lives," *Los Angeles Times*, April 14, 2000, 3.
11. Aaron Hoover, "TVs, Computer Monitors Contain High Lead Levels, Study Finds," *Science Daily Magazine*, December 7, 1998, www.sciencedaily.com/releases/1998/12/981204091724.htm.
12. Hoover, "TVs, Computer Monitors Contain High Lead Levels." Also see Matthew Powers, "The Trash Folder," *Harper's Magazine*, January 2005, 64–65. Powers indicates, "The half-billion computers rolling toward obsolescence in America contain 6.3 billion pounds of plastics, 1.6 billion pounds of lead, and 630,000 pounds of mercury, along with cadmium, barium, arsenic . . . and other hazardous elements."
13. Electronic Waste Recycling Act of 2003, California Integrated Waste Management Board, (2003), available at www.ciwmb.ca.gov/electronics/act2003/.
14. Dave Ryan, "Americans Encouraged to Recycle Old Computer Equipment, Televisions, Other Electronics in New Campaign Launched by EPA, Industry, Retailers, Recyclers, *Environmental News*, January 10, 2003, http://www.eiae.org/whatsnew/news.cfm?ID=71.
15. HP charges $17 to $31 to recycle and then gives customers a $50 coupon that they can use on purchases of $60 or more when they return old computers and monitors for recycling. Ian Fried, "HP: Don't Trash That Old Computer," *CNETNews.com*, February 5, 2003, http://news.com.com/2100-1040-983548.html.
16. Dell Initiates Industry's First Computer Collection Event Grant Program, Greenbiz.com, November 19, 2003, www.greenbiz.com/news/news_third.cfm?NewsID=26017.
17. "EIA Electronics Recycling Fact Sheet," Electronics Industries Alliance (October 2001), www.eia.org/news/pressreleases/2001-10-15.5.phtml.

18. About 40 million components are scrapped each year in the United States, and it's projected to increase to 100 million by 2010 according to figures from the International Association of Electronics Recyclers and Image Microsystems of Southern California. See Renee Koury, "Electronic Recycling Event Attracts Crowds to Stanford," *San Jose Mercury News*, October 11, 2003.

19. Karen Kaplan, "The Cutting Edge: Computing/Technology/Innovation," *Los Angeles Times*, December 20, 1995, 4.

20. Ross Kerber, "Southern California Enterprise: Yesterday's Computers a Boon for Junkyards Technology," *Los Angeles Times*, November 14, 1994, 2.

21. Briggs and Haldane, "Recycling Gives Old Computers," 3.

22. HMR's Los Angeles facility specializes in deinstalling, buying, selling, recycling, and demanufacturing used or obsolete equipment and computers, and claims to have the only license for the "de-manufacturing of cathode ray tubes," which is the part of the monitor that contains the most lead. See the company's LA Web site at www.hmrla.com.

23. The situation is reminiscent of the expropriation of American Indian reservation lands in the Southwest for the dumping of nuclear waste and landfills.

24. See report, "Exporting Harm: The High Tech Trashing of Asia," February 25, 2002, www.svtc.org/cleancc/pubs/technotrash.pdf.

25. In one sequence, the camera zooms in to reveal that some of the labels and tags on the computer equipment found in Guiyu came directly from California (University of California).

26. See, for instance, David Naguib Pellow, *Garbage Wars: The Struggle for Environmental Justice in Chicago* (Cambridge: MIT Press, 2002).

27. Todd Gitlin, *Inside Prime Time* (New York: Pantheon, 1983), 76–80.

28. Since that time, so-called trash TV has flourished, moving moral conservatives to turn off their TV sets and spurring political officials like William Bennett and Joe Lieberman to give "golden sewer" awards to television shows with violent and sexual content. I thank Henry Jenkins for pointing this out to me.

29. For a discussion of "trash TV," see Kevin Glynn's book *Tabloid Culture: Trash Taste, Popular Power and the Transformation of American Television* (Durham, N.C.: Duke University Press, 2000).

30. By 2001, the show had become TLC's highest rated prime-time series, attracting an average of 5 million viewers per week. TLC used the show to try to attract male audiences ages 25 to 54 who were loyal to the Comedy Central series *Battlebots* and the TLC Show *Robotica*. T. L. Stanley, "*Junkyard Wars*: It's More Than the Sum of Its Parts," *Los Angeles Times*, July 4, 2001, F15.

31. *Junkyard Wars* was created by Oxford graduate Cathy Rogers, who had developed science documentaries for the London-based company RDF Media, and, after seeing the film *Apollo 13*, decided to conjure up a show that would involve building "something out of a bucket of rubbish." Charles Strum, "In Britain It's a Challenge; Here It's War," *New York Times*, December 31, 2000, 4.

32. It is important to note that the junkyard was created for the show as a set and is stocked in advance with parts.

33. T. L. Stanley, "Junkyard Wars," F15.

34. TLC Fan Web site for *Junkyard Mega-Wars*, http://tlc.discovery.com/fansites/junkyard/episode/season11_format.html.

35. In February 2004, the show spun the formula of technological repurposing and warring factions in yet another direction with a special called *Junkyard Mega-Wars: At the Movies*, showcasing crews from Industrial Light and Magic, Jim Henson's Creature Shop, and KNB (special effects designers for *Kill Bill*). This was, without a doubt, the most ostentatious episode of the series. Teams were asked to construct a giant robot that would shoot aliens lurking in the alleyways of a film studio set before crossing the finish line. The contestants were also evaluated on the basis of the aesthetic value of their robots. The episode exposes the ludicrous excesses in American media industries and allegorizes the competition between media companies in an age of conglomeration.

36. At one point a retired British commander serving as the judge is invited to maneuver an amphibious vehicle through a river. Film footage of the beaches of Normandy during World War II appear, and when we cut back to a close-up of his face, he has tears in his eyes related to being behind the wheel of a machine that in his mind was connected to the end of World War II.

37. There is a link on the show's Web site to "How Stuff Works" where you can learn about "How Junkyard Wars Works" and one can follow links to find out "How Landfills Work," "How Television Works," and "How Cable TV Works." Available at http://stuffo.howstuffworks.com/junkyard-wars.htm.

38. John Hartley, *The Uses of Television* (London and New York: Routledge, 2000), 18.

4 New Lamps for Old: Photography, Obsolescence, and Social Change

Michelle Henning

Remediation and Equivalence

Sunlight falls through a slit in the curtains into a darkened room, and on the opposite wall tiny figures move in a faint and inverted image of the street. Light falls through a polished lens onto paper prepared with egg albumen, with cyanide or silver nitrate, ingredients from the kitchen, from the murder mystery, from the pharmacist. Light triggers thousands of minuscule photosensitive cells linked to a miniature electrical circuit made in a factory in Taiwan.

Unlike a number of well-established imaging technologies, such as the X-ray machine, the digital camera is photographic in the sense that it responds to visible light. Yet, since the first digital cameras, many writers have been preoccupied with whether digital photography counts as photography and how it affects our perception of photography and of the world itself. Do the classic theories of photography, which saw photography in relation to bigger questions about truth, desire, and death, hold true for digital photography too?[1] Writing on new media attributes great significance to the differences between analog and digital technologies, viewing the distance between the digital photograph and the photograph as greater than the distance between, say, the Polaroid pack films of the 1960s and the daguerreotype. The perception that digital photography means the "death" of photography is perhaps encouraged by the domination of twentieth-century photography by a standard chemistry of silver iodide prints and celluloid roll film, and by the extinction of what one writer terms the "photodiversity" of the late nineteenth century, when an enormous number of techniques and photographic materials proliferated.[2] Nevertheless, from the perspective of the ordinary users of photography, for whom chemistry and process matter little, digital photography may now seem just a novel format, less foreign than the nineteenth-century pictures on glass and tin. If home printing and processing via computer had made digital photography seem complicated and out of the ordinary, the availability of digital printing in small photography stores and drugstores—now equipped with digital "mini-labs" and digital "photo kiosks"—returns it to familiar terrain.

This familiarization of the new or novel is one aspect of what J. David Bolter and Richard Grusin have termed "remediation." This concept draws attention to the ways old and new intersect and coexist. It refers to both the content and the processes of media, to the reworking of one medium's content in another medium, the transformation of one medium into another, and the hybridization of media. It allows us to see how the new processes of digital photography do not simply supplant but instead transform chemical photography. For instance, photo developers offer digital processing for 35mm film. Film is chemically developed, then scanned and turned into prints or stored on CD. Indeed, photography could be said to be remediated from the moment it was reproduced in magazines and newspapers, on posters and billboards. New digital techniques refashion, rather than replace, old media forms. The billboard poster, a well-established medium, is now produced using digital reproductive processes. The photographic techniques, the graphic techniques, the printing technology, and even the machinery and methods involved in producing the paper and inks may be very different from the processes used in the past. Computer painting on vinyl billboards was researched as early as the 1970s, began to be widely used in the 1990s, and now the paper billboard shares the street with other kinds of billboards such as backlit plastic ones, ones made of rotating vertical strips, three-dimensional ones, and gigantic vinyl "skins."[3]

Remediation also describes how digital photography reproduces the formal characteristics and content of chemical photography. Digital cameras mimic aspects of film photography that really have nothing to do with digital photography, such as ASA/ISO settings—"film speed" settings, even though they use no film, and, as I discuss in more detail later, they are also are increasingly built to resemble traditional cameras. That there is no physical or technical reason for this is evidenced by early digital cameras, which resembled "normal" cameras only minimally—the Apple Quick Take 100 and the Kodak DC40 camera, which were introduced in 1994 and 1995, respectively, were shaped more like binoculars. Bolter and Grusin identify remediation as "a defining characteristic of the new digital media," but they also see it as a defining characteristic of media in general: "Each act of mediation depends on other acts of mediation. Media are continually commenting on, reproducing, and replacing each other, and this process is integral to media. Media need each other in order to function as media at all."[4] Elsewhere they state that, "What is new about new media comes from the particular ways in which they refashion older media and the ways in which older media refashion themselves to answer the challenges of new media."[5]

I want to suggest, however, that those things that are so often identified as changes in media, or problems brought about by "new media," are actually social transformations. The changes we see in billboards are both transformations in the means of production of billboards as a cultural form, and transformations in working practices and working relationships. Indeed, we might view remediation quite differently, if, following Raymond Williams, we recognize that technological changes are never just material changes or even changes in ways of seeing, but "altered processes and relationships in basic material production."[6] Williams's work provides a critical perspective for writings on new media, which tend to separate the medium from the social relationships that surround it, produce it, and are transformed with it. For Williams, the very concept of a medium is misleading because it presumes a preexistent intention or thought that is communicated *through* a particular form, for example, with language: "Words are seen as objects, things which men take up and arrange into particular forms to express or communicate information which, before this work in the 'medium,' they already possess."[7]

Williams's attack on the concept of the medium is informed by the 1930s work of the Soviet linguist V. N. Voloshinov, who argued that not only does a concept not exist outside its formation in language, but also that language is a "material social practice."[8] For Volosinov, communication is the material exchange of signs, and signs do not exist *except* as matter (even if this takes the form of vibrations in air) and only in the continually changing context of social interaction.[9] Using Volosinov, Williams explains that the concept of the medium confuses the means or material of a given practice with the practice itself. For Williams, practice "has always to be defined as work on a material for a specific purpose within certain necessary social conditions."[10] Williams's concept of media as material social practice opposes the tendency of media theory to treat the medium as a thing with its own properties.[11]

Photography, then, may be considered not so much a medium as a social practice that makes use of a range of different materials and means. This suggests that remediation should be considered in terms of changes in social relations and in the manipulation of physical "stuff" rather than as a property or tendency of new digital media. From this perspective, remediation cannot be separated from human social practice and especially from working practices and the exchange of commodities. To illuminate the relationship between old and new media, media studies needs to engage in questions about value and exchangeability, and to view media as practices and technical means that are part of an industrial, commodity culture. It is only in this context that remediation takes place. New media theory that does not take account of this context ends up fetishizing media. One antidote is provided by social anthropologist Arjun Appadurai, who advocates what he calls a "methodological fetishism":

> [W]e have to follow the things themselves, for their meanings are inscribed in their forms, their uses, their trajectories. It is only through the analysis of these trajectories that we can interpret the human transactions and calculations that enliven things. Thus, even though from a *theoretical* point of view human actors encode things with significance, from a *methodological* point of view it is the things-in-motion that illuminate their human and social context.[12]

By treating the things of media—the artifacts, the technical apparatuses, the material texts, as if they, like living things, have lives and therefore potential biographies, we can trace their paths as they pass across social classes and from newness into obsolescence. The relationship between new and old media is best understood in terms of value and via the actual exchanges and transactions give them their social significance. The exchangeability of cameras, technical equipment, and consumables, as well as the value of the skills needed to practice with these, relates to the "regime of value" in a given society, which is politically determined. Appadurai argues that societies establish equivalences between objects, and rules governing their circulation.[13] Remediation may similarly be conceived as the setting up of equivalences.

Another writer who can help us shed light on this process is Evan Watkins. In a study of social obsolescence, Watkins discusses the production of what he terms a "field of equivalence," which allows one thing to be viewed as replacing the other.[14] Rather than one kind of media superseding another automatically, the two first have to be established as equivalent in their use. Watkins argues that obsolescence is an ideologically produced designation. To study the *production* of obsolescence necessarily means to attend to social and cultural processes, to the production of a "field of equivalence." We can see exactly this process with digital photography. It is the production of "a field of equivalence" that makes digital photography a replacement for celluloid and silver-based photography.

The Production of Obsolescence

The technical means of digital photography did not develop from chemical photography. Their history is more closely linked to the development of magnetic recording technologies and computer memory technologies. The light sensors or charge coupled devices (CCDs) used in digital cameras were not initially a substitute for film but were first made in the 1960s as the memory chips for computers. CCDs respond to light as photosensitive emulsion does, but instead of the image being able to be fixed by a chemical process, their thousands of tiny photocells become electrically charged but then quickly lose that charge. We can think of the difference between this and all other photography in terms of capture and storage. The technology of digital photography developed out of what were essentially storage media. In a sense, all media are storage media, as the German theorist Friedrich Kittler has persuasively argued.[15] Even so, the distinction between capture and storage is sharper in some media than others. Until digital photography, all photography used photosensitive emulsion that allows the image to be stored on the same material the light has hit, so storage and capture occur at the same moment, on the same material. In the most common process this forms the negative, from which multiple positive copies can be made. Photography is thus characterized by its reproducibility and its ability to preserve a likeness, but also, and importantly, by its direct relationship with nature or the real.[16] In digital photography, light is sensed in what is again an analog process, but this is quickly translated, encoded, and stored, so that the same sensors are available to be used to record the next image. This is a fundamental difference between digital and chemical photography. It does not mean that digital photography is any less photographic than chemical-analog photography is. It means that it deploys very different processes, which had to be first established as equivalent to photographic processes, before digital photography as a practice became possible.

Although some technical innovations do become the material of new social practices, the vast majority are simply inserted into existing practices, sometimes increasing the reach and viability of that practice in very significant ways, sometimes simply replacing a now-outmoded technology. This is evident in the development of digital camera technology out of magnetic recording. Magnetic recording was first developed to store a performance already captured or mediated through broadcasting. One of its earliest uses was in the radio broadcast of prerecorded performances in the 1930s and 1940s, although the basic principle of magnetic signal storing had been known since the 1890s. In wartime Germany, and then in the United States in the late 1940s, magnetic recording was not a replacement for another technology but a solution to the problem of live performance. In America, Bing Crosby financed the mass production of tape recorders developed by Jack Mullin and Bill Palmer and based on the German Magnetophon. This enabled him to replace his live radio shows, which he had to repeat for the West and East Coast audiences, with prerecorded ones. In 1951, Bing Crosby Enterprises developed a video tape recorder, which allowed Crosby to do the same thing with his TV shows. From this technology were developed the first video stills cameras.[17]

Initially, therefore, magnetic tape and video technology were not considered replacements for another extant technology. They answered a need that resulted from the particular economic arrangements of media production of the period and from the demands of a centralized radio that crossed time zones. Thus the basic technology of the digital camera first developed out of a set of needs and desires related to the storage and replay of information, and it was then established as a technology equivalent to photographic cameras. This has been a slow process: as early as the 1970s, filmless electronic cameras were being developed by electronics firms (such as Sony), technical instruments firms, and companies that specialized in photography (Kodak), but they were not in popular use until the mid-to-late 1990s, and in 2004, new film-based cameras continue to be introduced.

Theoretically, the most significant advantage of digital cameras would be in reportage photography, where images need to be sent rapidly from one part of the world to another, yet early digital cameras were basically compact snapshot cameras, unsuitable for professional use.[18] The mass marketing of the digital camera addressed expectations shaped by the marketing of the Polaroid, one-hour processing, and domestic computer use. It played on, and helped create, new frustrations at the length of processing time or the difficulties of manipulating conventional photos. The practices of popular or domestic photography include not just the taking of the photos but also their circulation and display. The digital camera enables people to quickly circulate pictures to their friends, although only within a technically changed practice of circulation using Web pages and e-mail. Thus the story of the development of the new technology of the digital camera is part of the story of the transformation of a mundane, popular, and long-established practice—the everyday practice of photography. It is also a story of the production of obsolescence. This is true to some extent of all new technical developments. Even Crosby's investment in tape and then video recording was driven by a desire to make certain working practices obsolete (as well as, apparently, to increase the time available to play golf). In the case of digital photography, the successful marketing of consumer-model digital cameras depends on getting consumers to accept a relation of equivalence between them and traditional cameras.

The key point here is that photography was not being outpaced or becoming obsolescent, it had to be *made* obsolete, and its obsolescence had to be presented as inevitable. Remediation is therefore necessary to the production of obsolescence. The digital camera is marketed as an ordinary camera with added value: you can now do without film, view your images immediately, use an LCD instead of a viewfinder, and quickly transfer images to your and others' computers. By 2004, digital cameras were being made by the major names in photography, such as Minolta, Canon, Olympus, and even Leica. The next stage in the production of obsolescence was the increased focus on getting professional and hobbyist photographers to take digital photography seriously. This was hampered by the gimmicky associations of digital photography (cell phones with cameras, and so on). As digital photography aspired to become photography proper, the digital cameras at the top end of the market were made to look and work very like conventional cameras. The black and chrome body of the Leica Digilux 2 looks from the front like a reassuringly familiar, and expensive, old-style Leica. It gives an analog feel by returning control of the camera to rings mounted on the lens (rather than buttons), and it is marketed as an "analog digital camera."

The technology of the digital camera is being constructed as a replacement for analog/chemical technologies instead of as an alternative technology with its own specificity. In many ways, this is a loss. Downmarket and early digital cameras can produce digital effects different to film, and their small lenses and CCDs mean that they produce greater depth of field, but these qualities are never used as the basis for marketing them, and they are played down in camera design. The more upmarket cameras remediate photography so successfully that these distinctive characteristics are hardly noticeable. The "remediation" of photography by digital photography is not to do with tendencies inherent in media, but to do with the process of finding new markets for innovations and with the cultural and ideological production of obsolescence.

Although the new is (increasingly) not new for long, obsolescence is seldom produced overnight, as the example of the digital camera shows. Social and cultural change is uneven, with elements of the new coexisting with older forms and practices. Williams pictures culture as a dynamic process, so that at any point we have not only a dominant culture (by which he means the culture which best serves the interests of the dominant social group) but also "residual" and "emergent" aspects of culture. By attending to the recently outmoded and the obsolescent, culture as process becomes more evident. Yesterday's new thing does not simply disappear, nor does it just get refashioned by new technical processes or as new media content. Instead it continues to exist at the margins of culture. To consider residual media is also to consider the residues of past social arrangements and relationships, which continue in the present, although perhaps in fragmentary forms. As a stage in the social life of a thing or practice, obsolescence is determined not by the qualities of the thing itself but by the culturally produced regime of value. Things and practices do eventually become completely defunct—by being broken or by becoming impossible to sustain because of the unavailability of materials. We can see this in the case of certain old media, where hardware components become impossible to acquire, or, in the case of film and photography, where film stock is no longer manufactured. But certain highly valued objects and practices are culturally protected from being obsolete even if they lose their original function. Nor is relegation to the obsolete necessarily the final stage in the social life of a thing, as things may be restored to commodity status by being revalued as collectibles or antiques. Also, the cultural and social standards of "exchangeability" may produce the deterioration of exchange value in the obsolescent object, which may continue to have a high use-value. Its low exchange value makes it available to people who could not previously afford it. As Watkins points out, for many people, "[O]bsolescence . . . is a precondition for the availability of technologies."[19]

For Watkins, technical obsolescence is, in contemporary society, linked to the ideological production of certain populations as obsolescent, as leftovers irrelevant to contemporary society. Watkins's argument centers around what he sees as a shift in dominant ideology from "natural coding" to "technoideological coding."[20] In the well-known formulations, ideology is understood in photographic terms, as a frozen and inverted image of the status quo. Ideology naturalizes; it makes the existing social hierarchy seem inevitable and stable by an appeal to a static, eternal nature.[21] This is something ideology shares with myth. Roland Barthes, who used the two terms almost interchangeably, expressed it in the well-known phrase "myth turns history into nature."[22] Technoideological coding, according to Watkins, appeals to technology instead of nature. According to this model, to endure is to fail to move with the times. Survival is not "survival of the fittest" as in social Darwinism, but characteristic of remaindered and outmoded populations. Dominant ideology represents obsolescence as useless survival, rather than as something actively produced in the present.[23] Watkins insists that technoideological coding does not naturalize obsolescence. In coding people as obsolescent, just as an object or technology becomes obsolescent, it allows for the possibility that social positions and practices are not naturally given or simply the products of fortune. Technoideological coding accepts that an obsolescent social position or outdated practice can be forced on one by the choices made by others. However, it doesn't allow for the fact that the obsolescent are not irrelevant but in fact systematically or structurally necessary to a society that is based on social inequality.[24] Watkins shows how a discourse of technological progress gets applied to populations, but also how the two things—obsolescent technology and obsolescent people—become linked. Marginalized and exploited populations that are characterized as obsolescent are surrounded by obsolescent material culture, and out of necessity, they engage in obsolescent cultural practices.

Using the work of Williams, Appadurai, and Watkins, we can conceive of the production of obsolescence as a process that involves putting the old object or technology in a relationship of equivalence with the new, unyoking it from the relationships of equivalence that established its exchange value within a particular society. It is also a process connected with the marginalization of social positions and practices. Attributions of value affect the trajectories of objects and technologies through society. Residual media are material social practices, remainders of past social formations. Nevertheless, their residual character is also a product of the present, necessary to the production and maintenance of existing social hierarchy. On becoming residual, a cultural form does not become inactive or culturally irrelevant (although we might be persuaded to see it that way—as so much trash) but becomes available to those social groups also designated as "trash."

The Obsolescent and Oppositional Culture

If the obsolescent becomes the culture of the dispossessed and marginal, so the new becomes linked to positions of dominance. Watkins argues that the production of distance between the new and the old is a means of maintaining dominance through distinction from others:

> The positional "profit potential" of innovation and change is a matter of extorting the maximum of distance and distinction from each shift that occurs in the social field. Thus, rather than simply producing the conditions of change, whether new technologies or whatever, dominant social positions must depend on an emergent distance between what henceforth can be temporarily marked as "old" and "new" in the reconfigurations of positionality. That distance functions like a kind of "surplus value" over and above the fact of change and the labors of innovation, and is available to be realized as the social capital of distinction from "the others" condemned to a now rapidly disappearing configuration of what used to be.[25]

In academia and academic publishing, the study of "new media" (however critical) is engaged in this game of distinction. More worryingly, because the disciplines of cultural and media studies were founded on the basis of an interest in the mundane and in working-class culture, the emphasis on new media may restrict the focus to the culture of those with this kind of social capital and to dominant narratives about the direction of cultural change. The study of obsolescent media and outmoded cultural practices shifts this emphasis because of their availability for oppositional culture. Williams distinguished between those elements of residual culture that operate as alternative and oppositional and those that are co-opted for the purposes of the dominant class. Like Volosinov, he saw culture (and therefore value) as struggled over, as part of the social struggle between classes. This struggle involves the "dominant culture" (something Williams treats as perhaps more monolithic than it is) incorporating oppositional practices and oppositional groups appropriating from the dominant culture. Williams recognized that factors such as "the distribution of power or of capital, social and physical inheritance, relations of scale and size between groups" put "pressures and limits" on the possibilities for appropriation.[26] However, historically, there have always been opportunities for resistant appropriation. Williams pointed out that educating people to engage in one communications practice (such as reading the Bible) cannot prevent them using those skills for different purposes (such as reading the radical press).[27]

Although obsolescent materials may be more available for appropriation, their use-value for oppositional culture varies.[28] In the case of obsolescent hardware and outmoded media equipment, certain characteristics lend themselves to uses other than those intended by manufacturers or pushed by dominant financial interests. Sometimes these characteristics get written out as novel versions of the technology appear. As many technologies become increasingly miniaturized, increasingly multifunctional (or "convergent"), they may lose something of their capacity to be used "against the grain." Similarly, some obsolescent material and practices may be more easily incorporated than others. Writing about emergent (rather than residual) culture, Williams described advanced capitalist societies as increasingly able to expand into all aspects of experience and to seize on all new (as distinct from merely novel) cultural elements, but nevertheless,

> even here there can be spheres of practice or meaning which, almost by definition from its own limited character or its profound deformation, the dominant culture is unable in any real terms to recognize. Elements of emergence may indeed be incorporated but just as often the incorporated forms are merely facsimiles of the genuinely emergent cultural practice.[29]

Similarly, the dominant culture tends to incorporate residual culture, but only in limited ways. The nostalgia industry is one means of incorporating the residual and defusing its political significance. Evan Watkins argues, "There's every reason dominant ideological productions work very hard to endlessly construct itineraries of the obsolete as survival narratives, to flood so-called mass culture with nostalgic reproductions of a fading past—because obsolescence when reproduced as nostalgic object is no longer dangerous."[30] Nostalgic reproductions could be commodities (fake brass and mock-hardwood ceiling fans, distressed pine furniture) or media recreations of "obsolete" groups (such as the working class or Native Americans). These reproductions are "mere facsimiles" of the obsolete.

Whereas Williams sees both emergent and residual culture as having oppositional potential, Watkins sees obsolescence as key to oppositional politics.[31] He is interested in the availability of obsolete material, identities, and practices to be "repaired" and put to use for different ends.[32] Similarly, I am interested in the political effectivity of the appropriation of obsolescent and "lo-fi" materials and techniques for photography. However, photography is a good example of a social practice in which the obsolescent is currently being renewed through being incorporated into a commercialized "alternative" culture. The current fad for lo-fi cameras—especially the Lomo—may be understood in this context. The Lomo is a 35mm "point-and-shoot" compact camera with a glass lens, first produced in the 1980s by a Soviet optics company.[33] It was adopted by artists and photographers who were attracted to the saturated colors and unusual effects its lens produced. Now the product is marketed in terms of an "attitude," a way of taking photographs involves shooting "from the hip," and owning a Lomo connects you to a shared lo-fi aesthetic and a whole fan culture via the "Lomographic Society International."[34] A 1980s camera on the point of obsolescence has been revived as a fashion statement. Its origins in the Soviet Union just prior to perestroika became mythologized, translated into positive connotations of Soviet austerity and scientificity. Although Lomo cameras could be used in oppositional ways, its obsolescence is no guarantee of this. Through its renewed fashionability, the Lomo becomes, not obsolete or residual, but "new" in the sense that it enables one to stay ahead in the game of social distinction, and it allows one to feel "alternative" and different without necessarily taking an oppositional or critical stance. This thin line between obsolete and revived technologies points to one of the difficulties in applying Watkins's argument to obsolete media.

We also need to be cautious about seeing obsolescent populations as necessarily engaging in oppositional practices, since this includes the most destitute and disempowered in society. This is a population with nothing to lose but which also can be easily "bought off," which is one of the reasons Marx was so scathing about the *Lumpenproletariat*. Furthermore, extreme social exclusion tends to encourage cultural practices more mythical than oppositional. Take, for instance, the cargo cults of Oceania and Melanesia. The cargo cults developed out of the sudden impact of colonial relations and the perception of European wealth as somehow magically acquired. They involve the ritual imitation of European social forms and the promise of desirable goods suddenly arriving by ship or dropping out of the sky from airplanes. Appadurai says,

> Cargo cults . . . represent a particular mythology of production of European finished goods by natives embroiled in the production of commodities for the world trade. . . . Cargo beliefs are an extreme example of the theories that are likely to proliferate when consumers are kept completely ignorant of the conditions of production and distribution of commodities and unable to gain access to them freely. Such deprivation creates the mythologies of the alienated consumer.[35]

But perhaps even mythical practices may work critically on the dominant culture, exposing its limitations and broken promises. Cargo cults combine cultural practices and beliefs designated as obsolescent with entirely new ones to produce a hybrid culture, which may not be oppositional but does have a utopian aspect. By reinvesting a modern technical object, the airplane, with mythic significance, the cargo cult evokes the utopian potential of flight. Planes do drop food occasionally, but they also drop bombs. Early imaginings of human flight had anticipated much more generous possibilities, such as Leonardo da Vinci's vision of snow being collected from the mountains and dropped on hot city streets.[36] The cargo cult takes the utopian image for the reality, and in doing so it exposes the broken promise of technological modernity.

To summarize: obsolescent media become politically significant in a society where newness has become linked to social distinction and dominance. However, the obsolete can also be pulled into that game of distinction and stripped of its radical potential, through nostalgic reproduction or through being marketed as "alternative" culture, although it is never fully exhausted by this process. Obsolescent materials and practices may be deployed for mythical purposes as well as oppositional ones, but even mythical practices may have utopian potential. To see obsolescent material and residual practices as the material for utopianism rather than (necessarily) outright opposition is to make a more modest claim for them, without excluding the possibility of resistance. Indeed, imagining the world as other than what it presently is, in ideal form, may be a necessary precursor to radical social change.[37] We can see the outmoded and obsolescent as a means to catch a glimpse of other possible worlds, to imagine how the world might be "otherwise."

Trash Aesthetics

Materials and technologies are necessarily ambivalent, able to be pulled this way and that, and brought to life through their use in social practices. This does not mean they have no transformative power or specific effectivity of their own. This ambivalence and this transformative potential are captured in Walter Benjamin's writing of the 1930s. In his essay "The Work of Art in the Age of Its Technical Reproducibility," his optimistic characterization of the liberatory potential of film technology is framed by observations about the devastating effects of technological modernity and the fascist use of film and photography to spectacularly aestheticize war and violence.[38] Contrasted with this is the modernist project of the twentieth-century avant-garde and of early photographers and filmmakers, who exposed (literally, photographically) the emergent potential of photographic film. Benjamin described film's ability to enable people to go traveling, freed from their "narrow streets." In "A Small History of Photography" (1931), he described how photography brings things closer and makes perceptible every step of a movement, producing an "optical unconscious."[39] Early twentieth-century avant-garde photographs explored these qualities of photography, using extreme close-ups and fast shutter speeds as a means to represent everyday objects and scenes in unfamiliar or uncanny ways (photographer who used such devices in the 1920s and 1930s included Jacques-André Boiffard, László Moholy-Nagy, and Man Ray).[40] For Benjamin, the technology of film had critical potential in both its artistic and its popular uses. This potential was stymied by the attempts to hook that new technology to older notions of art and the artistic, as in studio portraits made to look "artistic" with props and backdrops imitating paintings, by artists' attempts to establish an art photography that deployed painterly symbolism, or in commercial attempts to use photography "creatively."

For Benjamin, the revolutionary potential of film technology is tied to its reproductive power and the concomitant destruction of the aura of the art object. His perspective offers an insight into another problem with the revival of obsolete photographic techniques, in those cases where the old technology becomes a means of returning photography its aura, giving an instantly "creative" look to anything photographed. We can find examples of this in Lomography. There are also a number of toy cameras—most notably the Diana and the Holga—that have become popular with photographers because of their unpredictable results and tendency to leak light.[41] This outdated equipment can be used to produce images that counter the emphasis on a glossy high-resolution world of appearances, and even has utopian potential in allowing us to see the world differently. However, grain, low resolution, distortion, and light leaks have become associated with art photography, to the extent that these cameras are increasingly marketed on this basis. Diana cameras fetch exorbitant prices on eBay, while Holgas are now sold in Lomographic kits with booklets and accessories. There is even a flexible lens called a Lensbaby that can be used to give digital photographs the appearance of having been taken by a Holga.[42] Instead of the technical means of photography allowing us to understand the world anew, as Benjamin saw it, they can become simply a means to endow the everyday with aura. As Benjamin wrote of commercial photography, "*The world is beautiful*—that is its watchword. Therein is unmasked the posture of a photography that can endow any soup can with cosmic significance but cannot grasp a single one of the human connections in which it exists."[43]

Although Benjamin disparaged attempts to make photography imitate older media such as painting, he was as much a theorist of the outmoded and discarded—the detritus of modernity—as he was a theorist of the avant-garde or of "new media."[44] He viewed the outmoded products of the recent past as revealing the false association of "technological change and social betterment."[45] His incomplete work, the *Arcades Project*, is both a study of the outmoded and a critique of the concept of progress. Like Raymond Williams later, Benjamin saw oppositional potential in both emergent and residual culture. Also, like Williams, he carefully distinguished between the new and the novel. He attacked that which passes for progress in capitalist modernity as "transiency *without* progress, a relentless pursuit of novelty that brings about nothing new in history."[46] For him, the outmoded contains the unrealized promises of modernity and the past's dreams of the future. Emergent technologies have potential that passes unrecognized because of the insistence on seeing the new as taking over from and improving on the old: "Technology, not yet 'emancipated' is held back by conventional imagination that sees the new only as a continuation of the old which has just now become obsolete."[47]

We can see this with digital camera technologies, the full potential of which are disguised in the attempt to make them appear just a new, improved means for the same old practice of photography. As I have already argued, the ideological production of the new as replacement for, and improvement on, the old is part of the production of a "field of equivalence." Benjamin shows how the production of equivalence is not just about establishing hierarchies of value, creating markets for new commodities, and producing obsolescence, but also about the construction of a linear progressive model of history. The obsolescent can be used to smash this ideology of progress. In photography, old technologies can be deployed in ways that disrupt common perceptions of the relationship between present and past. Alongside recent attention to the digital image has emerged a literature on scientific imaging techniques, wire photos, and other unusual and marginalized photographic practices and techniques. Numerous art photographers have returned to techniques such as pinhole photography and photograms or reinvented techniques that had been obsolete, such as the ambrotype and the cyanotype. Some artists, writers, and curators using and advocating these techniques want to return to a handmade art photography, or to reinvest photography with aura in the face of the ubiquity of the silver print, the advent of the digital image, and the institutionalization of the critical, conceptual art photography of the 1970s and 1980s. Ironically, many artists using old techniques share with the advocates of new media a kind of technical fetishism.[48] Yet there are others who are working with these old processes in more unorthodox ways, such as Justin Quinnell and Steven Pippin, who are among a number of photographers reinventing pinhole photography using everyday modern objects such as dustbins and washing machines as cameras. Whether old techniques are revived or reinvented, they are, in my view, most interesting when they manage to denaturalize the relationship between an era and the technical means by which it is pictured. Old techniques might also destabilize notions of social and technical progress by offering us an image of the now as past.

Benjamin's own writing uses a similar strategy. In 1927, when he began the *Arcades Project*, the covered shopping arcades of the late nineteenth century were still extant but outmoded. Benjamin described them as prehistoric: "[T]oday the Passages lie in the great cities like caves containing fossils of an ur-animal presumed extinct: The consumers from the pre-imperial epoch of capitalism, the last dinosaurs of Europe."[49] The prehistory the arcades represent is an earlier stage of capitalism, made obsolete by the department stores. The description of the arcades as prehistoric also allows Benjamin to attend to them as if they were natural phenomena, describing them like museum and zoo exhibits, geological formations, traces of ancient flora and fauna.[50] By reading recently outmoded cultural artifacts as natural-historical ones, he intended to overturn (even satirize) the prevalent faith in human history as evolutionary progress. His natural history emphasizes the transitoriness of contemporary reality and the rapidity with which it, too, will become ruin or fossil.[51] Benjamin shows how our experience of time and the relationship between new and old is tied up with aesthetics and embedded in material things. Yesterday's latest thing is today's trash, and history can be seen not as an onward march but viewed backward, as the production of obsolescence, the piling up of detritus. In his "Theses on the Philosophy of History," a much-cited passage inspired by Paul Klee's *Angelus Novus* picture inverts the conventional representations of historical progress. History is depicted as disaster, and progress as a storm that drives into the future leaving a growing pile of debris in its wake.[52] For Benjamin, the task of the historical materialist was to sort through this debris, like the ragpickers of nineteenth-century Paris. In the trash are commodities that have lost their allure, which can be used to undo the phantasmagoric appearance of modern reality. Against the glittering spectacle of modernity-as-progress, the obsolescent and discarded reveal the destructive side of modernity.[53]

Benjamin's writing draws attention to another way in which we might think about obsolescent photography: in terms of discarded artifacts as well as old processes. By linking obsolescence with waste, we reveal the connection between the production of obsolescence and discourses of hygiene. Even today, obsolescent goods such as cameras, video players, and TVs are seldom just thrown in the garbage: people store them, give them away, or sell them. Often, people find it emotionally difficult to rid themselves of the objects they have accumulated.[54] Even so, once those objects are disposed of they become associated with dirt, even if they have never been near a trash can. Hence thrift stores are considered by some to be dirty, and there are taboos about what used goods can be bought, whereas newness carries associations of cleanliness. The dispossessed cannot afford the same notions of hygiene. They have to depend on the material culture discarded by others, either for their own use or for economic reasons (modern-day ragpickers are the people who earn money sifting through rubbish dumps—the *pepenadores* of Mexico City, for instance, or the homeless in the United States who collect discarded bottles). If obsolescent goods become available to those populations ideologically produced as obsolescent, their reuse of them also compounds the sense of them as "dirty." Disorder and dirt go hand in hand. The belief that secondhand stores and markets are unclean is connected to their disorderliness. In discourses of cleanliness, dirt is outside classification.[55] So although historical progress appears orderly and rational, behind it is a disorderly heaping up of leftovers torn from their historical contexts. Another way of looking at this is that decontextualization itself renders things obsolete. For instance, one woman's photographs of her holidays may have been her most valued possessions, but they have no exchange value and no significance to anyone else except her immediate family and friends. The outdated technologies of media may retain some exchange value because they have a meaning and use for a wider community, but the cultural products of amateur media practice are usually destined to become trash once they become detached from the narrow group of individuals who had made and kept them.

Even this trash can become renewed and reactivated for critical purposes when it is reinserted into a social practice. The appeal of the objet trouvé for the historical avant-garde, especially the surrealists, was based in their interest in "profane illumination" or "materialist revelation." Their passion for flea markets and found objects was one that Benjamin shared.[56] The chance encounter offered an opportunity for the disruption of everyday life, an opportunity to be jolted from the rhythm of daily existence and to glimpse other possible realities. A similar interest in reviving discarded things can be found in the current trend for collecting found photographs. These are photographs discovered lying on the street, in dumpsters, or bought in flea markets and thrift stores (often sold only because the album in which they are kept is considered valuable). A number of Web sites are dedicated to these collecting practices. Several contemporary artists have also made found photographs the basis of their practice, such as Patrick McCoy, who has used photographs collected from the streets of Dublin, and Joachim Schmid, who since 1982 has been collecting, cataloguing, archiving, and assembling sequences of photographs found on the street.

Some of the criticisms that have been leveled at the surrealists for their interest in the chance encounter also hold for the current collectors of found photographs.

For one thing, they risk mythologizing these encounters, treating chance as inherently meaningful, and found objects or images as merely the stimulus for imaginative wandering.[57] This can be seen from the Web sites, where collectors speculate about the lives of the people in the photographs and sometimes treat the chance encounter as magical evidence of unseen forces, even implying that the mystery of the photograph would be lost if more were known about it.[58] Here, too, the engagement with obsolescent media can be nostalgic and a means of reinstating aura—which Benjamin defined not just in terms of the uniqueness and mystery of the art object but also as something that provides a profound sense of connection with the past.[59] Decontextualization makes the photograph available for aesthetic contemplation.[60] Benjamin saw the "true collector" as understanding his collection historically, giving things back their historical place and makes them "present" and meaningful once more.[61] But he also asserted that real collectors, including, and especially, children, re-enchant objects because by taking these decontextualized objects and giving them a new context in the collection, they release them from the existing regime of value. Through their aesthetic and imaginative attachment to objects, they provide a glimpse of another world not dominated by exchange value. If the torn or trodden photograph lying in the street speaks of a tragedy or a broken relationship, and the photo album found in a house clearance sale, of a forgotten and unmourned life, then the collector becomes a loving, if surrogate, mourner. In patiently giving order to the obsolete, collectors attempt to mend the rift that modernity violently produces between the present and the past. In doing so, they produce counternarratives to the ideology of progress. There is no reason, in theory, why digital prints might not also be found torn on the street and lovingly collected. But the collector of photographs might well feel, like Roland Barthes in *Camera Lucida*, that the particular poignancy of the found photograph is to do with its truth claim and based specifically in its analog-chemical nature (even while they are happy to digitize the photographs and display them on Web pages).[62] My point here, though, is not about one kind of photography over another but about the critical potential of the overlooked and obsolescent. Media theorists, like certain artists and collectors, would do well to turn their attention to the recently outmoded.

Of course, collectors can also be seen as the vanguard of the antiques trade, and their practice as the beginning of the recuperation of the obsolescent back into commodity status. The collector as bargain hunter or speculator, looking for economic gain, is not the collector Benjamin had in mind. But it is a short step from one kind of re-enchantment to the other, to the restoring of commodity value and the setting back in circulation of the obsolescent, cleaned up and scrubbed of its revelatory power. Today, not only do objects, cultural practices, and media texts become obsolescent increasingly quickly, they remain obsolescent for less time, before being seized on as collectibles, renewed as commodities, revived, replayed, and repeated. To return to the concept of remediation, it is worth noting that this word also has another meaning—the cleaning up of contaminated land for reuse. The collectors, the allegorists, and the ragpickers operate in that contaminated space, sorting through the dirt before the cleanup operation begins. Their time is limited: the space for appropriation of the obsolescent gets smaller and smaller. But it can never quite disappear. For capitalism relies on the production of obsolescence.

Notes

1. Geoffrey Batchen, in particular, has addressed this. See his *Burning with Desire: the Conception of Photography* (Cambridge, Mass., and London: MIT Press, 1999). For writing on photography and desire, see Victor Burgin, ed., *Thinking Photography* (London: Macmillan, 1982). On photography and death, see Susan Sontag, *On Photography* (New York: Farrar, Straus and Giroux, 1973/1978). On photography and truth, see André Bazin, "The Ontology of the Photographic Image," in *What Is Cinema?* (Los Angeles: University of California Press, 1967/1974). On the relation of photography to truth, desire, and death, see Roland Barthes, *Camera Lucida: Reflections on Photography* (New York: Hill and Wang, 1981).

2. Lyle Rexer, *Photography's Antiquarian Avant-Garde: The New Wave in Old Processes* (New York: Harry N. Abrams, 2002), 16.

3. This information is from the Web site of the Outdoor Advertising Association of America (OAAAA) at www.oaaaa.org/outdoor/sales/history.asp (February 9, 2004).

4. J. David Bolter and Richard Grusin, *Remediation: Understanding New Media* (Cambridge Mass.: MIT Press, 1999), 5, 55.

5. Ibid., 15.

6. Raymond Williams, *Marxism and Literature* (Oxford: Oxford University Press, 1977), 162–63.

7. Ibid., 159.

8. Ibid., 165.

9. V. N. Volosinov, *Marxism and the Philosophy of Language* (Cambridge, Mass.: Harvard University Press, 1984).

10. Williams, *Marxism and Literature*, 160.

11. Ibid., 159. Interestingly, a recent book argues the case against Williams's media theory, seeing him as a humanist who treats the materiality of media and technology as subordinate to their human use and place in human social interactions. Here I am using Williams, alongside the work of Appadurai and Volosinov, to argue for an approach that takes more account of the materiality of media and its existence as social practice simultaneously. See Martin Lister et al., *New Media: A Critical Introduction* (London: Routledge, 2003), 296–314.

12. Arjun Appadurai, "Introduction: Commodities and the Politics of Value," in *The Social Life of Things*, ed. Arjun Appadurai (Cambridge: Cambridge University Press, 1986), 5.

13. Ibid., 14–15.

14. Evan Watkins, *Throwaways: Work Culture and Consumer Education* (Stanford: Stanford University Press, 1993), 27–28.

15. See Friedrich Kittler, *Gramophone, Film, Typewriter,* trans. Geoffrey Winthrop-Young and Michael Wutz (Stanford: Stanford University Press, 1999).

16. See Henry Fox-Talbot, *The Pencil of Nature* (Longman: London, 1844).

17. See Heinz Ritter. *An Introduction into Storage Media and Computer Technology* (Ludwigshafen: BASF, 1988) and Peter Hammar, "John T. Mullin: The Man who put Bing Crosby on Tape," *Mix*, October 1, 1999, http://mixonline.com.

18. As TV companies and press compete to rapidly visualize events "as they happen," the digital camera shortens the time between an event and its visual reproduction, although this "speeding up" is not an inevitable effect of modern life but a social and economic imperative.

19. Watkins, *Throwaways*, 26.

20. This is a shift not just in dominant ideology, but in the very mechanism of ideology.

21. Evolutionary theory views nature in terms of change, but in the ideological use of the concept of nature it is conceptualized as static.

22. Roland Barthes, *Mythologies,* trans. Annette Lavers (New York: Hill and Wang, 1957/1984).

23. Watkins, *Throwaways*, 3–7.

24. Ibid., 58.

25. Ibid., 32.

26. Raymond Williams, *Television: Technology and Cultural Form* (London: Fontana, 1974), 130.

27. Raymond Williams, "Communications Technologies and Social Institutions," in *Contact: Human Communication and Its History,* ed. Raymond Williams (London: Thames and Hudson, 1981), 230.

28. Economic and social factors also determine (in Williams's sense of determination as "pressures and limits") the built-in functions of the technology.

29. Williams, *Marxism and Literature*, 125–26.
30. Watkins, *Throwaways*, 39. See also 84–87.
31. Watkins is skeptical about the political potential of the emergent and of avant-gardism. He sees avant-gardism as merely another means of social distinction, a way to stay one step ahead. I tend to view the avant-garde use of the obsolescent more positively.
32. Watkins, *Throwaways*, 158–59.
33. See www.lomoplc.com.
34. See www.lomography.com.
35. Appadurai, "Introduction: Commodities and the Politics of Value," 52.
36. Susan Buck-Morss discusses this passage in her *The Dialectics of Seeing: Walter Benjamin and the Arcades Project* (Cambridge, Mass., and London: MIT Press, 1989), 245.
37. See Vincent Geoghegan, *Utopianism and Marxism* (London: Methuen, 1987).
38. This is the title of the most recent version of this essay, in Walter Benjamin, *Selected Writings*, vols. 3 and 4 (Cambridge, Mass.: Harvard University Press, 2002 and 2003). The most-used English translation is under the title "The Work of Art in the Age of Mechanical Reproduction," in *Illuminations*, trans. Harry Zohn, ed. Hannah Arendt (London: Fontana, 1992), 211–44.
39. Walter Benjamin, "A Small History of Photography," in *One Way Street and Other Writings* (London: NLB, 1979), 243.
40. For a list of Benjamin's connections with the avant-garde, see Eduardo Cadava, *Words of Light: Theses on the Photography of History* (Princeton: Princeton University Press, 1997), xix.
41. The Diana camera is a toy camera, produced during the 1960s and 1970s in Kowloon, Hong Kong, by the Great Wall Plastic Company. The Holga is the successor to the Diana, produced from the 1980s, and also entirely plastic. Both provide an inexpensive means of taking medium-format photographs. Photographers who use the Diana and the Holga include David Burnett, Cris McCarthy, and Jonathan Bailey. For more information, see Rexer, *Photography's Antiquarian Avant-Garde*, and Robert Hirsch, "The Diana Camera" in *Photographic Possibilities* (Boston: Focal Press, 1991), 141–43.
42. See www.lensbabies.com.
43. Benjamin, "A Small History of Photography," 255.
44. Observations on the outmoded are scattered throughout Benjamin's writings, linked to discussions of childhood, memory, and collecting, as well as to the surrealists, and shaped by an early interest in kabalistic ideas of revelation and redemption as well as by Marxism.
45. Buck-Morss, *The Dialectics of Seeing*, 92.
46. Ibid., 96.
47. Ibid., 115.
48. See, for example, some of the work included in Lyle Rexer, *Photography's Antiquarian Avant-Garde: The New Wave in Old Processes*. Although Rexer puts forward a cogent argument for their value, some of the images seem to be about little more than using (antique) technique to produce a stylized and aestheticized appearance. There are, of course, some interesting exceptions in this volume.
49. The translation I have used here is from Buck-Morss, *The Dialectics of Seeing*, 64. A different translation appears in Benjamin, *The Arcades Project*, 540. As Buck-Morss relates, this notion of the passages as prehistoric owes something to Marx's conception of history as beginning after the revolution.
50. Buck-Morss, *The Dialectics of Seeing*, 66.
51. Ibid., 68. Benjamin's use of images of petrified nature derived from his study of baroque allegory. See Walter Benjamin, *The Origin of German Tragic Drama*, trans. John Osborne (London: Verso, 2003). It also derived from those modern allegorists, the montage artists and writers of the historical avant-garde. He was influenced by Louis Aragon's *Paris Peasant*, where Aragon describes Sacre Coeur as an ichthyosaurus. See Buck-Morss, *The Dialectics of Seeing*, 257.
52. Walter Benjamin, "Theses on the Philosophy of History," in *Illuminations*, trans. Harry Zohn, 245–68.
53. Buck-Morss, *The Dialectics of Seeing*, 92.
54. See Gavin Lucas, "Disposability and Dispossession in the Twentieth Century," *Journal of Material Culture* 7, no. 1 (2002): 8.
55. See Mary Douglas, *Purity and Danger: An Analysis of Concepts of Pollution and Taboo* (London: Routledge and Kegan Paul, 1966). Gavin Lucas argues out that where we do classify waste it is either to recoup it from being waste (e.g., in recycling schemes) or as a consequence of the classification of distinct spaces within the household. Lucas, "Disposability and Dispossession in the Twentieth Century," 8.

56. Buck-Morss, *The Dialectics of Seeing*, 238.

57. See Peter Bürger, *Theory of the Avant-Garde* (Manchester: Manchester University Press, 1984), 66.

58. Web sites hosting collections of found photographs include www.spillway.com and www.renewal.org.au.

59. See Walter Benjamin, "On Some Motifs in Baudelaire," in *Illuminations*, 152–96. This argument is put elsewhere in Benjamin's writing too, notably in Walter Benjamin, *Charles Baudelaire: A Lyric Poet in the Era of High Capitalism* (London: Verso, 1983).

60. Benjamin wrote, in reference to Marcel Duchamp's work with *objets trouvés*, "contemporary man would prefer to feel the specific effect of the work of art in the experience of objects disengaged from their functional contexts [*crossed out:* torn from this context or thrown away] . . . rather than with works nominated to play this role." Rosalind Krauss, "Reinventing the Medium," *Critical Inquiry* 25 (1999): 293.

61. Benjamin, *The Arcades Project*, 201 [H4a, 1; H2, 7; H2a, 1], 207.

62. Barthes, *Camera Lucida*.

PART II

Residual Uses

5 "Automatic Cinema" and Illustrated Radio: Multimedia in the Museum

Alison Griffiths

> Technology may be a conduit for information, but it is not the instrument that defines a "valuable experience."
> —James R. Beniger and Georgia Freedman-Harvey, "High Tech—the Dilemma for Museums"

At a 2001 Tate International Council Conference entitled "Moving Image as Art: Time-Based Media in the Art Gallery," the opening speaker, British Research Council video artist David Curtis, complained that museum gallery–based film and video art had not been subject to the same sort of historical scrutiny as other media forms. Although Curtis's decision to flag this marginalization is worthy, his neglect of museum-based screen and audio culture before 1960s/1970s video art is striking.[1] The conference organizers doubtless felt that their focus on video art precluded the discussion of earlier or alternative uses of media in the museum, but the fact that there was not even a single mention of film's role in the museum gallery in any incarnation other than video art was revealing. If the conference theme and Tate Modern venue served as sufficient justification for the occlusion of film and video as it has emerged within the gallery, the event nevertheless underscores the persistent truncation of the historical trajectory of museum-based screen media among art historians and museum scholars.

I would like to thank Mary Jo Arnoldi of the National Museum of Natural History for her input on the "African Voices" exhibit. Research for this chapter was supported by reassigned time from the City University of New York Professional Staff Congress and the Weissman School of Arts and Sciences at Baruch College, CUNY reassigned time and travel committee. Charles Acland and William Boddy also offered valuable feedback at critical stages of the writing. This chapter is dedicated to Soren Francis Boddy.

At the same time, nontheatrical film exhibition generally has attracted scant interest among film scholars, with remarkably little discussion of audiences' experiences of film and video across the diverse public sites beyond the movie theater, including museums of science and natural history. Also unexplored is the impact of the moving image on the architectonics of the museum space and on the museum visitor's understanding of exhibit goals—both in terms of what exhibit planners *hoped* motion pictures would accomplish as discursive aids to learning and what educators and visitors have *thought* of the technology. Digital media have infiltrated the contemporary museum, including touch-screen computer interactives, electronic orientation centers, MP3 random-access handheld audio guides, eDocent (handheld computers used by visitors to download information and bookmark Web pages as they move through galleries),[2] 3D animation and virtual reality (VR) installations. A casual stroll through the gallery spaces of the American Museum of Natural History (AMNH) in New York City offers striking contrasts between the high-tech video walls and touch-screen interactives of exhibits such as the Hall of Bio-Diversity and the neoclassical architecture and traditional display of museum artifacts in glass cases in the North American Indian Hall that transport us back to a bygone era. That museums have become high tech in the last twenty or so years goes without saying. But exactly how cinematic, electronic, and digital media crept into the museum is less clear. When did museums first start supplementing the display of historical or scientific artifacts with film, recorded sound, and interactive computer programs? How do these media technologies affect the experiences of museum visitors, and what are the stakes involved for institutions engaged in such modern- ization efforts? Precisely what is new about "new media" in the museum, and what has motivated curators to introduce media technologies? What impact have new media had on the status of the museum object, which, in French theorist Didier Maleuvre's words, is already deprived of "experiential content," imprisoned by the museum in a similar way to a corpse lying in an ossuary?[3]

To answer these questions, this chapter constructs a genealogy of audiovisual techniques in museums, tracing contemporary uses of video screens and computer interactives to their mechanical predecessors. The first part examines some of the earliest examples of media use in museums of science and natural history, considering how the gramophone, radio, and 16mm film were enlisted in a multitude of ways, such as herding large numbers of visitors through gallery spaces or furthering an interactive model of learning. Rather than focus on how traditional film screenings became customary in museum auditoriums, I want to better understand how the protocols of film spectatorship were adapted to meet the needs of the physical and social context of the public gallery. For example, in designing film's place in the gallery, how did curators respond to the ambulatory modes of film spectatorship anticipated in museums? What strategies were used to mitigate (or accommodate) the distracted gaze of the visitor? Finally, how were the conventions of nonfiction film exhibition and reception destabilized in the museum by this aleatory and highly mobile spectatorial gaze?

The second part of the chapter grounds these ideas in a case study of the African Voices exhibit that opened in 1999 at the National Museum of Natural

History (NMNH) in Washington, D.C.[4] Three key issues drive my discussion here: first, how the design of new media in the museum is influenced by a broadly postmodern shift in museological discourse, where knowledge can be democratized (and contextualized) via interactive kiosks; second, how the changes affecting new media, in the form of so-called augmented reality devices, affect the very nature of the museum experience; and third, the extent to which new media, while providing opportunities for enriching the experience of an exhibit, may in fact end up shortchanging its visitors by compromising more ambitious pedagogical objectives for the sake of "edutainment."

More pressing in relation to this anthology, however, is the question of the residual nature of the relationship between museum and media, and how the term "residual" signifies something quite distinct at different moments in the history of media use in the museum. In the first part of the chapter, the museum itself is the residual element, its galleries permanently altered by the increasing presence of media-based modes of exhibiting. And yet the obsolescence of such gimmicks as "illustrated radio," "listening circles," and photologues reminds us in quite stark terms that although some curators felt that radio and film threatened to supplant the museum, as an institution it remained largely unaffected by these transitory schemes. From a contemporary perspective, however, the term "residual" refers to the ways in which museums have continued to experiment with film, video, and multimedia interactives and the legacy of such earlier audiovisual experiments as "illustrated radio" on computer interactives. Indeed, one could argue that screen culture in the contemporary museum is rapidly shedding its "residualness" and becoming an essential, de facto component of most temporary exhibits and even playing a central role in the redesign of permanent exhibits as we see in the case of "African Voices."

"Go Ahead Museum People": Introducing Media into the Museum

Museums have been eager to attract and retain the interest of a wide swath of patrons since their institutionalization in the nineteenth century. Competing with such paratextual spaces of cheap amusement as dime museums, fun fairs, circuses, and freak shows, nineteenth-century museum directors were frequently highly cognizant of what differentiated their experience from that of rival institutions. Museums have always been seen as agents of transportation, metaphorical time machines, where objects "free of the drudgery of being useful"[5] (to cite Benjamin) become overdetermined symbols of singularity or preeminent achievement in the arts, natural history, and science.[6] The exhibition of magic lantern slides and motion pictures were de rigueur in large public museums by the mid-1910s, vital forces in the discourse of armchair travel pervading museums. Museum visitors in institutions large and small, urban and provincial, came to expect illustrated lectures, and some upper-class patrons possibly encountered cinema for the very first time in such museum settings in the late 1890s. Although curators have long been interested in the possibilities presented by modern communications media, there have always been dissenters and skeptics who argued that use of such media failed to nurture the intellect of museum goers, especially children, who were seen as vulnerable to distraction. British writer F. G. Kenyon, in an essay on children and museums from 1919, wrote that "The Museum can provide the objects, and furnish labels, guidebooks, picture-postcards, photographs, lantern slides; but it cannot provide the alert and prepared mind."[7] The winning over of the mind of the child visitor became the holy grail of early museological discourse, with countless articles and editorials devoted to appropriate display methods and the development of the children's museum as a separate institution. Some museum officials recommended altering the tenor of all museums, not just children's museums, by focusing more on "the subjects that will affect intimately the life and interests of the people." For example, AMNH director Herman Carey Bumpus suggested in 1924 that in the museum of the future, "the abandonment of the technical label and the abandonment of strict scientific arrangement of exhibits would be projected . . . with salutary results."[8]

However, despite the widespread use of lantern slides and motion pictures in early twentieth-century museums, there was little professional discussion of the utility of these or related technologies outside of the lecture format or separate screening venue. One of the earliest mentions of an audiovisual aid in a museum dates from 1908, when a gramophone provided commentary on photographs and objects on display in a hugely successful exhibit about tuberculosis (TB) at the AMNH (Figure 5.1). The exhibit attracted 10,000 visitors on its opening day, a total of 753,954 over the course of its seven-week run, 72 percent of the museum's recorded attendance for the year. Staying open for thirteen hours a day on weekdays and until 8 P.M. on Sundays, the museum did its utmost to cater to the phenomenal public interest in this epidemic, fulfilling a major public service in terms of health education. A report on the exhibit from *Harper's Weekly* noted that "at every stopping-place a talking machine delivered short lectures of warning and advice."[9] A light on another exhibit would also go out every two minutes and thirty-six seconds to show how often someone died from TB in the United States.[10] Although there is no extant discussion of the precise rationale among curators for using a gramophone in this exhibit, one might speculate that it would have played a vital role in imparting public health information to the greatest number of visitors unable to read English or even get near the wall labels. In addition, the gramophone might have improved the circulation of people through the exhibits by drawing them away from the crowded walls, and, as a novelty with respect to museum display technologies, the gramophone fulfilled the role of a public address system, perhaps in the process underscoring the seriousness of the epidemic. However, this innovative use of sound technology in an exhibit failed to spawn immediate imitators because one of the next references I have come across to gramophones being used in an exhibit is from 1930, when "automatic gramophones" giving the public information about exhibits were installed in the Deutsches Museum in Munich.[11] It is unclear whether these gramophones played continuously or were visitor activated.

What the appearance of the gramophone in the museum gallery did anticipate, however, was "illustrated radio." The idea was initially conceived as a form of museum extension work (early long-distance learning) offering people in

FIGURE 5.1 *Poster for the wildly popular tuberculosis show at the American Museum of Natural History, New York City, 1908.*

remote regions the opportunity to "listen to the lecture [on radio] with their eyes on the picture the lecturer is talking about."[12] Variously referred to as "Roto-Radio Talks" (talks given in partnership with the Buffalo *Courier-Express* newspaper)[13] or "Radio Photologues," the basic premise was the same: listeners would tune into stations broadcasting lectures while looking at photogravure pictures printed in local newspapers. The *Chicago Daily News* began carrying photogravure pictures in its Saturday paper: for example, a radio photologue on archaeological excavations at the Mesopotamian city of Kish included maps, numbered and captioned images, and even a "portrait of the lecturer as well as views and excavations and striking finds to be contemplated while the loud speaker is in action."[14] The most common subjects for radio talks were painting, sculpture, and architecture, although the immediate locale of the listening audience also proved a useful resource, such as the series of talks given by officials from the Buffalo Museum of Science and the Buffalo Historical Society on the birds of the region and Native American history in the region (respectively).[15] Schools were quick to pick up on the idea, lantern slides projected on auditorium screens substituting for the newspaper images.

Of course, this substitution of actual museum artifacts by their mass-produced visual simulacra troubled a few curators, who, despite being generally sympathetic to the democratizing impulse behind the scheme, had reservations about the technique. One writer in 1926 cautioned that

> The use of radio and stereopticon facilities aid materially in reaching a larger audience, but it should be borne in mind that photographs are at best but a substitute for actual objects and that radio talks are usually directed to unselected audiences.[16]

Admitting the value of expanding museum audiences, the writer not only predicted the eventual phasing out of radio museum talks but also identified the key flaw in yoking the new broadcast medium to public institutions such as museums; people wanted to *experience* objects at firsthand and could just as easily borrow a book on a museum-inspired subject from a local library as sit and listen to a radio-broadcast lecture while looking at the Sunday newspaper (the same argument is made today about bricks-and-mortar museums versus their associated Web sites). Radio's status as a mass medium was also regarded with some suspicion—the talks "directed to unselected audiences"—suggesting anxieties about the diverse and distracted nature of the broadcast audience. Although some may question the extent to which museum-themed radio broadcasts qualify as early examples of interactive media or audiovisual techniques in the museum—the visitor did, after all, remain at home while listening to the lecture—they nevertheless underscored the conditions of possibility for media's role in the museum explored by the institutions themselves. Such museum radio talks prepared visitors for interactive technologies in several ways: first, they smoothed over the potentially disorienting effect of a disembodied voice imparting an object lesson (visitors were used to attending lectures but not to listening to machines address them); second, they required interaction on the part of visitors, who would need to locate the images to fully follow the lecture; third, in similar ways to contemporary computer interactives serving as sponsorship vehicles for large corporations, they functioned as promotional tools for museums and broadcasters hoping to capitalize on the free publicity; as a reviewer of the 1925 book *Broadcast over Britain* argued: "Broadcasting will undoubtedly affect the attendance and utility of our museums.... Many of these lectures deal directly with material on exhibition in museums, and the general public is not slow to follow up on such lectures by an inspection of the exhibits themselves.... Thus there is a link between the 'unseen speaker' and the silent museum exhibit."[17]

Broadcast museum lectures might serve not only to shore up the idea of radio as a civilizing apparatus but to provide a concrete instance of the possible synergy between a very old mode of communication, exhibits in glass cases, and a uniquely modern technology. Such initiatives supported the notion of radio as a democratic apparatus, with the civic and pedagogic prestige of the esteemed institution of museums ideally rubbing off on the new medium. Likewise, radio's perceived democratizing influence and educational potential were, not surprisingly, welcomed by many museum administrators. Radio might also come to the rescue of the museum lecture, which some critics argued was inexorably threatened by the onslaught of modernity; an editorial in the British *Museums Journal* cautioned that "the conditions in our modern cities, with their clamor of bright advertisements, sky signs and cinemas, may have the effect of training the faculty of visual attention at the expense of the auditory, so that people may be beginning to find it harder and more fatiguing to listen than to look."[18]

That the illustrated radio lecture fell prey to a decreased audience interest and the imperatives of emerging commercial broadcasters is unsurprising, but the museum radio nevertheless anticipated experiments in gallery-based radio listening and gramophone sound displays in the museum. For example, in 1929, "listening circles" were established in museums and libraries, an arrangement in which a small group of visitors would sit in comfortable chairs around a loudspeaker and listen to a broadcast lecture while looking at the artifacts being described. According to C. A. Siepmann, head of adult education at the British Broadcasting Company (BBC), "With the exhibits ready to their hand, the wireless talk and subsequent discussion would have for listeners a new significance."[19] The mention of comfortable chairs in this experiment with museum-based radio was picked up the following year in a discussion of seating provision for visitors to technical museums; although the recommendation for the inclusion of stools is no match for comfortable chairs, it is interesting that the suggestion that "anything that will induce visitors to rest at intervals is to their benefit" is nevertheless connected to the role of music and radio in the museum. In the absence of chairs, however, music was likely to generate a quite different response from visitors, as evidenced in this *Daily Telegraph* notice of Strauss waltzes being banned in the London Science Museum gramophone concerts as a result of people "being inclined to use the polished floors of the museum for dancing." It was concluded that for the twice-daily "gramophone concerts" to be a success, "the music has to be of a serious character."[20]

The gramophone and radio were thus conscripted as both enabling and productive as well as disciplinary agents, ways of capturing visitors' attention *and* controlling their behavior. For those visitors with specialized interest in exhibits, stools made perfect sense as a way of taking the weight off one's feet while sketches were made or specifications noted.[21] Such technological aids to relaxation in the museum—furniture and music used as a "true accompaniment" to the predominantly visual experience on offer—were part of a growing cultural reconceptualization of the museum as an enveloping space, as an anonymous contributor explained in *Museums Journal* in 1931: "The moment the sound of music reaches the ear the objects of vision seem enveloped in a new architectural form of space and are seen as it were in a different light . . . aesthetic significance is enormously heightened when the music is so related, and *ear and eye walk as it were together through enchanted space* (emphasis added)." This enchanted space seems a precursor to the interactive multimedia space that we find in contemporary museums. Museums have constantly redefined preexisting desires to create novel contexts for objects, be it enveloping them with sound or transmogrifying the space from an ocular-centric one into a more sensually rich encounter with object and its context. That cinema should emerge in the gallery around the time of these experiments with the gramophone and radio should come as no surprise, and as film slowly crept into the museum gallery, so too did museums slowly creep into film, appearing as dramatic settings in films across the globe. Indeed, such was the demand for images of museums in French films in 1932, that one critic noted it "has been necessary to draw up a detailed tariff applicable to all manner of films, whether purely documental or for scenarios" (public monuments were also subject to tariffs).[22]

Back in the early 1920s, though, steps toward full adoption of motion pictures in the gallery were taken with ambivalence. For example, falling short of fully endorsing the use of film in the gallery, the 1922 author of "Movies and Museums" tempered optimism with caution:

> Many lessons are best taught by motion-pictures . . . all the processes can readily be explained in a lecture-room with the help of motion pictures. . . . [However], motion-pictures, to have an educational value, must be edited by someone in close touch with the work of the school, and must be presented in the right way and at the right time. . . . From the museum's point of view, the chief difficulty is the preservation of the films.[23]

The thorny issue of the suitability of films for museum use (as I've documented elsewhere, film use in the museum generated considerable debate among curators), their proper presentation to audiences, and their storage in the museum emerged in ongoing discussions of film throughout the 1920s.[24] Concern about the "effects" of motion pictures inspired the Payne Fund Studies, a series of thirteen social psychology research projects conducted between 1929 and 1932, some of which were concerned with the deleterious effects of motion pictures on the emotional makeup of young people. Even leaflets produced to accompany lecture series in museums were not immune from criticism, lumped together with motion pictures as yet another example of modernity's mind-numbing tendencies; as one reviewer put it several years before the Payne Studies began, "[T]he danger of a multiplicity of lectures, films, and leaflets is that which has been urged against cinema, viz., that it may tend to produce a state of mere 'passive receptivity' in the child's mind."[25] Similar criticisms were directed toward broadcasting in general, with arguments made that listeners "may listen and accept the broadcast . . . in an entirely passive and uncritical frame of mind."[26] Barbed comments about the lowbrow status of cinema were voiced across the Atlantic too. For example, the secretary of the progressive Imperial Institute was attacked in the pages of *Museums Journal* for exploring retail-inspired shop-window advertising and considering free cinema shows. Worrying that Keatinge's suggestions "might send a crowd of

FIGURE 5.2 *The Dramagraph ("Automatic Motion Picture Projector") showing "Pottery Making on the Rio Grande" in the North American Indian Hall at the American Museum of Natural History, New York City.*

curious to our museums," the author equated Keatinge's vision with that of P. T. Barnum: "too much from the point of view of the business manager of a waxworks show" than of a high-brow institute of cultural enlightenment.[27]

The mid-to-late 1920s and early 1930s mark a turning point with regard to the adoption by museums of audiovisual media as we begin to come across references to projectors appearing in the gallery.[28] One of the earliest discussions of gallery-based film is from 1925, when a "new venture" that had proven to be extremely popular with visitors was tried at the Imperial War Museum, London, namely, "miniature cinematograph machines which show portions of films such as destroyers in action."[29] A similar apparatus was described five years later when *Scientific American* reported the invention of the Dramagraph, a "miniature cinematograph machine," a projector housed inside a wooden cabinet with a small screen (Figure 5.2). Designed to alleviate the problems of using brittle 16mm film for "nickel-in-the-slot" projectors, which "must operate hour after hour and day after day, thus requiring the films to run through the mechanism thousands of times," the apparatus employed thin steel-clad 16mm film.[30] A longer description of the Dramagraph appeared in a spring 1931 issue of *Museums Journal* and is worth quoting in full:

> On the front of the projector cabinet is a push button and a framed card announcing the subject of the picture and inviting the visitor to "Press Button Below." When the button is pressed the motion picture appears on the translucent window built into the front surface of the cabinet. After the picture has been shown (100 feet of 16 millimeter film—4.5 mins. of duration) the mechanism automatically stops, automatically resets itself, and is ready for the next showing when the button is again pressed. There is no re-winding of film, no need for an attendant or fire-proof booth, and no breaking or scratching of film. The apparatus is operated by connection with direct electric current.[31]

For museum officials, automation was clearly an appealing feature of this apparatus, which ostensibly required no attendant to oversee its operation. This was not the first time, however, that an automated device had been used to illustrate museum artifacts. As early as 1922, the Art Institute of Chicago established in its museum instruction room an "automatic lantern which gives a continuous lecture of forty-six slides." An early version of a tape-slide apparatus (a device that allows recorded narration to be synchronized with 35mm slides), the lecture lasted nine minutes and took the form of alternating pictures and captions (each slide was projected on the screen for twelve seconds). "The machine is arranged with two parallel magazines, each consisting of two compartments—a delivery compartment and a receiving compartment. The magazines present their slides in alternate rotation. When the forty-six slides have been presented, the lecture is repeated without an interval."

The Dramagraph was clearly building on preexisting practices of automated visual instruction in the space of the museum, although unlike the automatic lantern, its novelty would have been considerable, especially for younger visitors who were not old enough to remember the kinetoscope parlors or mutoscopes of the 1890s. Constructed by the New York City–based Dramagraph Motion Picture Corporation, the projector found an early home in the North American Indian Hall at the AMNH (something of an irony given the current complete absence of screen-based media in this hall, which stands as living testimony to an earlier era of exhibitionary practices),[32] showing a film on Pueblo Indian pottermaking, and at the Museum of Science and Industry (in the RCA Building at Rockefeller Center) in New York City, where a Dramagraph was enlisted to illustrate an exhibit on the development of weaving. Each machine was equipped with an automatic counting register to check the number of times it had been used; initial surveys indicated that machines in use at the AMNH in 1930 were operated on an average about eighty times per day, increasing to about a hundred times daily on weekends.[33]

Discussions of gallery-based motion picture projectors were also taking place in the UK at the same time, including a reference in a paper delivered to the British Museum Association's 1930 annual conference by English curator Dr. E. E. Lowe, who invited a Kodak representative to demonstrate a "small experimental projector" (possibly a Dramagraph or certainly an instrument modeled on the same principle) to illustrate his talk:[34]

> We should not have a separate cinema hall or room in the building, but that in at least one of the more dimly-lighted corners with which most museums are provided we should fix up a projection apparatus which a visitor could start into action just as he starts a machinery model moving in the Science Museum, by pressing a button or switch. The screen would be small but of sufficient size to be visible in dull daylight to twenty or thirty people, if so many had gathered.... By pressing a switch the visitor starts the show and gives himself and his party a performance which lasts say five minutes and then repeats itself.[35]

The space of the museum would be transformed by the operation of the motion picture projector, arresting the fluid movement of spectators as they are invited to gaze at the screen for the duration of the film. Conversely, because there is no way of knowing the nature of the movements of a 1930s museum patron, one must also consider the possibility that a five-minute film might have had the opposite effect of indeed quickening a visitor's pace. It is difficult to know whether film in the gallery was likely to be part of the conditions of the contemporary, impatient, glance, or mitigate against it. Given the novelty of motion pictures in a space where the only movement one would expect to see is that of another visitor or, possibly, a machine demonstrating an industrial process, I would argue that film *did* have the desired effect of attracting spectators, perhaps even more so than today, when visitors are so familiar with video walls and plasma displays found in large retail stores that even if they stop for a moment, they usually do not stand for the duration of the screening, unless the visuals are extremely captivating. Of course, there are bound to be exceptions, instances where museum-based videos transfix visitors, performing a similar hailing function as diorama life groups (taxidermy specimens or mannequins arranged in naturalistic tableaux).

The architectonics of static display thus exist in tension with the time-bound space of performance because the visitor is expected to divert her or his attention away from the surrounding exhibits toward the motion picture screen. The pattern of patron engagement with the Museum of Modern Art's new multichanneled video gallery suggests the contemporary playing out of this tension. That the projector's position in the gallery warrants a careful selection of lighting conditions (a "dimly-lighted" corner in "dull daylight") suggests a negotiation between the needs of motion picture projection on the one hand and exhibitionary space on the other. Although the "projection apparatus" evokes the self-service quality of the early mutoscope in terms of the agency it afforded the spectator, unlike its peephole predecessor, there is a very public feel to this screening. In contrast to the mutoscope, the viewing is, or at least has the potential to be, a collective experience, should another visitor decide to stop in front of the screening space. The British Science Museum referent in the previous description bridges the discursive leap between the concept of interactivity as understood in the Science Museum, where visitors activate machinery by flicking switches, and gallery-based motion pictures that demand a similar level of involvement. Recommending that the film subject be changed every month and that, where possible, guide-demonstrators be enlisted to use the projector in conjunction with gallery talks, Dr. Lowe concludes his proposal with a discussion of the costs incurred in the purchase and upkeep of the film apparatus.[36]

Dr. Lowe's paper appeared in a special issue of the *Museums Journal* devoted to film in the museum; in the issue's editorial, entitled "Museums and Movies," science and children's museums were singled out for praised for their success in capturing the interest of audiences through the use of illusionistic dioramas and reconstructions. But now, the journal editor notes, "The cinematograph has come to change all that and to give us real living pictures, to bring all the life and movement of the world onto a few square feet. . . . It is not surprising that a few go-ahead museum people have seized on this as a new and greater aid to their educational activities."[37] In the same way that taxidermy and dioramas could vivify specimens and function as hailing devices in the cluttered gallery spaces, so too could the novelty of seeing film projected in the gallery cinema engender appeal in and of itself. The fact that the screening was "automatic and repeating" provided the museum visitor with an experience not dissimilar to that engendered by today's video screens and interactive kiosks.

Similar experiments with gallery motion pictures were taking place at the Deutsches Museum in Germany, where an apparatus, designed to be "set in motion by the visitor," was on display in 1930. In an announcement of the device from the *Museums Journal*, the author noted that "Ordinary museal illustration of physical and chemical principles is unattractive, but the visitor whose active cooperation is enlisted, if only to the extent of pushing a button, is sure to have his interest aroused."[38] A notice posted by the secretary of the British Museums Association in a 1934 issue of *Museums Journal* requesting feedback from curators on the inclusion of "mutoscope machines" in their institutions indicates a measure of institutional acceptance of such apparatus by the mid-1930s; the information being sought included how curators regulated the times of shows, in other words, "did the projectors function only when set by visitors, at fixed intervals, or continuously?" Curators were also asked about the types and source of machines employed (whether mutoscope or projection screen), whether the apparatus had proved reliable, and finally, the types of films shown.[39]

Also notable in this special issue is a discussion of film both as an illustrative aid in the exhibit halls *and* as a part of the museum's programs in public auditoriums. In a section on the exhibition of motion pictures at the Imperial Institute in London, where films were shown daily in four separate programs, Sir William T. Furse noted the difficulty of finding suitable film subjects, although he reported that with the formation of the Empire Marketing Board, headed by John Grierson, the situation was improving rapidly.[40] That the Dramagraph or similar devices represented a breakthrough in museum motion picture exhibition is indisputable because the technique, if not the name or technological base, has endured as a means of providing contextual material for exhibits, spawning hands-on computer interactives, and more recently, digital screen-based exhibits.

Interest in "mechanical aids to learning" increased exponentially during the 1930s, as evidenced by a British Institute of Adult Education three-day exhibition at the London School of Economics in 1930. The event showcased "visual, aural, and other aids to learning," including such subjects as "television, educational talking films, and the use of the cinematograph in schools." A contributor reporting on the exhibition in the *Museums Journal* reminded readers that "in view of the growing importance of the cinema in museum work, curators should make a point of seeing this exhibition which will include a wide range of immensely interesting instruments and apparatus."[41] Excited by the fact that most of the leading firms connected with the production of the cinematograph, the gramophone, the lantern and epidiascope, and the microscope would have stands at the event, the author visited the exhibition in the hope of "finding there a cinematograph instrument of the *press-the-button type* [emphasis added] suitable for use in museums. There were, it is true, at least two types of continuous cinematograph on exhibition, but, alas! there were difficulties connected with both which render their use in museums out of the question at present."[42] If this response to "press-the-button" type projection devices is in any way typical, it would appear that in spite of the assurances offered by Dramagraph convert Dr. Lowe, some curators nevertheless remained skeptical of the suitability and durability of these "continuous cinematograph" devices.

Notwithstanding these technical drawbacks, by the late 1930s, the dual functions of film in the museum had reached a level of intellectual and logistical maturity; writing on the establishment of a National Film Library in the UK, Ernest H. Lindgren offered the following assessment:

> Already some museums are using films to illustrate activities in their particular field, from the short cyclic band of film which can be seen by pressing a button, to the showing of special films, with or without a lecturer. I am convinced that the day will come when no museum will be

without a projector. There is no other way which can so infuse with life and movement the apprehension of a series of objects which ordinarily tend to remain . . . static, isolated articles in a showcase.[43]

One can't help notice the full circle we have come with regard to the selection of suitable titles for museum exhibition; the following list of films from Furse's 1927 Christmas screenings are virtually identical to recent Imax releases exhibited in museums of natural history: *The Epic of Everest, With Captain Scott in the Antarctic*, and *My Polar Adventures with Shakleton in the Antarctic*. In our current era of neoimperialist "regime change" and imagined communities of "freedom-loving" citizens, perhaps even *Outposts of Empire*, a film listed in the 1927 lineup, is not too outlandish a title for Imax to consider.[44] There is, of course, an Imax screen currently operating at the Science Museum in London, the reference point in the 1930 Dramagraph example discussed earlier, home also to over sixty interactive exhibits in the high-tech Wellcome Wing (including "Comment," an award-winning exhibit with a track network of 27,000 LED units allowing messages to be sent through a maze suspended over all four floors of the gallery space).[45] A separate admission charge is made for Imax (currently about $12) and special shows ($10), although general admission to the museum is free.

What Lindgren could not anticipate, however, was the extent to which such visual media would become the fully elaborated means of a highly interactive and immersive experience for contemporary museum goers. Let us now turn to more recent developments in museum practice to gauge what has changed and what has remained the same in museum uses of interactive moving image technologies via an examination of the "African Voices" exhibit at the NMNH in Washington, D.C. A brief digression into the history of interactivity in the museum is our first port of call; however, with video an intrinsic part of so many interactive units, it would be shortsighted to overlook the merging of computer and video monitor in the gallery. My goal in the next section, then, is to think through some of the enduring features of multimedia use in the museum, to consider how the idea of "immersion," for example, a staple of Imax promotion and a long-standing trope of illusionistic exhibits such as dioramas and period rooms, continues to define museum going. Another point of intersection concerns how notions of branding operate differently in each era and how branding works to lend rhetorical unity to exhibits in a multitude of ways (not just in the form of corporate underwriting but in the trafficking of marketing ideas across retail and museums).

"African Voices": An Exploratorium of Sound and Image

To a large extent, there is nothing new about the idea of interactivity in museums; if the concept (and its overuse as a 1990s buzzword) smacks of high-tech digitized galleries where every exhibit features cool-looking plasma screens and interactive kiosks, the reality suggests that museums have taken the idea of visitor interaction seriously for quite some time. Museums have *always* been interactive insofar as they have invited visitors on occasion to touch artifacts and seek specific information. What *has* changed, however, is the discursive construction of the term, which over time has come to mean a lot more than simply touching an object or manipulating an exhibit to affect an outcome. Today, at least in the world of museum education, interactivity connotes agency, a more dialogic vision of visitor-centered learning, and some awareness of the politics of exhibiting where including contextual material thickens visitors understanding of the phenomena or processes at hand. Although it was rare in the nineteenth century for museums to allow visitors to handle museum artifacts for fear of breakage or wear and tear, objects were occasionally removed from their cases (the "hands" part in the "hands-on" exhibit concept). For example, cabinets with removable drawers facilitating the handling of objects were in use by 1910, and by the early 1930s, exhibits had incorporated electrical features such as a 1933 "Electrical Questionnaire for Young Visitors," which consisted of a "box with a number of compartments, in each of which is an exhibit which is not labeled. Against these are lists of possible names, one of which the child can select by pressing a pointer against a disk. An electrical contact is made, and a white light appears for a correct answer and a red light for a wrong one."[46] Where "hands-on" referred quite literally to handling, manipulating, or exploring a process or mechanism, interactivity evoked a more distinctly dialogical experience, what museum writer Christopher Nash defines as one allowing the user to "reach out in time and space to access and manipulate resources not currently in their immediate environment . . . to contextualize objects more richly, to provoke thought and debate about issues of importance to their communities, and to reach out to more diverse audiences and client groups."[47]

Interactivity has become a near prerequisite for contemporary museums hoping to attract visitors, especially children. As galleries constructed in the 1950s and 1960s undergo renovation, new museums such as the International Spy Museum in Washington, D.C., integrate interactive elements at every level of the experience. Visitors explore the world of espionage by adopting a "cover," "crawling through an air duct designed for eavesdropping, deciphering code, and identifying disguised spies."[48] With an opportunity to experience interactive exhibits about "disguise, surveillance, [and] threat analysis,"[49] museum backers hope the $36 million building will attract at least fifty thousand visitors a year (at a cost of $12 per adult ticket; $10 for children). Gimmick-driven museums like the International Spy Museum are not the only institutions turning to interactive media as selling points; an element of interactivity extends even to memorials commemorating individuals lost to acts of terrorism, such as the Oklahoma City National Memorial that opened five years after the event and includes a park as well as a museum offering background material on the Murrah building, testimony from victims, memorabilia, and photographs of those killed. However, not all visitors are appreciative of the edutainment theme running through the exhibit; reviewing the memorial in the *New Yorker*, Paul Goldberger observed that the museum provides a "much more American experience" than the outdoor memorial, "since it is grounded in the belief that almost anything, including the most horrendous events imaginable, can be made entertaining."[50]

Notwithstanding this embrace of interactivity as a structuring principle of the exhibit/memorial, museums still have a reputation for being late adopters of new technology (although large urban museums such as the Whitney Museum of American Art and the American Museum of the Moving Image in New York City and the Walker Art Center in Minneapolis have done a great deal to dispel this image).[51] At the 2002 Whitney Biennial, kiosks allowed visitors with Palm electronic organizers to download a digital dancer (male or female) from James Buckhouse's digital artwork "Tap"; eDocent, being tested at the American Museum of the Moving Image, enables visitors to download information into handheld computers and bookmark Web pages. Finally, the Walker Art Center, as part of a $90 million renovation and expansion, is including a wireless network that will give visitors a chance to access far more of the collection, what design director Andrew Blauvelt calls "social computing," where groups of visitors sitting at "smart" tables touch screens embedded in tabletops, giving them access to the Walker's collections and resources.[52] Not losing sight of the artwork or exhibition content is critical for curators, even those with a large stake in the success of interactive technologies such as Whitney director Maxwell L. Anderson, who says that "Museums are not in the hardware business, we're not in the software business—we're in the content-development business."[53]

Finding the right balance between traditional and more cutting-edge exhibitionary techniques is the goal of most curators and certainly a key consideration in "African Voices," an exhibit that opened at the NMNH in December 1999. At the same time as "African Voices" was in development, the NMNH was itself taking big strides to develop "group-oriented interactive-media programs" for its Discovery Center, in part because of the large number of visitors each year (about 9.5 million annually). Entitled "Vital Space," the interactive exhibit located in an auditorium tells the story of an astronaut who is contaminated by a sample of Mars dust; with fifty touch-screen consoles that can each accommodate two users, audience members influence the narrative by selecting icons such as whether to search for the contaminant either in the brain or the spine. As many as a hundred audience members "collaborate and compete with one another in order to influence the plot of a computerized, fictional drama projected on a high-definition [three-panel, 150-degree wraparound] screen."[54] There is a competitive, video-arcade dimension to the exhibit, with the highest scoring participant—whoever has prevailed in the fight against the larvae and eggs using electrical, laser, and chemical weapons—winning a prize. Produced for free by Immersion Studios (which receives a share of ticket-sales revenues), the sixteen-minute film and interactive exhibit were the first in a series of collaborative projects with Immersion; according to Immersion CEO Stacey Spiegel, "by creating an experiential environment around real learning and real science, learning must be taking place."[55] If "African Voices" is a far cry for the frenzied (according to some accounts) "hitting, poking, and slapping" at touch screens that occurs in "Vital Space," it nevertheless points to a shared belief in the value of experiential learning and interactivity as central tenets of modern museological design.

With a price tag of $5.5 million, "African Voices" consists of a 6,500-foot rectangle divided by a colonnade that effectively splits the space in half lengthwise.[56]

Replacing the "Hall of African Cultures," which had closed in 1992 following accusations of anachronistic display techniques and primitivist representations of native peoples from the African continent, "African Voices" uses video, sound elements, and interactive displays such as the "Money Talks" installation, where visitors turn a wheel to rotate examples of African currencies, to disseminate the exhibit's intellectual themes and "stories." In addition to computer interactives found on the gallery floor, there is also a Learning Center at one of the exits in which eight computers running Microsoft's Encarta Africa CD-ROM are available for use, and a viewing gallery where two short documentaries play continuously on a large video screen.[57] To create a "more personal and immediate story of contemporary and historical Africa," curators strove to convey a strong sense of the polyphony and dynamism of Africa, using African voices as text or in recorded excerpts drawn from literature, songs, poems, proverbs, scholarly essays, and interviews.[58] Greeting you upon entering "African Voices" is a loud, bass-heavy soundscape emanating from 10-foot-tall video walls situated at either end of the exhibit (Figure 5.3). The music is bright and contemporary, blending African world music with hip-hop beats and urban rhythms; the soundscape, which includes a female narrator talking about African life, is arresting, drawing our eyes to the wall of twelve video screens that represent Africa as modern, inviting, knowing, and a space of multiple modernities and cultural diversity.

On the face of it, this is an Africa far removed from the images of poverty, starvation, civil war, and political unrest that we've witnessed for decades on our television screens; this high-tech signifier of Western commercial culture,

FIGURE 5.3 *Video wall in the "African Voices" orientation section, National Museum of Natural History, Smithsonian Institution, Washington, D.C. Photograph by James F. DiLoreto.*

along with the ubiquitous reminders of new technology in the large number of video installations and computer interactives, clearly places new technology at the service of a recuperative discourse that overcompensates for the primitivist presuppositions it expects visitors will bring to the exhibit. Multimedia as it is used in "African Voices" is therefore doing a lot more than simply delivering content; it functions as a mnemonic, a reminder to visitors that this is an Africa looking to the twenty-first century rather than an allochronic construction from a mid-nineteenth-century colonial past. It creates a context that stakeholder audiences—diasporic Africans or African American visitors—as well as other constituencies will recognize as empowering and contemporary, the kind of Africa that might be invoked in a documentary on global music on VH1 or the Discovery Channel. The economic hardship and political realities fracturing the African continent are not addressed immediately; they are instead dealt with in the "Africa Today" section of the exhibit. And yet one can't overlook the irony in the fact that our first impression of Africa in this exhibit comes prepackaged in the semantic architecture of mass-media "shoppertainment," an iconic form associated with single-brand superstores such as NikeTown, the consumerist excesses of Western industrialized capitalism, and the exploitative labor practices of the retailing multinationals who doubtless have manufacturing plants in the struggling economies represented on the screens.

We therefore view Africa through the cultural lens of corporate branding because Africa *is* quite literally branded for us at this moment. As a phenomenon, branding has assumed preeminence in contemporary advertising; once aimed at the customer, it has come to imbue the very heart of the corporation, turning products into iconic, fetish objects that function as a form of cultural (global) shorthand. To possess a brand is, therefore, to belong, as British brand analyst Wally Olins explains: "Displaying the brand . . . turns you into a perambulating advertisement . . . [which] in turn relates to membership of a niche group, a club with a wide variety of nuances relating to social, cultural, and economic status."[59] By helping products break through the clutter of an incredibly noisy marketplace, and offering consumers consistency, empathy, and branding for self-definition, the brand has become the lingua franca of the marketplace *and* the museum. This branding of Africa as hip, modern, and globally savvy is clearly recuperative, an undoing of the colonialist legacy of Western stereotyping. Branding Africa in this way is an act of reclamation but also a way of connecting with younger museum goers whose views of the continent may be especially shaped by televised images of political unrest, genocide, and starvation. But just as point-of-purchase video installations offer the "promise of symbolic movement across identity categories, of inhabiting different *social positions*, as well as spaces," to quote Anna McCarthy,[60] so too does this video wall, the key difference being that the spatiotemporal shape-shifting invited by the "African Voices" installation has an overtly humanist and pedagogic dimension far removed from the starkly consumerist underpinnings of the equivalent installation found in NikeTown.

The video wall therefore succeeds in undergirding the central theme of "African Voices": Africa's diversity, dynamism, long history, contemporary relevance, and global reach.[61] In the words of the curators, "The kaleidoscope of images of people, objects, and places set against a background of contemporary African music and voiced narration alerts the visitors that they have entered a new and exciting section devoted to the present-day Africa."[62] Sharing "authoritative tasks" (McCarthy's term) both similar and distinct from television in public spaces, the natural history museum may be seen as yet another institutional location for an ambient TV experience, a location that is itself invested in constructing an immersive, educational environment. Curators often find themselves in something of a double bind, though, always looking over one shoulder at the uses of audiovisual technologies in point-of-purchase video, theme parks, and discovery centers, knowing that what works in these commercial spaces can easily be adapted for use in the not-for-profit space of the museum.[63]

As an orientation device, the video wall is increasingly a standard feature of contemporary science and natural history exhibits, supplanting the extended wall text or "celebrity artifact" as the traditional contextualizing device upon entry into the museum or gallery. The National Museum of Australia in Canberra, for example, includes a forty-seat orientation center where four thin plasma video screens crawl on hydraulic lifts across a rear-projected backdrop depicting Australia's landscape and human impact on the environment. According to *New York Times* journalist Peter Hall, at the end of the show, "the entire theater rotates 90 degrees to face a new presentation on the theme of nation, while a fresh batch of visitors watches the *Land* performance."[64] In the same way that the use of gramophones in the AMNH TB exhibit in 1909 may have addressed the problem of circulation, so too do the revolving theaters ensure the smooth flow of visitors in manageable batches of forty or fewer. When asked about maintaining the right balance between the theme park–styled multimedia and traditional object-driven exhibits at the NMA, museum director Dawn Casey, an Australian Aborigine, defends her approach, arguing that the blend of "scholarship and entertainment . . . contemplation and involvement" is just right and that "everything—from the multi-narrative exhibitions to the building—intends to convey that there are several, sometimes conflicting stories of the country's history."[65]

Putting the architectonics of the exhibition space to work in the service of a more discursively complex articulation of ideas that eschew a linear historiography sounds a great deal like the curator's vision for "African Voices." Embodying revisionist models of history that challenge audience assumptions about grand narratives is a bold one, but also one that demands recognition of the possibility that not all visitors will come away from the exhibit having grasped the "big ideas." For example, "African Voices" uses a color-coded, hierarchically organized "informational architecture"[66] consisting of "Gallery texts," "Main texts," and "Subtexts" that impart information ranging from the general to the specific (highlighting main gallery themes and overarching messages at one end and additional information about a topic at the other). Although the hierarchy served as a logical structuring and suturing device during the planning stage, its usefulness for audiences is uncertain, with anecdotal evidence suggesting that most visitors are oblivious of the hierarchized organization of texts, not even noticing the color coding of labels that distinguishes the different thematic galleries.[67] Indeed, based on formative research conducted with museum visitors, the exhibit planners

FIGURE 5.4 *Somali aqal, showing the virtual Somali American docent on the full-size vertical screen, "African Voices," National Museum of Natural History, Smithsonian Institution, Washington, D.C. Photograph by Donald Hurlbert, Department of Anthropology, Smithsonian Institution.*

found it necessary to simplify many themes, knowing that some of their work would have to be corrective, including a frequent visitor confusion about Africa's status as a continent rather than a country.[68] Curators also found themselves defending the cacophonous audiovisual media in the exhibit, arguing that "it uses audio and video components selectively to enliven the space and emphasize the main message of diversity, dynamism, modernity, and global connections . . . The decision to fill 'African Voices' with energizing sounds and moving images was a calculated one, meant to offer an approach to African history and culture that is anything but mute and passive."[69]

On a quieter note, the Somali *aqal* exhibit installed diagonally across from the video wall furthers a similar message about modern African life, combining video and a partially reconstructed Somali *aqal*, a circular house displayed with the outer covering peeled back to reveal the interior to underscore the tradition/modern dialectic (Figure 5.4). In contrast to videos that play continuously in conventional rectangular black boxes in exhibition halls (Figure 5.5), this 6-foot vertical screen with green LED supertitles located directly above the image—reminiscent of the supertitle translations found above the proscenium arch in American opera houses—at first reminds us of a hologram, but its 2D flatness quickly registers as we realize we are looking at the equivalent of a virtual docent, or rather two virtual docents, Somali-Americans Abirahman

FIGURE 5.5 *Black box video playing alongside Chewa Kasiyamaliro mask, "African Voices," National Museum of Natural History, Smithsonian Institution, Washington, D.C. Photograph by Donald Hurlbert, Department of Anthropology, Smithsonian Institution.*

Dahir and Faduma Mohammed. Growing up nomadic in Somalia, the pair ended up working in the United States as part of the extended team of exhibit planners along with the production crew and filmmakers from Northern Light Productions, the company responsible for the audiovisual component of the exhibit.

The decision to exhibit the *aqal* in the first place had been a fraught one for the curators; the reifying tendencies and fetishizing impulses of the life group diorama, with its illusionistic mannequins and allochronic representations of a petrified culture, made the curators uneasy about working with the life group trope. And yet Dahir and Mohammed recognized the fact that the Somali house had the potential to meld the features of a contemporary nomadic existence with a centuries-old architectural form that is not a relic of an ancient way of living, but rather a compelling example of cultural continuity and vibrancy.[70] Supporters of the *aqal* idea thus convinced the "African Voices" team that the "primitive housing" connotations of the *aqal* could be transcended if its centrality in the Somali imagination and its value as an object of cultural memory could be conveyed.[71] In addition to the video, above the *aqal* are two large photomurals, one of a group of *aqals* in northern Somalia, the other of the camp's camel herders and text rail at the front of the house that includes labels, Somali proverbs, and excerpts from poetry. The team thus succeeded in transforming the *aqal* from a static diorama into a "dynamic dialogic display."[72] The question of agency is sovereign here: by giving voice to both male and female diasporic Somalians through an embodied gaze out at the museum spectator, African subjectivities are foregrounded and the primitivist tendencies of the ethnographic diorama repressed.

As close cousins of video installations found in contemporary art museums, the audiovisual exhibits in "African Voices" depend to a large extent on the presence of the visitor to close the circuit of meaning; for example, in the absence of spectators, the *aqal* docents resemble virtual shop-floor sales associates making a pitch for the newest kitchen aid as they address an invisible spectator. But with visitors standing in front of them, they are immediately transformed, constructing a quite different notion of "presence," namely, a heightened sense of African subjectivity and agency. Being physically present in front of the exhibit is therefore necessary to engender affect and connectivity between the Somalians represented on screen and the visitor, especially given the size of the video image. But while the life-size *aqal* video screens succeed in arresting our attention and inviting us to listen to the monologues, videos embedded in black boxes seldom compel us to stand around long enough to watch the entire screening of three to five minutes. The location of the video monitors, whether at eye level, as in the large rectangular photo mural of a Makola market plasma screen, or 3 feet off the ground, as in the mudcloth artist Nakunte Diarra's interactive kiosk, which allows visitors to select videos illustrating the mudcloth designs, and drawing power of its visuals doubtless determine the degree of interaction, which brings us to the larger question of how documentary footage is resignified by the gallery and how the protocols of nonfiction film viewing are shaped by the overdetermining effect of the exhibition context.

Documentary film's discursive work seems threatened by the contingent nature of museum film spectatorship, in which we are invited to construct our own fragmentary versions of narratives and piece together meaning based on elliptical flashes of objects, images, and sounds that ideally contribute to a meaningful narrative. We are promenading spectators, absorbing images, sounds, and texts as our feet shuffle through sometimes crowded, sometimes deserted galleries, creeping in for a closer look, backing out carefully to allow other visitors to take our place. The experience is elliptical and thoroughly somatic as our bodies, tired of standing or wandering, cue the mind to switch gears and move on in search of somewhere to sit down. That mini screening rooms are often located at the end of exhibits (as is the case in "African Voices") should come as no surprise because our "museum-fatigued" bodies are far more willing to submit to video at the end of an exhibit than at the start. In the likelihood that most gallery spectators are not sutured into the conventional subject positions offered by documentary cinema, we must therefore rethink how spectators negotiate the meaning(s) of electronic media in the museum. As a result of the low sound levels and fidelity of certain interactives, which also have to compete with the ambient sound of the gallery, we absorb most information visually. "African Voices" uses electronic media to create what Ralph Appelbaum, founder and president of a company responsible for the $210 million Rose Space Center at the AMNH and the U.S. Holocaust Memorial Museum in Washington, D.C., describes as an immersive environment that foregrounds an "emotional" as well as an intellectual interactivity.[73] Exhibits don't just deliver content; they shape that content into emotional, sensual, and memorable experiences that clearly affect visitor expectations. The fact that electronic images are now being summoned to perform this affective work alongside objects is significant.

For example, the refurbished Milstein Hall of Ocean Life, formerly known as the Hall of Oceanic Life and Biology of Fishes at the AMNH, now features plasma screens above all of the upper level marine dioramas and a continuous video on ocean life playing on a large screen on the lower level. According to the news release, the $25 million renovation set out to transform the existing exhibit into a "fully immersive marine environment with video projectors, interactive computer stations and new ocean dioramas." Lighting, video, and sound are central to this exhibit—as *New York Times* art critic Virginia Smith comments, the former silence has now been replaced by whale sounds—our eyes darting from 3D models housed in glass cases to flat, kinetically charged video images of the specimens on display. The shift in spectating registers seems normal for most visitors; there are no buttons to press to trigger the video images that hang like videographic paintings on the cavernous walls of the Oceanic tomb. However, like Smith, I, too, miss the magical sense of wonder that the former dioramas used to evoke when they weren't competing with video, and second her lament about spending time "staring mostly at a flat-screen monitor showing seals swimming" rather than absorbing the systematic display of information. Although justified pedagogically, the trade-off is still that, a trade-off, "sounds for silence, information for contemplation, facts for poetry."[74]

If the pervasive influence of screen and Internet culture in the increasingly corporate-looking architecture of museums is disquieting to visitors such as Smith, nostalgic for the idea of museums as serene spaces of quiet reflection rather than vibrant heterotopias, new media technologies are creeping into the most unlikely of museological spaces, such as the Pope John Paul II Cultural Center in Washington, D.C., where visitors are encouraged to "design a stained-glass window out of available motifs on a digital kiosk, practice bell-ringing, and create a giant collaborative collage by floating designs onto a video wall."[75] Even the Vatican, it would seem, is getting hip to interactivity. Like the shopping mall, airport, restaurant, and sports bar, the museum is a public spectatorial space, where bodies either stand shoulder to shoulder or drift through galleries drawn toward objects, sounds, and haptic sensations. That the museological community hasn't fully formulated a theory of museum spectatorship with regard to interactive media and video installation—despite the fact that film and audio have been in the museum for quite some time—is apparent from the haphazard way in which the spatiotemporal logic of the screen gets configured in the museum. Despite the fact that some research suggests that the presence of computer interactives influences the way visitors attend to and process information, with more visitors likely to become involved with an exhibit, stay longer, and recall information better if they have obtained that data from a screen-based source,[76] there is always the nagging fear that museums may never fully understand how visitors interact with the technology or what it means to view video in the gallery versus the movie theater or home.[77] We are reminded by museum curator and theorist Suzanne Keene that multimedia productions are not only the dominion of the computer screen: there is, she argues, an entire spectrum of delivery means from "handheld portable devices to wall-sized screens and even installations where the interface is the whole space occupied by the user."[78] Although Keene is right to broaden the platforms of interactive and video delivery in the gallery, her argument that the costs and benefits of gallery interactives can be judged in a "straightforward manner" oversimplifies the complexity of visitor interaction with screen and audio-based technologies.

Given the difficulty of knowing with certainty how visitors responded to the Dramagraph in the 1930s, there were inevitably dissenters who disapproved of the developments (as we saw earlier, some curators were de facto skeptical of all forms of audiovisual technology or even hyperillusionist display methods), one cannot help wonder whether visitors in the late 1920s and 1930s might have shared Virginia Smith's trepidation when they first encountered film and audio in the museum. To evoke the term metaphorically, galleries were being branded as signifiers of a popular culture; the Dramagraph, for example, looks very much like early television sets that were publicly launched at the 1939 New York World's Fair. If interactivity today encompasses both the hands-on sophistication of many Exploratorium-type science centers as well as the simple push-button kiosks found in "African Voices," it also refers to the way in which exhibit spaces are being branded not just literally, via corporate tie-ins or sponsors, but through the incorporation of marketing-like strategies for exhibitions. This may appear to be in stark contrast to an earlier era of media use in the museum, but it would be inaccurate to assume that retailing and museums have only recently formed synergistic ties.

Despite the fact that branding conjures up distinctly late twentieth-century notions of corporate marketing techniques promoting the emotional properties of consumer goods, it has precedents much earlier in the century, with exhibit designers turning to department store display windows and audiovisual technologies to support exhibit themes. There is a long history of museums and department stores appropriating one another's display techniques in pursuit of a common didactic, disciplinary role regarding visitor behavior. Philadelphia department store magnate Rodman Wanamaker in 1906, for example, organized educational lectures in his store, and museum curators employed the visual rhetoric of advertising in their exhibits.[79] There's even discussion of corporate underwriting from 1908, when British curator Dr. Hoyle suggested to his colleagues that they should follow the American model of corporate sponsorship by persuading "rich Corporations or individuals to start [i.e., sponsor] one or two [diorama] cases."[80] The principles of commodity fetishism thus cut across museums and marketing, with museum stores quick to exploit consumer desires to own the objects on display by selling replicas of them in their shops. Experientially, too, we can trace several continuities across these distinct periods in museum history. Part of the appeal of multimedia exhibits has been their ability to immerse visitors in reconstructed environments while attempting to give equal status to both cognitive and emotional registers. Display techniques using video-walls are thought to be especially effective/affective in this regard because not only do they draw visitor's attention to an exhibit—especially if a large screen or video wall is used—but in the case of multimedia interactives, they increase the likelihood that at least some learning goals will be met.

However, to talk about the collapse of the representational technologies of museums, theme parks, and retail is to slight the very real differences between these institutions, including how the museum's interpretative techniques mediate the presence of a screen- or audio-based culture. In other words, although our ability to make sense of computer interactives and video monitors clearly comes from competencies acquired in distinct cultural spaces—the home, the office, and the shopping mall, to name a few—these skills don't fully prepare us for the cognitive mapping exercises we are expected to perform if we are to grasp the meanings circulating between object and image in the increasingly noisy gallery. Digital culture is certainly superimposing itself on the traditional spaces of galleries, with most visitors (especially younger ones) finding the iconography of the computer interface enticing; indeed, a 1999 London Science Museum survey on visitor use of computer interactives confirmed this assumption, arguing that "techno-fatigue" rather than "techno-phobia" was more likely to be an issue with (some) adult audiences and that lessons from the 1930s about the desirability of providing seating and sound in the gallery still hold true today. Visitors are clearly drawn to computer interactives (as they may have been to the Dramagraph eighty years ago), finding the computer screen "several billion times more interesting than any printed text around the screen."[81] But while we seem to have reached a consensus on the efficacy of screen-based media and the feedback on visitors' uses of computer interactives is positive, we have a long way to go before we fully grasp just how a digital interface is unraveling the discursive underlay of the early twentieth-century museum, forming heterotopias where meanings are not ontologically bound to the aura of the object or the museum face, but are co-constructed in the interstices of object, image, sound, and interactive keyboard. "Automatic cinema" and its practical culmination in post-1970s video in the museum was an inevitable by-product of early twentieth-century modernist tendencies that looked to technology as a way to kill several birds with one stone: utilize a technological novelty to draw audiences into a gallery, reinforce educational themes by providing contextual material on the exhibit objects themselves, and continue in the long-standing tradition of integrating popular culture into museum display techniques.

Notes

1. A Webcast of the conference can be found at www.tate.org.uk. All quotes are taken from this site.

2. For more on recent developments in new technology, see Karen Jones, "New 'Smart' Galleries, Wireless and Web-Friendly," *New York Times*, April 24, 2002, G27.

3. Didier Maleuvre, *Museum Memories: History, Technology, Art* (Stanford: Stanford University Press, 1999), 16.

4. The Web site for "African Voices" can be found at http://www.mnh.si.edu/africanvoices.

5. Walter Benjamin, *Reflections: Essays, Aphorisms, Autobiographical Writings*, trans. Edmund Jephcott (New York: Schocken Books, 1978), 155.

6. For more on museums and the politics and poetics of collecting, see the following anthologies: Robert Lumley, ed., *The Museum Time-Machine* (New York: Routledge, 1988); Peter Vergo, ed.,

The New Museology (London: Reaktion Books, 1989); and John Elsner and Roger Cardinal, eds., *The Cultures of Collecting* (Boston: Harvard University Press, 1994).

7. F. G. Kenyon, "Children and Museums," *Museums Journal* 19, no. 1 (July 1919): 10.

8. Herman Carey Bumpus, cited in Anonymous, "Human Interest to Shape Exhibits of Future," *Museum News* 1, no. 1 (January 1, 1924): 2.

9. Anonymous, "Weapons in the War against Tuberculosis: The Exhibition in the New York Museum of Natural History," *Harper's Weekly* (November-December 1908), 7; cited in Geraldine Santoro, "To Stamp Out the Plague of Consumption': 1908–09," *Curator* (March 1993): 13–27.

10. For discussion of the exhibit, see the March 1908 and February 1909 issues of the *American Museum Journal*, the magazine of the AMNH. Other reviews include John Martin, "The Exhibition of Congestion Interpreted," *Charities* 20 (April 4, 1908): 27–60, and Everett B. Mero, "An Exposition That Mirrors a City," *American City* 1 (November 1909): 100.

11. *Museums Journal* 31, no. 7 (January 1931): 290.

12. Anonymous, "Illustrates Talk by Radio," *Museum News* 2, no. 4 (June 15, 1924): 2.

13. In conjunction with the Buffalo Historical Society, Buffalo Fine Arts Academy, and the Buffalo Museum of Science, printed photographs of museum objects appeared in the "Rotogravure" section of the Sunday paper. Listeners tuning in to WKBW at 6 P.M. on Sunday evenings would "hear a talk on the pictures shown therein." Anonymous, "Museum and Newspapers Co-operate in Roto-Radio Talk," *Museum News* 6, no. 21 (April 15, 1929): 1.

14. Anonymous, "Radio Photologues," *Museum News* 3, no. 10 (November 1, 1925): 2.

15. Anonymous, "Museum and Newspapers Co-operate," 1.

16. "Museums and Radio," *Museum News* 4, no. 10 (November 1, 1926): 2.

17. Review of J. C. W. Reith, *Broadcast over Britain* (London: Hodder and Stoughton, 1925) in *Museums Journal* 24, no. 9 (March 1925): 212.

18. Anonymous, "Lectures and the Public," *Museums Journal* 25, no. 8 (February 1926): 215.

19. C. A. Siepmann, "The Relation of Broadcast Education to the Work of Museums," *Museums Journal* 29, no. 5 (October 1929): 120.

20. Unattributed *Daily Telegraph* quote from Anonymous, "Science Museum: Radiogramaphone," *Museums Journal* 35, no. 9 (December 1935): 358–59.

21. Anonymous, "The Dissemination of Information by Exhibition and Display," *Museums Journal* 30, no. 6 (December 1930): 218.

22. Anonymous, "Museums and Cinematographers," *Museums Journal* 32, no. 2 (May 1932): 76.

23. "Museums and Movies," *Museums Journal* 21, no. 9 (March 1922): 200.

24. For more on the emergence of film at the AMNH, see Chapter 6 of my book *Wondrous Difference: Cinema, Anthropology, and Turn-of-the-Century Visual Culture* (New York: Columbia University Press, 2002).

25. Review of "Museum Stories for Children," *Museums Journal* 24, no. 6 (December 1924): 145.

26. R. A. Rendall, n.t., *Museums Journal* 32, no. 11 (February 1933): 419.

27. Anonymous, "Brighter Museums," *Museums Journal* 27, no. 9 (March 1928): 292.

28. The validation of motion pictures in museum work may be gleaned from the topic of the sixteenth annual meeting of the American Association of Museums held in Cleveland in 1921: "Motion Pictures as a Phase of Educational Work in Museums." Reference in *Museums Journal*, "The Activities of Museums in America," 21, no. 1 (July 1921): 9.

29. Review of J. C. W. Reith, *Broadcast over Britain*, 220. By 1936, the Imperial War Museum housed seven hundred official war films (over 450,000 feet) representing the navy, army, and air force on various fronts. Copies of certain films were available for hire and, in collaboration with the Admiralty, War, Office, and Air Ministry, had been used by commercial film companies. Information from H. Foster, "Imperial War Museum," *Museums Journal*, 36 no. 5 (August 1936): 221.

30. Anonymous, "Unbreakable Movie Film: Steel-Sheathed Film Reduces Replacement Costs; Has Great Educational Value," *Scientific American* (April 1930): 299.

31. "The Dramagraph," *Museums Journal* 31, no. 1 (April 1931): 24.

32. Film was first exhibited at the AMNH in 1908. In 1923, the AMNH formed a Motion Picture Library, "an international depository for the very best nature films, with a view to preserving valuable records for study and research." The films included those shot in Africa by Paul. J. Rainey, Carl E. Akeley, and James Barnes and the Asiatic films of Roy C. Andrews. *Museums Journal* 23, no. 4 (October 1923): 115.

33. Ibid.

34. Reference to the demonstration projector appears in Anonymous, "In the Magazines: Movies in Museum," *Museum News* 8, no. 1 (May 1930): 7.

35. E. E. Lowe, "The Cinema in Museums," *Museums Journal* 29, no. 10 (April 1930): 343.

36. Lowe estimated that the cost of a camera outfit capable of projecting black and white films would be £25 (about $37) and for color £120 (about $180). The cost of changing the film each month would be about 35s (about $5).

37. "Museums and Movies," 334.

38. Anonymous, *Museums Journal* 31, no 7 (January 1931): 439. At the same time as interactive film exhibition technologies were being experimented with in the museum, curators still contended with the dearth of suitable subjects for screening in museum lecture theaters; one British author complained in 1933 that "films for Museums are at present practically non-existent," although such organizations as the Central Information Bureau for Educational Films did publish leaflets on the availability and distributor information of educational films. Mary Field, "The Film in the Museum," *Museums Journal* 33, no. 10 (January 1934): 350; *Museums Journal* 33, no. 8 (November 1933): 239.

39. Reginald V. Crow, "Correspondence. The Film in the Museum: An Inquiry," *Museums Journal* 34, no. 8 (November 1934): 336.

40. Sir William T. Furse, "The Panoramas and the Cinema at the Imperial Institute," *The Museums Journal* 29, no. 10 (April 1930): 339. The cinema building at the Imperial Institute was about 120 feet long and contained 329 seats. Finding films of suitable subjects and that were not too long were just two of the major difficulties the Institute faced; according to Furse, "We beg, borrow, and occasionally hire these films, and we are much indebted for the loan of films to the High Commissioners of the various Dominions and India, to many big firms and corporations which have established industries in various parts of the Empire, and to private individuals."

41. Anonymous, *Museums Journal* 30, no. 2 (August 1930): 65.

42. "Exhibition of Mechanical Aids to Learning," *Museums Journal* 30, no. 5 (November 1930): 206

43. Ernest H. Lindgren, "The National Film Library," *Museums Journal* 37, no. 12 (March 1938): 549.

44. Furse, "The Panoramas," 335.

45. Anonymous, "Interactive Service," *Design Week*, March 28, 2002, 16. Another notable interactive is the Human Genome Interactive, which introduces visitors to the concept of genetics by letting people "'see the genome—an artifact that is as much an information abstraction as a physical reality. We made the interactive controls like an aircraft. Users can fly into the genome, seeing more detail as they got deeper into it.'" Exhibit founder David Small; cited in ibid.

46. "Electrical Questionnaire for Young Visitors," *Museums Journal* 33, no. 6 (September 1933): 236.

47. Christopher John Nash, "Interactive Media in Museums: Looking Backwards, Forwards, Sideways," *Museum Management and Curatorship* 11, no. 2 (June 1992): 173, 178.

48. Karen McPherson, "International Spy Museum," *Pittsburgh Post-Gazette*, May 28, 2002, D1.

49. See www.spymuseum.org for more information about exhibits.

50. Paul Goldberger, "Requiem: Memorializing Terrorism's Victims in Oklahoma," *New Yorker*, January 14, 2002, 90.

51. Marjorie Schwarzer, "Art Gadgetry: The Future of the Museum Visit," *Museum News*, July/August 2001, 36.

52. Karen Jones, "New 'Smart' Galleries: Wireless and Web Friendly," *New York Times*, April 24, 2002, G27.

53. Ibid.

54. Jessica Ludwig, "Smithsonian's Interactive Exhibit Allows Museum-Goers to Collaborate and Compete," *Chronicle of Higher Education,* http://chronicle.com.

55. Ibid.

56. Initial corporate sponsors for the exhibit included the Shell Oil Foundation and the Coca-Cola Foundation.

57. The two videos, made by Northern Light productions who were also responsible for the *aqal* video and computer interactives, are *Struggle for Freedom* (20 minutes) and *Struggle for Justice* (18 minutes). They are both expository in style, combining archival footage and narration and interviews with Africanist scholars.

58. Mary Jo Arnoldi, Christine Mullen Kreamer, and Michael Atwood Mason, "Reflections on 'African Voices' at the Smithsonian National Museum of Natural History," *African Arts* 34, no. 2 (Summer 2001): 16.

59. Wally Olins, "How Brands Are Taking Over the Corporation," in Majken Schultz, Mary Jo Hatch, and Mogens Holten Larsen, eds., *The Expressive Organization: Linking Identity, Reputation, and the Corporate Brand* (New York: Oxford University Press, 2000), 63.

60. Anna McCarthy, *Ambient Television: Visual Culture and Public Spaces* (Durham, N.C.: Duke University Press, 2001), 2, 4, 14.

61. Arnoldi, "African Voices," 21.

62. Ibid., 26.

63. McCarthy, *Ambient Television*, 14.

64. Peter Hall, "Now Showing: Something Dazzling," *New York Times*, May 2, 2001, H17 [special section on museums].

65. Ibid.

66. The term is Appelbaum's and refers to the mutual imbrication and support of the museum's architecture and the exhibition. Because this exhibit is one space in a larger institution, the term "architecture" in this context refers to the topographical design of the gallery space, how the different areas are partitioned, and the overall design schema. Ibid., 89.

67. Arnoldi, "African Voices," 21.

68. During a visit to the exhibit in May 2002, I overhead two women talking about wanting someday to visit Africa, but fearing for their safety because the "country" seemed so politically volatile. The plan to include an exhibit on the history of African paper currencies was also modified when it transpired through visitor testing that many people didn't realize that as members of global economies, Africa actually uses modern paper currencies (Arnoldi, "African Voices," 24).

69. Arnoldi, "African Voices," 26.

70. Ibid., 29.

71. Ibid.

72. Ibid.

73. Ralph Appelbaum, "Designing an 'Architecture of Information'—The United States Holocaust Memorial Museum," *Curator* 38, no. 2 (June 1995): 87.

74. Virginia Smith, "Swimming in the Sea of Memory," *New York Times*, June 15, 2003, 3. For an insightful discussion of museum and corporate sponsorship, see "The Sponsored Museum; or, The Museum as Sponsor," in Mark W. Rectanus, *Culture Incorporated: Museums, Artists, and Corporate Sponsorship* (Minneapolis: University of Minnesota Press, 2002).

75. Hall, "Now Showing," H17.

76. Susanne Keene, *Digital Collections: Museums and the Information Age* (London: Butterworth-Heinemann, 1998), 31.

77. Kristine Morrissey, "Visitor Behavior and Interactive Video," *Curator* 34, no. 2 (1991): 109–18.

78. Keene, *Digital Collections*, 52.

79. For more on these connections and ensuing debates, see Chapter 2 of *Wondrous Difference*.

80. Dr. Hoyle, "Egyptological Collections," *Museums Journal* 8, no. 5 (November 1908): 158.

81. Ben Gammon, "Visitors Use of Computer Interactives: Findings from Five Grueling Years of Watching Visitors Getting It Wrong," Science Museum Welcome Wing Exhibition Development unpublished report, May 1999, 2–3.

6 The Residual Soul Sonic Force of the 12-Inch Dance Single

Hillegonda C. Rietveld

FIGURE 6.1 *Twelve-inch dance record. Photograph by Daniel Rubinstein.*

Elephant & Castle, South London, England, 2004: the MC (master of ceremonies or, in the UK, mike controller) is pirate king of the ether, the beat is often broken, and vinyl rubs shoulders with CD-R in the practices of the pirate radio and club disc jockey (DJ). Being lighter, more compact, and cheaper than vinyl, digital sonic media like MP3 and compact disk (CD) offer clear advantages to the jet-setting as well as to low-earning DJs and to DJs who have no access to vinyl distribution. Music genres that are geared toward home entertainment are hardly available anymore on vinyl, with the exception of a recent return of vinyl albums in London's larger megastores. As a result, the vinyl 12-inch format looks like an oversized anachronism. Nevertheless, vinyl is still the sonic medium of preference to London-based dance DJs who specialize in drum and bass, hip hop, garage, deep house, techno, and related dance genres. Even underground trance DJs, known for early experiments with digital audiotape (DAT), are often seen to hang on to their heavy record bags. In local dance music record stores in London's Soho, I have had many conversations, wondering why collectors and DJs alike spend huge amounts of time and money on the acquisition of large slices of vinyl, dance music on 12-inch and, sometimes, 10-inch record discs, that are heavy and cumbersome to store compared to their digital alternatives. The buying crowd here ranges in age from sixteen to fifty and is dominantly male with varied ethnic backgrounds, although African Caribbean descendents from Jamaica as well as English and continental Europeans are well represented, in addition to a sprinkling of Japanese enthusiasts. The collectors usually have a disposable income, able to afford vinyl while maintaining a working-day life, but the DJs are the real vinyl junkies, "giving up food for funk," as an old expression goes, unanimously exclaiming that vinyl is "nicer."[1] It is this intense passion for dance vinyl that has inspired the discussion here.

Vinyl records have been the source material for increasingly complex DJ mixed dance soundscapes; which are characterized by spiritual repetition, hedonistic flow, and antagonistic rupture. Created by the manipulation of vinyl on twin turntables, by the structuring of sound from the building blocks provided by one's record collection, a new musical narrative is produced. Mixing a variety of danceable musical sources, possibly from different times and places, results in a meaningful aesthetic. The contemporary dance DJ thereby gains the aura of an auteur, whose practice has looped back into the recording studio, to produce dance floor–friendly material on 12-inch vinyl, the dance single. Although there was an expectation that vinyl would be replaced by digital sonic media during the 1990s,[2] the persistent preference by dance DJs for vinyl is especially because of its tactile, hands-on analog qualities. First, these provide pleasures in record manipulation; this specific DJ craft that has influenced formats of current digital sonic media, making the CD and MP3 within a dance context derivative media despite their unique "radical potential" as new media. Second, because of the characteristics of its analog grooves and the bias of the needle, vinyl provides a sonic spectrum that is not only audible but can be literally felt when played through large sound systems; it thereby provides a sense of presence to disembodied studio music. Third, there is a fetishization of the collectable object, the 12-inch vinyl disc,[3] which also has kept vinyl in business outside of the dance market.

This chapter draws on culturally historical and geographical links to the development of 12-inch dance single as a creative sonic medium in the context of professional dance DJs, in particular within disco music production. By employing a genealogical approach, insights into the residual relationship between analog vinyl and digital sonic media can be gained, showing that the development of postdisco dance music is intricately bound up with particular, historically based, engagements with technological forms. Cyberfeminist N. Katherine Hayles makes a similar point,

The new cannot be spoken about except in relation to the old... whatever the form, it can be expressed only by articulating its differences from what it displaces, which is to say the old, a category constituted through its relation to the new.[4]

This chapter will address historical formations of dancing to records and of various vinyl-based DJ techniques to gain an understanding of vinyl's residual legacy in the context of dance DJ-practices, such as CD mixing, sampling, re-edits, remixes, and the development of new dance music genres. In brief, the specific forms of dance music that favor the 12-inch vinyl dance single are historically tied to a musical aesthetic forged on New York dance floors in the 1970s. Partly rooted in downtown's arty underground dance music scene and partly in midtown's glitzy discotheques, this was eventually tagged as disco music and later as club, garage, and house music. Disco DJs developed seminal mixing and programming skills that were passed on to, and developed in parallel fashion by, hip hop DJs in the uptown areas (who worked in tandem with rap

FIGURE 6.2 *Hillegonda Rietveld DJs at London's Big Chill Bar. Photograph by Daniel Rubinstein.*

MCs). In the 1990s, break beat–driven drum & bass and club crossover UK garage have, like rap, not only cultural links with disco but also with reggae and its Jamaican sound system public address practices of toasting and MCing. In the late 1960s, the Jamaican reggae sound system environment provided the concept of the unique one-off remixed disc (the "dub plate," initially a 7-inch disc) that was specifically made for the dance floor. Independently, the New York disco scene created the right climate in the mid-1970s for the development of the "disco single," as the 12-inch vinyl dance single was called at the time, offering dance floor–specific reedits and remixes that were longer than a 7-inch single would permit. Hip hop's ruptured mixing techniques ultimately inspired the 12-inch vinyl-based DJ scratch art of turntablism.

During the 1980s, affordable production technologies made the remix and studio production accessible, which was successfully exploited by dance DJs in the African American Chicago house music scene as a type of DiY (do-it-yourself) disco,[5] providing a template for further developments in dance floor–specific music. Although the cassette tape made such productions immediately available to the DJ, the 12-inch vinyl dance single was the preferred format for DJ performances. In the 1990s, some trance DJs explored the cassette-based DAT, a digital format adopted during their travels from Europe and the United States to hot climates of Goa and Thailand because vinyl buckles in heat and is bulky, while DAT allowed the DJ to have access to new studio material before it was released; yet club DJs that play trance do use 12-inch vinyl in addition to CDs. Only since the mid-1990s, when CD mixers became available, has CD become a more serious contender in the DJ market, especially with the increased use of CD-Rs to copy and promote new material. Importantly, CDJ practices have been normalised in those areas where poverty promotes the use of (bootlegged) CDs, for example in Bangkok or Havana. The discussion here will show that dancing to records was initially a sign of poverty, a taste of necessity, in which Black Atlantic[6] musical forms became dominant. Yet, ironically, as a specialist DJ tool, the 12-inch single was not destined for a mass market; in the current context of recording studio–based music as well as the encroaching CD and MP3 market, the 12-inch dance single has become a sign of distinction, a limited issue artifact, a mechanically reproduced sonic sculpture that requires connoisseurship.

The support from DJs and hi-fi buffs in globally affluent areas has kept the few geographically dispersed vinyl pressing plants going that would otherwise perhaps have disappeared in the 1980s. Although a mixture of formats is increasingly prevalent, when comparing the "new," the CD, with the "old," vinyl, the latter is mainly cherished because of a reported analog and tactile quality of the vinyl 12-inch. Most of my DJ interviewees point out that vinyl enables direct manipulation through hand–eye coordination, because the grooves on the record are an analog representation of the sound and therefore look different when the sound textures change in a track. There is an intimate relationship with the recording when the needle is put in the groove, as opposed to the coordination of numbers on a digital screen. Those DJs that focus on sound quality point out that vinyl's audio representation has a "warm" bias that is especially important for the physical impact of the bass in dance music. There are some valid technical reasons for this tactile effect. Dance music CDs are sonically more defined; yet this comes at a price as certain surplus frequencies and carrier noises are eliminated in the (finite) form of digital formats. In contrast, the analog transfer of sound waves through the needle into the groove adds extra frequencies, which increase the sense of "presence" of a (infinite) sound that can certainly be perceived as being "fuller." This is especially noticeable in the textures of large powerful sound waves that are produced when the sub bass is loudly amplified on a well-constructed sound system, as is the case in a dance club. The sonically embodied presence of analog recording fills a lack, the absence of the performer, which is difficult to satisfy with an ephemeral digital code. This is also an incentive for collecting vinyl records; as one DJ interviewee put it: "Vinyl is a fetish."

Brian Winston has created a useful model to frame the development and impact of media technologies, showing a complex interaction between technological potential and social necessity.[7] Institutional interests may suppress the radical potential of a new medium, but unexpected spin-offs may also occur while a similar device or technique may be developed in parallel fashion, following slightly different technical trajectories in several localities in response to similar socioeconomic contexts. This chapter will show such paralellism in comparable, yet independent, creative responses to an interaction among recorded music, DJ performance, and market competition. For example, the reggae dub version is a sound engineering spin-off produced in the later 1960s in response to the competitive needs of Jamaican sound systems; the disco remix is a spin-off from the specific use of vinyl on twin turntables by DJs in the mid-1970s club scene of New York. An increasing global synchronicity implies that choices of "zero moments" and "seminal inventions" in linear historical narratives may be mostly a matter of cultural politics and economic resources, a matter of whose voice is documented, distributed, and heard.

As a sonic medium, the vinyl 12-inch is used for both the disco single and for long playing (LP) formats. The latter was designed for home listening, in turn inspiring the CD format of the 1980s. The 33⅓ rpm vinyl LP format was introduced by Columbia (CBS) in 1948, as result of an exploration by CBS of vinyl as a sonic medium from around 1944, when war activities in Southeast Asia caused an interruption in the supply of shellac.[8] DJs for small American radio stations were the first to use recordings on this new plastic, broadcasting 30-minute prerecorded music programs distributed on 16-inch vinyl transcript discs. Because the finer grooves in the vinyl pressings provided a much better sound quality than the characterstic surface noises of previous shellac discs, radio DJs could pretend they had musicians playing live in the studio.[9] The actual format of the LP goes back even further, to before World War I, when it was used by the film industry. Yet single audio recordings of up to five minutes, distributed on ephemeral fragile shellac, had so far been better for music business.[10] The arrival of television changed this sense of complacency because domestic entertainment technologies and industries had to compete even more intensely; as a result, the vinyl LP was launched to maintain the attention of a home audience. In 1949, GE introduced the extended play or 12-inch EP at 45 rpm and

RCA-Victor came up with the 45 rpm 7-inch single,[11] maintaining a pop format of three-minute songs. Singles were the backbone of a buoyant music market, exploited on pop radio and at dance parties. New York's professional dance DJs emerged from the crossroads of a range of developments that required public dancing to records rather than to live bands, followed by a competitive and creative improvement of musical skills by dance DJs, leading eventually to the introduction of the 12-inch dance single as a professional DJ tool.

The idea of public dancing to recorded music goes back to at least the 1930s. For New Yorkers, precedents were set in rent parties and working-class jukebox joints, as well as in a more elegant jazz discotheque concept imported from France. The gramophone-based jukebox was patented in 1899 but was not popularized until the late 1920s, with the introduction of electric amplification. The jukebox became a hit in the United States after 1933 when Prohibition was repealed[12] and the Depression made musicians too expensive as public entertainment. A utilitarian tool, the jukebox was a signifier of poor taste, a taste produced by necessity. According to Chanan,

> The machine and its music carried low-life connotations: the derivation of the term is the creole (Gullah) word "juke," as in "juke house" for brothel. As for the music, the jukebox particularly stimulated "race" and "hillbilly" recordings for the black and white markets, respectively; and the record companies found distinct advantages in using jazz musicians to help them cut the cost of production.[13]

The presence of such jazz musicians can be noticed, for example, in the jazz concept of the "middle eight" or break in a pop song, in which session musicians would improvise with rhythms and alternative melodies. Before World War II, the distinct absence of black radio stations in the United States meant that the jukebox became an alternative public medium for African American music. As a consequence, the development of African American music as dance music, the jukebox, and the aesthetic of the pop single were intimately intertwined. Ethnomusicologist Kai Fikentscher points out that disco is, in fact, part of a cultural lineage in which "underground dancing can be understood as part of an American continuum of social dance styles that has been marked by a pervasive African American imprint since the 1920s."[14]

The notion of "underground" here indicates a cultural space for social groups that are repressed or marginalized because of their ethnicity, sexuality, or political beliefs; although hillbilly country music could be heard as listening music on the radio before World War II, for African Americans an enforced underground provided a space hidden from hegemonic surveillance in which idiosyncratic dance styles could develop. In African American dance, the body is often used as a highly skilled expressive musical instrument, "in which sound can be experienced physically,"[15] as a feeling. In the interaction with an otherwise mechanically reproduced dead artifact, the record, a living interactive musical event is produced though dance, which later became of importance to DJ-related dance styles.

In Winston's terminology, "supervening social necessity" creates the right moment to utilize and develop available technologies and prototypes in a socially and culturally specific manner.[16] Recorded music was logistically easier to manage, was cheaper than a dance band with musicians, and also provided a potentially wider repertoire of music that might not normally be locally available or performed by a limited set of musicians. Following the cultural and economic logic that developed around juke joints, specific DJ aesthetics were given shape in the 1970s with "few economic assets and abundant aesthetic and cultural resources."[17] Representing, like the jukebox, a threat to the livelihoods of traditional music performers, the DJ was not accepted by the music unions as a musician and was therefore more easily exploited by entertainment establishments. Together with the increase in studio recorded music, such circumstances led the DJ to become an attractive alternative musical entertainer, who creatively competed with the jukebox to produce a specific soundscape built from an archive of recorded sounds to suit the mood of a dance event.

The use of two turntables was essential in the production of such a soundscape, if only to shorten the gap between records. For example, as early as 1946, in a Leeds club, North English working-class teenager Jimmy Savile exposed his fragile shellac collection of swing records to a proto version of twin turntables. Being an able entertainer on the microphone, comparable to mobile DJs at school and office discos as well as weddings and birthday parties, he eventually became a national radio and television personality with the children's wish-fulfillment program *Jim Will Fix It*.[18] Vinyl was sturdier in transport than shellac, so when this was introduced it made even more sense to play one's records in public spaces. The introduction of vinyl coincided with big R & B dance events organized by rock radio DJs in the United States, such as Alan Freed's notorious 1952 Moondog Coronation Ball, a mixed band and DJ-led extravaganza, which, thanks to radio exposure, drew an African American crowd of nearly 25.000 "hepcats" and "chicks," double the capacity of the designated Cleveland Arena.[19] As dance record parties became more popular in the United States, twin turntables were reinvented—for example, by disco sound system designer Bob Casey's engineer father in 1955.[20] Twin turntables offer the possibility to mix and juxtapose records, allowing a powerful sense of a rush when feeding the sonic flow with a seemingly endless supply of new material, satisfying a displaced and unattainable urge to gain control over an unspoken finite: the end of the record, of the party, and, ultimately, on a deeper level, of life. A hedonist "time is now" attitude of the dance party is enhanced by a temporal suspense that exists in the twin turntable aesthetic.

The success of rock and twist parties in the 1950s shows how radio DJ activities can complement dance DJ-ing and that the use of speech over the music can be utilised as a marketing device. A similar device was used by Jamaican sound system DJs,[22] who in turn took their cue in the 1950s from the jive talk that was heard over R & B records on southern African American radio stations based in New Orleans and Miami.[23] Their selection of dance music suited the Jamaican experience of "rapid post-war urbanisation."[24] Keyes has shown that many of Jamaica's local swing musicians had left the country for

Britain after the latter had changed its immigration policy in 1948, inviting a new workforce to rebuild the war-damaged country. DJs filled the gap, initially playing on sound systems that were based on the PA (public address) systems used for political rallies.[25] In effect, a radio DJ practice was transferred to a dance floor setting, combined with the macho stage bravura associated with political speakers, providing a type of communal dance "radio" experience in dance halls and open-air yards. This may explain why, in Jamaica, the title "DJ" is reserved for the speaking performer, like a jive radio DJ who "rides the discs" while the "selector" takes care of programming the music, usually on a single turntable that required the talking DJ to maintain the attention of the crowd. Such home-built sound systems became complex constructions, each competing with a unique sound, each the brainchild of a sound engineer who would also be an essential member of the sound crew.

Two important DJ tools were created in this context: the "dub plate" device and the "rewind" technique. The latter is based on a teasing lifting up of the needle and placing it back to the start of the track while the DJ rouses the crowd to create a frenzied sense of anticipation. As the record was interrupted, the flow of the soundscape was ruptured. That way, the speaking DJ—the toaster—became a larger-than-life personality. Such an interruption forced the crowd to listen even more intently to the reintroduced music selection. Such assertion of power and charisma effectively rallied the crowd to support the sound system they attended. With the emergence of hip hop, in the 1970s, the Jamaican dance hall aesthetic seems to have inspired a similar assertive attitude toward recorded music, in terms of sonic rupture, the format of the MC, and the occasional "sound clash." In the 1990s, UK break beat DJs used the rewind literally, winding back the introduction of a 12-inch record on a DJ turntable deck (preferably with a magnetic drive) while the needle remains in the groove, accompanied by cheers from the crowd.

The dub plate became a crucial tool in the arsenal of the sound system—this is a one-off uniquely mixed cut on acetate, a material used for test pressings of vinyl records. Sound systems operated in competition with each other, and stories of sound clashes abound.[26] As well as a unique repertoire, sound quality was crucial, such as the depth of the (sub)bass and the finesse of the tweeters. A master in this, sound engineer King Tubby introduced the dub plate around 1968, as part of the race to produce a unique music experience for the crowd. Remixing master tapes for Coxsone[27] and Duke Reid, Tubby would create new mixes by taking out the vocals or instrumentation from multitrack recordings to allow sonic spaces for his toaster, U-Roy, to dub lyrics live during a sound event.[28] He thereby used a full arsenal of sound-molding effects, especially doubling up instruments with reverb and echo units, blasting their sounds into unique abstract directions.[29] Although R & B star James Brown put straight instrumentals of songs on B sides in the 1960s, the concept of the dubbed-up instrumental was new. Eventually, during the 1970s, reggae artists recorded special instrumental versions, also called "riddims," providing bass and drum rhythm sections to enable a variety of vocal performances.

In the 1990s, the dub plate appeared in 12-inch format in the practices of British break beat DJs; their genres, jungle, drum & bass, and UK garage, are historically linked to reggae as well as to the 12-inch vinyl mix techniques of disco and hip hop. According to several of my interviewees, the dub plate seems to have been a reason why British break beat DJs have long stuck to the use of vinyl. An affordable vinyl cutter was even introduced in 2001 by DJ equipment manufacturer Vestax, allowing home vinyl cutting of new mixes. Nevertheless, home-burned CD-Rs combined with a dedicated CD player currently present a viable alternative, as illustrated by experimental club night CDR, successfully organized in London, offering music hot from the CD burner of the music recording studio. MP3 formats are seeing an increase in popularity with young break beat DJs, who store digital audio files on a laptop and manipulate these with a digital 12-inch interface disc on their DJ turntable, for example via the Traktor Final Scratch system. The success of such digital equipment is based on the 12-inch vinyl record disc; this format therefore sustains its importance as a residual medium.

The British vinyl junky is also culturally linked to the Rare Groove collector of the 1980s and the Northern Soul fan in a North West English dance scene who, from the late 1960s onwards, danced exclusively to imported American soul music. In the latter case, the fast-paced stomping Motown sound of Detroit was preferred, with its heartbeat bass drum and its snare and tambourine on the 4/4 count. Mostly working-class fans buy, sell, and exchange rare imports at amphetamine-driven all-night dance events.[30] Obscurity raises the authentic status of a mechanically reproduced record; for some rarities, a log is available showing details of past sales, in terms of ownership and sales value. In one extreme case, Frank Wilson's 7-inch "Do I Love You" was sold for £15.000 in 1989.[31] Such exclusive knowledge highlights the librarian research activities and connoisseurship of the vinyl collector and DJ. The notion of archival knowledge was also tied up in the word "discotheque" and in the increased prestige of the sound-recorded disc as artifact.

FIGURE 6.3 *Cueing a 12-inch dance record. Photograph by Daniel Rubinstein.*

The French *discothèque* (literally, disc library) was established before World War II in Paris, where people would listen to jazz recordings while enjoying a social drink. Flirting with the "otherness" of jazz as a "double consciousness" of modernity, to use Gilroy's terminology,[32] African American artists, including dancer Josephine Baker, were celebrated. During the Nazi occupation, jazz was forbidden, thereby undoubtedly making this genre even more attractive to the anti-occupation Resistance. However, during the war, it was impossible to bring over bands to western Europe from across the Atlantic, so jazz records filled the gap at secret parties.[33] Thus a European connoisseurship of African American music added to a sense of intimate conspiracy of a war-based anti-Nazi underground movement. The discotheque reappeared after the war as a 1950s exclusive club concept.

In 1960, the first French-styled cosmopolitan discotheque, Le Club, opened in New York City. It provided a blueprint for the aspiring glitzy aspects of many discos that followed, and demonstrated to Americans that dancing to records was not necessarily a sign of poverty but a "cool" cultural activity during an economic peak for affluent seekers of good times. By 1965, even the rich, famous, and beautiful people were dancing to records, at midtown clubs like the "really chic" Arthur's.[34] However, as the 1960s drew to a close, New York's white middle classes got bored with the discotheque and, instead, some of America's social margins appeared on the disco dance floor; hidden in the obscurity of the late night, a range of identity configurations made up of African Americans, Italian Americans, Latinos, working classes, gays, and women entered the discotheque to shake their booty in a sexually liberated manner. New cultural crossroads in the hidden space of the "unfashionable" discotheque in the economic recession days of the early 1970s created a new energy necessary for the development of fresh ideas toward disco.

Also in the mid-1960s, DJ David Mancuso experimented with party concepts at his loft home, which were seminal for an innovative underground disco concept. Inspired by a combination of rent parties, Puerto Rican loft parties, countercultural hangouts, and the Electric Circus, as well as Timothy Leary's ideas around LSD trips, Mancuso's loft featured a fantastic hi-fi sound system and a type of music programming that took the psychedelic gathering on a lengthy shamanic sonic journey. Mancuso was, and still is, a perfectionist about sound, in which vinyl plays a central role. After a serious personal crisis, Mancuso professionalized his parties in the 1970s to turn into the Loft, which for nearly two decades inspired influential underground dance DJs and clubs, such as Nicky Siano's The Gallery or Paradise Garage with DJ Larry Levan, as well as Chicago's The Warehouse with New Yorker Frankie Knuckles.[35] The latter two clubs forged particular DJ aesthetics associated with the 12-inch vinyl dance single that in the 1980s would be marketed as garage and house music, respectively.

There is a close link between underground dance music (UDM) and disco, which Fikentscher defines as:

> a particular performance environment in which technologically mediated music is made immediate at the hands of a DJ, and in which this music is

responded to via dance by bodies on the dance floor. I call this concept the disco concept. To understand the disco concept is to understand the conceptual and historical foundations of UDM.[36]

Italian American DJ Francis Grasso is regarded as the first modern DJ who, to maintain a dance groove through lengthy dance sessions that in his late teenage years could last in excess of twelve hours a night, utilized DJ techniques such as slip cueing, the "slow mix"[37] (a layered soundscape), and "beat spinning" (matching the beats of two or more recordings). His DJ sets at The Sanctuary, in 1970, were typical of disco's promiscuous genre bending for a gender-bending crowd. Grasso experimented with a monitor headphone and achieved what seemed impossible: to play two records at the same time, creating a new soundscape, or "third record." In his seminal book *Disco*, music journalist Albert Goldman observed, "He would super the drum break of 'I'm a Man' over the orgasmic moans of Led Zeppelin's 'Whole Lotta Love' to make a powerfully erotic mix . . . that anticipated the formula of bass drum beats and love cries . . . now one of the clichés of the disco mix."[38] In addition to such a funk/rock/soul/R & B mix of dance music, DJs would "seek out exotic imports from unusual world music sources" in order to satisfy an "ever-demanding crowd,"[39] showing disco's flexible and adaptable musical attitude. The legacy of musical adventures with Latin dance music can still be heard in, for example, the dominance of salsa clave rhythms in the riffs of (post-disco) house music.

With its specific mix of Black Atlantic and Hispanic musical cultures, New York City was a hotbed for the development of dance cultures. Uptown, hidden from the global music business in its first six or seven years of development, hip hop emerged from the convergence of disco DJ techniques and an MC rap style that was reminiscent of the Jamaican toaster, the American jive radio DJ, and poetry from the black arts movement.[40] Mobile sound systems provided a welcome alternative to the discotheque for young African Americans, West Indians, and Puerto Ricans, for whom Manhattan's midtown discos were "too nervous . . . too fast . . . too gay"[41] and the local disco too dangerous as capitalist local politics produced self-policed gang-ruled ghettos.[42] The disco DJ style mutated in a block party sound system culture, on which soul historian Brian Ward comments,

> As dazzlingly dexterous black club deejays worked the twin turntables of their huge sound systems, they transformed the stark electronic hardware into a flexible musical instrument in its own right . . . Using the second turntable to overlay snippets of old and fresh funk, lost-found soul classics, and even splashes of reggae, rock and rhetoric, they created seamless collages of new danceable black noise . . . the seeds of the rap music and hip-hop culture which flowered in the 1980s.[43]

Hip hop shows similar musical diversity to disco but breaks the beat by utilizing the "zigue zigue" scratch sound, heard when slip-cueing a vinyl record as a rhythmical device and by sequencing small isolated rhythm sections, the breaks from recorded songs, on a disco twin turntable set-up. In an early and influential publication on the subject, David Toop described the break beat as

> the important part of a tune in which the drums take over. It could be the explosive Tito Puente style of Latin timbales to be heard on Jimmy Caster records; the loose funk drumming of countless '60s soul records by legends like James Brown or Dyke and the Blazers; even the foursquare bass-drum-and-snare intros by heavy metal and hard rockers like Thin Lizzy and The Rolling Stones. That was when the dancers flew and the DJs began cutting between the same few bars on two turntables, extending the break into an instrumental. One copy of a record—forget it.[44]

Grandmaster Flash and Afrika Bambaataa, two early commercially successful hip hop DJ-producers, have pointed to Kool Herc as their source of inspiration. He developed the break beat when, as a teenager, he set up his legendary sound system in 1972, five years after he had arrived from Jamaica. Flash has remarked that Herc's mix style was quite inexact because he did not use headphones—instead, he cued his breaks by sight,[45] relying on the visual position of the label and imprint of the grooves on the vinyl records. On reflection, Herc seemed to use the breaks like makeshift "riddims," the bass and drum instrumentals that underpinned Jamaican reggae sets at the time. He was known in particular for using the drum breaks of James Brown's 'Funky Drummer' and The Incredible Bongo Band's 'Apache' for its conga break.[46] Importantly, the latter funk band was just as popular in the downtown underground dance music and disco scene as it was in the uptown hip hop scene, showing that these scenes had more in common than several of its chroniclers wish to admit. Grandmaster Flash was more a perfectionist than Herc, however, applying disco's beat mix technique to create perfect break beat loops between two copies of the same record.[47]

In the use of vinyl as a musical instrument, the equipment is used in a manner that is alien to the home hi-fi, creatively breaking the rules and discarding the manual. The crowd is moved through the loud and forceful manipulation of recorded sound and juxtaposed lyrics while in hip hop the MC maintains the flow. The artistic aura is thereby firmly on the side of the DJ/MC team, rather than the consumable recordings, as fingers, even stickers, are put on the grooved surface to find the right splicing moments. In the 1990s, international DJ competitions in turntablism take the activity of scratching and cross-cutting between records, developed mainly by hip hop DJs, to an abstract extreme. Although the dance DJ's main function is to enable a good dance party, at such events the serious self-absorption of the DJ virtuoso into the technique of ruptured vinyl manipulation has become an end in itself, without much regard for the fun of dancing at a party. Interestingly, Technics has long sponsored such DJ competitions. At the end of the 1970s, this company introduced a record (or turntable) deck, the SL1200, with a long and sensitive pitch control slide and a unique magnetic (nonfriction) drive for the turntable, which has become a classic DJ tool that has not changed since 1980.[48] Pioneer, the prominent CD mixer manufacturer, has now set up a countercompetition for CDJs with some success.

Although an MC or a spoken link between records can provide a ruptured sense of continuity, the New York disco DJ would communicate exclusively through the order of music programming. On the disco dance floor, the idea was to float on the groove, to dance until a giddy sense of unshackled freedom and timeless abandon would indicate lift-off from the gravity of mortality and the niggling things in life. The same idea lies at the heart of house, techno and trance.[49] The format of disco music was adapted to enable such a flow, emphasizing the rhythm section, especially in the "intro" of a track, so that harmonies and textures of various sound sources would not clash when two tunes would be blended. Tom Moulton was the first to professionally to produce re-edits on tape and 7-inch vinyl formats that were specifically geared toward dance clubs, taking parts of records that worked best with dancers. In this manner, the practice of the dance DJ was extended into the studio, producing a lengthy disco mix from existing recordings in which the dancer could really lose herself.

The music industry traditionally used radio for marketing purposes, but in the 1970s, the larger discotheques in New York became popular enough to enable seemingly obscure recordings to reach high sales without radio support. Attempts were made by the music industry to control this new market, by providing dance floor–friendly disco mixes on the B sides of selected 7-inch singles. In 1974, Moulton accidentally stumbled upon the 12-inch disco single format. His sound engineer had run out of 7-inch acetates when Moulton did a master cut for Al Downing's 'I'll Be Holding On.' This was resolved by using a 12-inch blank on which the grooves were spread, so it would look better. This extra space also meant that the cut could be louder because this requires wider grooves. Although the resultant 12-inch test pressings sounded impressive on the dance floor, this track was eventually commercially available as a 7-inch. Six months later, in June 1975, the very first 12-inch single finally saw the day as a DJ tool: a re-edit of Bobby Moore's 'Call Me Your Anything Man.'[50] This is a good example of Winston's notion of a spin-off, where a technology is not used as initially intended. With its bulky size and the usual three-minute pop song format that had previously been dictated by the size of the shellac disc, the 12-inch had not been regarded as an attractive medium for a single recording. Yet the specific needs of the dance floor for longer grooves with an excellent loud sound quality led eventually to the adoption of the 12-inch vinyl disco single and the further development of its dance floor–specific aesthetic. In 1976, Salsoul released the first commercially available underground disco remix. Walter Gibbons, known for his DJ use of rhythm breaks, deconstructed the original three-minute recording of '10 Percent' by Double Exposure to a percussive dance groove of nine minutes. Digitally sampled in many later dance club productions of deep house and garage, it lives on in the fabric of current house music. Being distributed in 12-inch form, the extended disco mix went beyond the function of the pop song; disco successfully enabled interactive immersion in total sound. As a DJ tool, each 12-inch vinyl release is like an exclusive event, which draws its main economic support from dance DJs, rather than from their dance crowds.

Although DJs had different styles, an increased preference for a continuous disco groove foregrounded the four-to-the-floor (4/4) beat of rock and R & B, as this was easiest to mix on the night. Salsoul's drummer Earl Young has been credited with crystallizing this into a signature disco shuffle, using a pounding bass drum as a framework around which a variety of rhythms could be arranged, as can be heard on the 1974 seminal disco classic *Love Is the Message* by M.F.S.B.[51] However, the organic speed differences displayed by human drummers are difficult for DJs to blend during a set. Because the metronomic predictability of drum machines is easier in the mix, the disco aesthetic moved toward electronic tracks that provided minimal funky syncopation in the background. A landmark in this respect is a 17-minute Euro disco epic, 'Love to Love You Baby,' which Italian electronic composer Giorgio Moroder produced with Pete Bellotte for the label Casablanca, featuring Germany-based African American diva Donna Summer. With its thudding bass beat and climaxing soft porn vocal sounds, it exploited disco's sexually experimental sensibilities and established a commercial formula that crossed over into pop radio in 1975.[52] It set a standard that was particularly popular at New York's flagship Euro-American gay club The Saint. Visits to this club inspired British northern soul DJ Ian Levine in the early 1980s to introduce hi-NRG at the London gay club Heaven, a stripped version of pounding phallic 4/4 drum box beats combined with black diva vocals. Echoes of such "Euro beats" are still heard in versions of "gay" disco, hard house, and epic trance, which have long held on to the 12-inch vinyl format even though also here moves toward CD mixing have been noted.

The foregrounding of the 4/4 beat in disco has led to accusations of a musical ethnic cleansing. For example, African American critical theorist Houston A. Baker argues that "what was displaced by disco, ultimately, was R & B, a funky black music as general 'popular' entertainment," in addition to patriarchal "black manhood rights (rites)," which hip hop reclaimed from gay disco.[53] Such arguments not only hide the existence of black and Latino (gay) disco, such as influential underground dance clubs Paradise Garage and The Warehouse, but also obscure the fact that the emergence of hip hop is historically intertwined with the development of disco's DJ tools, techniques, and musical references. All dance DJs, whatever their sense of identity, used a mixture of groove, syncopation, and rhythm breaks in a varying balance between flow and rupture. In addition, competitive hip hop DJs explored music beyond local boundaries, such as German electronic outfit Kraftwerk, who produced electronic interpretations of American soul music that have become a classic reference point for house music, techno, trance, and hip hop related electro. Kraftwerk had a disco hit with 'Trans-Europe Express' in 1977 that provided a main component in the successful 1982 electro rap release 'Planet Rock' by Afrika Bambaataa & Soul Sonic Force, with producer Arthur Baker, on 12-inch vinyl for hip hop label Tommy Boy.[54]

Like electro for hip hop, house music showed a new electronic phase for disco; the practice of vinyl manipulation was partly replaced and partly supplemented by the digital sampler and electronic drum machines and synthesizers. House music has its generic roots in Frankie Knuckles's disco DJ sessions at the Chicago club The Warehouse to which he was invited from New York in 1977. This was the same year as the global success of the film *Saturday Night Fever* in which the working-class Italian American straight hero is rescued from a disco lifestyle.[55] By the end of the 1970s, the business of disco had become worth $4 billion in the United States, making it more important than the broadcasting, film, or sports industries,[56] much to the chagrin of rock radio DJs. Chicago was the heartland of reactionary Middle America, when, in 1979, white rock radio DJ Steve Dahl instigated the "Disco Sucks" campaign; at Comiskey Park stadium, in front of about 70,000 baseball fans and broadcast on prime-time television, a riotous crowd of white men detonated a huge pile of vinyl dance records, mostly gay and commercial disco mixed with R & B, accompanied by great cheers and a hysterical rush onto the game field. Yet this was not the end of disco or of the 12-inch vinyl disco single. On the contrary, house music became, as Knuckles put it, "disco's revenge."[57]

Whether electro or house music, the production of different-sounding and even new material was a way of staying one step ahead in a competitive DJ–based party scene. The arrival of cheap Japanese electronic music technologies helped the production process of new material. The drum box was used live during DJ sets to enable the continuous groove and to enhance the bass drum—in particular, the Roland 808 was popular in the early 1980s. The digital sampler was initially developed as a sound engineering tool, to "repair" pop recordings during studio recording sessions by digitally recording parts of a track, which can then be digitally manipulated and duplicated. A spin-off was created by the DJ practice using the samples of breaks and of danceable parts of records. Especially in conjunction with a sequencer, the sampler enabled DJs to produce re-edits for their remixes without high studio costs, recording their tracks onto reel-to-reel as well as a new cheap sonic medium, the audio-cassette tape, which had gained popularity across the DJ world in the early 1980s. As an extension of the particular DJ practice of the house music DJ, the samples were just as choppy as for hip hop but superimposed on a drum machine track that retained the steady disco groove with its characteristic 4/4 bass foot. Eventually, in 1984, the house music adventures of Chicago DJs found their way onto DJ-mix-friendly 12-inch vinyl.

Although it was still a special privilege in the 1970s, through the 1980s an increasing number of DJs developed from being remixers to becoming producers in their own right and, with the help of computer technology, even studio

FIGURE 6.4 *"Needle in the Groove at Black Vinyl Boat Party on the Thames." Photograph by Hillegonda Rietveld.*

musicians who could play their own recordings to their crowds. Because the major music industry tried to enforce a consumer change to CD, vinyl plants were already hard to find in the 1980s. Yet because of the need to use these recordings in a DJ mix, the 12-inch vinyl remained the medium of choice for these DJs. In Chicago, one pressing plant, run by Larry Sherman, seemed to have a near monopoly in that city; producing pressings for his Trax label on recycled vinyl resulted in wobbly discs with a collectable trademark crackling sound. In a time of mechanical reproduction and sonic simulation, the performance with one's own studio material is the closest an audience can get to the source of studio-generated music. The CD-R in combination with a CD mixer could achieve the same aim and is currently a popular option—at such occasions, the presence of the DJ-composer-producer makes up for a "lack" of presence in the digital medium.

As a concept of DiY disco, house music was embraced by the British youth marketing machine in the late 1980s, providing a driving beat and community-minded messages of love and togetherness for a budding rave scene. In 1993, the UK postrave house music–based dance club business was estimated at between 1 and 2 billion pounds a year, more than the total value of the UK-based publishing business.[58] In Germany, a similar success of the disco format can be measured by its yearly Love Parade in Berlin in the late 1990s, where 1 million young people danced annually to mobile techno sound systems to celebrate the anniversary of the reunification of Germany. In its Anglicized and Europeanized forms, dancing to the "third record"[59] returned to the United States[60] in the 1990s, around the same time that MTV finally embraced rap and the hip hop aesthetic. This inspired kids in America's heartland, like their counterparts in other privileged economies, to take up record mixing instead of playing the guitar. The image of the DJ, with headphones, Technics decks, a mixer, and 12-inch vinyl records has become iconic, even if only simulated as a residual sign in a computer game, on MTV, or as intertextual shorthand in a TV commercial.

In conclusion, the vinyl 12-inch dance single has been developed in the context of a resourceful cultural response based on social necessity in public dance music provision. The promise of simplified logistics as well as a more flexible provider of repertoire made the DJ an attractive cheaper solution to club entertainment. As a result of such economic considerations, traditional musicians were replaced by entertaining record collectors, who eventually developed a new form of musicianship, which in turn provided the cultural conditions for the development of the 12-inch dance single. Being forged within a specific musical aesthetic language for the dance floor, the 12-inch vinyl single is residual in the practices of the current digital DJ and as such remains as a status symbol for DJs. Analog vinyl has qualities that make it attractive because of, and beyond, the cultural authenticity forged by the history of dancing to records described here, especially its analog sound quality, its hands-on manipulation, and its fetishization as collectible object. However, digital music formats, such as the CD and MP3, are cheaper and easier to obtain, making them attractive alternatives to vinyl records, especially for less affluent DJs. In economically challenged countries, such as Thailand and Indonesia, a 12-inch single would be at least three times the daily income of most of its people, so social necessity dictates here that CDs are the preferred medium for resourceful professional DJs. In such countries, only expensive cosmopolitan clubs and cocktail bars can offer their music on 12-inch vinyl, indulging their wealthy clientele in deep house as a form of cool new jazz that has been imported via Internet retail. To some extent such establishments are comparable to the French discotheque concept half a century earlier. Meanwhile, American and British DJs who work on an international level are also moving to the use of CDs; being lighter and a more compact medium than vinyl, CDs are easier and safer to transport by plane, while pre-releases increasingly occur on CD-R. Nevertheles, despite the move to digital music formats, the global mediation of dance music is historically centered in New York and London (followed by Berlin, Paris, San Francisco, LA, Melbourne, and Tokyo),[61] where the 12-inch vinyl dance single has been perceived and marketed as a signifier of underground dance music, of soul sonic good taste in dance music, of connoisseurship, and as a residual standard for its digital sonic alternatives.

Notes

1. Based on a series of interviews conducted between September 2003 and January 2004, with many thanks to Dan and JP at Vinyl Junkies; Andrew and Anthea at Flying Records (a dance vinyl specialist which closed in 2004); Goldie and Clarkie at Blackmarket; Jake at Westend DJ; Lerato "Lakuti" of Sud; Gavin Alexander of CDR; Charlie Dark of Blacktronica; Maureen "Sisamo" Schipper of Lady Bugz/Les Fleurs; Patrick Hagenaar at Ministry of Sound; Egbert of JazzSole; and James Nice at LTM. Also thanks to all other vinyl junkies, CD DJs, record distributors, sound engineers, and dance enthusiasts whose additional expert opinions I have had the pleasure and honor to share.

2. For additional insights into this issue, see Will Straw, "Value and Velocity: The 12-inch Single as Medium and Artifact," in *Popular Music Studies*, eds. David Hesmondhalgh and Keith Negus (London: Arnold, 2002), as well as Sarah Thornton, *Club Cultures: Music, Media and Subcultural Capital* (Cambridge: Polity Press, 1995), 58–76.

3. The traditional spelling "disc" is specifically applied here to analog music recordings, as in "disc jockey" (DJ) and "discotheque"; the American alternative spelling, "disk," is used for all other flat round objects.

4. N. Katherine Hayles, "The Life Cycle of Cyborgs: Writing the Posthuman," in *Cybersexualities*, ed. Jenny Wolmark (Edinburgh: Edinburgh University Press, 1999), 158.

5. DiY is a cultural form of DIY or "do it yourself." For further expansion of this concept, see also Hillegonda Rietveld, "Repetitive Beats: Free Parties and the Politics of Contemporary DiY Dance Culture in Britain," in *Boom Creation! DiY Culture and the New Protest*, ed. George McKay (London: Verso, 1998).

6. Paul Gilroy, *The Black Atlantic: Modernity and Double Consciousness* (London: Verso, 1993).

7. Brian Winston, *Media, Technology and Society. A History: From the Telegraph to the Internet* (London: Routledge, 1998), 1–15.

8. Winston, *Media, Technology and Society*, 134.

9. Bill Brewster and Frank Broughton, *Last Night a DJ Saved My Life: The History of the Disc Jockey* (London: Headline, 1999), 30.

10. Winston, *Media, Technology and Society*, 134.

11. Michael Chanan, *Repeated Takes: A Short History of Recording and Its Effects on Music* (London: Verso, 1995), 93.

12. Chanan, *Repeated Takes*, 83.

13. Ibid., 88.

14. Kai Fikentscher, *You Better Work!* (Hanover and London: Wesleyan University Press, 2000), 25.

15. Fikentscher, *You Better Work!* 67.

16. Winston, *Media, Technology and Society*, 6.

17. Tricia Rose, *Black Noise: Rap Music and Black Culture in Contemporary America* (Hanover: Wesleyan University Press, 1994), 61.

18. Jimmy Savile; cited in Broughton and Brewster, *Last Night*, 25.
19. Brewster and Broughton, *Last Night*, 37.
20. Ibid., 51.
21. Cited in Tim Lawrence, *Love Saves the Day: A History of American Dance Music Culture, 1970–1979* (Durham, N.C., and London: Duke University Press, 2003), 207.
22. David Toop, *The Rap Attack: African Jive to New York Hip Hop* (London: Pluto, 1984), 39.
23. Dick Hebdidge, *Cut 'n' Mix: Culture, Identity and Caribbean Music* (London: Comedia/Routledge, 1987), 62–70.
24. Brewster and Broughton, *Last Night*, 106.
25. Cheryl L. Keyes, *Rap Music and Street Consciousness* (Urbana and Chicago: University of Illinois), 51.
26. Hebdidge, *Cut 'n' Mix*, 63.
27. Ibid., 83.
28. Lloyd Bradley, *Bass Culture: When Reggae Was King* (London: Viking, 2000), 295.
29. Bradley, *Bass Culture*, 295–97, 311–16.
30. Russ Winstanley and David Nowell, *Soul Survivors: The Wigan Casino* (London: Robson Books, 1996); Pete McKenna, *Nightshift* (Lockerby: S.T. Publishing, 1996).
31. Brewster and Broughton, *Last Night*, 96–98.
32. Gilroy, *The Black Atlantic*.
33. Hillegonda C. Rietveld, *This Is Our House: House Music, Cultural Spaces and Technologies* (Aldershot: Ashgate, 1998), 101.
34. Alan Jones and Jussi Kantonen, *Saturday Night Fever: The Story of Disco* (Edinburgh and London: Mainstream Publishing, 1999), 17.
35. Lawrence, *Love Saves the Day*.
36. Fikentscher, *You Better Work!* 22.
37. Rietveld, *This Is Our House*, 109.
38. Cited by David Harvey, "Behind the Groove," in *Collusion #5*. Publication date unknown (early 1980s).
39. Jones and Kantonen, *Saturday Night*, 19.
40. Rose, *Black Noise*, 27–34.
41. Bill Adler; cited in Keyes, *Rap Music*, 44.
42. Rose, *Black Noise*, 27–34.
43. Brian Ward, *Just My Soul Responding: Rhythm and Blues, Black Consciousness and Race Relations* (London: UCL Press, 1998), 429.
44. Toop, *Rap Attack*, 14.
45. Ibid., 62.
46. Keyes, *Rap Music*, 57.
47. Nelson George, *Hip Hop America* (London: Penguin, 1999), 17.
48. Ulf Poshardt, *DJ Culture*, 234.
49. Hillegonda Rietveld, "Ephemeral Spirit: Sacrificial Cyborg and Communal Soul," in *Rave Culture and Religion*, ed. Graham St. John (London and New York: Routledge, 2004), 53–54.
50. Lawrence, *Love Saves the Day*, 212.
51. Brewster and Broughton, *Last Night*, 173.
52. Jones and Kantonen, *Saturday Night*, 48–50.
53. Houston A. Baker, *Rap and the Academy* (Chicago: University of Chicago Press, 1995), 86.
54. Keyes, *Rap Music*, 76. In addition, the bass line for Planet Rock was based on another track by Kraftwerk, *Tour the France*.
55. The soundtrack was dominated by white pop stars the Bee Gees, who contributed six tracks. However, the soundtrack also included the Trammps, with the classic *Disco Inferno*, as well as disco groove originators M.F.S.B.; the latter two are credible within the discursive cultural values of UDM, or underground disco, a scene well described by Fikentscher (2000) and Lawrence (2003).
56. Brewster and Broughton, *Last Night*, 185.
57. Rietveld, *This Is Our House*, 99.
58. Ibid., 61.
59. Dan Sicko, *Techno Rebels: The Renegades of Electronic Funk* (New York: Billboard Books, 1999), 118.
60. Mireille Silcott, *Rave America: New School Dancescapes* (Toronto: ECW Press, 1999), 17–46.
61. Based on research travel between 1998 and 2002, supported by the Faculty of Humanities and Social Science, London South Bank University.

7 Reporting by Phone

Collette Snowden

The ringing of the telephone woke me early one Sunday morning in Adelaide, South Australia, and reacting with a sense of urgency, as most people do, I answered its insistent sound without question. Calling from London, where it was Saturday evening, a journalist from one of the daily newspapers was searching for local information. I answered his questions, but more importantly provided a telephone number for a more useful local source. My caller was assigned to a major breaking news story, and working the phones was just a necessary and routine part of his task as a reporter. The telephone was simply the technology that allowed him to reach across the globe. No doubt another local media contact would have received a call if I hadn't answered. The next day the information he obtained appeared in his report and was reprinted and broadcast by other news media outlets. More than a century after its commercial launch, the use of the telephone to conquer distance, to tap into a network of contacts, to talk to strangers and then retransmit the information received through the mass media continues to be remarkable, if too often unremarked upon.

This example demonstrates just one typical use of the telephone by media professionals, and every day in news media around the world the telephone continues to be a critical communications technology that contributes to the production of news media content on a persistent and pervasive basis. The widespread availability of the telephone provides rapid communication over distance that allows media professionals to access privately held knowledge and expertise and transform it into publicly available information. Media organizations also depend on the telephone system for organizational and administrative functioning. Yet there is a paradox in the use of the telephone by media professionals, for although the telephone is thoroughly normalized in the everyday working practices and processes of the news media and media professionals to the extent that it can be regarded as ubiquitous, it is also curiously invisible in analysis of the news media and in our understanding of the content we receive via the news media.

Studies of practical aspects of media work rarely consider the use of the telephone as a primary tool for media professionals. For example, most existing material on interviewing techniques for media professionals is based on the premise that interviews take place between co-located individuals. Although the National Council for the Training of Journalists in the United Kingdom has now introduced a short course on interviewing that includes aspects of telephone interviewing techniques,[1] such attention as there is in the professional and academic domain to the use of the telephone is brief and superficial. Sedorkin and McGregor argue that

> While the face-to-face interview is the ideal way of gathering information from primary sources, it is steadily becoming a less common method used by journalists. Today more interviews are being conducted by fewer staff who are turning to faster methods of news gathering. The two most common time-saving methods used today are telephone and email interviews.[2]

Ideal or not, the face-to-face interview is receding in the daily practices of media professionals with anecdotal accounts of media professionals and the limited empirical material available,[3] indicating that a vast amount of material is obtained by media professionals via telephone. Rather than something new and emerging, this dependence by media professionals on the use of the telephone has taken place over many decades, and over that time the use of the telephone has moved from the margins of the practice of media professions to a more central position so that a substantial part of the work of news gathering and production is based around gaining access to people and information through the telephone system wherever and whenever possible.

Today news and information is conveyed by telephone everyday on what a quarter of a century ago was called "the largest electrical communication network in existence."[4] With the addition of the Internet and mobile phone networks, and the increasing integration of voice and data communication, it is now claimed that "the global telephone system is the largest and most complex machine in the world."[5] The general use of the telephone for communication ranges in individual daily use from the mundane (*"Please could you pick up some milk on your way home?"*) to the profound (*"Help! I think my child is dying. We need an ambulance!"*). It also has social and political significance from the local level (*"It would be useful if you could attend the meeting."*) to the global and international (*"The prime minister spoke to the president by phone late last night to negotiate a settlement."*). On a global scale the telephone is fundamental to the functioning of the global economy, with the amount of international telephone traffic generated by and between nations regarded as a key indicator of economic activity. The most cursory consideration of the telephone reveals that in many areas of contemporary life the telephone is an instrument of pervasive utility, and the communication it allows has the ability to alter and influence human behavior significantly. How then can such a powerful technology be considered residual? Therein lies the central paradox of the telephone, that is, in its very ubiquity and acceptance the particular and special properties of the telephone have become so commonplace that we no longer perceive them or think about them as extraordinary. The properties of the telephone thus reside in the background of our understanding and use of communication technology, particularly those forms that have developed out of and along with the telephone system.

The residual nature of telephone communication lies not in the withering away or replacement of the telephone by new communications technology. Rather, there are two critical elements of telephone communication that persist and leave a strong "residual" trace in other forms of media. First, there is the continuing formation of communication at a distance on the same model as the telephone; that is, electronic communication continues to be largely dyadic and highly personalized. Second, the foundation of telephone communication in orality and conversation gives particular properties to the information that is conveyed and produced, which is then reproduced in content for other media, where the effect of the original telephone communication persists. That is, following Hutchby, the telephone has particular affordances, by virtue of the manner in which it has been developed and deployed, which "impose themselves upon humans' actions with, around or via that artefact."[6] Such affordances that the telephone possesses "are functional aspects that frame, while not determining, the possibilities for agentic action."[7] Although a telephone continues to be an instrument or artifact by which people communicate with others verbally, rather than face to face, particular affordances are associated with the use of the telephone.

Turning from the global use and effects of the telephone system to its more specific application in the news and entertainment media, the telephone is continually present as an underlying enabling medium, from live action "reporting by phone" on radio and television, to reports phoned into newspapers and online news services by the public. The telephone is an irreplaceable and crucial tool for media professionals because it has incredible spatial reach and the global network that it provides access to enables information to be acquired, checked, and verified quickly and economically. The telephone is a critical communication technology for creating the vast amount of material produced and consumed in the contemporary news media.

Yet few scholars have turned their attention to the specific subject of the telephone, and those that have tend to share the view of James E. Katz that "no technology is more ubiquitous than the telephone, and none less examined."[8] Avital Ronell in *The Telephone Book*, "a study of the history, psychology, and unnoticed politics of the telephone," declares that,

> Lodged somewhere among politics, poetry and science, between memory and hallucination, the telephone necessarily touches the state, terrorism, psychoanalysis, language theory, and a number of death-support systems.[9]

Despite the paucity of academic work and analysis of the telephone, there is no suggestion that the telephone is completely absent, rather that because "the miracle of telephone conversation is too readily forgotten by laypeople and scholars alike,"[10] the telephone is glossed over, especially in favor of newer, emergent forms of communication. It is in this sense particularly that the telephone can be regarded as a residual medium. For although the use of the telephone is persistent and pervasive, the qualities of telephone communication are not always considered as peculiar or specific, despite their effect on the communication associated with the telephone.

If we consider the use of the most pervasive of all communication technology in and by the news media and consider how it influences and shapes the way the news media works, the residual effects associated with the use of the telephone become more apparent. These effects are both historic and continuing, and they relate in part to the basic conditions of telephone-based communication as a two-way (dyadic) oral communication between individuals, but are also related to the way that telephone networks have been developed, managed, and regulated. In assessing the use of more recently developed communication media to report and deliver news and information, and those that are emerging, it is useful to consider how they continue to retain elements of the much older telephone and telephone system.

Consideration of the use of the telephone in the production of news reveals that its use contributes an enormous amount of information to the daily output of news and information, but also that the specific affordances of telephone communication influence how the news media structure content originating from telephone communication. The telephone is used so extensively by the news media that it could almost be regarded as a medium itself, through which vast amounts of news and information are passed from the private to the public domain, from the individual to a mass audience.

Yet the telephone for the most part is an indirect medium, compared to traditional news media, that is, newspapers, radio, television, or new media such as online news services. Until quite recently, the telephone was rarely used directly to distribute media content, although new wireless telephone technology now makes it possible. Nonetheless, even in its older conventional form, the telephone can be regarded as a news medium because its use indirectly supports and allows other news media to function as we expect them to. Via the telephone the news media make wide-ranging connections that frequently allow news and information to be discovered and reported. Although the introduction of newer technology, especially e-mail, is creating changes in these practices, the long-standing conventional work processes associated with the use of the telephone continue to shape and influence the contemporary media. In a survey conducted to assess media professionals' use of mobile communications technology, the majority of respondents asserted that the most frequently used and important communication technology they used is a simple fixed-line telephone.[11] Interviews with media professionals supported this finding with the assertion, "I simply couldn't do my job without a phone," made repeatedly in one form or another. Despite the advent of new forms of media, the telephone remains the most important communication tool available to journalists, as it has been for decades.

The residual effects of the original person-to-person voice-only telephone instrument remain important even as the technical convergence of the telephone with other media increases. For example, improved satellite coverage now permits direct live telephone links to reporters in the field for radio and television, but many of the traditional practices of telephone usage persist even in this new application. In such circumstances the telephone is a medium on which all other news media have a strong reliance, even a growing dependence. Audience members, too, are able to engage directly in the news reporting process through telephone interaction. Consequently we see and hear so-called eyewitness reports phoned in to news media outlets, and additional information and tips are provided directly by readers, listeners, and viewers and contribute to the production of news and information.

The contribution of the telephone in shaping contemporary media practices is a marginal field of study, but pioneering communication scholar Harold Innis noted in his work that,

> the telephone supplemented the telegraph for local news and the reporter saw his work divided between the leg man who visited the sources of news and the rewrite man to whom he telephoned his information and who prepared it for the paper.[12]

Later, Marshall McLuhan also noted the particular qualities and importance of the telephone to the media, although in the analysis of his wide-ranging exploration of other media the observations he made about the telephone have been somewhat overshadowed. One of McLuhan's most pertinent observations about the use of the telephone by the news media was that,

> [T]he American newspaperman in large degrees assembles his stories and processes his data by telephone because of the speed and immediacy of the oral process. Our popular press is a near approximation of the grapevine.[13]

The concept of the press as a grapevine, as McLuhan proposed, is consistent with the use of the telephone in the everyday practice of media professionals, among whom a key measure of professionalism is the quality and extent of contacts. Such contacts are frequently maintained by telephone despite the increasing use of e-mail.

Media professionals rely heavily on the telephone to provide them with fast, cheap, and efficient access to their contacts, as well as access to new sources of information. There is also a need for them to trace and contact people in distant locations, and the global telephone system allows them to do so with relative ease. Telephone contact provides an immediate response in many situations, with the added feature of being an acutely personal form of communication. It is also possible to follow and track people by telephone, literally chasing them electronically. Additionally, media professionals rely on tips and leaks provided to them via telephone from inside sources, and in many areas of reporting such as crime, courts, politics, and business, the grapevine of contacts and sources that journalists are connected to is fundamental to the conduct of their work. It is through these networks of contacts that off-the-record briefings are made; gossip, hearsay, and rumors are passed on, discussed, and analyzed; and important face-to-face meetings and interviews are arranged. These forms of communication are fundamental to the operation of the news media, and through them media professionals frequently acquire and check data and information from which they assemble stories and reports. The telephone is thus an enabling technology that is critical to the professional practices of media professionals.

Psychologist Robin Dunbar argues that gossip is a form of grooming that contributes to social cohesion, and proposes that one of the purposes of language development was to facilitate the social grooming afforded by conversation.[14] Sociologists Jack Leven and Arnold Arluke also acknowledge the importance of gossip in their work, *Gossip: The Inside Scoop*, with its own telling use of a common media term in the title.[15] Leven and Arluke consider gossip necessary for social bonding, for the distribution of socially and culturally valuable information, and the establishment and maintenance of social networks. As the telephone increases social grooming by increasing the number and extent of contacts, so the distribution of socially and culturally valuable information also increases. For media professionals, an extensive contact network that can be reached by telephone is therefore a professional necessity.

Such social networks have a remarkable resemblance to telephone networks, with the powerful at the center, and the powerless either at the margins or with limited access, epitomized by the term "out of the loop." The capacity of the telephone to facilitate gossip has been noted throughout its development and use. News media exploit this function of telephone communication most successfully, and the content of news media also reflects this form of communication, with the extensive use and presence of gossip, hearsay, and unsubstantiated information in the news media indicative of both the reliance on telephone communication by media professionals and the persistence of the role of gossip for purposes of social grooming, as Dunbar argues.

The questions of how the particular form and conventions of telephone communication affect the way in which information is conveyed and how the discourses and agenda of news media are influenced are closely allied to understanding the technologization of oral communication via the use of the telephone.[16] As the telephone system has expanded, telephone communication has also influenced general behavior and expectations and had a powerful effect and influence on the manner of communication and on social behavior. However, the importance of the telephone in shaping forms of modern communication has not been widely acknowledged, partly as a result of its normalization and acceptance in modern societies. Similarly, our understanding of the production of news media content owes very little to understanding how information is affected because it has been received and filtered via telephone. The latent attributes of telephone communication may persist in all forms of media to which telephone communication contributes and where the particular affordances created remain useful, yet our understanding of those qualities is still largely unexamined.

McLuhan returned repeatedly to the importance of the telephone in expounding his theories of media effects on the processes of communication, individual consciousness, and perception more generally. In particular, McLuhan used the example of the telephone to illustrate his theory about the manner in which electronic media disembody the user. In a lecture at Columbia University in 1978 he explained,

> When you're on the telephone, you don't have a body. You're a discarnate, disembodied intelligence. You have an image, acoustic, but no physical body. When you're on radio or TV, you have no physical body. What do you think the effect of not having a body is on the user of the telephone or the radio or television? The effect is always a hidden ground; it's never part of the figure. The thing you see is figure; the thing that affects you is ground, and that's what I mean by the medium is the message. The medium is hidden; the message is obvious. But the real effect comes from the hidden ground, not from the content, not from the figure. It comes from the ground which is never noticed. Nobody in the history of the telephone has ever mentioned that on the telephone you do not have a physical body.[17]

Indeed, considering the telephone as "the hidden ground" in news media reveals its vital role. Often news reports and other media stories are actually the recounting and reinterpretation of what were originally telephone conversations or interviews. Consequently, the telephone has, throughout its history, shaped reporting conventions and influenced the priorities and values of news media. Its importance persists in the news media in the reliance on information obtained in telephone interviews, and increasingly in the conversational style of delivery of the broadcast media, and even in the tendency to a less formal colloquial, conversational style of prose in the print media.

According to Ian Hutchby, "There is a particular relevance in thinking about the nature of ordinary conversation for our understanding of how technologies for communication function in everyday life."[18] The conversational properties of telephone communication and the pervasive orality of the medium of the telephone persist in the subsequent reporting of conversations and in the conveyance of information obtained by telephone. By delivering the power of oral communication quickly and relatively cheaply, the telephone has vastly increased the connectivity between news media and audiences while the specific properties of telephone communication in turn have shaped and influenced news media conventions and processes.

The use of the telephone to enable conversation or oral communication, as established by the limited technology available when it was introduced, is a key residual element of the telephone. Despite the possibility for many more communication capabilities, the telephone is still most often regarded as a device for assisting conversation over distance.

Telegraphy and the Media

Before the telephone, the use of the telegraph firmly established many of the conventions and priorities we associate with modern communication technology and the news media specifically. The invention and adoption of the telegraph was powerful in managing the problem of communication at a distance and in increasing the speed of communication. Lewis Mumford drew attention to the importance of the telegraph and its critical importance in allowing communication over distance in his early work, noting that

> With the invention of the telegraph a series of inventions began to bridge the gap in time between communication and response despite the handicaps of space: first the telegraph, then the telephone, then the wireless telegraph, then the wireless telephone, and finally television.[19]

The importance of the telegraph to the news media, specifically the newspaper industry, was also explored by Innis,[20] who recognized there was a relationship between news media and the dominant form of technology used in news production processes and argued,

> [T]he newspaper has been a pioneer in the development of speed in communication and transportation. Extension of railroads and telegraphs brought more rapid transmission of news and wider and faster circulation of newspapers; and newspapers, in turn, demanded further extension of railroads and telegraph lines.[21]

Innis was prescient in recognizing that the news media increased both power and profitability by accessing and controlling particular means of communication, and also noting that particular forms of communication eventually affected the content produced by the news media. Although Innis did not pay particular attention to the telephone or the emerging global telecommunications system, his theories about the role of communications systems and the specific bias inherent in particular communication media remain relevant in assessing the use of the telephone by news media.

Building on the observations made by Innis, James Carey later asserted that telegraphy separated communication over distance from transportation for the first time in human history, giving it a critical role in the shaping of the modern world and impacting many elements of human activity and society.[22] For this reason Carey expanded Innis's thesis to argue that telegraphy also influenced our perception of space and time by "freeing communication from the constraints of geography"[23] and by requiring and allowing the introduction of standard time zones.[24]

One of the most significant consequences for the news media, and for the modern era, arising from the successful invention and eventual diffusion of telegraphy, was the development of newswire services. Introduced just four years after the first telegraph service, the newswire services quickly became successful and profitable. Perhaps more significantly, the success of the newswire services introduced an expectation that news reporting could and would be provided as close to real time as possible. The use of telegraphy also laid the foundation for the twenty-four-hour news cycle by expanding the time in which news organizations operated,[25] with many large circulation newspapers producing several editions each day to manage the new phenomenon of "breaking news." Carey also attributed changes in the nature of language to the use of telegraphy by the media, arguing that the demands of the newswire services "led to a fundamental change in news," specifically a new style of writing arising from the need for the content produced to be understood in many different locations and to be concise because of the high cost of telegraph services. The language of the media continues to conform to such requirements, with the essence of journalistic prose regarded as the use of the active voice and simplicity of expression.

The use of telegraphy increased the ability of reporters to submit reports to their newspapers from distant locations, and it also allowed material to be used in newspapers that was not the product of the direct participation or observation of their own reporters. Telegraphy allowed the reproduction of news from the wire services and the inclusion of information from additional external sources. The telephone extended the ability of the news media to access and use widely dispersed sources of information, and it increased the speed at which the news media could operate, beginning with reporters calling in stories to newspapers by phone, to the current practice of live satellite telephone links from war and disaster zones.

If the impact of the telegraph on the development of media practices and the news media was profound, the impact of the telephone, which, in comparison, has endured and is more widely dispersed, must surely be greater. The expectation for fast, instantaneous communication established by the telegraph was more than matched by the use of telephony. Critically, the telephone did not require specialized skills to be used successfully and therefore enabled access to fast, instantaneous information, or to a source, with the additional feature of being more direct and more discreet without the mediating presence of a telegraph operator (although clearly the discretion of telephone operators in the early years could not always be guaranteed). Use of the telephone thus enhanced the potential for private communication to be transformed into news for public consumption quickly and discreetly, but it also introduced properties of orality and conversational affordances that were not present with the use of telegraphy. These properties include the use of direct speech and vernacular language, instant rather than reflexive responses, and the construction of communication on a strong dyadic model. The limitations of the telephone have not been accounted for in contemporary analyses of the news media. In particular, the loss of nonverbal clues and the privileging or misinterpretations of paralinguistic cues, such as tone or rate of speech, have not been considered. However, media professionals themselves are acutely aware of the limitations and constraints of telephone communication.[26]

Telephone Journalism and the Media

Soon after the official patenting of the first telephone and the establishment of the first telephone networks, the companies that invested in the technology and established the first telephone companies began to engage with the media. Regardless of the purpose of the inventors in developing the technology or the capabilities and potential of the technology, as dot.com companies were also to discover in the 1990s, if they wanted to profit from a new, untried technology, the early telephone companies quickly had to find applications people were willing to pay for. As heavy users of telegraph services, newspaper offices were an obvious target for the early telephone companies, as Michele Martin notes,

> [N]ewspapers were not initially overly enthusiastic about the telephone. At the end of the 1870s, company management lamented the fact that newspapermen were devoting very little attention to telephonic experiments. In fact, it was this indifference that forced Bell Telephone Co. to adopt definite policies for publicising its product in the press. Company management reached an agreement with some widely distributed papers that entailed the exchange of a free telephone in their offices for advertisements and insertion of notices at no charge. The company considered the deal a "profitable arrangement," since it guaranteed the "friendliness of the papers" and produced more numerous and more positive news reports on the company's activities (BCA, d 26606, 1887, 24).[27]

However, once they began to use the telephone, newspaper companies quickly realized the practical benefits it provided. The telegraph provided timely communication, but the telephone was even more immediate. The first recorded news despatch by phone "was from Salem, Massachusetts, to the Boston Globe, a distance of 16 miles, on 12 February 1877."[28] By 1880, Kate Field, a journalist with the *New York Tribune,* was in England promoting Bell's telephone;[29] and "news of the bombardment of Alexandria by British warships in 1882 was the first important announcement to be phoned from Reuters' head office in London to the Press Association for onward distribution to the UK press."[30] The telephone bettered the speed of the telegraph and also meant that journalists and anyone with access to a telephone could contact a newspaper office. Offering both a practical advantage and economic efficiency, the telephone soon became integral to the running of the modern newspaper office and a tool that journalists quickly came to rely on.

In 1889, Jules Verne wrote a short story set a thousand years in the future, *In the Year 2889,* in which "telephonic journalism"[31] was the norm. Verne explains telephonic journalism as "a system made possible by the enormous development of telephony during the last hundred years."[32] The future, however, came much faster than Verne anticipated with a similar service introduced in Hungary by 1893. The *Telefon Hirmondo,* as reported in *The Electrical Engineer* in 1895,[33] and which Caroline Marvin has also documented, lasted for twenty years.[34] Throughout its existence it provided a daily telephone broadcast of news, musical performances, an events calendar, and stock exchange reports. A service modeled on it, *The Telephone Herald,* was established in Newark, New Jersey, in 1911, but it proved to be an even shorter lived experiment in the use of the telephone as a mass medium.[35] Services now being offered via mobile telephone connections are remarkably similar to those provided by this early use of the telephone.

Representations of journalists in popular culture from the early years of the twentieth century also began to reflect the use of the telephone by journalists, and by the 1930s, journalists in films were frequently depicted rushing for a telephone, receiving "just in time" telephone tip-offs or frantically trying to submit stories by telephone. Although it took longer to be accepted by the general population, for media professionals the telephone became entrenched as a critical professional tool early in its development.

However, despite its successful integration into the daily work of the news media, Brian Winston records the failure of the early telephone technology to be fully realized, arguing that the radical potential of the telephone as a mass communication or broadcast medium, as in the example of the *Telefon Hirmondo*, was actively suppressed. Instead the telephone system developed as "a child of commerce, specifically of mid-nineteenth century commercial developments."[36] As a result, the technical structure of the telephone system that was adopted in its early decades focused instead on the provision of point-to-point communication rather than on the development of its potential as a broadcast medium, imposing limitations on how the telephone worked in practice, how telephone communication became defined, and how the particular affordances of telephone communication were normalized. The properties of the telephone that persist are in part a result of the technical limitations that arose from the commercial and regulatory decisions made in the early period of its development. Thus the telephone has been largely configured as a personal point-to-point, person-to-person form of communication. For a long time electronic communication in general reflected this form of communication, and even media content originating from telephone communication has shared these properties. For example, talk-back radio using telephone calls is primarily dyadic, and the rules and conventions of conversation of telephone communication, including accepted understanding about turn taking and the absence of silence or reflective pauses, persist.

Nonetheless, with the use of telephone communication and some judicious legwork, media professionals were able quickly to make connections between disparate individuals and the wider public. The use of the telephone also meant that information could be provided to the news media by individuals with greater protection of privacy, and often with anonymity, while information could be mediated via the actions of journalists. The range of uses of the telephone system by media professionals varies enormously, as a few examples illustrate:

> My big break in journalism was *minding a phone box* outside the Longbridge car plant, while working on the Birmingham Evening Mail in the mid-seventies. These were the days before mobiles, when getting the story back to the office was more important than getting the story in the first place. There was only one phone box within a mile of the shop stewards' office and it was my job to occupy it and repel all borders from 10.30 AM until the industrial editor, Tom Condon, came running out at lunchtime with the result of the latest strike ballot.[37]

Greek authorities banned all calls from jailed November 17 guerrilla suspects to the media on Tuesday *after a broadcast phone interview* with one prisoner sparked outrage.[38]

Supreme Council of the Islamic Revolution in Iraq (SCIRI) member Akram al-Hakim, speaking *on the phone from* London has told Al-Jazeera TV that information coming out of Basra had noted a "limited popular move" in the city.[39]

Queen Noor of Jordan was discussing her new memoir *by phone* from Washington, when, as is usual in media interviews, the allotted 25 minutes were up just as the conversation was getting somewhere.[40]

The next day the women's magazines started to phone. Some of the reporters were "enormously pushy," he says. "They rang non-stop. *Continuous calls.* One company phoned me two or three times by 6.30 in the morning, after phoning me about 15 times the night before. . . . I said I'd call them back and before I could they phoned me again six times.[41]

The importance and use of the telephone by the news media is vividly illustrated in the Watergate story, which Bonnie Brennan argues

> codifies an ideology of journalism which has framed an understanding of the role of the press in the United States and Western Europe since the 1970s. *All the President's Men* can be read as an ur-text—a seminal text that illustrates a specific structure of feeling regarding the construction of contemporary journalistic practices.[42]

The telephone is a critical element in the entire Watergate saga, especially its use by *Washington Post* reporters Woodward and Bernstein, in tracing and contacting the key players, tracking events through an examination of telephone records, in gaining information from sources and in liaising with their senior staff of the newspaper. In every respect the reporting of the Watergate scandal would not have been possible without the telephone. But its use in this famous example also demonstrates the extent to which a connection to the telephone system creates pathways and connections that media professionals are able to trace and tap into to acquire information and sources of information.

Over time, however, the use of the telephone has replaced some of the actual work of reporting in the first person. Its fundamental attribute of allowing communication over distance and being able to reach people in both public and private domains led to the situation in which many journalists now spend much of their time, "reporting by phone." Technological developments such as long-distance dialing, improvements in international telephone connectivity, as well as the increasing diffusion of the telephone throughout the population in many countries, meant that media professionals could contact people almost anywhere by telephone with increasing ease.

A study of specialist journalists in 1971 reported that "specialists are often to be seen sitting at their desks engaged in rambling telephone conversations and taking a shorthand note,"[43] and a journalist interviewed by Tunstall reported that "sometimes there are so many calls that the left ear becomes positively painful."[44] More recently, a journalist interviewed by this author about the use of the mobile telephone echoed this in saying, "Sometimes I am on the phone for so long my ear gets hot." In other recent research a case of dysesthesia of the scalp associated with mobile phone use in a journalist has been documented.[45] McLuhan's framing of the extensive use of the telephone by media professionals as a grapevine is therefore accurate but also indicative of how the telephone plays a major, if unheralded, role in the construction of news media content.

The media grapevine functions not only to pass on news and information; it also serves to create a homogeneous or "pack" view of journalism. Those on the grapevine or in the loop, so to speak, are privileged with access to information and are able to take part in agenda setting. Veteran journalist Phillip Knightley gives an example involving the International Consortium of Investigative Journalists, which produced reports about U.S. president George W. Bush's involvement with the company Harken Energy. Although the reports were published in 2000, Knightley argues that during the U.S. presidential campaign,

> a lazy American mainstream press failed to notice them, and it was not until 2 July [2002] that the *New York Times* columnist, Paul Kruger, mentioned them in the context of the deepening crisis in American business, that the Center's phones started ringing, and didn't stop for days.[46]

The use of the telephone and development in telephone technology increasingly allowed media professionals to work without leaving news production centers, saving time and expense in traveling. Media organizations rely more on the use of communication technology, particularly the telephone, to reduce overhead costs in relation to staffing and have come to rely more on desk-based telephone reporting. Tapsall and Varley argue, citing Walsh[47] and Richardson,[48] that this has created "the *battery hen* model of news, with journalists cloistered in their newsrooms, which they seldom leave, and fed on a high-fibre diet of paper press releases and phone interviews they regurgitate into space-filling news."[49] In Britain, many experienced journalists "deprecate the new cost-conscious, accountant-dominated Fleet Street" and "decry the modern, far-flung office blocks where teetotal journalists peer forlornly into their computer screens."[50] Subsequently, "journalists increasingly come to rely on *dial-a-quote* sources—spokespeople who are used repeatedly because they can be counted on to quickly deliver a promising sound bite on the topic of the day."[51]

The widespread use of the telephone by the media is indicated in the print media in the use of the phrases "interviewed by phone," "in a phone interview," or "speaking by phone" in text reports, and in the broadcast media, especially radio, by the constant references to "reporting by phone." Radio in particular is highly dependent on the use of the telephone. From news reports, to interviews, to listener talk-back, call-in competitions and audience calling, the interaction between the two media of radio and telephony is constant and with the use of mobile communications is becoming further entrenched. In each of these situations, the use of the telephone allows news media to convey information with immediacy but also often with a high degree of intimacy. We have come to expect from the news media the same instantaneous quality afforded by the telephone, and the telephone continues to be an important delivery mechanism.

With nothing more than a telephone and a good connection, a journalist can report from most parts of the world, and even without a typewriter or computer, he or she is able to contact sources and file stories for other media. In 2003, Australian radio journalist Nonee Walsh was caught up in one of the most dramatic news stories of the year as the train she was a passenger on derailed.

> Suffering personal injuries and still trapped in a carriage waiting for rescue, Walsh made her first attempt to record news via mobile phone from the scene. Later, with the guiding hand of the ABC's Michelle Brown, she gave several versions of events [by telephone] for ABC Radio News. One part witness and one part reporter, she gave the ABC the scoop on the cause of the crash when she flagged the issue of the speed at which the train was travelling before the accident.[52]

In numerous similar situations, the telephone continues to be important to media professionals as a persistent background tool of communication. Recent developments in mobile communications, particularly the addition of audio and visual communication technology, also have the potential to maintain the telephone as the technological tool of choice of media professionals. At the same time, the use of the telephone with and alongside other forms of technology also has some negative aspects for the news media, especially when the limitations of the telephone are used to perpetuate deceptive and false impressions about news and information.

In 2003, the *New York Times* revealed that one of its journalists had routinely fabricated details of news reports, quoted people he had never spoken to, reported from locations he had not visited, and plagiarized material from other publications. The cunning use of various communication technologies, but especially the telephone and cell phone, was critical in enabling this extensive deception to occur. The reporter, Jayson Blair, used the capacity of telephone communication to enable communication at a distance and to claim that he was reporting by phone. The practice was further exposed when another *New York Times* reporter and the Pulitzer Prize winner, Rick Bragg, also resigned after it was discovered "that an unpaid assistant had done virtually all of the reporting for a story on oyster fishers in Florida for which he took full credit."[53] Here, too, the deception was executed and assisted via the practice of telephone journalism, specifically in Bragg's use of the telephone for contacting stringers and interns whom he relied on and guided to obtain information. Both the Blair and Bragg cases were seen as aberrant, and they were met with a high degree of moral indignation, especially by other journalists. With the widespread use of the same technology throughout the news media and the same pressure for fast, continual turnaround in the production of material, such incidents are found industrywide but have not received the same level of attention. In an interview about the issues raised by the Blair/Bragg saga, veteran American journalist Jimmy Breslin offered a solution: "I would take half of the phones out of the city room, put out a bunch of bus passes and get reporters out on the street."[54]

The Blair and Bragg examples also draw attention to the particularities of the use of the telephone by journalists and other media professionals. In the abuse of the technological power of the telephone in these two incidents, the common practice of using a telephone to communicate at a distance is highlighted. The Blair example also involved the specific newsroom practices and processes of the *New York Times*, and it is instructive in revealing the particular "biases" of telephone communication, most notably the capacity for distancing communication. Innis notes in discussing the bias of particular forms of communication, "We can perhaps assume that the use of a medium of communication over a long period will to some extent determine the character of knowledge to be communicated."[55] Consideration of the use of the telephone over a long period by the news media indicates that the "character of knowledge communicated" is influenced markedly by the specific qualities or biases of the telephone. These qualities include the way in which the telephone allows users to transcend space or distance and to engage in a personal and quite intimate form of communication. However, although telephone communication has these qualities, there is also a loss of other features of communication. On the telephone, the illusion of being close may substitute for real proximity, and the illusion of intimacy created by dyadic conversation substitutes for actual intimacy. The intimacy of the telephone therefore can create a false consciousness in the communication that it produces, including news media reports. The telephone can also be perceived, as Ronell asserts, as "a synecdoche for technology,"[56] standing for all the communication technology carried on what was originally the telephone network. The affordances of the telephone are sufficient to convince us that we are engaging in close, meaningful communication, yet face-to-face communication contains so much more than is possible via the telephone. However, as the Jayson Blair incident demonstrates, there is much that can be withheld and obscured in telephone communication. When the news media presents telephone journalism, not as the next best thing to being there, but as if the reporter is or was actually there, the report is at best misleading, if not false.

The use of the telephone by the media also requires the identification and acknowledgment of the importance of orality in the media. Despite the focus on text in media mythology, in vocational training, and in academic discourses, the strong reliance on the telephone by media professionals reveals the importance of conversation and oral communication. The level of reliance on oral communication conducted by telephone by the news media suggests that much media content contains characteristics associated with oral communication, even where the medium of delivery is primarily text based. The secondary orality[57] of the telephone pervades the media, influencing and changing its practices and priorities. Just as the form, technical limitations, and cost of the telegraph profoundly affected methods of reporting and introduced new uses of language, so too the telephone has influenced the way media professionals work and their use of language.

The most notable use of the telephone by the media is in transforming private communication into public communication, for so much of telephone communication is experienced as a private act between individuals. Although it would be going too far to assert that the telephone is solely responsible for this transformation, its critical role in the everyday practices of the media and its widespread diffusion more generally have played a key role in developing many elements of the modern social practice of communication. In particular, concepts of intimacy have been influenced by the widespread use of the telephone. In the modern telephonic age, there is an overwhelming expectation that what is talked about is reported. Consequently, in many areas of media communication, intimacy has been commodified, especially in the reporting of gossip, in the revelation of personal details of public figures and especially in the self-revelatory conversation of radio talk back, which is also engendered in television talk shows. In this respect, the sociability and social grooming afforded by the telephone is exploited constantly by the media. The barrier between communication that was at one time considered private, or at most shared via gossip in a small social networks, is now accessed with the aid of the telephone by the media until there is virtually no field of human activity that is not available for consumption. The application of specific protocols and rules of etiquette that once applied to maintain social hierarchies and to establish contact with individuals or social networks can be quickly bypassed or contravened with the use of a direct telephone contact. Taboos separating the private and public domains also have been altered and sometimes discarded as a consequence of the use of the telephone, and in Western countries especially, ideas of propriety and individual availability appear to correspond closely to the ways in which the telephone is used as a vehicle for communication.

The revelation of private information permeates modern culture so that it sometimes seems that everything is talked about, and everything is available for media appropriation. The telephone has played a significant role in promoting the sociability of strangers, and in acclimatizing people to "intimacy at a distance."[58] Additionally, the notion of immediacy, a communication affordance directly associated with and residing in the medium of the telephone, is critical in the news media, whether as text-based print media, radio and television, or the newer online and mobile media. The interactive properties of the telephone have also influenced media by allowing direct audience participation by large numbers of people, through telephone polling and in the medium of radio, especially talk back. A tendency of the contemporary media content to use the vernacular or conversational language rather than literary language also reflects the origin of much communication in telephone communication.

Innis and other communication scholars rightly argued for the bias of communication technologies to be understood and for the effects of those biases to be analyzed. To understand the modern news media, media organizations, and new communication technologies thus requires a more complex analysis of the use of the technology of the telephone because its use is so comprehensively ingrained within the news media. Any particular biases of the telephone therefore have consequences for the news media. With an emphasis on orality, and the capacity to provide immediate, personal communication over distance, the telephone has been instrumental in introducing and developing acceptance of disembodied communication and instantaneous accessibility regardless of location. Understanding the qualities of the telephone, the affordances of telephone communication, and their application by the news media is therefore critical to understanding the contemporary news media and how the application of newer technologies to news media practices and processes is influenced by past and existing uses of the telephone.

Notes

1. The National Council for the Training of Journalists, "Short Courses—Interviewing Skills," (2004), http://www.nctj.com/courses_london.htm#intskills.

2. Gail Sedorkin and Judy McGregor, *Interviewing: A Guide for Journalists and Writers* (Crows Nest: Allen and Unwin, 2001), 131.

3. Don Middleberg and Steven S. Ross, *The Eighth Annual Middleberg/Ross Media Survey: "Change and Its Impact on Communications,"* (2002), http://www.middleberg.com/toolsforsuccess/fulloverview_2002.cfm; Collette Snowden, *News and Information to Go: Mobile Communications, Media Processes and Media Professionals*, Ph.D. thesis, School of Communication, Information and New Media, University of South Australia, 2005.

4. Colin Cherry, "The Telephone System: Creator of Mobility and Social Change," in *The Social Impact of the Telephone*, ed. Ithiel de Sola Pool (Cambridge, Mass.: MIT Press, 1977), 112–26.

5. Bruce Sterling, *The Hacker Crackdown* (New York: Bantam, 1993).

6. Ian Hutchby, *Conversation and Technology: From the Telephone to the Internet* (Cambridge, UK, and Malden, Mass.: Polity Press, 2001), 33.

7. Ibid., 194.

8. James E. Katz, *Connections: Social and Cultural Studies of the Telephone in American Life* (New Brunswick, N.J., and London: Transaction Publishers, 1999).

9. Avital Ronell, *The Telephone Book: Technology, Schizophrenia, Electric Speech* (Lincoln: University of Nebraska Press, 1989).

10. James E. Katz and Mark Aakhus, "Introduction: Framing the Issues," in *Perpetual Contact*, eds. James E Katz and Mark Aakhus (Cambridge: Cambridge University Press, 2002).

11. Snowden, *News and Information to Go: Mobile Communications, Media Processes and Media Professionals*.

12. Harold Innis, *The Bias of Communication* (Toronto: University of Toronto Press, 1951).

13. Marshall McLuhan, *Understanding Media: The Extensions of Man* (London: Sphere Books, 1964).

14. Robin Dunbar and Ian McDonald, *Grooming, Gossip and the Evolution of Language* (London: Faber and Faber, 1996).

15. Jack Leven and Arnold Arluke, *Gossip: The Inside Scoop* (New York: Plenum Press, 1987).

16. Walter Ong, *Orality and Literacy: The Technologising of the Word* (London and New York: Methuen, 1982).

17. Marshall McLuhan, *Voyager Live* (1978), http://www.voyagerco.com/catalog/mcluhan/indepth/live.html.

18. Hutchby, *Conversation and Technology*, 8.

19. Lewis Mumford, *Technics and Civilization*, 9th ed. (London: Routledge and Kegan Paul, 1934/1967), 239.

20. Harold Adams Innis, *Political Economy and the Modern State* (Toronto: Ryerson Press, 1946); and Innis, *The Bias of Communication*.

21. Innis, *Political Economy*, 32.

22. James W. Carey, "Technology and Ideology: The Case of the Telegraph," in *Communication as Culture: Essays on Media and Society*, ed. James W. Carey (London: Unwin Hyman, 1983), 201–31.

23. Ibid., 204.
24. Ibid., 227.
25. Ibid., 228.
26. Snowden, *News and Information to Go: Mobile Communications, Media Processes and Media Professionals.*
27. Michele Martin, *Hello, Central?: Gender, Technology, and Culture in the Formation of Telephone Systems* (Montreal and Buffalo: McGill-Queen's University Press, 1991), 36.
28. Peter Young, *Person to Person: The International Impact of the Telephone* (Cambridge: Granta, 1991).
29. Ibid., 11.
30. Ibid., 49.
31. Jules Verne, *In the Year 2889*, 1889, (2002), http://wondersmith.com/scifi/2889.htm, 2.
32. Ibid., 2.
33. *The Electrical Engineer* (2003), http://earlyradiohistory.us/1895th.htm.
34. Carolyn Marvin, *When Old Technologies Were New: Thinking about Electric Communication in the Late Nineteenth Century* (New York: Oxford University Press, 1987).
35. Ithiel de Pool, ed., *The Social Impact of the Telephone* (Cambridge, Mass: MIT Press, 1977).
36. Brian Winston, *Media, Technology and Society, A History: From the Telegraph to the Internet* (London: Routledge, 1998), 54.
37. Richard Littlejohn, "Why I'll Never Give up the Day Job," *British Journalism Review* 13, no. 3 (2002): 65–70.
38. Karolos Grohmann, "Greece Bans Nov 17 Guerrilla Phone Calls to Media," *Reuters News*, September 17, 2002.
39. Al Jazeera Television, "Iraqi Oppositionist Notes 'Limited Popular Move' in Basra," BBC Worldwide Monitoring, March 26, 2003.
40. Mary Delach Leonard, "Queen Noor's Memoirs Extend Hope for Peace," *St. Louis Post-Dispatch*, March 18 2003, E1.
41. Sally Jackson, "Innocents Abroad in a Media Storm," *The Australian*, April 24–30, 2003, 3.
42. Bonnie Brennen, "Sweat Not Melodrama: Reading the Structure of Feeling in *All the President's Men*," *Journalism* 4, no. 1 (2003): 113–31.
43. Jeremy Tunstall, *Journalists at Work: Specialist Correspondents; Their News Organizations, News Sources, and Competitor-Colleagues* (Beverly Hills, Calif.: Sage, 1971), 152.
44. Ibid.
45. B. Hocking and R.Westerman, "Neurological Changes Induced by a Mobile Phone," *Occupational Medicine* 52, no. 7 (2002): 413–15.
46. Philip Knightley, *Reflections of a Warhorse*, ABC Radio, The media report, Mick O'Regan, December 27, 2003, http://www.abc.net.au/rn/talks/8.30/mediarpt/stories/s1008070.htm, 6.
47. M. Walsh, personal interview with S. Tapsall, 20 November, Sydney, 1998, cited in Suellen Tapsall and Carolyn Varley, "What Is a Journalist?," in *Journalism: Theory in Practice*, eds. Suellen Tapsall and Carolyn Varley (South Melbourne: Oxford University Press, 2001), 3–20.
48. J. Richardson, personal interview with S. Tapsall, 19 November, Sydney, 1998, cited in Ibid.
49. Ibid., 12.
50. Stephen Glover, "Introduction," in *The Penguin Book of Journalism*, ed. Stephen Glover (London: Penguin Books, 1999), xiii.
51. Ibid.
52. Walkley Awards 2003, http://www.walkleys.com/2003/winners/walsh&brown.htm.
53. Gary Younge, "Reporter's Plagiarism Claims Scalp of Editor as New York Times Becomes the News," *Guardian Unlimited,* June 2003, http://media.guardian.co.uk/presspublishing/story/0,7495,971692,00.html.
54. Jimmy Breslin; cited in Joe Strupp, "Hard Times: Journalism's Credibility Problem," *Editor and Publisher* online, 2003, http://www.editorandpublisher.com/eandp/ news/ article_display.jsp?vnu_content_id=1909835.
55. Innis, *The Bias of Communication*, 34.
56. Ronell, *The Telephone Book*, 13.
57. Ong, *Orality and Literacy.*
58. Hutchby, *Conversation and Technology*, 83.

8 Vaudeville: The Incarnation, Transformation, and Resilience of an Entertainment Form

JoAnne Stober

In 1906, the Empire Theatre in Edmonton, Alberta, opened with this announcement, "Don't be afraid to bring your wife, sweetheart, or children . . . as we cater particularly to that class." The storefront theater opened as a vaudeville house, and although it was said to have connections to the Sullivan-Considine vaudeville circuit, it opened with locally recruited talent. The notice for the opening continued:

> Empire Theatre: Two Doors South of New Post Office. Open Monday, June 25th. In Refined Vaudeville, Motion Pictures, Illustrated Songs. Four Shows Each Day. Afternoon 2.30 and 4 o'clock. Each Performance One Hour and Twenty Minutes. Change of Program Each Week. Matinee Prices: 10c and 15c; Night: 15c and 25c. Reserved Chairs for Ladies and their Escorts without Extra Charge.[1]

The opening of the Empire in Edmonton is typical of smaller city vaudeville houses of this period. The Empire attempted immediately to draw in families and to establish itself as "clean and classy." Vaudeville's roots in what was considered "rowdy" entertainment were not ancient history and sometimes led to it being misunderstood, so every attempt was made to ascertain the wholesomeness of vaudeville as an entertainment form. Vaudeville underwent a number of transformations to get to this point. Variety entertainment had its roots in the popular entertainment institutions of live theater, minstrel shows, and burlesque. Consequently, vaudeville also had roots in the practices associated with these entertainments. In the late 1850s and early 1860s, variety was the performance of brief entertainment acts that became concentrated as part of concert saloon entertainment aimed principally at audiences of working-class men.[2]

So-called small-time vaudeville made its debut in ramshackle theaters, often doubling as saloons. The Empire in Vancouver was famous for its variety acts but even more renowned for housing a brothel on the second floor.[3] Some of the cities' "ten-cent" houses presented inoffensive shows but were operated sporadically, and the nature of the acts was often uncertain.[4] However, by 1900, vaudeville was appealing to a mass audience in North America by guaranteeing clean, inoffensive entertainment with strict standards for vaudevillians and audiences. This outcome took its course over a half century, from the 1850s to 1900, and everything that vaudeville became was a reconstruction of earlier popular entertainment forms.

Vaudeville's influence was pervasive, and its presence was increasingly seen as a sign of urbanity and success. Even smaller cities and towns attempted to offer some form of vaudeville entertainment either by booking the big-time circuits and offering shows two or three times a week or by offering a more local bill of music and comedy performers. A number of popular entertainment forms that had been relatively independent of any particular presentational venue, including magic lantern shows, puppetry, and magic illusions, were absorbed into vaudeville. "Vaudeville shows compressed spoken drama into short playlets, derived production numbers from musical comedy and even presented capsule versions of recent stage hits. Though often scorning or mocking 'legitimate' theater, ballet, opera and classical music, vaudeville also emulated their status" as symbols of high culture.[5] This emulation was necessary because variety acts had a long-running association with burlesque and, as Robert C. Allen argues, burlesque and vaudeville were actually "negative reflections of one another."[6] Implicit in the cultural construction of what vaudeville became as an entertainment form were continual shifts and changes in both composition and performance practices. That is to say, to understand vaudeville as an entertainment form is to think about it as a composition of new and old media and entertainment practices.

In this chapter, I survey the existing research on the history and evolution of vaudeville.[7] I propose that tracing the transformation of vaudeville as it begins to incorporate entertainment practices already in existence and new media technologies allows us to gain insight into the emergent and residual forms of "variety" entertainment. Raymond Williams encourages us to read continuity into cultural practices and experiences. To help us do so, Williams identifies residual forms as "experiences, meanings and values which cannot be verified or cannot be expressed in terms of the dominant culture" because they are the residue of a "previous social formation."[8] Inherent in cultural practice is change and as Williams describes "emergent forms [create] new meanings and new practices and experiences."[9] To be sure, the emergence and consolidation of vaudeville occurred over more than a half century and traces of vaudeville are present well into the era of synchronized sound cinema and on radio and television as these new technologies emerge. Certain components of the vaudeville presentation and its relationship to technology, programming practices, exhibition sites and practices and audience formations, create layers in the entrepreneurial structural logic of vaudeville as an entertainment form. The incarnation, transformation and resilience of vaudeville is intrinsically linked to residual and emergent forms of media and pursuant practices.

The expansion of vaudeville can be seen as an emergent and defining aspect of its consolidation while certain components of its presentation and its relationship to film exhibition and programming practices in theaters became residual forms. The sediment of this popular entertainment and its corresponding social formations are part of its continuity. In addition, I focus on the resilience of vaudeville, which is intrinsically linked to new media and exhibition history, including distribution, sites of exhibition, audience formations, practices and discourses of popular entertainment. Vaudeville's resilience during the transformation from silent cinema to synchronized sound cinema can be seen specifically in Montreal, Quebec. Contrary to dominant perceptions of new media replacing old, vaudeville made its return on the very heels of synchronized sound and in Montreal, the first Canadian city wired for sound movies in 1928, vaudeville had been a dominant cultural form. The presence of vaudeville as an enduring and popular entertainment form in the city is complex and related to the changing dynamic and cultural patterns of film exhibition during the period when silent cinema and sound cinema are settling their differences.

The residual and emergent forms of popular "variety" entertainment seem to accompany the introduction of new technologies and changes in the practices associated with these technologies. When discussing the discourses and activities that fall into the realm of commercial entertainment, I find that, as Harold Innis argues, one of the effects of technology as it was introduced into society was that it would influence patterns of perception. Specific shifts to new media of communication have been characterized as creating profound disturbances in the current relationships and patterns of living that existed previous to the establishment of the new media.[10] Innis's insight into the moment when a new media is introduced as something that would, in its use, destabilize the current relationships and patterns of living that existed is a powerful legacy, simply for the reason that it draws our attention outward from the media as an object to the effects of its destabilization on social and cultural life. To be sure, rather than disappearing when new technologies are introduced, vaudeville adapts and shifts into forms that follow dominant entertainment media, and in each shift there is a destabilization of surrounding social relations. I contend that these relations and intersections, examined together, stress the cultural and social impact of the arrival of new media, entertainment forms, and the resistance of established forms and formation of hybrid forms.

In asking about the ongoing contexts of the emergent and residual forms of variety and vaudeville, I begin in 1870. In the early 1870s, burlesque and minstrel shows were uniting in a hybrid form of entertainment.[11] Allen says this "marriage" was "facilitated both by economics and by an underlying structural logic emanating from the homology between the blackface minstrel and the burlesque performer."[12] A consequence of the union was to reroute the course burlesque had been taking as an autonomous popular entertainment form for mainstream bourgeois theaters and audiences and put it on the minstrel show's course, destined for a working-class audience. As burlesque became a working-class form of entertainment, variety, too, began to emerge as a "separate popular entertainment form."[13] Within the dense downtown of New York City, rising from the concentration of opera houses and dime museums, a retargeting of audiences from working-class men to women and families took place. Tony Pastor's Opera House in Bowery Row of New York City was one of the first entertainment houses to target a more familial audience in 1865.[14] By 1890, Pastor "played a major role in establishing a new audience for variety and, in New York at least, in loosening variety's tie to prostitution and rowdy, working-class leisure."[15] Pastor positioned his variety theater amid New York's theater and retail shopping district and offered his audience a steady stream of acts and performances without the vice of alcohol and cigarettes. Pastor had managed, through the elimination of smoking, drinking, and explicit acts, to separate the "concert saloon's connection with liquor and sexuality" from variety entertainment, thereby succeeding to attract an audience of both men and women.[16] In addition to the rise in urban entertainment, the railway networks that were spreading across North America were making it possible for burlesque and variety shows to travel. Both the spread to smaller towns and the solidification of sites and an entertainment structure in cities were influencing the entertainment dynamic of the time and setting a course for future mass amusements.

Having seen the results of Pastor's variety theater, Keith endeavored to create an enterprise of clean, legitimate, and popular variety entertainment. He converted his New York dime museum (located in Boston) into a continuous performance variety house with a schedule of performances that repeated throughout the day, and he consequently reported a rise in attendance.[17] Unlike New York, Boston's liquor licensing laws did not allow for theatrical performances to take place in taverns, so there was not as strong a connection among booze, rowdy behavior, and entertainment.[18] In this sense, Keith did not have to struggle with variety's association to alcohol and immorality and in turn was free to tackle what he saw as the next task: breaking variety's link to working-class leisure. Keith introduced an eccentric twist of variety and comic opera performances into an environment where the patrons already felt comfortable, hoping to attract a refined middle-class patron.[19] By actually separating variety from entertainments associated with the traditional dime museum (freaks and curios and the lower-class patrons entertained by curiosities, medical mishaps, and strange sites), Keith had astoundingly reintegrated the previously fragmented American audience into the transformed dime museum.[20] To prevent any slippage toward the old norms of variety in dime museums and concert saloons, Keith enforced a strict code of conduct and decorum, which he applied to both performers and patrons. To ensure proper bourgeois behavior, uniformed ushers handed out fliers requesting gentlemen not to stamp their feet, ladies to remove their hats, and no talking, smoking, or raising of voices to respond to performers.[21] Gilbert Douglas also claims Keith lectured his audiences personally on behavior, and he had bouncers onsite to ensure patrons did not smoke, spit, whistle, or crunch their peanuts.[22]

The environment for Keith's new clean and sophisticated brand of entertainment was propagated in 1894 when he partnered with Edward F. Albee, a former circus man, to construct his new vaudeville theater, B. F. Keith's New Theatre. As an entrepreneur, Keith saw an opportunity to incorporate variety further into the norms and behaviors traditionally associated with bourgeois theater. The legacy of Keith's desire took variety and turned it into vaudeville as a structured entertainment institution complete with the characteristics now commonly associated with vaudeville: clean entertainment, appeal to family and children audiences, no interaction between audiences and entertainers, traveling circuits, continuous performances, lavish theaters, and quality entertainment scheduled through a central booking office.[23] Vaudeville's entrepreneurs made a discursive choice by consolidating what had been considered "variety" entertainment into "vaudeville." As Allen indicates, "The terminological shift from variety to vaudeville signifies not so much a change in performance structure as changes in the form's institutional structure, social orientation and audience."[24] Robert Snyder argues that Keith wanted to "beat the stigma attached to variety" and so "christened his performances as 'vaudeville.'"[25] The B. F. Keith New Theatre was proclaimed to be the first vaudeville palace. As an architectural site, it featured a custom-designed original exterior and an interior featuring accoutrements like leather sofas, telephones, writing desks, and messengers in the lobby, signaling an attempt to incorporate business-class working patrons into the regular audience. Albee's decoration was inspired by the legitimate theaters but also by the public buildings of Europe. He incorporated ornamental ironwork, stained glass, marble pillars, mirrors, gargoyles, and original paintings into the theater's interior.[26] Vaudeville "grandiloquence," as Robert C. Allen labels it, swept through North America, and theaters for big-time vaudeville arose in major cities and even smaller centers that benefited from the vaudeville acts on traveling circuits. Of all the opulent theaters constructed in cities across North America, the Palace in New York City was the "throne room of the Kings of Vaudeville" and the "Queens of Comedy for 19 years."[27] Keith went to great lengths to make vaudeville high class, including advertising interior decoration and amenities in his theaters to appeal to higher tastes.[28] When Keith opened the Colonial Theatre in Boston, he announced publicly that $670,000 had been spent on decorations and $89 alone for a red velvet carpet.[29] He extended big-time vaudeville to smaller towns, and using the technology of the railroad and the telegraph created traveling circuits coordinated from central booking offices.

Part of the sustaining ability in vaudeville was what Raymond Williams identified as the allure of the individual performance. Of the music hall he wrote, "There emerged a generation of solo performers whose line has since been unbroken: the performers we call entertainers and comedians. Their songs, monologues, sketches and routines have precedents in a long tradition of comic acting, but their presentation as individuals was in effect new."[30] Always a delicate balance between material that was seen to belong to low culture and an appeal to high drama, vaudeville was built on articulations of spectacle, the "tinsel and plush kind . . . sequins rather than diamonds."[31] The adaptations and resilience of vaudeville was more than the evolution of an entertainment form. Even as vaudeville was institutionalized into Benjamin Keith's cultural schema, the roots of variety held in older entertainment forms like circuses, sideshows, and burlesque were visible.

Each phase in vaudeville's development is also composed of a process of dynamic contradictory relationships in the interplay of dominant, residual, and emergent forms.[32] Vaudeville carried with it the living traditions and institutions that had contributed and been amalgamated to shape its existence. The articulations of vaudeville are derived from everyday life and comedic situations whose relevance reflects technological change, the history of cultural practices, the concepts of high and low culture and the various incarnations of the entertainment industry as it vied for control and standardization of the market and audiences. Whatever degree of standardization we assume in terms of vaudeville programs and audiences, the emergence of the entertainment paradigm cannot be grasped as a simple middle-class experience. We can see the formation of the vaudeville audiences as a process of multiple and uneven transitions just as Miriam Hansen advocates for the transitional period when audiences of live entertainment were becoming audiences of classical cinema.[33] There are striking similarities among the market interests, industry attempts to create respectable audiences in vaudeville's consolidation as an entertainment form, and the later formation of cinema spectators that Hansen is treating. Hansen's argument that the cinema, "in its emancipation from existing live entertainment outlets . . . grafted itself onto surviving structures of working-class culture" can also apply to vaudeville as it struggled to disassociate itself from lower-class entertainment offerings. Vaudeville grafted itself onto higher forms of entertainment and amusement like the legitimate theater and ballet, even adopting the institutional structure of grand-scale theaters. The similarities between the consolidation of vaudeville and the transition from silent to sound cinema are striking and point to the need to develop an understanding of the transitional period as both an industrial shift and as an integration of mass entertainment and consumer culture.

It is possible to account for the history of film exhibition as a tale of novelties bridged together by allegiances of audiences and exhibitors. Prior to industry integration and control of cinema chains, the consumer and the showmen had a relationship based on appeal and appreciation: the showmen would attempt to appeal to the audiences, and the audiences might appreciate their efforts or not; thus was the cycle of consumption. Vaudeville's transitions to this point demonstrate what Roy Rosenzweig refers to as a transfer of allegiances from the existing cheap entertainments and entertainment venues. Rosenzweig situates the transfer of loyalty from the saloon to demonstrate that in the small industrial town of Worcester, Massachusetts, the prosperity of commercial entertainment was linked to incomes, industrial and leisure time, and conceptions of amusements as either high or low culture.[34] For working-class patrons in small towns, the movies were amenable as entertainment, offering cheap prices and short programs that allowed even the overworked and underpaid a chance to participate. Rosenzweig claims that the movie theater became a central working-class institution like the saloon had been. Along with the saloon patrons, women, children, and immigrants were a part of the new audience. This was part of the phenomena for vaudeville as audiences formed in both urban and rural locations.

The men at the helm of the consolidation of vaudeville, Keith and Albee, may not, as Allen notes, have been the entrepreneurs solely responsible for affecting the shift from variety to vaudeville, but "their strategies set the tone for this nascent show business industry."[35] Many theaters were built explicitly for touring vaudeville, including the Orpheum in Vancouver, the Sherman Grand in Calgary, and the Palace in Montreal and Toronto.[36] "As a high-class vaudeville theater, the Orpheum (Vancouver) required a staff of about 100, including the stage crew, maintenance people, musicians, and four projectionists (two for each shift). But the majority was front-of-house staff: one outside and three inside doormen to greet 'milord' and 'milady,' ticket sellers, the . . . 'French' maids, and 65 usherettes."[37] The Sherman Grand was the most luxurious theater in Calgary, and its manager Martin Beck was touted as holding his patrons in the highest regard by featuring Orpheum vaudeville. "There is nothing in the world so cosmopolitan, and so delightfully so, as modern vaudeville."[38] The opinion that vaudeville was high-class entertainment seemed to have spread across North America.

In terms of performance, vaudeville was a compilation and a continuation of the previous entertainment forms from which it took attractions. Robert Allen argues that "Vaudeville existed only as a distinctive, presentational, environmental and institutional form; in terms of content, vaudeville was nothing and everything."[39] Big-time bills would often feature larger groups of singers and dancers as well as sketch comedy and physical or slapstick turns. A program from the Palace, New York's premier vaudeville house, shows the organization of a typical vaudeville show. Continuously available, vaudeville acts ran from early morning until late into the evening and embodied the serious and the trite. The juxtaposition and uneven transition from show to show was the opposite sort of effect that now goes into television programming: the "flow" in entertainment was avoided in vaudeville; instead, performances were programmed to be disjunctive in an endless stream of novelty.[40] The vaudeville program may have been loosely knit together in terms of content, but acts were tailored to audience interests, timed for shifts in mood to maintain excitement and incorporate all six to nine acts plus the headliner in a dynamic tempo. Programs were printed and distributed to the audiences, and acts were scheduled and selected well in advance based on the traveling schedules of the vaudeville circuits, which were based in the United States and traveled cross border from south to north. The circuits were organized in the most conscientious time-saving way to eliminate unnecessary travel and costs. Local acts would target their routines to the audience, sometimes using ethnic characterizations carried over from minstrel and so-called coon shows.[41] Humor based on ethnic characterizations was a major component of many vaudeville routines, which also included comic sketches, joke routines, songs parodies, and acrobatic and dance turns.[42] Just as the terminological shift from variety to vaudeville resulted in wider changes in the institutional structure, social orientation, and audience, so too did technological shifts in film, audio recording, broadcast, and eventually sound films, which more than one historian has proclaimed to be the technology that dealt the final blow to the popular entertainment form of vaudeville.

Vaudeville was solidly anchored in the tradition of itinerant spectacles like the minstrel show and burlesque, circuses and Wild West shows. Following a pattern set down by P. T. Barnum in which Barnum had mastered the "rhetoric of moral elevation, scientific instruction and cultural refinement in presenting his attractions," B. F. Keith carefully packaged mass entertainment to appeal to middle-class patrons.[43] Vaudeville established itself firmly on the map of popular amusement in three concrete ways: the development of commercial circuits, the stabilization of a programming mode, and the establishment of permanent vaudeville houses in the big-city centers.[44] Vaudeville's centrality and dispersal over a vast territory was a means of disseminating entertainment. The infrastructure for the movie industry was laid with the early vaudeville shows.

When moving pictures took up residence in vaudeville houses in the early 1900s, their place on the program was typically as an intermittent break from the live vaudeville acts. The emergence of nickel movie houses established a new trend; many vaudeville houses were converted into nickelodeons where motion pictures and illustrated songs were the main attraction.[45] As Russell Merritt points out, nickelodeons began by showing "a miscellany of brief adventure, comedy, or fantasy films that lasted about an hour," and they willingly tailored exhibition techniques found in vaudeville and used "sing-alongs, inexpensive vaudeville acts, and illustrated lectures" to augment their programs.[46] The introduction of the nickel theaters and the continuation of the vaudeville houses resulted in a coexistence and an interaction between the two sites and forms. Rosenzweig claims that in Worcester "those with only ten cents to spend could sit in the gallery of Lothrop's Opera House and watch melodramas and minstrel shows or visit the nearby Front Street Musee for burlesque or vaudeville."[47] The mixing of live entertainment in the form of either vaudeville or musical acts with motion pictures became a prevalent entertainment model and thrived as an entertainment program in various sites from the nickleodeon to the vaudeville theatre to converted music halls and church basements throughout the teens and into the 1920s.[48] In nickledodeons, which stood in clean contrast to vaudeville and attracted middle class audiences that had previously not gone to the movies with any notable frequency, moving pictures were the main attraction and the illustrated song was typically an intermittent break from the moving pictures. In *Moving Picture World*, George Craw speculated, "The formative stage through which the film theaters are passing will probably make the illustrated song an institution and will eventually discard cheap vaudeville."[49] In giving the illustrated song priority, Craw argued against vaudeville's economic viability and its cultural standing, including the "half-baked 'actors' who smirk and wiggle and clog-prance about the stage."[50] He claimed the expense of the number of acts required for a vaudeville program would never allow it to compete effectively against box office draws. This shift in vaudeville's popularity, shortly after it had reestablished its presence alongside the nickelodeon theaters, cannot be read as simply disinterest in the entertainment or persistent accusations of vulgarity in the form. Rather, the "democratic leveling" of the nickelodeon's pricing fostered a new sense of proprietorship in the movie house and business independence to which the working class responded positively. The nickelodeon, in the eyes of community advocate John Collier, writing in 1908 for *Charities and Commons*, had successfully bucked its legacy of all that is evil and still represented in vaudeville. Collier wrote, "Five years ago the nickelodeon was neither better nor worse than many other cheap amusements are at present. It was often a carnival of vulgarity, suggestiveness and violence."[51] Vaudeville remained lumped in with the bad guys—penny arcades, saloons, and melodrama—painted as reliant on illegitimate methods for success and only of "limited interest for the great, basic, public of the working and immigrant classes in New York."[52] The nickelodeon, in contrast, had shimmied up the ladder of lowbrow entertainment to occupy a new rung of social and moral respectability starting fresh with motion pictures, offering a cheap, varied program with families in mind.

By 1909, small-time vaudeville had combined several reels of film with an abbreviated vaudeville program in the same sort of legitimate theaters to which nickelodeon owners had moved their shows. The emergent hybrid form of entertainment saw vaudeville once again on the bill and the perception of the variety show bounced back to being one of clean fun. Sometimes, in the case of the integration of the nickelodeon and vaudeville, the emergence of a hybrid form of media is more telling than either the birth or death of the prior media. Seeing beyond the technology to what de Certeau called the "functionalized space in which consumers move about"[53] enables us to focus on the way vaudeville entrepreneurs viewed and competed against the popularity of the nickelodeon. The reemergence of vaudeville as a family-oriented entertainment is what de Certeau would call *strategy*; media technologies use in standardizing, institutionalizing practices of space. The incorporation of film as a prevalent part of the vaudeville program is what de Certeau might call a *tactic*; media technology and the everyday practices associated with them are operational and depend on practice.[54] Vaudeville, as a cultural form, is a composition of "strategic representations offered to the public as the product of these operations."[55] To ensure its survival and prosperity, vaudeville reemerged as a cultural form in which variety and film had a place. In addition, the social and cultural standing the nickelodeon had achieved in the community was not lost; rather, it was transferred onto the emergent vaudeville form and ensured that the public significance of vaudeville was augmented.

During the transitional period, the vaudeville palaces continued to emphasize opulence, grandeur, and the value of the "show" over specific acts or films. The vaudeville acts that took their place in these upscale houses were more akin to the theatrical world of live entertainments than the burlesque shows and bars of vaudeville's principal incarnations. As the vaudeville house continued to rise in stature, it concentrated on establishing itself as different from the "sensational" houses that catered to "the element which craves 'action' pictures, that is, not only western subjects, but others that may be called melodramatic in the extreme."[56] By emphasizing the "show," the act of going to vaudeville theaters remained essentially a theater experience as opposed to a film experience. As we see in the late 1920s in the transition from silent cinema to sound cinema, the same emphasis is placed on the "show" as an event rather than specific films or stars. As Richard Koszarski argues, the belief in a "balanced program" was almost mystical among silent picture palace managers, who saw this part of their business as closer to the vocation of vaudeville manager.[57] As head of Loew's Inc., the theater chain MGM, Marcus Loew's motto, "We sell tickets to the theater, not movies,"[58] emblematized the transition from silent to sound cinema, whereas the first evidence of selling the experience rather than the actual acts or performances had its roots in the establishment of vaudeville.

As I have argued thus far, the disjointed variety-style program never really vanished; it just reshaped itself into different forms that we can trace in the sediment of popular amusement from burlesque and variety to the consolidation of vaudeville as an entertainment form that is institutionally regimented and recognizable. By the mid-1920s, vaudeville's popularity and the well-established mixed programs of live and filmic entertainment in cinemas was being encroached on by the industry experiments with sound-on-disc systems and other methods of synchronizing sound. Vaudeville managers had sought solace in the silent screen, believing that as long as cinema was silent, vaudeville could not be supplanted. It was with the secure sense that silent films could not trump live delivery that vaudeville houses had added films to their programs. The widescale booking of films began to interfere with the traditional circuits of traveling vaudeville shows. In addition, the theater chains that were controlling the production and distribution of films were now creating longer films, and to book these films vaudeville houses were shaving their live acts to the bone.[59] Congruently, movie palaces were playing some live vaudeville acts, thereby profiting from vaudeville's popularity and at the same time lessening the hold of the vaudeville houses on the entertainment form.

In 1926, a dubious feeling accompanied the coming of sound and, as with the arrival of most new technologies, opinions differed about the introduction of synchronized sound to the existing mix of silent cinema and live vaudeville. As Donald Crafton points out, the popular tale of Hollywood's shift from silent to sound cinema has become a sort of urban legend: "The components of the popular retelling of sound always represent it as a dividing line between the Old and New Hollywood . . . sound divides the movies with the assuredness of biblical duality."[60] The sound to silent border has become an organizational axis for the study of film and continues to divide film into two separate worlds where sound would be the victor and silent would be relegated to the back shelf.[61] The talkies quickly became one of the inventions in the evolution of film that led "inexorably to the modern movie industry."[62] Jacob Lewis's account of the transition to sound film that had swept through the American film industry was written in 1939 and is a good example of the degree to which sound film's institutionalization and the creation of the studio system was understood and for many cases is still understood:

> Suddenly in 1927 the progress of motion picture technique was brought to an abrupt halt by the invention and adoption of sound. The incorporation of spoken dialogue as a permanent element of motion pictures

caused a cataclysm in the industry. Technique lost its sophistication overnight and became primitive once more; every phase of the movie medium reverted to its rudiments. The interest in artistic film expression that had been stimulated by the superior foreign films, now having reached a climax, was stifled in the chaos that the advent of sound produced. The new film principles that were just beginning to crystallize seemed destined for the dump heap, and directors, stars, writers, musicians, and foreign talent who had succeeded in the era of the "silents" found themselves unwanted. Movie art was forgotten as the studio doors were flung open to stage directors, Broadway playwrights, vaudeville singers, and song-and-dance teams. Voice, sound, noise, were all that now mattered. Diction schools sprang up; everyone took singing lessons; voice tests became the rage; speech filled the ears of the movie capitol.[63]

Standard film histories widely accept that the introduction of synchronized sound to cinema quashed vaudeville and all the practices associated with the exhibition of silent cinema whether they were established or not. In stating the consequences of the coming of sound in film exhibition and reception, Kerel Dibbets emphasizes a radical break: "Sound changes not only the film, but also the film's presentation and its relation to the viewer. In fact, the roots of silent film culture had to be demolished to give room to the rise of talking pictures."[64] He continues, "In the first place, the transferral of the orchestra from the pit to the sound-track marked the end of cinema as a multimedia show with live performance, giving way to the cinema as a single-medium event."[65] The conception of film exhibition in the transitional phase of silent to sound cinema created unpredictability between what was understood as a live performance (variety acts, vaudeville, and performance combined with film presentation) and what became a more streamlined cinematic exhibition, momentarily phasing out live performance. Cinemagoers were subject to a shift in technology and presentation that imposed a new form of viewing on the public. Not only did the cinema undergo a dynamic innovation with the introduction of synchronous sound, films were viewed differently. As Miriam Hansen points out, films were likely to have "a wide range of meanings depending on the neighbourhood and status of the theater, on the ethnic and racial background of the habitual audience, on the mixture of gender and generation, and on the ambition and skills of the exhibitor and the performing personnel."[66] Lying at the heart of the transition from silent to sound cinema was a change from what Hansen has characterized as a disjointed presentation of live and filmic performance to a streamlined, all-filmic program.[67] The adoption of an all-filmic program meant that the live character once lent to the exhibition was vanquished, therefore rendering exhibition practices and programs seamless and homogeneous from milieu to milieu and city to city.

Film historians have documented the differences in exhibition practices from city to city and particularly the difference between going to the movies in the city as opposed to the experience in small towns.[68] Although the same film may have been the feature on a playbill, it would have been programmed differently depending on the location, the exhibition capability of the theaters, and the audiences. Douglas Gomery points out that well into the late 1920s, film production companies were still making two copies of films, one talking and one silent, to be shown in theaters not yet wired for sound. Gomery demonstrates that the uneven adoption of new technologies, and certainly the study of specific locations, is an integral part of understanding the experience of going to the theater in these transitional phases. Moreover, the forms of entertainment that exist prior to the innovation and adoption of new technologies do not just vanish; rather, they are integrated and incorporated into the emergent forms. Such is the case with vaudeville. After theater managers and critics believed the thrill of sound had worn off, they quickly turned back to a program style proven popular prior to the adoption of the all-filmic program that accompanied synchronized sound. The return to programs featuring live acts was an effort to return to the entertainment style audiences were accustomed to before the takeover of the synchronous sound feature film.

The New York City exhibition of *The Jazz Singer* (Alan Crosland, 1927) corresponded to Warner's plunge into exhibiting talking films and its merger with Western Electric. The film marked the grand departure on the part of the major studios to go ahead with talking film and with their investment: "Warner's broke the logjam that had blocked the introduction of sound equipment to the nation's movie theatres."[69] Although the premiere of *The Jazz Singer* in New York City marks a key moment in film history, it was not shared by other cinemas, cities, or countries. Locating the premiere of the film in Montreal, Quebec, puts researchers into the year 1929, skipping over the period and cultural practices underscoring the transition to sound exhibition in that city. Prior to the business deal that brought *The Jazz Singer* to the screen in New York City as a part talkie, synchronized sound technology was being demonstrated and tweaked in exhibitions for audiences in the United States and Canada.

In Montreal, vaudeville was firmly established in the city's cinema programs, but with the introduction of synchronized sound, a shift in practice was taking place. Experiments with synchronized sound had been demonstrated in theaters in the city since 1926, and in 1928, the Palace was the first theater wired for synchronized sound in Canada. The Palace, Loew's, and the Capitol spent thousands converting to synchronized sound, and with great triumph their owners announced to the moviegoing public that their theaters had been wired and completely refurbished in grand elegance.[70] The first indication that live vaudeville might be in jeopardy was at the Palace's opening night, which featured an all-filmic program with the exception of the Palace Symphony Orchestra.[71] Gradually vaudeville was phased out and replaced by filmic entertainment except at Loew's, where it remained at the heart of its program until late in 1930. By September of that year, long after other cinemas in the city relied on a bill of entirely film to draw audiences, Loew's announced vaudeville would be discontinued, and in its place, patrons could expect "a splendid program of talking and singing pictures."[72] In turn, Loew's offered a new program of talking films and shorts at reduced prices of 25 cents to 50 cents, but even the bargain did not change the irony of the disappearance of the last standing vaudeville accompaniment to film exhibition in Montreal just before long-standing houses like the Capitol took it up again. In what at first seemed like a dead cat bounce for vaudeville, the Capitol brought back its live stage show within a month of the discontinuation at Loew's.

The Capitol announced its inaugural gala show as "the most important event in Montreal's theatre history."[73] In almost a revival theme, the Capitol "took its place among America's Finest Theatres—presenting the same programs as the world's biggest theatres provide."[74] The "greater new show idea" at the Capitol brought several features to the program for prosperity week, including stage productions, organ novelties, musical surprises, a concert orchestra, and the "Greatest Talking Pictures!"[75] The theaters in Montreal were reacting to a plunge in attendance by rejuvenating the screens with added attractions, a resurgence of live acts to accompany the talkies, and new bargain prices, making filmgoing even more accessible to the public. Moviegoers paid only 25 cents before twelve-thirty in the afternoon and were able to see a complete show. Other theaters also introduced bargain prices and thrift matinees. The Palace theater even brought in a special French-language movie to include French-language spectators in their target audience. Theater managers brought back the sort of entertainment audiences were accustomed to before the takeover of the synchronous sound feature film. Montreal is not the only case of vaudeville enjoying a resurgence in Canada. Doug McCallum notes that at the Orpheum in Vancouver, vaudeville was the main attraction from 1927 to 1932, even in the mixed program of live acts and film. He comments, "For some people the movies were the main drawing card, while others considered them decidedly less important—at least at the Orpheum. Some even sat out the movies in the foyer, returning to see the vaudeville twice."[76] This action by the audience points to a steady appreciation for the live acts that were now considered as accompaniments to the movies. It also indicates the resilient expectation audiences had for the continuous mixed show as the dominant form of programming.

For years, audiences at the Orpheum had seen the who's who of the vaudeville stage, including George Burns and Gracie Allen, Jack Benny, and even Bob Hope. The performers passed through on a circuit that moved west from Chicago to Vancouver, and a weekly bill consisted of five or six acts, fifty-two weeks a year.[77] By this time, vaudeville circuits had been disrupted because of decreased booking. The Orpheum, not wired for sound until 1937, was unable to offer acts from vaudeville circuits. Under the direction of Ivan Ackery, the Orpheum continued to present a mixed program of filmic and live acts while drawing live performance from the local community and select touring acts. Although vaudeville and silent cinema are believed to have taken a leap hand in hand from the entertainment stage, there is evidence that the shift to synchronized sound was uneven and inconsistent. Even the infamous Palace in New York City did not see its last vaudeville turn until 1932 when it became a motion picture house.[78] What this demonstrates is that not only do we have to push the dates of the conversion of sound forward beyond what is commonly believed to have taken place as early as 1928 and to have been largely completed by the 1931–32 season; we also have to imagine that theaters not yet wired for sound were still featuring live acts in addition to films, shorts, and orchestral programs. Because there is evidence that refutes a homogeneous and rapid conversion to sound, we must, in turn, refute a rapid and sudden purging of vaudeville. Live stage shows may have been an obvious reminder of the old days after the motion picture industry adopted synchronized sound, but we can argue that the power of the stage show was its reintegration into film programs in spite of the novelty of sound. Not only is this a reminder of vaudeville's popularity as a way to add novelty to filmic programs, it is also a reminder to examine the period when new technologies are introduced for residual media forms.

In September 1930, the *Montreal Daily Star* announced that the public was fed up with sound, claiming, "The novelty of sound was thrown as a sop to the public. So far it has worked. But the news now is that it has ceased to work."[79] The last two years "have seen the talkie take hold on the imagination and pocketbook of the theatregoer."[80] The article in the *Montreal Daily Star* was critical of the film industry's use of sound to ward off the encroaching radio entertainment and to fill otherwise emptying theaters. Referring to the initial adoption of sound by Hollywood as an effort to invigorate the amusement industry, the article says that once again the industry must deal with a public bored by the novelty of sound. We can read this "once again" boredom as part of a pattern in the movie industry to implement a novelty that will be received as the next best thing in entertainment and draw in the audiences. The *Montreal Daily Star* claimed, "People are fed up on lame products offered with a sugar coating of sound and theatres are beginning to worry."[81] It was announced that, "In an effort to restore waning interest the big local show houses have restored the stage prologue, with its lavish acts and pretty dancing girls discarded two years ago."[82] Just a month after cutting its stage show, Loew's jumped to restore vaudeville in October 1930, announcing, "Loew's theater have completed arrangements whereby they can definitely assure the public of the weekly selection of the very best acts available in this particular field of theatrical."[83] Other cinemas in the city followed suit. In an effort to attract crowds, theater managers were attempting to recreate the flush of excitement that had shrouded the first exhibitions of talking film and the format of incorporating film into a wide range of entertainment and live musical acts.

Around the film world, a debate was raging. Would the talkies take a nosedive right off the screen? Other forms of entertainment, particularly the popularity of "midget golf" (mini-golf), and a severe summer slump were blamed for the decreased attendance at the cinemas.[84] As Crafton notes, "By summer even the Fox, Publix, and Warner organizations were turning their unprofitable theaters into miniature golf courses.[85] In an industry just beginning to realize that novelty had great power in terms of reception and bringing in audiences, the wearing off of a novelty should not be taken lightly. The industry looked for ways to bring back the crowds, and despite Warner's confidence in the talkies, the extent to which silent cinema could be brought back was not ruled out as an option.[86]

Louis Mayer of MGM suggested pantomime artists as a solution to declining attendance; Radio Keith Orpheum's William Le Baron foresaw the proper formula for exhibition as one-half dialogue and one-half pantomime.

It appeared that the return to silent films was not the only threat to the somewhat disenfranchised talkies. Morgan Powell of the *Montreal Daily Star* wrote, "Movies and talkies are here to stay. [They] are an interesting and educational form of cheap entertainment and will continue to be patronized extensively by the public; but, they can never take the place of spoken drama."[87] According to Powell's article, people wanted "plays and music rendered by real flesh and blood people"—they also wanted "plays teeming with human interest, wit, educational value and entertainment." They wanted "wholesome entertainment that was a true representation of life not an exaggeration—except for the innocent exaggeration of life that harms no one by its merriment and lifts everyone for a few cheering hours from life's drab realities."[88] While the public tastes were deemed fickle, the major studios were housecleaning by cutting their contract layers down.[89] If they had not been drawing a crowd they were out, and no one was secure any longer because studios had adopted a "one-picture contract" rather than a long-term or lifelong contract they may have agreed to in the past. The excuse used by an industry that needed to employ cost-cutting measures because of lower box office returns was that the public was calling for an industry malleable to the ever-changing whim of the moviegoer.[90]

The new economic pressure of the Depression inspired lower prices as theater owners struggled to pull in audiences. The thrift matinee and the bargain matinee and programs around the city had begun to add stage shows and other novelties like the organ, symphony orchestras, and even so-called freak shows. The Godino Siamese twins and their brides appeared on the stages of "Five United Amusement Theaters in Addition to Their Regular Double Film Programs!"[91] Some new technological innovations began to clamor for the public's attention, including the film *Cimarron* (Wesley Ruggles, 1931), to be presented on a gigantic screen at the Palace theater.[92] The changes brought about in exhibition were not entirely based on the need for something novel; although the studios and theater managers blamed the fickle attitude of audiences, the pattern that had been established in the industry was part of the equation. We see, in examining the lineage of entertainment practices and technological transitions, a consistent pattern of overlapping progressions and recessions as entertainment forms shifted. In the lead up to the 1930s, the resilience of the live stage show I demonstrate is both a reminder of the residual and emergent practice in entertainment and an opportunity to think about how audiences may have identified with each form. The audiences, or interpretations and assumptions about the audiences, played an integral role in the hybrid entertainment forms that developed as the use and incorporation of new sound technologies was negotiated.

Similar to the criticism Staples notes of vaudeville when it began closely to mimic Broadway revues, audiences complained about Hollywood's persistence in producing films that tell stories about "gold-diggers, unfortunate chorus girls, successful thugs and underworld characters, and all the rest of the stuff that has been served up ad nausea in the past."[93] Some critics still thought vaudeville could survive in spite of the talking picture. Alexander Bakshy harshly criticized vaudeville as it was in 1929, "In spite of the fifty thousand vaudeville artists in this country, the programs even in the leading American theaters seldom contain more than one or two really satisfying numbers. The rest are the veriest junk which only the utter degradation of vaudeville standards has permitted to be performed."[94] He continued to say that vaudeville had lost its class, but in spite of the current state of entertainment, vaudeville was needed to rival the talking film. Bakshy felt, "There will be one weapon, however, which will never be found in the armoury of the talking picture—the power of the direct and personal appeal to the audience which distinguishes the art of vaudeville more than any other form of stage entertainment."[95] Predictions of what cinema would become after the talkies ceased to draw a sole spectator demonstrated the vulnerability of the synchronized sound film as a new invention. Talkies fell prey to the hype that surrounds a new innovation and consequently to the predictions of demise that follow. The talkies were different in that they were not just a novel addition to cinema programs. They threatened both the established practices of silent cinema and greatly destabilized the live theater as an entertainment form.

In the wake of this destabilization speculation, an interview with seventy-seven-year-old theater producer David Belasco forecasted the "death of the talkies" and claimed good silent pictures would sweep the country.[96] Many filmmakers and theorists agreed. Charlie Chaplin had already pushed the limits on silent film in 1931 with *City Lights*, yet he surprised people by doing a silent film, *Modern Times*, in 1936. In other words, the movement away from sound was not only on the basis of aesthetics. This stance came as no surprise; however, I maintain that the changes in exhibition brought about by the initial success of the feature-length sound film and the competition among theaters for the public presence contributed to the decreased attendance at the theaters more so than the single idea that the novelty of the talkie had worn off. The variety-style program was popular with audiences, and when theaters ceased to present live acts and musical performances, opting instead for a program of only film, attendance was affected. Managers competed to entice moviegoers to the theater, returning to programs that had proved popular previously—a mixture of sound film, vaudeville acts, and short features. They deliberately made use of vaudeville artists and performances (sometimes even local amateur acts) as part of their overall presentational packages.

The "return" of vaudeville after synchronized sound film was adopted in the entertainment industry is not historically unique. A flash forward to the advent of television demonstrates that in 1948 as networks were bringing in seven-day-a-week programming, time needed to be filled, and former vaudevillians were recruited and signed to perform live and be broadcast on television. By this time, former vaudevillians were the epitome of star talent—established performers with reputations in the entertainment world. These stars could command serious salaries and be a considerable audience draw for television.

Joe Cohen wrote in *Variety* magazine, "The comeback of vaudeville is television's hottest development. Both talents and networks foresee as many variety forms used in this medium as in the days when the Keith-Albee, Pantages, and Orpheum circuits flourished."[97] In prior demonstrations of television, vaudeville had also been the easy answer for test programming. In September 1930, *Saturday Night* announced that in London, a public screening of live television would take place at "one of the West End cinema theatres" wherein "short performances comprising of simple turns and the televising of personalities singing or talking" would be watched by the moviegoers as a special screen was mounted on wheels and moved onto the exhibition platform.[98] This early demonstration mimicked a vaudeville stage show in content, and similar to early trials of synchronized sound, it featured performers speaking and singing to demonstrate the technological prowess of the media. The surge in popularity and sales of television sets in the late 1940s and early 1950s was attributed to the everlasting popularity of the variety show, termed "vaudeo" by film and television critics. Major stars of the vaudeville stage were contacted by the television industry in hopes that they would make a return to the small screen. Joe Laurie Jr. cited television as the future of vaudeville in his birthday tribute to the Palace theater in New York. These were his final words in the article: "Who knows, maybe television will be the Palace of tomorrow?"[99]

Bob Hope had been one of the most successful stars of vaudeville, and long after theaters no longer featured live acts he toured internationally. In 1931, amid the wake of controversy over whether the talkies would drive vaudeville under for good, Hope secured his first contract at the Palace in New York. *Variety* magazine reviewed his performance: "He is a nice performer of the flip comedy type, and he has his own style. These natural resources should serve him well later on."[100] Hope's return to the stage and move to the small screen of television was relished by the National Broadcasting Corporation. On June 29, 1949, John F. Royal wrote a letter to Hope:

> I have been involved in this television racket for a long time, and have been amused by the hurrying and scurrying of late regarding a lot of rough, slapstick comedy, and I want to take a little bit—in fact, a goddam big bet—that the first time Hope gets into television, he will do to this industry what Jolson did to talking pictures.[101]

This reference to Al Jolson and the fame and mythic presence of *The Jazz Singer* (1927) is emblematic both of the legends built in the media industry and the exhilarating desire to revive the vaudeville acts and the heroes of the stage seen as possible though television. Hope's response to Royal was cordial and signaled not only Hope's persistent adoration of vaudeville but his feeling that the habitual practice of taking in the vaudeville shows as an amusement habit would need to be re-created in the new media of television if a vaudeville revival was going to be a success. Hope wrote,

> It was nice to receive your vote of confidence with regard to television, but Berle can have the medium all to himself for the next year. Then I shall have my head blocked and we'll all go back into vaudeville!

> Without a doubt television will really be going in a couple of years and we will have to put on our very best manners and do a nice half-hour show every week. I don't think any less than that will do, as television will have to become a habit, but, nevertheless, an interesting one.[102]

The following year, the vaudevillian revival was magazine cover material. *Look* magazine paid tribute to the return of the vaudeville stars with a cover photo featuring vaudevillians on television including Fred Allen, Jack Benny, Bob Hope, Groucho Marx, Eddie Cantor, Ken Murray, Ed Wynn, Bobby Clark, George Burns, Gracie Allen, and Jimmy Durante.[103] In general, the rehashing of vaudeville stars as new television stars can be seen as a competitive shift instigated by the instability of the new media of television and broadcast. The variety show format is consistently the entertainment form that exhibitors and broadcasters return to in the case of a perceived instability in attendance. It was believed, on numerous occasions, that vaudeville was dead: the form was not alleged to be interesting or sophisticated enough to withstand the various innovations in site, sound, synchronization, and streamlining that occurred in entertainment programming. However, as I have shown, by following vaudeville past the first certain date of its decline and beyond its impending doom at the advent of synchronized sound, we can see that the entertainment form had a resilience not often noted in the history of film.

By 1930, the model of the all-filmic program that accompanied the transition from silent cinema to synchronized sound was economically not viable for many exhibitors trying to draw in Depression crowds more concerned about the national bread prices than the Saturday matinee. In what could be characterized as a desperate move, exhibitors brought live vaudeville programs back to the cinema. The cost of converting to sound made making additional large-scale adaptations to theaters impossible for most exhibitors. Having wired their theaters and already booked sound films would have made it impossible to do anything but show the films. The movie industry was also financially committed to producing and distributing sound films. Although vaudeville had been eulogized numerous times, it made a return on the very heels of synchronized sound and the feature film–based program, the new media that was said to banish live entertainment. This move backward to a mixed program was brought about by what theater managers perceived as audience's declining taste for the novelty of sound, the perception that audiences were not responding to all-filmic programs, and the economic uncertainty of the Depression. Contrary to dominant perceptions of new media replacing old, the unexpected historical return of vaudeville to filmic exhibition in the early 1930s offers a vivid example of vaudeville's hold on the popular imagination and its resilience as an entertainment form. At the very least, the use of vaudeville to augment programs and the recruitment of vaudeville artists by theater managers and, later on, television producers demonstrates that within technological transitions, vaudeville brought a sort of resilience and reliability to the stage and the screen. Moreover, examining the resurgence of live performance over the course of several periods of change shows that a form of entertainment as enigmatic and effervescent as vaudeville actually had a much broader relevance in the concepts of technological adoption, the history of cultural practices, and the concepts of high and low culture. A closer examination of the cultural life of vaudeville reveals that resurgences are associated with periods in which the entertainment industry (in its various incarnations) is vying for control and standardization of the market. The residual and emergent forms of vaudeville over the first few decades of the development of the film industry demonstrates patterns indicating that more than anything, the introduction of new media proves, as Harold Innis argues, to destabilize the existing relationship of communication media and patterns of living.

Notes

1. *Bulletin*, June 26, 1906, 1. For a history of the Empire theater and other theaters in Edmonton, Alberta, see John Orell's *Fallen Empires: The Lost Theatres of Edmonton, 1881–1914* (Edmonton, Alberta: NeWest Publishers, 1981), Glenbow Archives, Calgary, Alberta.
2. Robert C. Allen, *Horrible Prettiness: Burlesque and American Culture* (Chapel Hill: University of North Carolina Press, 1991), 178.
3. Doug McCallum, *Vancouver's Orpheum: The Life of a Theatre* (Vancouver: Social Planning Department, 1984), 6.
4. Ibid.
5. Ibid.
6. Allen, *Horrible Prettiness*, 179.
7. There is not shortage of historical work on the history of vaudeville and most histories of early film, silent film, and the coming of sound era will mention vaudeville. For an excellent history of burlesque and vaudeville see in particular, Robert Allen's *Horrible Prettiness: Burlesque and American Culture* (Chapel Hill: University of North Carolina Press, 1991) and Robert Snyder's *The Voice of the City: Vaudeville and Popular Culture in New York* (New York: Oxford University Press, 1989) For more on the relationship between vaudeville and film see Robert C. Allen, *Vaudeville and Film, 1895–1915: A Study of Media Interaction*, Ph.D. dissertation, University of Iowa, 1977. For more on the history of vaudevillians and the vaudeville aesthetic see Shelly Staples, *Male-Female Comedy Teams in American Vaudeville, 1865–1932* (Ann Arbor: UMI Research Press, 1984) and Henry Jenkins, *What Made Pistachio Nuts?* (New York: Columbia University Press, 1992).
8. Raymond Williams, "Base and Superstructure," in *Marxist Cultural Theory, Problems in Materialism and Culture* (New York: Verso, 1980), 38.
9. Ibid.
10. Harold Adams Innis, *Bias of Communication* (Toronto: University of Toronto Press, 1951), 188.
11. Allen, *Horrible Prettiness*, 177.
12. Ibid.
13. Ibid.
14. Shelly Staples, *Male-Female Comedy Teams in American Vaudeville, 1865–1932* (Ann Arbor: UMI Research Press, 1984), 33. Pastor even tried ham and turkey giveaways to lure more women into the Opera House.
15. Allen, *Horrible Prettiness* 180.
16. Ibid., 179.
17. Robert C. Allen accounts the detailed tale of Keith's move from dime museum manager to variety, as does Robert Snyder in *The Voice of the City: Vaudeville and Popular Culture in New York* (New York: Oxford University Press, 1989).
18. There were also strict regulations governing the number of saloons in the city, which may have been a factor. Allen, *Horrible Prettiness*, 182.
19. Ibid., 183.
20. Although vaudeville immediately differentiated itself from the curio side of popular entertainment, a residual presence of live stage shows featured freak shows and odd characters like bearded ladies, dwarfs, and fat ladies. In 1925, Lester (Bob) Hope and George Byrne were booked on a tour in which the headliners were eighteen-year-old Siamese twins Daisy and Violet Hilton. The Hilton sisters' show featured the twins telling stories of their lives, playing saxophone and clarinet

duets, and dancing with Hope and Byrne. Brochure for "Siamese Twins" Daisy and Violet Hilton, ca. 1925, Bob Hope Collection, Motion Picture, Broadcasting & Recorded Sound Division, Library of Congress. From "Vaudeville: Bob Hope and American Variety," Library of Congress online exhibition, http://lcweb.loc.gov/exhibits/bobhope/vaude.html.

21. Allen, *Horrible Prettiness*, 182–85.

22. Douglas Gilbert, *American Vaudeville: Its Life and Times* (New York: Dover Publications, 1963), 204–5.

23. These characteristics are most commonly associated with vaudeville, but there are of course variances. For instance, the illustrated song was very popular on Canadian programs, and audiences interacted with performers in these cases. For a detailed account of illustrated song, spectatorship, and the role of exhibitors in Quebec, see Pierre Verronneau, "The Reception of 'Talking Pictures' in the Context of Quebec Exhibition 1894–1915," *Film History* 11, no. 4 (1999).

24. Allen, *Horrible Prettiness*, 179.

25. Robert Snyder, *The Voice of the City: Vaudeville and Popular Culture in New York* (New York: Oxford University Press, 1989), 27.

26. Allen, *Horrible Prettiness*, 185.

27. Joe Laurie Jr., "Happy Birthday," *Variety*, March 10, 1948. Accessed as part of the Guy Weadick Collection, M1287, file 34, Glenbow Archives, Calgary, Alberta.

28. Ibid., 205–6.

29. Allen, *Horrible Prettiness*, 206.

30. Raymond Williams, *Television: Technology and Cultural Form* (London: Wesleyan University Press, 1974), 59.

31. Ibid.

32. Raymond Williams, *Marxism and Literature* (London: Oxford University Press, 1977).

33. Miriam Hansen, *Babel and Babylon: Spectatorship in American Silent Film* (Cambridge: Harvard University Press, 1991).

34. Roy Rosenzweig treats the emergence of nickel theaters in Worcester, Massachusetts, and the popularity of the movies in *Eight Hours for What We Will: Workers and Leisure in an Industrial City, 1870–1920* (New York: Cambridge University Press, 1983).

35. Ibid., 185. Robert Snyder similarly describes what Allen calls "vaudeville grandiloquence." In New York, Frederick F. Proctor was starting the construction on Proctor's Pleasure Palace. The Romanesque auditorium had a roof garden, German café, library, and barbershop, Turkish bath, flower stand, and a smaller auditorium called the Garden of Palms. Oscar Hammerstein constructed the Olympia Theatre in 1895. The Olympia trumped Proctor's Palace like the Chrysler Tower did the Empire State Building. The Olympia had two large auditoriums, a musical hall, concert hall, café, roof garden, billiard hall, smoking lounges, a bowling alley, and a Turkish bath. See Robert Snyder, *Voice of the City*, 82–87.

36. Canadian theaters booked the vaudeville acts that toured North America, usually originating in the United States. These theaters also played local and amateur acts.

37. McCallum, *Vancouver's Orpheum*, 15.

38. "Orpheum Bulletin," Sherman Grand Theatre, Calgary. December 19–21, 1912, Barron Enterprises Collection, M7268, Glenbow Archives, Calgary, Alberta.

39. Allen, *Horrible Prettiness*, 185.

40. Raymond Williams's concept of flow was first identified in relation to television programming and a comparison of British and American programming schedules.

41. Blackface minstrel shows were sometimes seen as part of vaudeville programs. Al Jolson was a popular vaudeville entertainer before he began his career in movies, and his character in the *Jazz Singer* (1927) was based on his popular vaudeville routine. Other ethnic acts featured comic characters that were commonly Irish, Jewish, German, and Italian.

42. For sample programs and examples of vaudeville sketches and turns, see "Vaudeville: Bob Hope and American Variety," Library of Congress online exhibition, http://lcweb.loc.gov/exhibits/bobhope/vaude.html.

43. Shirley Staples, *Male-Female Comedy Teams in American Vaudeville 1965–1932* (Ann Arbor: UMI Research Press, 1984), 76.

44. Robert C. Allen, *Vaudeville and Film, 1895–1915: A Study of Media Interaction,* Ph.D. dissertation, University of Iowa, 1977.

45. Rosenzweig, *Eight Hours for What We Will*, argues that by 1904 the moving pictures were no longer sporadic novelties for view on busy corners or in makeshift movie houses but began to be given more permanent homes in Worcester's vaudeville theaters. See Roy Rosenzweig, *Eight Hours*

for What We Will: Workers and Leisure in an Industrial City, 1870–1920, 192. For a complete picture of early vaudeville and its relation to film, see Robert C. Allen, *Vaudeville and Film, 1895–1915*.

46. Russell Merritt, "Nickelodeon Theaters 1905–1914: Building an Audience for the Movies," in *The American Film Industry*, rev.ed., ed. Tino Balio (Madison: University of Wisconsin Press, 1985), 85.

47. Rosenzweig, *Eight Hours for What We Will*, 194.

48. For an excellent overview of the contexts of film distribution and exhibition in relation to understanding the function of movie theatres and sites for movie consumption as public spaces, see Gregory Waller's edited collection: *Moviegoing In America* (Oxford: Blackwell Publishing, 2002).

49. Craw wrote a weekly column in *Moving Picture World* advising theater managers on the box office business from programming, to audiences and how to make their theatre popular. George Rockhill Craw, "Swelling the Box Office Receipts," *Moving Picture World* 8 (May 13, 1911): 1059–60.

50. Ibid., 1060. Vaudeville was also argued to be civically damning, and the regulation of motion pictures was thought to be a civil authority in North America as illustrated in Boyd Fisher, "The Regulation of Motion Picture Theatres," *American City* 7 (1912): 520–21. Fisher argues that vaudeville should be ruled out as a cheap expedient for profit and should be barred from theaters. He is explicitly referring to small vaudeville shows, which were often thought of as cheap entertainment: "Vaudeville cannot be profitably furnished unless it is either immoral and cheap or simply inferior and cheap. To rule it out of all except vaudeville theatres is to safeguard the moral and intellectual quality of picture theatres."

51. John Collier, "Cheap Amusements," *Charities and the Commons* 20 (April 1908): 73–76, reprinted in *Moviegoing in America*, ed. Gregory Waller (Oxford: Blackwell Publishing, 2002), 46–47.

52. Ibid., 46.

53. De Certeau identifies trajectories or what he calls "wandering lines" (lignes d'erre) which are the tracings of audiences as they move within functionalized spaces. He warns against simplifying or "flattening out" the path of practice because we would miss the complex combination of elements that form these trajectories which are, to De Certeau, the combination of discursive elements of practice and use. Michel de Certeau, *The Practice of Everyday Life*, trans. Steven Rendall (Berkeley: University of California Press, 1984), xviii.

54. Ibid., xviii.

55. Ibid.

56. Harold B. Franklin, *Motion Picture Theatre Management* (New York: Doubleday, Doran, and Company, 1928), 27; reprinted in Waller, *Moviegoing in America*, 116–23.

57. Richard Koszarski, *An Evening's Entertainment: The Age of the Silent Feature Picture, 1915–1928* (Berkeley and Los Angeles: University of California Press, 1990), 9.

58. Ibid.

59. Staples, *Male-Female Comedy Teams*, 301.

60. Donald Crafton, *The Talkies: American Cinema's Transition to Sound, 1926–1931* (New York: Simon & Schuster/Macmillan, 1997), 1.

61. Alexander Walker, *The Shattered Silents: How the Talkies Came to Stay* (New York: William Morrow, 1979). Walker's work supports the legend that claimed the "Art of the Silent" cinema as victim of the new sound technology.

62. Ibid., 27.

63. Lewis Jacobs, *The Rise of the American Film* (New York: Harcourt, Brace and Company, 1939), 334.

64. Karel Dibbets, "The Introduction of Sound," in *The Oxford History of World Cinema*, ed. Geoffrey Nowell-Smith (Oxford: Oxford University Press, 1997), 214.

65. Ibid.

66. Miriam Hansen, "Early Cinema, Late Cinema: Transformations of the Public Sphere," in *Viewing Positions: Ways of Seeing Film*, ed. Linda Williams (New Brunswick, N.J.: Rutgers University Press, 1994), 147.

67. Ibid.

68. See Gregory Waller, *Main Street Amusements: Movies and Commercial Entertainment in a Southern City, 1896–1930* (Washington, D.C.: Smithsonian Institution Press, 1995), for an examination of midsized American cities; for Manhattan, see Robert C. Allen, "Motion Picture Exhibition in Manhattan: Beyond the Nickelodeon," *Cinema Journal* 19, no. 2 (Spring 1979), 2–15; Russell

Merritt, "Nickelodeon Theatres, 1905–1914: Building an Audience for the Movies," in *The American Film Industry*, ed. Tino Balio (Madison: University of Wisconsin Press, 1976), 59–79; and Melvyn Stokes and Richard Maltby, eds., *American Movie Audiences from the Turn of the Century to the Early Sound Era* (London: British Film Institute, 1999).

69. It is well accepted in film history that the 1927 decision of Warner Bros. to add synchronized sound dialogue to their film *The Jazz Singer* (1927) marked the move that shut the door on silent cinema. See Koszarski, *An Evening's Entertainment*, 90. Film historians have also acknowledged the legend of *The Jazz Singer* and attempted to work around the mythic proportions of the film, which has become a temporal marker in film history. Notably, Al Jolson, one of the most famous vaudeville stars, commanding a salary of $100,000 per year, helped the film go down in a privileged historical place. Accounts of the film in media sources and popular writings from the period are consistently intertwined with stories about Jolson. See Donald Crafton, *The Talkies: American Cinema's Transition to Sound, 1926–1931* (Berkeley: University of California Press, 1999), 520–30.

70. The Palace theater's interior had been completely restored and wired for Fox Movietone and Vitaphone sound systems. Theater manager George Rotsky announced in the French- and English-language press that $100,000 had been spent to make the theater the most luxurious in Montreal: *Montreal Daily Star*, September 1, 1928, 22; *La Presse*, September 1, 1928, 69.

71. Dane Lanken, *Montreal Movie Palaces: Great Theatres of the Golden Era 1884–1938* (Waterloo, Ontario: Penumbra Press, 1993), 103. Not long after it reopened, the Palace's famous orchestra led by Maurice Meerte lost its spot, and the Palace shows were entirely film. Meerte turned up soon after at the Capitol theater. For more on the transition to synchronized sound in Montreal, see my master's thesis, *That's Not What I Heard: Synchronized Sound Cinema in Montreal, 1926–1931*, Concordia University, Montreal, 2001.

72. Advertisement for *Way Out West* (1930) at Loew's, *Montreal Daily Star*, September 13, 1930, 24.

73. Advertisement for the "Capitol's Greater New Show Idea during Prosperity Week," *Montreal Daily Star*, October 11, 1930, 23.

74. Ibid.

75. Ibid.

76. McCallum, *The Orpheum*, 17.

77. Ibid., 16.

78. Snyder, *Voice of the City*, 152.

79. "Public Getting 'Fed Up' with Lame Movies Despite Sound Novelty," *Montreal Daily Star*, September 1, 1930, 14.

80. Ibid.

81. Ibid.

82. Ibid.

83. Ibid.

84. "Photoplay Executives Give Options on the Future of the Talkies," *Montreal Daily Star*, October 11, 1930, 24.

85. Crafton, *The Talkies*, 263.

86. "Photoplay Executives Give Options on the Future of the Talkies," *Montreal Daily Star*, October 11, 1930, 24.

87. Morgan Powell, "About Ticket Prices" [editorial], *Montreal Daily Star*, October 25, 1930, 23.

88. Ibid.

89. "List of Movie Players Will Undergo a Weeding Out Process Very Soon," *Montreal Daily Star*, November 2, 1930, 6.

90. Crafton, *The Talkies*, 182.

91. Advertisement for the appearance of the Godino Siamese twins at five United Amusement Theatres, *Montreal Daily Star*, March 4, 1931, 6.

92. The *Montreal Daily Star* reported that *Cimarron* cost $1 million to make.

93. Morgan Powell, "How About the One-Reeler?" *Montreal Daily Star*, March 7, 1931.

94. Alexander Bakshy, "Vaudeville: Vaudeville's Prestige," *The Nation* 129, no. 3342 (July 24, 1929): 98–100.

95. Ibid.

96. "Belasco Predicts Death of Talkies," *Montreal Daily Star*, July 24, 1930, 6.

97. Joe Cohen, "Vaude's 'Comeback' Via Vaudeo: Talent, Agents Hopping on TV," *Variety*, May 26, 1948, 43–44.

98. "Progress of Television," *Saturday Night* 45 (September 13, 1930): 15. This demonstration of early television featured the Baird system, which allegedly cost between £50 and £60 and did not then have a wavelength to broadcast. The article claimed this was the first public exhibition of television.

99. Laurie, "Happy Birthday."

100. Bob Hope Collection, Motion Picture, Broadcasting and Recorded Sound Division, Library of Congress. From "Vaudeville: Bob Hope and American Variety," Library of Congress online exhibition, http://lcweb.loc.gov/exhibits/bobhope/vaude.html.

101. Letter from John Royal to Bob Hope, June 29, 1949. Typed manuscript. Bob Hope Collection, Motion Picture, Broadcasting and Recorded Sound Division, Library of Congress. From "Vaudeville: Bob Hope and American Variety," Library of Congress online exhibition, http://lcweb.loc.gov/exhibits/bobhope/vaude.html.

102. Letter from Bob Hope to John Royal, July 13, 1949. Typed manuscript. Bob Hope Collection, Motion Picture, Broadcasting and Recorded Sound Division, Library of Congress. Within months of this correspondence Bob Hope agreed to host his own show, and fellow vaudevillian Milton Berle continued to host Berle's *Texaco Star Theater*. Hope's first national television appearance was on April 9, 1950, as a regular host of *Star Spangled Revue,* a variety show produced by Max Liebman. Broadcast live from New York, the program was much like Hope's radio show with a monologue, skits, and musical performances. "Vaudeville: Bob Hope and American Variety," Library of Congress online exhibition, http://lcweb.loc.gov/exhibits/bobhope/vaude.html.

103. "TV's Old-New Stars," *Look* magazine (April 10, 1951). Library of Congress online exhibition, http://lcweb.loc.gov/exhibits/bobhope/vaude.html.

PART III

Collecting and Circulating Material

9 Every Home an Art Museum: Mediating and Merchandising the Metropolitan

Haidee Wasson

> Catalogues published by the museum, photographs of all objects belonging to the museum, photostats of books, photographs, and prints. Postcards, color prints, etchings, and casts are on sale at the Fifth Avenue entrance. Lists will be sent on application. Orders by mail may be addressed to the Secretary.
> —*The Metropolitan Museum of Art Bulletin*, October 1927

> There are relations between the department store and the museum, and here the bazaar provides the link. The amassing of artworks in the museum brings them into communication with commodities, which—where they offer themselves en masse to passersby—awake in him the notion that some part of this should fall to him as well.
> —Walter Benjamin, *The Arcades Project*

In 1949, the Metropolitan Museum of Art, New York (Met), and the Book of the Month Club (BMC) announced a collaborative project. Together they conceived, designed, and circulated 2 by 2–inch poster stamps to book club members featuring color reproductions of paintings and other objects held by the museum. Entitled "The Metropolitan Museum of Art Miniatures," these stamps arrived along with an album in which they would be mounted. Each

---

I would like to acknowledge the dedicated research assistance of Mark Frank and Steve Groening. Considerable credit is also due to Joan Acland for providing astute commentary on an earlier draft of this essay. Funding of this research was generously provided by the McKnight Landgrant Foundation, University of Minnesota.

album was identified by a letter of the alphabet and had its own distinctive theme. Album E contained stamps depicting the development of seventeenth-century Dutch painting. Album W supplied twenty-four examples of twentieth-century American painting. Album L illustrated "Art in the Middle Ages." Album O was solely dedicated to van Gogh, comprised of works held by the Met and also by several Dutch museums. Other albums included pictures of ornamental vases, sculptures, and gewgaws originating across historical periods and places: ancient Greece, fourteenth-century China, and nineteenth-century France. Artists ranged from the canonical Titian, Courbet, Bruegel, and Cézanne to anonymous creators of German armor and Roman marbles.

The Met's book club series promised "Your Own Museum of Art in Miniature," and each individual reproduction was sent attached to others by perforated seams and preapplied adhesive, ready for its back side to be moistened and its front side to be displayed on a designated square. Every image was thus dated, affixed alongside the artist's name (when available) and country of origin. Additionally, beneath the mounted image, or the blank space provided for the diminutive objects, paragraph-long descriptions indicated the item's significance in the general history of art appeared. Each album provided a kind of home schooling in art history, sanctioned by the museum, at the low, low cost of $1 per lesson. Yet these miniatures were more than just quaint collectibles, weekend hobby, or didactic amusement. They were extensions of museological space and authority, further reorganizing and transforming the already decontextualized and abstracted museum originals they meagerly indexed. Bolstered by the Met's imprimatur, these tactile shrunken assemblages transformed the hallowed halls of the museum into handheld satellites, orbiting far beyond the bricks-and-mortar museum site.[1]

Selling more than eight million in their ten-year run, museum miniatures arrived in homes across the United States by mail, providing but one example of an impressive range of modes by which the American art museum was at midcentury working to maintain a presence in the everyday life of ordinary and geographically dispersed peoples.[2] Typifying this phenomenon—like art catalogs, framed reproductions, and postcards before them—museum minis reproduced the discourses of an otherwise staid museum, in this case the Met, on inexpensive paper sent by bulk mail. The miniatures also confirm the Met's collaboration with a largely placeless cultural institution, the BMC, whose members were scattered across the country. Through the art albums, each institution worked to enliven their own respective missions. The BMC marketing literature accompanying the minis claimed to generate "a museum in every home," linking the acquisition of books with the apparent excitement of collecting and curating art. Club correspondence proclaimed that the BMC was proudly helping the museum by "bringing its priceless art treasures into the homes of cultivated people."[3] The club further likened the miniatures to the actual physical spaces of the Met, calling the albums "guided visits," promising a kind of virtual travel connecting home to museum, viewer to art. BMC solicitations implored potential subscribers to understand the uniqueness not of the mini album but of the phenomenon the album sought to replicate, reminding customers that the Rembrandts reproduced delivered readers a unique experience available only at the Met. "Few institutions in

FIGURE 9.1 *Advertising insert for the Metropolitan Museum of Art Miniatures, suggesting that over time subscribers will build a miniature Museum of Art for their home, linking museum, art, and domestic spaces through inexpensive reproductions.*

the world have in their custody such a representative variety of Rembrandt's art as the Metropolitan. They are beautiful, beautiful works—every one of them; people come from far and away to see them, and it is seldom that there is not a group of art lovers before each one, enjoying their beauty."[4] Appealing to those interested in "art and their children," letters and sales pitches to club members worked hard to connect the home and family with the faraway ideals of institutional sanction, aesthetic achievement, unique genius, and cosmopolitan culture. All of this was a method by which books together with museums might maintain and perhaps improve their prominence and pertinence within the shifting cultural landscape of the 1950s.

Clearly, the Met's minis were fantastical journeys into middlebrow taste, reproducing familiar discourses of beauty and reverence. But, they were more than this. These publications were also designed to be stored on shelves alongside other objects of erudition and cultural capital: literary classics, encyclopedias, dictionaries. Art became small pieces of stored data—handled, classified, mounted, and filed away in the home library. In short, the minis transformed the museum into a domestic object; the Met became part of the family archive and library, and thus an element of the middle-class project to display its cultural capital through practices of interior design and decoration, a process Pierre Bourdieu termed "distinction."[5] Yet while the museum became one of many elements constituting bourgeois domesticity, its aura was paradigmatically resituated, complementing not only the sacred but also the banal objects typifying familial accumulation and home decor. The museum could conceivably and in practice sit beside a leather-bound issue of Shakespeare, a Pat Boone album for the hi-fi, or a county fair snow globe featuring pig wrangling.

In all of this, the museum boldly claimed the home as one part of its expanding remit, proposing very particular ideals of domestic life. According to the Met, the tasteful and well-read home emulated museological practice: acquiring, arranging, and displaying old and usually European masterpieces. The "Home Museum of Art" linked quotidian domesticity to the timeless ideals of cultural institutions (literature, art), affordable books to priceless objets (classics and masterpieces), and interior decoration to auratic aesthetics.[6] As the museum rearticulated a domestic ideal of supposed high aesthetic purpose, domesticity also rearticulated the art museum. The grand halls of the Met were intentionally shrunk, its marble floors turned to paper, its foundation stones transformed into cheap binding, its galleries became as two-dimensional as its Goya. The museum's "priceless art" was recast as lickable fun, deemed appropriate for decorating "lampshades, coasters, ash trays and boxes."[7] The American art museum was not only a household name, it was a household presence—an unlikely shrine to the old masters, to be dusted alongside other decorative objects and acquired tchotchkes. It was thus also an intimate index to future travel, a memento of a place never seen or yet to be visited, a reference to exotic cultures and historical events. In short, the museum became a distinct kind of material object as well as a domestic fantasy, linked to a network of other middlebrow cultural forms, proliferating across homes urban, suburban, and rural.

Using museum minis as a starting point, this chapter documents an earlier period in which the American art museum was already a mediated museum, seeking to shape the private and domestic spheres as much as the wider public spheres so often deemed to be its primary remit. The following will focus on one of the most prominent, authoritative, sizable, and well-endowed such institutions, the

Metropolitan Museum of Art, New York, and map the institution's shifting address during the 1920s, a period crucial for understanding the art museum's most basic structure and identity throughout the twentieth century. The interwar period allows us to trace the American art museum's paradigmatic turn toward consumer culture and domesticity as modes of curation and as sites for furthering the museum's presence in everyday life. The ongoing commodification of reproducible art as well as the reproducible museum coalesced during this period, manifesting concretely in the birth of the museum gift shop. Among its many influences, the world of retail and advertising provided a powerful model for, in this instance, Met administrators who were tasked with attaining visibility for the museum in a crowded cultural field and thus burdened with finding methods for effectively circulating its things and ideas. Mediating and merchandising the museum functioned as a necessary mechanism for mobilizing art beyond the museum's walls, inserting both art and the museum into the ebbs and flows of everyday life. Inspired in part by its close relations to the department store, the mediated museum and its gift shop are indexes to—and symptoms of—the general conditions of culture and art under capitalism. Helping the art museum maintain relevance and also assert influence, the mediated museum is a salient stage on which debates about cultural value have long been played out. This dispersed sometimes ethereal and sometimes material museum reshaped objects old and new, of high and low status, sending them out into the world, forever changed and forever changing.

The Expanded Museum

The idea of an expanded and mobile museum was especially prominent during the interwar period, buttressed by an enthusiasm for modern technologies. In 1932, for instance, the *New York Times* celebrated the increasing use of lantern slides, motion picture films, and film projectors to circulate museum exhibits. The production of "portable art and science exhibits" for classrooms was seen as a laudable public service. The paper assured its readers that soon the "collections of the Field Museum of Natural History in Chicago, the American Museum of Natural History in New York, the Metropolitan Museum of Art and similar great institutions will ramble over whole states in special railway trains." Already in the 1930s, science and art museums were being imagined as mobile, their collections easily reproduced and benevolently dispersed. The conditions of possibility for this were of course linked to the emergence of photography and lithography in the mid-nineteenth century. Such technologies fundamentally transformed not just the location of art but our most basic ideas about what art is.

Museum minis are part and parcel of these wider shifts, further confirming the importance of other media in exploring the changing nature not just of art but of the museum as well. In other words, just as art changed under the conditions of reproducibility, so too did its institutions. Museums in the modern period were responses to modernity itself: industrialization, the rise of technology and capitalism, urbanization. Some museums formed as direct reactions against the overproduction of things and the commensurate regimentation of rapid industrial time, crafting themselves and their exhibited objects as distinct from, if not outside of, the ephemeral and the quotidian.[8] Other museums took modern life as an opportunity to engage with the objects and the cultures of cities, factories, assembly lines, and urban migration by forwarding workers' education, and arguing for the potential beauty of manufactured and mass-produced objects.[9] Of course, most American museums—in this case, museums of art—adapted to these conditions by incorporating elements of both these positions, organized to varying degrees around the timeless and the ephemeral, the sacred and the ordinary.

The American art museum, then, has a related and constitutive tension at its core, one that is endemic to privately funded yet publicly mandated—as opposed to princely—museums.[10] It has been tasked with managing the dual tendencies of, on the one hand, functioning as an adaptive site of public education and democratic access, and, on the other, serving as an enduring and sacral repository for precious objects. Or, as Lawrence Levine has phrased it, American art museums have from the beginning enacted a contest between the "didactic and the ideal."[11] The balance of this contest has shifted significantly throughout the public art museum's long life, from its origin in curiosity to its current imbrication with Vegas-style spectacle. Inside and outside of the museum, debates have persisted about the institution's goals, as well as the methods it should adopt to realize them. Like all institutional debates, these must be understood in relation to the historical, social, aesthetic, and political forces that provide the context in which such institutions function. Some art museums have clung tightly to their elite remit, and others have worked toward more populist aims. Within this spectrum, one fact remains clear: the American art museum itself has long been—to greater and lesser degrees—a mediated museum, integrated with a range of media technologies and systems, occupying a complicated and not simply antagonistic relationship to popular and consumer culture.

Considering the first generation of grand American art museums, emergent in the latter half of the nineteenth century, it is important to note that their lineage derives less from powers explicitly linked to the state, as it does for their earlier European counterparts. The American art museum was born of the declarative civic impulses of wealthy industrialists and a social elite seeking to punctuate America's growing political and economic might with what were usually objects from Europe's past. Inextricably connected to this was the effort to provide alternative and reformist amusement for the great mass of immigrant workers whose interests in amusement parks, saloons, vaudeville, and movie theaters were deemed unsettling. During this period, Levine has argued that American museums underwent—like other cultural institutions—a gradual yet identifiable change, morphing from their birth in eclectic and curiosity-based displays into exhibitionary palaces. In sharp contrast to populist forms of leisure, museums increasingly organized around a cultural ideal of studied contemplation and ritualized reverence. According to Levine, this transformation was instrumentalized most effectively by regulating the space of the museum itself. A growing assortment of museums employed grand architectural and neoclassical design, rationalized exhibition techniques, and exercised strict enforcement of dress and behavioral codes crafted to ensure that those who entered the museum were observant of proper rather than beer hall etiquette. The ethnic and class-based coding of these regulatory practices is palpable in discourses of the time. The idea of timelessness was a persistent element of art museum rhetoric and authority, according the institution a plane of temporal existence far removed from the grind of industrial and quotidien time.[12]

There is no question that the site of the museum and the knowledge it presents has been elemental to regulatory efforts exerted from above seeking to reform those deemed to dwell below. Naturalizing these efforts through resorts to rhetoric of timeless beauty and eternal value are foundational to museum art. Yet, as crucial as the impulse to identify and assess such institutions by examining what Tony Bennett has termed "the political rationality of the museum,"[13] it behooves the historian of the American art museum to also note that what we call *the museum* cannot only be understood as a unique and physical place in which art hung, time stopped (or at least slowed), and authority accrued. Recent work on the history and politics of the museum has gone a long way toward reminding us that museums are best understood as points of convergence. As Daniel Sherman and Irit Rogoff have written, we must understand the museum "as an amalgam of historical structures and narratives, practices and strategies of display.[14] Despite this, the overwhelming tendency has been to overlook the fact that museums have both practically and abstractly extended their museological functions—using a variety of technologies and systems—far beyond any one singular location. In short, the museum's authority as a producer of knowledge and manager of people has been sustained but also seriously restructured and perhaps transformed by its active role as a cultural programmer. Its enduring and persistent presence across a range of media forms and spaces confirms that just as the masses and the media provided the bad object against which many museums struggled, the museum has also re-formed itself, extending out, and working through the mass media.[15]

The fact of the art museum's expanded reach includes a predictable impulse toward what is conventionally termed museum education and museum extension, programs that take place either at the museum using lecturers, slides, and other forms of augmentation or that unfold at other locations such as schools, galleries, colleges, and art societies. Integral to this is an equally long-standing effort to publicize and merchandize the museum, assisted by ever-changing technologies of reproduction and an ascendant mass media. That is, just as Levine's art museum was discursively fashioned as a sacral space, unique and auratic like the originals it held, it was also slowly yet steadily migrating outward toward the chaotic and disordered world from which it worked so hard to differentiate itself.

At the turn of the century, already worried that its tendency toward becoming a lifeless warehouse might doom it to irrelevance and inefficacy, museums of art as well as science, natural history, and technology began to formalize their role as active and engaged educators. The term "docent" was coined to designate a new role for museum guides, who were conceived as both students and teachers. Docents guided visitors through collections and exhibitions, providing explanations and instruction about displayed objects. Additionally, traveling educators with lantern slides and elaborate lectures filled increasingly large museum auditoriums, built to augment museum operations and punctuate their civic function. Adults and children alike were served by regular presentations animated by photographic or painted slides. These were held in the evenings, Saturdays, and in the late teens and 1920s, increasingly on Sundays, to facilitate the attendance of working people and encourage family participation. Classes, formal gallery talks, group tours, and discussion groups also emerged as standard aspects of museum activities during this period. Lectures and programs were organized in conjunction with public schools as well. Children's clubs, children's rooms, and galleries, as well as after-school classes, all became standard elements of the museological institution. Groups that largely functioned at locations far away from the museum also sought use of its resources. YMCAs, YWCAs, 4-H clubs, Boy and Girl Scout groups made field trips during which museum workers attempted to engage children either with lectures but more often with games, quizzes, and treasure hunts orchestrated around whole exhibitions or individual objects and artworks. Such activities transpired in art museums as well as in museums of science and nature.[16]

The difficulty of travel to museums and the constraints of the museum's physical immobility soon gave way to the institutionalization of extension programs. In the first decade of the new century, museums such as the Met and the American Museum of Natural History actively began to assemble traveling exhibitions, packages of educational materials, printed matter, lantern slides, mounted objects and sometimes films, film stills, and film strips.[17] By 1916, for instance, the comparatively small Boston Museum of Fine Arts had 30 trained docents serving 4,300 annual visitors, with an additional 8,000 student attendees. They circulated 25,000 reproductions for classroom use.[18] At the Met, the extensions programs grew rapidly, and by 1925, as many as 3,427 different loans were made, consisting of 115,954 individual items.[19] Ten years later, these numbers had increased, respectively, to 4,369 and 179,239.[20] The sites of the museum multiplied; the museum audience proportionately spread.

During the 1920s, attendance to the museum itself was well over a million visitors per year. Yet the number of those exposed to its art and its discourses about art increased exponentially. In addition to its extension programs, the museum also attained a steady presence across media forms. With magazine sales exploding, radios proliferating, and film audiences growing, fulfilling the museum's educational mission required responding to and accommodating the rapidly changing environments in which the museum's public engaged the world. Eager to continue adapting to the modes of display and curation that held sway with the public, Met curators and educators established a surprisingly wide range of strategies to enliven their art and improve museum status.[21] One telling and unexpected example arises in the museum's use of film. The Met organized a film library as early as 1922 to complement its growing collection of educational loan items. And, in 1925, it inaugurated "a series of cinema films relating to various phases of art and illustrating the objects in its galleries."[22] In other words, the Met became a film producer, distributor, and exhibitor.[23] These films, along with those made by other educational organizations, were assembled and circulated to other museums, art schools, and societies.[24] In the spring of 1926, these films also appeared twice weekly in museum galleries, gradually expanding to four regular showings, two during the week and two on weekends.[25] Subjects of these films included travel, history, architecture, and—of course—art appreciation. Noteworthy among them were numerous tours of particular collections or museum wings, designed to augment and mobilize views of museum space, transporting its galleries and collections to a dispersed public. In *A Visit to the Armor Galleries of the Metropolitan Museum of Art*, viewers were treated to slow tours of chain-mail and gothic plate armor displayed on walls and in cases. The Met's filmed tour of its American Wing highlighted housing design and interior decoration. A prominent feature of the museum, the American Wing opened in 1924, to much national acclaim in the popular media and women's magazines.[26] Building on this, the forty-five-minute film surveyed the halls of the wing in chronological order, emphasizing important furniture and tapestry details using close-ups and long takes to punctuate and dramatize them. The film provided a "delightful introduction to the galleries" for those unable to make the pilgrimage to see the real thing.[27] The museum's film program can be thought of as one element in a larger strategy to animate museum space and to implicate the museum and its contents in a wide range of other discourses and modes of popular leisure: magazines, movies, advertising, shopping.

The Met films also reflect the size and scope of museum activities. For instance, other films commissioned by the museum emphasize less the quaint and feminized link between museum and home, but more the grand and problematic relations of the Met to the history of imperialism. *Firearms of Our Forefathers* promised a sweeping overview of weaponry owned by the museum "from the Indian's Bow and Arrow to the Modern Machine Gun." Some of these films constituted documents of museum excavations (and colonialist thievery), which were then edited into educational tours of, for example, Egyptian digs. Two such films were made of the Tutankhamen excavation, feeding what amounted to an international Tut craze. Other films were narrative shorts that used objects held by the museum as storytelling devices (i.e., Greek myths told by markings on a vase). Still others furthered the museum's more practical art education programs, instructing would-be students in the techniques of making pottery. Each type of film was designed to attract a different segment of its perceived audience. Despite or perhaps because of this range, the film collection grew throughout the 1920s and 1930s, resulting in the publication of a full cinema catalog in 1940.[28]

How were these films understood within the museum itself? By 1926, as the film program formed, motion pictures were being credited by a small but increasing number of museum officials and educators with paving the way to the very future of the art museum. The Met's bulletins hosted occasional celebrations of cinema as a necessary strategy to enliven art and to expand the museum's audience. The Met educational staff prophesied that "the motion picture screen can do for art as much as it has done for the drama, and even more."[29] Like in the natural history museum across the park, film was imagined not only to make dead fossils come to life, but to embolden the most basic mission of the museum, in this case, to make art "live again." Further, the reduction of these films from 35mm to 16mm in 1928 enabled an expanded audience both inside and outside of the museum. *Inside* the museum, reducing film prints to the smaller 16mm gauge allowed portable projectors using nonflammable acetate film to be set up in classrooms, galleries, and halls not equipped with the comparably prohibitive apparatus of 35mm. This granted the museum greater programming flexibility as well as conceptual latitude. For instance, a learning laboratory and theater was planned that invited museum goers to choose a film upon entering, and to project the film more than once if need be, "to fix it more firmly in memory."[30] These images depicted things both far and near, past and present. In short, the Met designed a controlled yet interactive learning center, constituted by an archive of images projected on demand.

It was also very clear in this same literature that these films were not intended simply to bring motion and an afterlife to ostensibly dead art objects. These films were also expected to bring the same to the museum, expanding the reach of its collections and curatorial practice to "places heretofore untouched."[31] Films served as a delivery system transporting images of paintings and sculptures as well as lessons in art interpretation designed and orchestrated by the museum.[32] Using a method of instruction deemed more relevant to modern times, the museum reaffirmed and extended its museological influence, making its films available to a growing but privileged and specialized film circuit that superexisted commercial cinemas. Taking root in affluent homes as well as a range of other institutional sites—museums, libraries, colleges and universities, art societies—the 16mm network represented the latest innovation in visual technologies and display to a select but expanded museum audience.

The frequently unspoken subtext of the museum's experiments with cinema was the relatively populist commercial movie theater, a phenomena that the Met could no longer ignore as it planned its reformist activities. Established along with the museum's film library was a standing Committee on the Cinema. In support of this committee, museum trustees issued statements that acknowledged two basic facts: the utility of the motion picture to art education and the cinema's influence on the recreational habits of potential museum goers.[33] In other words, the commericial movie theater served as a "hidden dialectic" in the museum's film projects.[34] In response, administrators designed what they believed was a radically distinct and reputable form of cinema, transforming film into a carefully controlled and edifying museological form of exhibition—a new kind of movie theater. Poised as a direct correction to the popular movie house, museum movie theaters signaled not only that the Met sought to continue distinguishing itself from lesser forms of leisure, but that the business of film, the commercial movie theater, and the unruly moviegoing crowd were among the museum's prominent antagonists. Huger Elliot, head of Museum Education, declared in 1926: "Thousands of people are looking for motion pictures which have artistic, instructive, and entertaining value, but the average motion picture producer does not show such films. He has overlooked his latent public and has sought to please only those who enjoy the type of pictures now generally shown, with their cheap love scenes and impossible plots."[35] Although Elliot officially derided popular melodrama and Hollywood formula, it is also clear that in preaching against commercial cinema he was admitting a certain defeat. Far more people were habituated to moviegoing than to the museum. Swallowing a bitter pill, the Met integrated films into its regular operations, attempting to craft its own kind of countercinema, a good or *useful cinema*.[36]

To be sure, film was but one of the many modes by which the Met struggled to maintain its relevance. The Met also actively supplied radio content early on in the development of the medium; in turn, radio rapidly became yet another standard element of museum outreach.[37] By 1928, weekly museum talks were broadcast on WOR, Wednesdays at 7:30 P.M., a prime family listening spot. Discussions about the museum itself were as prominent as discussion of the art curators sought to illustrate. Program topics encompassed current exhibitions and objects from its permanent collection: Egyptian Art, Armor of Famous Men, Art in Daily Use. Radio topics also included "General Introduction to the Museum" and "The Collections and How They May be Studied."[38] These radio spots grew in frequency. By 1929, the Met's annual reports list fifty-five radio talks delivered under the aegis of the Department of Educational Work.[39] And, by the middle of the 1930s, there were two to three such radio talks each week, broadcast by three different radio stations. Like other museums of the period, the Met, rather than reject the new and emergent mass media, embraced it for its ability to expand its influence through the ideals of educational programming. This entailed some short-lived and highly experimental modes of curation. For instance, in 1929, local high school classes viewed select images from the Met's circulating library of lantern slides. Scheduled to coincide with the class's meeting time, information about the projected images was broadcast by radio on a local educational channel (WNYC).[40] The Met literally entered the classroom by air, coordinating sounds traveling through the ether with projected images shown in spaces far removed from the museum itself. Such experiments had precedents in other museum-media undertakings.

Museums and newspapers, for instance, enjoyed a long courtship. On the one hand, museums of science and art held an ample supply of inexpensively reproduced images imbued with the aura of the venerable institution: quality content. On the other hand, newspapers had a comparably vast, dispersed public that daily demanded word of what was new in the immediate and wider world. Relationships of mutual benefit developed. During the 1920s, these culminated in attempts to coordinate newspaper schedules with radio as well as museum programming. For instance, Chicago's *Daily News* called these "radio photologues" and Buffalo's *Courier Express* called them "Roto-Radio Talks." In newspaper picture pages and in advertisements announcing special picture-page features, listeners were directed to look at particular images in the Sunday paper while the radio broadcast unfolded; sound effects as well as educational lectures augmented and transformed the otherwise blurry images. For instance, jungle noises were added to images of wild animals.[41] Or biographical information was supplied about an artist. A new form of virtual museological curation became apparent, coordinating airwaves, newspapers, museums, and a geographically expanded audience relocated to urban and suburban homes.

These experiments are closely related to the frequent reproduction of artworks on special photo pages in newspapers. Sometimes these art images occupied an entire page. At other times they shared space with bathing beauties, pie-eating contests, coronations, and presidential speeches. For its part, the Met regularly collaborated with newspapers. Reproductions of its art appeared often in the *New York Times* on the paper's picture pages or in its rotogravure section. Such images appeared as news of important acquisitions or as announcements of noteworthy exhibits. In 1924, an ongoing series of art reproductions was printed, with individual pieces selected by Edward Robinson, then-director of the Met. The grandiose series was entitled "From the Renaissance to the Present." In advertisements for the museo-newspaper exhibit, the paintings were described as representative of the Met's collection, "all varying in theme, but all bearing the mark of genius." Apparently important to the endeavor, the ad also announced that the masterpieces would appear in a "form suitable for framing," further indicating the museum's and the newspaper's deliberate attempt to transform these seemingly ephemeral images into domestic objects suitable for home decoration.[42] Magazines also participated. The omnipresent *Ladies Home Journal* regularly reproduced color versions of "great masterpieces" in its monthly issues, to be framed for the home. During the 1930s, many of these publications also sold higher quality versions of its published images as an ancillary business. The *New York Times*, for instance, made the photos published in its picture pages available for purchase "for framing or other private use," advertising its service in the paper itself.[43] Department stores also contributed to this vernacular art practice. Macy's established an exclusive arrangement with *Vanity Fair*, a then-new magazine that strongly advocated for basic art literacy, to sell art reproductions that appeared in its magazine, framed or unframed, singly or in a set.[44]

Museum-media collaborations also include early experiments with television that began as early as television itself. The year of the medium's debut at the 1939 World's Fair, the Museum of Modern Art (MoMA) broadcast the grand opening of its signature building on West 53rd Street.[45] The Met followed thereafter with didactic art programming during the late 1940s. Early tourism programs also featured museums. For example, "Treasures of New York" provided live televisual tours of New York area museums during the 1951 season.[46] Further, as Lynn Spigel has recently shown, MoMA actively engaged the potential of television during the whole of the 1950s, orchestrating and planning new genres of programming.[47] Indeed, museums were at the forefront of several television innovations. In 1954, NBC conducted its first color-compatible broadcast. The show was titled "A Visit to the Metropolitan Museum of Art" and featured its newly renovated building.[48] Collectively, these ventures into reconceptualizing the museum as a genre of cultural programming, fully integrated around mass media and the American home, represent the genealogy of what is now a commonsense aspect of contemporary art institutions. Museum publicity events, spiraling Web sites, satellite museums, mall-based retail outlets, television series, and countless coffee-table books now punctuate this foundational fact of museum authority and identity. Each of these reorients museum address away from an abstract idea of "the public" and *toward the individualized and mobilized consumer.*

The Museum's New Shopper and Her Home

Using reproductions to extend the reach of the museum was more important for art museums than for museums of science and natural history because they made the wider dissemination of otherwise fragile and expensive originals possible. Yet beyond democratic access or benevolent outreach, circulating reproductions was also always one element of a wider museum effort to establish legitimacy and authority, as well as generate profit. Almost as soon as it formed in 1870, the Met produced and distributed etchings of its fledgling collection to European museums as a way to garner recognition for its endeavors. It also sold these same etchings to members and nonmembers of the museum. As early as 1874, museum officials contracted with a local photographer both to make comprehensive records of the art it held and to sell copies of this art, taking a small profit after costs. The Met displayed these reproductions in the museum from this period onward, often in varied sizes, alongside art catalogs and folios, all of which were for sale. Prices ranged from 50 to 75 cents.[49] These photographs were understood by museum administrators as methods by which to circulate the museum's holdings widely and efficiently, serving the purposes of education (popular and specialist). The photographs also provided international publicity for museum holdings, advertising its collection to other museums, galleries, scholars, collectors, the press, and to consumers who might either visit the museum or obtain reproductions for personal collections or home decoration.[50]

In an effort better to orchestrate its production and distribution of art copies, the museum established its own in-house photography studio in 1905, fully institutionalizing the medium as both a bureaucratic tool and as an instrument of circulation, distribution, and display. Internal reports indicate great pride in the size and accessibility of the growing photograph collection, registering the belief that it was the largest program of its kind in the world.[51] By 1908, sales were brisk enough that a permanent sales shop (rather than sales desk) was installed to better organize and display the goods.[52] The shop featured photographs (framed and unframed), casts, folios, lantern slides, and catalogs generated from the museum's collection.[53] The number and type of available items grew to include postcards, calendars, and notepaper. Color reproductions were sold as early as 1920. As the museum's retail function expanded, the original space of approximately 400 square feet tripled in 1926. This expansion allowed for increased and improved merchandise display. It also ushered in a way of talking about museum operations that was relatively new to museum discourse, one that indicates a gradual conflation of the museum goer with the shopper.

By no means dominant but certainly visible, during the 1920s a small portion of the museum's internally generated discourses began to resemble those ascendant

FIGURE 9.2 *H. W. Kent, secretary of the Metropolitan Museum of Art, tells the readers of* Woman's Home Companion *that "we cannot relegate art to the museum alone any more than we can consider language as suitable only to grammarians." From* Woman's Home Companion *56 (January 1929): 11.*

in the world of advertising and retail. In 1924, the Met explicitly coarticulated its reproductions with the idea of gift giving, issuing a sales flyer that read,

> Reproductions, color prints, etchings and woodblock prints, books, casts, the Museum Calendar—good things in good taste—make inexpensive presents. . . . The Museum sells all of these, some of the pictures

framed in attractive styles, and you will find them displayed to great advantage at the Information Desk at the Fifth Avenue entrance, or if you prefer, it can send you what you wish to you and your friends.[54]

Museum offerings continued to grow. In August 1927, the Met's bulletin announced that available postcards had grown to number eight hundred. Some reproduced paintings; others, sculptures. Some illustrated prized museum wings such as the American Wing.[55] The collection's service to students, collectors, and appreciators of art was also foregrounded. The size and comprehensiveness of the collection were seen as foundational virtues; the wide selection and easy availability to the shopper rang loud and clear.

The invitation to shop at the museum was also linked early on to the holiday shopping season. Calls to purchase museum products escalated in November and December. In 1928, the widely distributed museum bulletin reminded readers that "Christmas remembrances of many kinds are waiting for the shopper at the Museum Information Desk."[56] This included images that thematically related to Christmas and thus could be used to decorate homes in the holiday spirit. The bulletin also announced holiday gifts, which had grown to include the previously mentioned framed reproductions and art books, as well as note cards and stationery sets, and the ever-popular annual museum calendars, which contained a different art reproduction for each month. For those who could not travel to the museum, a small pamphlet that functioned as a shopping catalog entitled "Christmas Suggestions" was promised upon request so that would-be shoppers would not be hindered by geography. They could shop at the museum by mail.[57]

By 1933, the pitch to increase museum sales became even more explicit as the museum poised itself against the din and disorder of conventional retail spaces, encouraging consumers to shop at the museum to "avoid the crowds" of other stores.[58] Plenitude and selection continued as standard ways in which merchandise was described. "Shelves and counters full of cards, calendars, prints and books" sat tastefully in waiting.[59] Uniqueness and distinction of merchandise also emerged as salient features.[60] Perhaps most indicative of the rising importance of retail for the museum's ideas about distribution and education was the renaming of the Information Desk. In addition to providing pamphlets, memberships, and advice on viewing the museum collection, it also served as a sales kiosk. As of 1939, the Information Desk was thus referred to in the monthly bulletin as the "Information and Sales Desk": "Questions answered; fees received; classes and lectures; copying, sketching and guidance arranged for; and directions given. The Museum publications—handbooks, color prints, photographs, and postcards—are sold here."[61]

The preceding is a rough sketch of a slow but sure tendency toward the expansion of museological space, from civic theater to shopping center. Museum art was being integrated with a range of cultural rituals and social structures. Images of art marked the passing months in the museum calendar and participated in Judeo-Christian rituals such as holiday gift giving. Museum officials presented its reproductions as elements of everyday knowledge through radio programs and daily newspaper articles. Museum art thus became a constitutive component of home entertainment, home education, and home decoration. The museum itself was figuratively and literally a place to shop.

Attempts by museums to insert art lessons into daily life proliferated throughout the 1920s and 1930s. The Met and others worked to exert influence over American taste, to maintain relevance as a cultural authority, and to secure a place on the increasingly complex stage of cultural value. Magazines, newspapers, radio, and cinema all became relevant for thinking both about the site of the museum and the modes by which the institution obtained its status as a commonsense idea. To be sure, what we think of as a museum had changed in structure and form. Yet the museum was also changing in content. Responding to the promises of the machine age, the marriage between the industrial and the aesthetic further blurred the boundaries demarcating the spheres of retail and the museum. Cemented and effectively popularized by the *Expositions des Arts Décoratifs et Industriels Modernes* in Paris (1925), the merger of design with industrial methods provided an international stage on which art eschewed handicraft and ornamental detail, embracing sleek lines and bold geometry. The objects this event yielded also became more widely available to an increasingly affluent middle class of consumers.

Despite America's nonparticipation in the 1925 *Expositions*, a bold endorsement of its modern styles was amply evident in the U.S. media. Women's magazines, long the site for promulgating the mutual entanglements of the home, consumerism, and good taste, energetically forwarded the emergent generation of dining room furniture, electric lights, and bedroom ensembles. For its part, the surge of interest in the modern conjoined happily with what was the Met's long-standing interest in industrial and decorative arts. As early as 1917, this affinity became explicit when the museum developed direct relations with American manufacturers, conducting annual displays of industrial design throughout the latter half of the teens and 1920s. In response to the Parisian *Expositions*, the Met selected and exhibited four hundred objects from the Paris show, which subsequently toured eight American cities.[62]

The ascendance of modern design in popular discourse provided a natural continuation of Met efforts to reach a segment of the public it deemed particularly important to ensuring art's place in daily life: women as homemakers, consumers, and as retail workers. These efforts took several forms. From January 1914 onward, the museum orchestrated a regular lecture series for department store buyers to shape their acquisition practices and for salespeople to help them counsel their customers. By 1917, these courses were also offered on weekends. Eventually classes in art history and design included Friday morning lectures. These were given in coordination with retail organizations, which allowed their department store clerks to leave work to attend. Special study rooms were designated at the museum; individual lectures targeted salespeople as well as artisans and manufacturers. Some classes were formulated explicitly for "girls and homemakers."

The education staff at the Met understood that if it was to exercise the widest possible influence and respond to changes in the most basic definition of art, that influencing retail environments and consumer practices was essential.

Thus museum professionals began writing advice columns and general art tracts in a range of women's and design magazines. Serving as museum publicity, crafted as a consumer-friendly form of education, such magazines provided lessons in modern design and advice on purchasing. H. W. Kent, secretary of the Met, serves as an example. In 1929, he authored an article in *Woman's Home Companion*, entitled "Modern Art Makes Itself at Home." Alongside ads for electrical appliances and home furnishings, Kent acknowledged the confusing and unfamiliar world of modern design. All the while he implored the housewife/consumer to accept her responsibility as the guiding force behind a revolution in the arts. He wrote,

> America is on the verge of the most remarkable, and in many ways the most promising, development in the long history of art, and the women of America will largely determine at the counters of the department stores and other retails shops exactly where this development is to lead and how fine a contribution it will make to the evolution of art.[63]

Addressing women magazine readers explicitly as consumers, brandishing the authority of the museum expert, Kent offered a kind of dramatic crisis in contemporary culture. It was not one in which consumerism had failed the museum, encroaching irretrievably on its sacred autonomy, but one wherein the museum might in fact fail the consumer. Indeed, the very mission of the museum had been reframed: to provide guidance for the housewife-shopper. It was this feminized shopper who was burdened with the ultimate responsibility for nurturing the next phase of art's forward movement, one that embraced the copy, the machine, and the everyday. The museum therefore both ceded and claimed authority, openly collaborating with industrial designers and department stores while assuming benevolent stewardship. In short, rather than a threat, consumerism and the expanding world of retail was seen as an opportunity for assuring museum purpose, providing a significantly expanded public created by the vast distribution, advertising, and display systems in place to fuel the buying of things.

It is in the spirit of museum–consumer consort that a surprising range of magazine articles about museums—especially about the Met—and home decoration began to appear. Commenting on industrial art and also supplying images of it, the Met endorsed a whole range of celebratory articles on this particularly modern form and its interface with home design. Articles in publications such as *Magazine of Art, Arts and Decoration*, and *The Architectural Record,* and women's magazines, such as *The Ladies Home Journal, House Beautiful*, the previously mentioned *Ladies Home Companion,* and *Country Life,* materialized. The Met, whether by resort to an idealized American past or to a modern machine age, addressed the American middle-class home as if it were a natural and necessary museum satellite.[64] Saturated with rhetoric that coarticulated good consumerism with the moral individual, a discourse endemic to the contemporaneous better homes movement, the Met explicitly and implicitly forwarded the idea that the moral rectitude of art was a happy complement to the moral rectitude of the consuming housewife.[65]

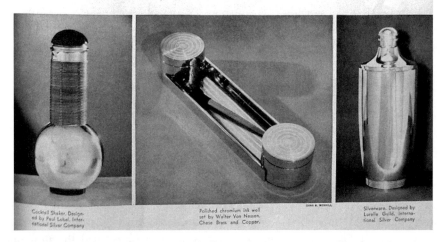

FIGURE 9.3 *The Met actively inserted itself into the ascendant discourses of consumerism and good design, which linked museum art, industry, and home. These select household items held by the museum were featured in* Arts and Decoration 42 *(December 1934).*

Let me provide a final example of the Met's search for retail influence and use of media. In May 1927, the museum sponsored an exposition of modern design held by Macy's, the well-known department store. The pairing was mutually beneficial, simultaneously satisfying the museum's determination to spread its influence and Macy's desire to assert legitimacy as a tastemaker sanctioned by the cultural elite. Macy's provided valuable retail space to museum-like exhibits of design objects. It also offered a lecture hall in which art experts—including those employed by the Met—delivered talks on the history of art and design. The museum supplied objects from its collection and retained visibility in all advertising for the week-long event.[66] The show provided content for thirty thousand printed catalogs, which highlight the Met's prominent role in planning the show by invoking the museum's name on every cover.[67] The daily lectures given at Macy's were also broadcast on the regular "Metropolitan Museum of Art Show" on WJZ, a local radio station. The actual exhibit itself hosted over fifty thousand visitors in its one-week run, an audience many times that which entered the museum's primary site just up the street. Using a dynamic strategy that included newspapers, catalogs, radio shows, and elaborate displays—with voluble museum presence (a practice today we would call branding)—the show's official purpose to link "art with everyday household life" fortified an enduring model both for museums and for retail.[68]

The Met effectively appropriated Macy's as a hybrid extension of the museum itself, reaching an audience considerably larger than otherwise possible. Macy's was simultaneously and perhaps temporarily imbued with museological authority, acquiring credibility as an informed and worldly cultural institution. Other department stores soon followed this lead, curating their own design shows, emulating museological modes, and collaborating openly with willing museums. For their part, art museum administrators and educators seem to have been happy and willing participants. The institution of art and its influence was boldly expanding. This institution did not just include objects of industrial design and consumer culture, but *it embraced their methods of distribution and mediated modes of display* as well. The museum goer was a proud consumer, frequently gendered and tasked with the moral caretaking and design of the better home. The commodified institution of domesticity found a ready ally in the art museum, which increasingly fashioned itself as a producer of good taste and moral consumption, positing its art as auratic everyday objects.

During the 1920s, American institutions of domesticity appropriated discourses of art and discourses of the museum; institutions of art actively appropriated discourses of domesticity. At the Met, methods by which art and the museum might be enlivened entailed using a range of so-called mass media and modes of consumption to cultivate a specific but expanding audience. The Met sent its reproductions and discussion of them to newspapers and women's magazines, radio stations, and department stores. It advertised itself as a gift shop, issuing sales flyers and mini catalogs. In other words, throughout the Met's educational and increasingly normalized consumer-related activities, museum goers—often configured as women—were addressed through a spiderweb of museological address, merging art education, art reproduction, home decoration, and shopping.

Commodifying its art as well as its own cultural authority, the Met and museums afterward continued to foster intimate relations with consumer culture. At MoMA, founded in 1929, these ideas shaped the new modern museum at the cellular level. Embodying the provocations of modernism in its charter, MoMA not only exhibited a range of modern art in its galleries—photography, design, posters, film—it used each of these to sell the idea and the importance of the museum itself. Some of its early catalogs looked more like those of Sears rather than the erudite reference texts we associate with museum publication programs, featuring retail prices and location.[69] Its 1934 Machine Art show took the concept of industrial design and consumer relations one step further than the Met's earlier forays. MoMA put plastic spoons and glass bowls, metal kitchen pots and machine gears, ball bearings, gas pumps and flower vases in its galleries. It placed an airplane propeller on West 53rd Street to help publicize the show. A boldly modern antiart display, the museum's bulletin described the show's purpose: "to provide a practical guide to the buying public."[70] Eschewing the convention of hanging art on walls to create a sense of distance, visitors were encouraged to handle the museum objects, which for the most part were exhibited not at eye level but at hand level. Reviews of the show likened MoMA's galleries to department store displays. The *New Yorker* compared the exhibits to those in a hardware store.[71] Still others reported that museum goers not only touched items but handled them and tested them, checking prices and attempting to make purchases.[72] Not deemed a mistake or a misfire by any means, MoMA's Machine Art inspired exhibits that followed. Beginning in 1938, the museum organized a show entitled "Useful Household Objects under Five Dollars." Orchestrated to augment the Christmas shopping season, the space of the art museum was given over to trinkets and tools arranged first according to retail cost and second according to aesthetic achievement. During the late 1940s and 1950s, it was this show that morphed into MoMA's national "Good Design" campaign. In full collaboration with manufacturing and retailing organizations, Good Design entailed museum sanction of design objects being sold across the country in upscale retail sites. Museum approval was indicated with an orange sticker that was featured prominently on these items.[73] The Good Design program later became a series of objects sold in the ever-expanding bookstore, culminating in the opening of the design store (1989), the museum online store launched in February 2001, and, most recently, the flagship design stores now with three separate New York locations.[74] In short, what began as an active collaboration with retail interests that existed outside of the museum in the 1930s was fully inverted. Some eighty years later, MoMA, the Met, and many other major museums are now self-proclaimed retailers themselves, sometimes even supplying their own branded goods to other museum stores.

New Institutions of Art

The discussion of museum miniatures that began this chapter situates the Met and the Book of the Month Club collaboration as a symptom of the steady and gradual transformation of the twentieth-century American art museum. Undergirding the handheld and mobile museum was a basic fact of twentieth-century American art museums: an increasing interdependence with a range of media forms and systems, and an expanded imbrication in a complex network of other cultural institutions (the press, retail/consumerism, the family, the nation-state). Through a wide range of projects to extend the museum via technological reproducibility and consumer culture, *the museum has long functioned as a tentacular system rather than a singular site*. Crafting discourses of timelessness and the sacral as its core values, the art museum has equally embraced the ephemeral and the prosaic modes of modern life, sending its particular ideologies of old and sometimes new art to spaces irretrievably undergirded by the flux of media flows. In short, to fully understand the art museum-as-space and as institution, we must also acknowledge that this same museum has been radically transformed by technological modernity and consumer capitalism throughout the twentieth century.[75] Particularly true of American art museums, beginning in earnest during the 1920s, institutions such as the Met, and a few years later MoMA, reimagined the museological institution, cementing cultural authority by inserting the museum and its art into a spiraling mediated web of cultural content—lithography folios, radio shows, illustrated magazine features, newspaper picture pages, greeting cards, photographic reproductions, art books, and films. The home represents one of the primary sites in which this amorphous and ephemeral museum was reconstituted, rematerialized, and thus integrated into the structures of everyday domestic life. In short, the museum forged a new frontier, colonizing living rooms, home libraries, recreation rooms, and coffee tables, providing a lasting articulation of the conditions in which art acquires meaning in consumer culture.

The interwar period in the United States marks a paradigmatic shift in the cultures of art. It crystallized ongoing changes to the conditions in which art was seen and circulated, cementing the great dissolve of high and low cultures. Contra Levine, the museum was a full participant not only in consecrating art but also in desecrating it. I am not arguing that the museum was a radical equalizer or an avant-gardist force. Nevertheless, this was a period that witnessed a fundamental reorganization of the museum and its relationship to mass media, which in turn requires us to rethink the institutions relevant to identifying the patterns, meanings, and modes by which art became an intelligible site for mediating twentieth-century life. By the end of the 1930s, the American art museum was fully habituated to the ebbs and flows of mass media and consumer culture. Museum staff merchandised exhibits, advertised the museum as a shopping destination; they also worked happily with department stores. Museum administrators hired full-time publicists and established publicity committees to manage public relations and press coverage. The results included active solicitation of public attention and ongoing extramuseological attempts to shape public opinion. Among the many strategies, art contests, celebrity judging panels, and subway marketing campaigns emerged.[76] Publicists also issued daily press releases, supplied illustrations to magazine stories about art and their exhibitions, gave away prizes, and polled audience opinion. In attempting to assert their relevance and maintain their authority, traditional and modern art museums infused the everyday with its art, a distorted example of the avant-garde's challenge to art tradition. Andreas Huyssen has written that postwar consumer culture rendered the avant-garde's plea for the elision between art and the everyday obsolete through its mass reproduction and circulation of abundant commodities.[77] If this is true, the interwar museum provides a primary mechanism for assessing one of the ways in which the proposition of aesthetic-politico revolution transformed into radical consumerism and hyperacquisition.

At first glance, there seems something fundamentally contradictory between the prosaic and ephemeral nature of the newspaper or radio and the sacral timelessness of the museum. Yet, throughout this period, museums, art, and media forged distinct relationships that complicate—among other things—the temporalities of art. Through their activities they demonstrated that the museum's ideal of timelessness was accompanied by a much more modern time ideal. The museum has long adopted the rapid pace and stable grids that shape everyday time in general: the twenty-four-hour clock, the five-day workweek, the tourist, holiday and shopping seasons. Museums may have continued to borrow the signs of endurance through time, but they did so in part to counteract the acceleration and diversification of time that was not just happening around them but inside and through them. The American art museum was increasingly not about the *then* of art but also about the *now* of modern commercial leisure, the time sensitivity of news, and precision grids of cultural programming. The resulting ephemeral museum was neither antithetical to nor discordant with the rhetoric of timelessness and endurance forwarded so frequently in museum literature. In short, the interwar period witnessed considerable *integration of museum time with media time*. Thus the ideal of timelessness must be paired with the realities of the ephemeral, the temporary, and the ever-changing schedules of publishing, programming, and broadcasting. In short, it's clear that the enduring value of art propagated by the museum also found a home in ephemeral media and modes of circulation whose materiality enacts a notably unmuseological relationship to art objects, refashioning art reproductions as participants in everyday and domestic life.

The affluence of the 1920s, combined with shifts in media technologies and the ascendance of advertising, mark a paradigmatic shift for museums that helped pave the way for the populist ethos of the 1930s. Such shifts also clearly predict the arrival of a multimediated museum like MoMA, which integrated a range of modern forms as content and as structure. Subtending all of this is the ongoing question of art's function and value under capitalism, and the ways in which art as a mobile commodity both acquires, maintains, and changes value throughout its life span. The museum has long been a full participant in these transformations, brokering the metamorphosis of its art from unique original to fridge magnet, from museum gallery fetish to the frayed edges of a tourist's postcard.

To better understand this museum, we must accept the fact that it is indeed a highly and purposefully mediated museum. The Met, the MoMA, the Louvre, the British Museum—among many others—have each achieved a kind of symbolic presence in our everyday lives as places where matters of significance and art of importance are always happening. Consumerist mechanisms have been part and parcel of this ongoing process for the better part of the twentieth century—especially in American art museums—providing a sober reminder of the challenges to making spaces for art free from market concerns. Fully assessing the fantastical, the democratic, the regulatory, and the prosaic nature of this proposition will do much to help us critically assess the new and the old, the enduring and the ephemeral of the museum and its art.

Notes

1. Contemporary parlance would effectively deem this a phenomena of branding, an example of the museum using its authoritative stamp to endorse objects and activities ostensibly far removed from its primary mandate and usually informed by corporate sponsorship, consumerism, and spectacular display. See, for instance, Michael Sorkin, "Brand Aid," *Harvard Design Magazine* (Fall/Winter 2003): 5–9.

2. For more on the Met's reproduction programs, see Regina Maria Kellerman, *The Publications and Reproductions Program of the Metropolitan Museum of Art: A Brief History* [pamphlet] (New York: Metropolitan Museum of Art, 1981).

3. Frank Magel, Book of the Month Club, "Dear Friend" [letter] (n.d.; ca. 1950), 1.

4. Francis H. Taylor, Book of the Month Club, "Dear Friend" [letter] (n.d.; ca. 1951), 2.

5. For a general history of the Book of the Month Club, see Joan Shelley Rubin, *The Making of Middlebrow Culture* (Chapel Hill: University of North Carolina Press, 1992), and also Janice Radway, *A Feeling for Books: The Book-of-the-Month Club, Literary Taste, and Middle-Class Desire* (Chapel Hill: University of North Carolina Press, 1997). For the standard history of the Met, see Calvin Tomkins, *Merchants and Masterpieces: The Story of the Metropolitan Museum of Art* (New York: E. P. Dutton, 1970). For a full articulation of the ways in which taste functions as a stage for class politics, see Pierre Bourdieu, *Distinction: A Social Critique of the Judgement of Taste* (Cambridge, Mass.: Harvard University Press, 1984).

6. Pamphlet insert, Metropolitan Museum of Art Miniatures, Album H (property of author).

7. See, for instance, "Metropolitan Miniatures," *Recreation* 41, no. 11 (February 1948): 538.

8. Andreas Huyssen provides one example of this idea in arguing that the modern museum is a symptom of the perceived loss of tradition and material certainty. See Andreas Huyssen, "Escape from Amnesia: The Museum as Mass Medium," in *Twilight Memories: Marking Time in a Culture of Amnesia* (London: Routledge, 1995), 13–35.

9. The Victoria and Albert Museum (V & A) is the quintessential example of this latter museum type. Established in 1857 and inspired by the success of the Great Exhibition of 1851, held in the Crystal Palace (London), the V & A was intended to advance the commercial life of Great Britain by providing a space wherein design objects (industrial and artisanal) could be displayed to stimulate the superior manufacture of such objects by British designers and manufacturers. The decorative arts were thought to be more accessible to the working classes, who were also offered education in the aesthetics of the everyday. This type of museum should be distinguished from those that explicitly designed themselves against industrial and consumer culture by offering grand quiet spaces in which contemplation of serious art, usually from the past, was idealized. For more on the influence of these two basic museum models on American museums, see Carol Duncan, "Museums and Department Stores: Close Encounters," *High High-Pop: Making Culture into Popular Entertainment*, ed. Jim Collins (BFI: London, 2002), 129–54. For a general history of the relationship between museums and the world of retail, see William Leach, *Land of Desire: Merchants, Power, and the Rise of the New American Culture* (New York: Pantheon Books, 1993).

10. Eilean Hooper-Greenhill argues that that these deeply contradictory functions are endemic to the public museum, born of the republicanism emergent with the French Revolution. These museums represent a sharp turn away from the powers of the cleric, the king, and the aristocracy and toward the functions of the modern state and its needs to manage populations. See Eilean Hooper-Greenhill, *Museums and the Shaping of Knowledge* (London: Routledge, 1992), 167–90.

11. Lawrence W. Levine, *Highbrow/Lowbrow: The Emergence of Cultural Hierarchy in America* (Cambridge, Mass.: Harvard University Press, 1988).

12. See ibid., 150–68, 184–86.

13. See Tony Bennett, *The Birth of the Museum: History, Theory, Politics* (London: Routledge, 1995).

14. Daniel J. Sherman and Irit Rogoff, "Introduction: Frameworks for Critical Analysis," in *Museum Culture: Histories, Discourses, Spectacles*, ed. Daniel J. Sherman and Irit Rogoff (Minneapolis: University of Minnesota Press, 1994), ix–xx.

15. For but one example of the links between the art museum and mass media, see Lynn Spigel, "Television, The Housewife, and the Museum of Modern Art," in *Television after TV: Essays on a Medium in Transition*, eds. Lynn Spigel and Jan Olsson (Durham, N.C.: Duke University Press, 2004), 349–87.

16. Grace Fisher Ramsey, *Educational Work in Museums of the United States* (New York: H. W. Wilson, 1938), 111. See also Alison Griffiths, *Wondrous Difference: Cinema, Anthropology, and Turn-of-the-Century Visual Culture* (New York: Columbia, 2002).

17. By 1915, this collection had grown to include 34,219 slides. This collection was lent to 876 borrowers in its first four years. Ramsey, *Educational Work in Museums*, 171–73. See also Grace Fisher Ramsey, "The Film Work of the American Museum of Natural History," in *Journal of Educational Sociology* 13, no. 5 (January 1940): 280–84.

18. Andrew McLellan, "A Brief History of the Art Museum Public," in *Art History and Its Institutions: Foundations of a Discipline*, ed. Elizabeth Mansfield (NewYork: Routledge, 2002), 19.

19. *Annual Report to the Trustees*, The Metropolitan Museum of Art (1925), 44.

20. *Annual Report to the Trustees*, The Metropolitan Museum of Art (1935), 51.

21. Alison Griffiths addresses similar attempts at the Museum of Natural History in this volume. See chapter 5.

22. "Cinema Films," *Metropolitan Museum of Art Bulletin* 20, no. 1 (January 1925): 2.

23. From 1923 forward, 16mm emerged as the standard gauge for extratheatrical film exhibition. The Met's entry in the field is early and suggests the emerging common sense both of the 16mm format and the idea that institutions seeking to keep current needed to respond to and integrate cinema into operations. For more on the importance of 16mm and cultural institutions during this period, see Haidee Wasson, *Museum Movies: The Museum of Modern Art and the Birth of Art Cinema* (Berkeley: University of California Press, 2005), especially chapter 3.

24. The Met also acted as distributor and exhibitor for films made by Yale University Press, entitled "The Chronicles of America Photoplays." In 1925, this entailed 15 films of a planned 33. Each covered a topic in American history, for example: Columbus; The Pilgrims, The Puritans; Alexander Hamilton; The Frontier Woman.

25. Charge for these films was $5 per reel. Converted into 2005 dollars, this equals approximately $56 per reel. "The Chronicles of American Photoplays," *The Metropolitan Museum of Art Bulletin* 20, no. 7 (July 1925): 186.

26. For instance, *Ladies Home Journal* declared the wing proof that "Americans are a home-loving race." Ethel Davis Seal, "The American Wing of the Metropolitan Museum," *Ladies Home Journal*, May 1925, 20.

27. "A New Museum Film," *Metropolitan Museum of Art Bulletin* 30, no. 7 (July 1935): 150.

28. Metropolitan Museum of Art, *Cinema Films; A List of Museum Films and Others, with the Conditions under Which They Are Distributed* (New York: Metropolitan Museum of Art, 1940).

29. Huger Elliot, "The Museum's Cinema Films," *The Metropolitan Museum of Art Bulletin* 21, no. 9 (September 1926): 216.

30. "Accessions and Notes," *Metropolitan Museum of Art Bulletin* 23, no. 9 (September 1928): 222.

31. Elliot, "The Museum's Cinema Films," 216.

32. This affirmed that "quiet contemplation of a thing of beauty" remained the keynote of aesthetic pleasure, and that the motion picture had little in common with this task. Its value to the quiet contemplation of art resided in its service to the art object, providing it a secure vessel and thus safe passage into the world. Ibid.

33. "Metropolitan Has Produced Motion Picture Films," *Museum News* 1, no. 5 (March 1, 1924): 3.

34. I am purposefully referencing Andreas Huyssen's phrase in his well-known book *After the Great Divide: Modernism, Mass Culture, Postmodernism* (Bloomington: Indiana University Press, 1986).

35. Elliot, "The Museum's Cinema Films," 216.

36. I am borrowing here from Tony Bennett's paradigmatic formulation of culture as useful. See Tony Bennett, "Useful Culture," *Cultural Studies* 6, no. 3 (1992): 395–408.

37. For more on this see Rubin, *The Making of Middlebrow Culture*.

38. *Metropolitan Museum of Art Bulletin* 23, no. 2 (February 1928): 59.

39. *Annual Report to the Trustees* (New York: Metropolitan Museum of Art), 27.

40. *Metropolitan Museum of Art Bulletin* 24, no. 8 (August 1929).

41. Ramsey, *Educational Work in Museums*, 195–97. See also "Museums and Newspapers Cooperate in Roto-Radio Talk," *Museum News* 6 (April 15, 1929): 1; "Lantern Slides Supplement Museum Radio Talk," *Museum News* 6 (November 1, 1929): 2; "Illustrated Talks by Radio," *Museum News* 1 (June 15, 1924): 2; "New Automatic Motion Picture Projector Installed in Three Museums," *Museum News* 9 (February 1, 1931): 4. Richard F. Bach, "Neighbourhood Exhibition of the Metropolitan Museum of Art," *Museum News* 15 (September 1, 1938): 7–8.

42. "Masterworks Reproduced," *New York Times*, February 24, 1924, E7. The *New York Times* advertised its Sunday picture pages throughout the week, addressing "lovers of art" and scrapbook makers with high-quality reproductions of art "well worth saving."

43. November 5, 1933 Display ad: "Photographs reproduced in The New York Times Sunday Rotogravure Picture Section and other portions of the paper, as well as in Mid-Week Pictorial, credited to The Times Wide World Photos, may be obtained on application to the Sales Manager, Times Wide World Photos, 229 West 43rd Street, New York, NY. Prices for prints for framing or other private use (not for publication) are: Glossy 8x10, $.50; mat finish, 8x10, $.75; 11x14 glossy or mat finish, $1.00, plus postage" [advertisement]. A dollar in 1933 equals approximately $15 in 2005. *New York Times*, November 5, 1933, E7.

44. [advertisement]. *New York Times*, May 5, 1935, X7.

45. For further discussion of this, see Spigel, "Television, the Housewife, and the Museum of Modern Art," 349–87. For a fuller discussion of MoMA and its relationship to mass media in the 1930s, see Wasson, *Museum Movies*, 2005.

46. For more on this, and the Met's refusal to participate in this particular program, see Saul Carson, "On the Air," *New Republic* 124 (June 25, 1951): 30–31. For contemporaneous writing on the Met and television, see Gilbert Seldes, "Television and the Museums," *Magazine of Art* 37 (May 1944): 178–79; and on the place of modern art on television during the 1950s, see Lynn Spigel, "High Culture in Low Places: Television and Modern Art, 1950–1970," in *Welcome to the Dreamhouse* (Durham, N.C.: Duke University Press, 2001), 265–309.

47. See Spigel, "Television, the Housewife, and the Museum of Modern Art."

48. Cited in Spigel, "Television, the Housewife, and the Museum of Modern Art," 367.

49. Indexed to 2005 U.S. dollars, this converts to a price range of $10.83 to $16.25.

50. The fact that some of these photographs were framed indicates their likely use as decorative objects rather than reference copies. For more on the Met's reproductions program, see Regina Maria Kellerman, *The Publications and Reproductions Program of the Metropolitan Museum of Art: A Brief History* [pamphlet] (New York: Metropolitan Museum of Art, 1981).

51. Ibid., 9.

52. Ibid., footnotes 9, 12. The museum was also an active producer of plaster casts, opening its own atelier in 1895, circulating them to other museums, schools, and libraries from 1897 onward.

53. The museum developed a relationship with other retail sites early on as well. In 1877, it had contracted with Tiffany's, allowing the jewelry company to make copies of the items held by the museum for resale. Tiffany's exhibited these at the Paris Exposition of 1878, selling the objects at its stores and to other museums. It featured these items in its sales catalog as well. For more, see ibid.

54. "Museum Suggests Gifts," *The Museum News* 1, no. 1(January 1, 1924): 2.

55. "Postcards," *Metropolitan Museum of Art Bulletin* 22, no. 8 (August 1927): 219.

56. "Christmas Suggestions," *Metropolitan Museum of Art Bulletin* 23, no. 11 (November 1928): 280, 282.

57. "To those who cannot come to the Museum to shop or who wish to have a more detailed account of our holiday wares, the little booklet, Christmas Suggestions, will be sent upon request." Ibid., 282.

58. "For Christmas Shoppers," *Metropolitan Museum of Art Bulletin* 28, no. 11 (November 1933): 200–1.

59. Ibid., 201.

60. *Metropolitan Museum of Art Bulletin* 33, no. 11 (November 1938) offers "unusual Christmas gifts." "These make a festive display at the Information Desk, a convenient place for Christmas shopping, and a list of them, with illustrations of the cards, may be obtained on request by anyone who wishes to order by mail" (254). A 1940 bulletin also announced items whose prices had been significantly discounted. In short, the museum started having sales. *Metropolitan Museum of Art Bulletin* 35, no. 11 (November 1940).

61. *Metropolitan Museum of Art Bulletin* 24, no. 3 (March 1939): 76.

62. Marilyn F. Friedman, *Selling Good Design: Promoting the Early Modern Interior* (New York: Rizzoli, 2000), 16.

63. H. W. Kent, "Modern Art Makes Itself at Home," *Woman's Home Companion* 56 (January 1929): 11.

64. "A Parade of Contemporary Achievements at the Metropolitan Museum," *Arts and Decoration* 42 (December 1934): 12–25; Richard Bach, "A Museum Exhibit as a Spur to Industrial Art," *Arts and Decoration* 18 (February 1923): 10, 62; Ethel Davis Seal, "The American Wing of the Metropolitan Museum," *Ladies Home Journal*, May 1925, 20, 21, 188, 191; Reginald T. Townsend, "The Metropolitan Museum of Art and Your Home," *Country Life* 40 (September 1921): 65–66; "The House in Good Taste," *House Beautiful* 57 (January 1925): 33–36.

65. Janet Hutchison, "The Cure for Domestic Neglect: The Better Homes in America, 1922–1935," *Perspectives in Vernacular Architecture II*, ed. Camille Wells (Columbia: University of Missouri Press, 1986), 168–78.

66. Friedman, *Selling Good Design*, 18–19.

67. *Catalog of the Exposition of Art in Trade at Macy's* (New York: Macy's, 1927).

68. Friedman, *Selling Good Design*, 26. This program continued, and one source estimated that in 1928 some 250,000 people had wandered through Macy's exhibit on art and industrial design (7). For a fascinating account of other contemporaneous collaborations between department stores and the Newark Museum, as well as the related progressive ideas of that museum's director, John Cotton Dana, see Duncan, "Museums and Department Stores."

69. See *Machine Art* (New York: Museum of Modern Art, 1934). For fascinating discussions of MoMA and their efforts to promulgate modern design in the 1930s and later, see Felicity Scott, "From

Industrial Art to Design: The Purchase of Domesticity at MoMA, 1929–1959," *Lotus International* 97 (1998): 106–43; Mary Anne Staniszewski, *The Power of Display: A History of Exhibition Installations as the Museum of Modern Art* (Cambridge, Mass.: MIT Press, 1998). See also Edward Eigen and Terence Riley, "Between the Museum and the Marketplace: Selling Good Design," *The Museum of Modern Art at Mid-Century: At Home and Abroad* (New York: Museum of Modern Art), 150–79.

70. "Machine Art," *Museum of Modern Art Bulletin* 3 (November 1934): 2.

71. "Talk of the Town: Machine Art," *The New Yorker*, March 17, 1934, 18; See also "Machine Art," *Parnassus* 6, no. 5 (October 1934): 27.

72. See, for example, Jane Schwartz, "Exhibition of Machine Art Now on View at Modern Museum," *Art News* 32, no. 23 (March 10, 1934): 4.

73. The Good Design program involved a complex strategy and included exhibitions, publications, symposia, advertising, consumer opinion polls, and other public programs. Edward Eigen and Terence Riley, "Between the Museum and the Marketplace: Selling Good Design," *The Museum of Modern Art at Mid-Century: At Home and Abroad* (New York: Museum of Modern Art), 152.

74. Like the Met, MoMA early on sold art catalogs, lantern slides, Christmas cards, as well as photographic copies of paintings and objects it held. Sales were conducted by mail and at the main museum Information Desk. By 1941, a full slate of such items was available including two hundred photographs. Gift catalogs circulated in the 1950s. For this brief sketch I am indebted to research notes made available to me through the Museum of Modern Art Archives, prepared by Rona Roob.

75. I am borrowing here from Walter Benjamin's well-known essay "The Work of Art in the Age of its Technological Reproducibility," in *Walter Benjamin: Selected Writings, Vol. 3, 1935–1938*, eds. Howard Eiland and Michael W. Jennings, trans. Edmund Jephcott, Howard Eiland, et al. (Cambridge, Mass.: Belknap Press, 2002) in which he famously grapples with the implications of reproducibility for questions of art under capitalism and fascism. On the compelling nature of museum miniatures, See Susan Stewart's brief but suggestive discussion of them in her wonderful book *On Longing: Narratives of the Miniature, the Gigantic, the Souvenir, the Collection* (Durham, N.C.: Duke University Press, 1993).

76. But one example of the ways in which urban space was used to fortify museum presence was a program launched by the Met in the early 1940s. Color reproductions of select works were posted on subway cars throughout the New York transit system. One estimate suggests that the artworks were seen by a total of 5,638,800 riders. This resulted in 1,023 mail-in orders and 1,600 on-site purchases of the reproductions. The subway program lasted for at least three years and fed the development of more elaborate projects to market reproductions. Museum officials were sufficiently impressed by such sales that they continued issuing series of art prints, reporting 16,008 pictures sold of a painting by Sir Thomas Lawrence in January 1945. See "Subway Prints: Masterpieces," *Magazine of Art* 38, no 3 (March 1945): 112.

77. Andreas Huyssen, *After the Great Divide: Modernism.*

10 Recovering a Trashed Communication Genre: Letters as Memory, Art, and Collectible

Jennifer Adams

At the dawn of the twenty-first century, people accustomed to the communicative speed of technologies like fax and e-mail disdainfully refer to correspondence sent through the postal service as "snail mail." Proponents of electronic communication argue that these messages trump the traditional letter because they arrive instantly, allow for easier response, are stored more rapidly and conveniently, and are more environmentally friendly. Whether or not these new technologies actually contributed to the dwindling use of traditional letters, the fact remains that the frequency of letter writing has been decreasing since the advent of the wire telegraph and the telephone.[1] Literary critic John Brown argues, "The health of the letter has been undermined and finally dealt a fatal blow by the telephone, the telegram, the cassette, the fax, and other technical innovations that have deprived it of its raison d'être."[2] This is not to suggest that the U.S. Postal Service suffers for lack of business; indeed, the opposite may be true. Many of us still find our mailboxes full, but they are full of business solicitations, utility bills, and credit card statements. Rarely do we find letters from loved ones penned by their hand.

Paradoxically, as fewer and fewer individuals use pen and paper to write personal letters, scholars across disciplines seem to become more and more interested in studying and preserving them. This growing interest in the epistolary art is reflected by an increase in both the publication of sets of correspondence written by famous or successful individuals and in the study of these types of letters by scholars engaging in a myriad of subjects from literature to history. The letters of both the famous and the anonymous may be found in archival libraries around the world, and these sources have led to the publication of countless collections of letters that are "read by a public less attracted to fiction than to biography and personal 'revelation.'"[3] As Brown aptly notes, "If the letter is a dying art, the tomb in which the defunct will repose—the elaborate editions of correspondences completes—is certainly an impressive one, and its mourners, largely drawn from the academic community, are legion."[4]

Despite the popularity of the letter as an artifact of analysis in academia, these studies are largely limited in scope. Often, it is not the letter as communicative medium that historians or literary critics find interesting; rather, it is the content or the significance of the author that makes these letters objects of study. Yet there is value in the mere *form* of letters when we consider them as the residue of a once-dominant channel of business and social communication over distance. In an effort to broaden notions of what the study of epistolary can reveal, I explore the role of historical personal letters as an object of memory, art, and collectible. Specifically, I first discuss the parameters of letter writing as a historical communication genre. I then consider the ways that letters contribute historical depth to the collective memories of our culture, redefine our conception of art, and become the objects that populate the collections of individuals and museums. Finally, I argue that any scholarly consideration of the historical personal letter is incomplete without understanding the layered and multifaceted character of the letter as memory, art, and collectible. My goal in this chapter is to expose those lasting qualities of the historical handwritten personal letter as the genre experiences profound change resulting from newer and faster communication and media technologies.

In recent modern history, letter writing was central as a mode of communication between individuals who found themselves separated by geographical distance. Postal historian Wayne Fuller notes, "Few, in fact, are the aspects of American past—military, religious, social, political, economic—with which the postal system has not been connected in one way or another."[5] The significance of the letter to the lives of everyday people in the early decades of the twentieth century was aptly described in a 1936 manual for writing social correspondence:

> Letters make and keep friendships; they bring to our doors invitations, acceptances, and regrets; they carry congratulations and greetings; they express our appreciation for gifts and favors; they make our apologies; they state announcements; they bear sympathy where it is most needed. Indeed, there is hardly an occasion when we do not gratefully turn to the letter as a means of expressing our feelings, opinions, and decisions.[6]

Despite the eventual decline of letter writing, personal letters remain a viable source of historical communication research.

Letters are unique as a genre of mediated interpersonal communication in that they paradoxically result from absence and yet retain presence by negotiating gaps in space and time. Letters are unlike most other relational communication in that the sender and receiver are separated by both geographical and temporal distance. In fact, the single most important factor resulting in the need for communication by personal letter is geographical separation; this was especially true before the advent of other more timely and efficient methods for long-distance communication. Cicero noted this unique quality in 53 B.C. when he wrote, "Letter writing was invented just in order that we might inform those at a distance if there were anything which it was important for them or for ourselves that they should know."[7] Further, unlike more contemporary communication technologies designed to conquer distance, letters span a period of time between when they are written and mailed and when they are received and read.

Yet despite the geographical and temporal isolation of the writer and reader of the private letter, their presence is represented by it. Even the earliest known letters, those composed in ancient Greece and Persia ca. 500 B.C., were said to embody the presence of the sender. During that era, personal messengers of an authoritative or powerful person delivered letters, and these messengers were said to carry the *parousia*, or projected presence, of the sender.[8] In the twentieth century, Derrida alluded to this notion of presence when he used the format of love letters to explore philosophical questions of language and time in *The Postcard: From Socrates to Freud and Beyond*; the unusual format of the envois seems to challenge reader's expectations of form, tone, and content in philosophical writings.[9]

Letters, especially personal letters, suggest a construction of their writers in that they represent a fragment of the life experience of that person. Authors of letters are identifiable in descriptions of their feelings, hopes, and activities. They are present in the use of the first-person pronoun "I" to refer to themselves as well as in the signing of the letter. Yet the "I" of the letter is not identical to its author, nor are the experiences, feelings, or opinions discussed necessarily whole. The "I" of the letter is to some extent a fabrication that has been created by the author as a particular type of representational self.[10] This "I" may be fabricated or insincere, or "I" may be a rhetorical construction. Furthermore, as noted by Martyn Lyons, "Private letters always reconstitute fragments of experience. They carry no guarantee of completeness."[11] Writers of letters, then, construct an identity and a pattern of experience that is then projected to the reader, who then interprets those constructs. Similarly, personal letters also represent addressees originally intended by the author; they are the "you" referred to throughout the body of a letter, and its contents typically cater to the preferences and interests of that person.[12] Furthermore, the relationship between the writer and the intended addressee can further contribute to the content of the letter as well as to the way in which that content is presented.

Despite a gap between the writer and the reader of the private letter in both space and time, letters retain the presence of both. In this way, the world of the here and now of the writer and the here and now of the reader are both invoked in the letter, enabling letters to allude to an intimacy more often found in face-to-face interpersonal communication.[13] In his famous muse on the love letter, Roland Barthes observed, "What I engage in with the other is a relation, not a correspondence: the relation brings together two images. You are everywhere, your image is total."[14] In some cases, the personal letter seems to substitute or retain physical presence for the individuals involved in their reading and writing and thus can provide an interpersonal experience despite gaps in space and time.

Paradoxically, however, absence is as much a part of a letter as is presence. After all, geographical or emotional separation is usually the primary reason for the composition of a letter. The very reason that presence is such an important part of a private letter is precisely because of the separation or absence from a friend or loved one. Although this absence may be construed as literal, it may often be the case that there is merely a feeling of separation between the writer and the receiver, as in cases where one writes what they feel they could not speak to another. Without this absence, there would be no reason to focus on establishing presence as intently as do the writers of personal letters. Barthes, in elaborating on this "preposterous situation" of the sustained "discourse of the beloved absence" notes, "The other is absent as referent, present as allocutory. This singular distortion generates a kind of insupportable present; I am wedged between two tenses, that of the reference and that of the allocution: you have gone (which I lament), you are here (since I am addressing you)."[15]

In many ways, it is the physical materiality of the letter that helps solidify this process as one in which the presence of both the reader and the writer is retained despite geographical and/or temporal distance. Porter observes, "After the letter . . . has been delivered to its addressee, after its addressee has interpreted the letter and its words according to what his or her personal perceptions will permit or encourage, the letter-object remains."[16] Letters as physical objects persist in time, potentially long after they are written, and they sometimes take on a life of their own by reappearing in contexts unintended or undesired by the writer. Writers throughout history have been acutely aware of this characteristic of the letter, and they have often requested that their letters be destroyed after being read by the intended recipient. For example, Franz Kafka and Emily Dickinson both requested that their letters be burned after their deaths; they were not, and ironically have become some of the most popular collections of letters read by both scholars and laypersons alike. Usually, a request that a letter be destroyed stems from the writer's awareness of its ability to persevere though time as a semipermanent object.

Letters as objects are comprised of many elemental pieces, including the objects and materials that people need to compose a letter. Letter writers select the paper, stationery, or card they will use to write on, and they select the type and color of the ink or pencil as well. Often, writers select their stationery carefully; some may use personalized or monogrammed letterhead on their paper, whereas others use standard lined notebook paper. Even the physical texture and weight of the paper may serve as a reminder of the writer who selected and wrote on the page. Although these choices may appear insignificant, they are quite important in the overall appearance of the letter as a physical object. In fact, the message contained in the letter may be enhanced, illustrated, or even obscured by the writer's choice of paper, writing instrument, or handwriting style.

The style of one's handwriting might reveal that a letter was written hastily or with great care; it may reveal the emotional state of the writer when considered in context with the words on the page. Tear stains or drawings on the pages might provide similar insight. Personal handwriting also is a factor in the feelings of presence invoked by private letters from loved ones. William Decker notes, "In eras when letters were always handwritten, the bodily trace of a correspondent stood before one on the sheet, so that the state of a partner's health might be read in the steadiness of his or her inscription."[17] Even something as seemingly innocent as handwriting, then, can suggest to some degree both the mental condition of the person writing the letter as well as his or her actual physical being.

In all these ways, letters as objects have a physical presence; they are semi-permanent documents that represent a private representation of a moment in time. As Carey Burkett writes in "The Joy of Letter Writing" for *Sojourners* magazine, "Letters are tactile pleasure unduplicatable by the ring of the telephone or the blinking neon of a computer: paper comes in all thicknesses; ink has a smell and a way of changing appearance as the mood or speed of the writer changes; even the lowly postage stamp adds a colorful and festive air to a letter."[18] The fact that the letter gives such a tactile pleasure often leads to the fetishization of the letter object. Letters from loved ones are often carried on one's person, to be read and reread. Love letters especially tend to be fetishized; they can be touched, stroked, and kissed (as surrogates for the presence of the lover), or they can be placed under one's pillow at night. Conversely, they can also be torn to shreds or shared with unintended others, resulting in the humiliation or embarrassment of the author. The sending of personal objects with letters also represents a kind of fetishism, as when one sends a lock of one's hair, common in the nineteenth century, or when one sends photographs, more common in the twentieth century.[19] Letters as objects usually give pleasure, then, not just in the reading, but also in the appearance and the feel of their physicality.

In sum, the letter is a communication genre that is defined by its ability to obtain the phenomenological experience of presence with another through the negotiation of temporal and spatial absence. Letters achieve this end partially because of their physical presence as object in the world, a characteristic that has allowed them to survive so that scholars and others may appreciate their historicity as memory, art form, and collectible item.

One of the ways that letters can be conceptualized is as repositories of cultural knowledge and memory. Although personal letters are rightly categorized as belonging to the private realm because they serve their intended communicative function, letters may enter the public realm when they are preserved as artifacts. When recipients save letters, their existence as objects may span decades; they may even span generations when letters become cherished heirlooms from ancestors. In this way, letters from the past can slip into the public realm, to be read by unintended and unimagined people living decades or centuries after they were penned. Such is the case with letters that become archived by public libraries, letters that are sold to private collectors, and letters that are published in books or featured in documentaries. These preserved private letters then become artifacts that represent traces of a larger public in which the writer and reader belonged. In this way, letters can be understood as receptacles of cultural memories.

During the nineteenth century in the United States, the familiar or personal letter was consciously recognized and promoted as a cultural object that carried with it not only an appropriate and socially polite form, but cultural knowledge and values as well. During that time, the letter was the primary means for long-distance communication, and the ability to write an effective letter was important to many citizens. As such, developing public schools and universities instructed students in the epistolary arts, a skill that was also featured in many nineteenth-century textbooks of composition. And, although it was the form of the letter that was overtly taught to these students, these lessons also served to indoctrinate the values of the upper middle class. According to Lucille M. Schultz, letter writing instruction in nineteenth-century schools provided knowledge about the proper form and content of the familiar letter at the same time it provided codes of manners and morals in the form of Puritan values like personal discipline, self-sacrifice, honesty, duty, faithfulness, and obedience.[20] Schultz argues that through epistolary instruction, the ideals and decorum of the hegemonic ruling class were preserved and perpetuated, ultimately alienating and excluding children from less privileged backgrounds. For example, Schultz notes that in writing sample notes like invitations and replies, students from working-class and rural children were alienated because they were not familiar with these sorts of social formalities typical of the upper middle class. The function of teaching such formal letter styles to these children was to instill in them certain modes of behavior as much as literacy. Seemingly, that school leaders and reformers recognized the way that learning about the appropriate form of the personal letter contributed to a student's understanding of proper morals and values underscores the letter's ability to carry with it cultural knowledge.

Arguably, we can look to personal letters that have survived from the past to deepen our historical, cultural, or collective memories. Collective memory differs from history in that it can be understood as popular public acceptances or recollections of history. The level of collective memory is where historical events and people help create and maintain our societal institutions and traditions as well as our personal identities as citizens of a particular culture.[21] Occasionally, letters may be used as epideictic rhetoric to celebrate the identity and values of a culture, and thus to reify a popular memory of the past. One recent example can be seen in the publication of a collection of letters edited by Tom Brokaw, titled *The Greatest Generation Speaks: Letters and Reflections*.[22] This volume contains many letters from members of the American Armed Forces who served in World War II. Many of these letters are anecdotal, describing the day-to-day events of the war, but Brokaw's stated purpose in publishing them was to pay tribute to that generation for "building the world we have today."[23] In publishing these highly personal letters written in the midst of an impersonal event such as war, Brokaw seemingly intends to make World War II something that Americans can relate to and remember as one of the defining moments of their culture. In sum, the letters written by soldiers during that war simultaneously personalize and memorialize that event in American history, making it something that many citizens of the United States can celebrate.

Although letters can be used to reinforce cultural memories of important historical events that are at the forefront of collective thought, they can also be used to remind us of facets of history that have been omitted or strategically forgotten. For example, women's experience in the United States has not traditionally been an important part of history. Generally, this is explained by the fact that most women were relegated to the private realm of the home while history was unfolding in more public forums. However, this explanation implies that women were of no consequence in American history or at least that their personal accounts of historical events were unimportant to the larger issues of historical scholarship. To remedy this oversight and reintroduce women's voices and perspectives into the cultural memory of historical events, feminist historians and sociologists have studied and published personal letters written by women.

One such example is Margaret Baker Graham's essay, "Stories of Everyday Living: The Life and Letters of Margaret Bruin Machette."[24] Says Baker about Machette, "She was neither famous nor influential during her life-time. Nonetheless, her letters documenting the people and events around her stand as a notable source about the everyday life of Missourians in the nineteenth century."[25] Unlike the patriotic letters collected in Brokaw's book, Machette chose not to write about national events in favor of local ones, yet her descriptions of religious practice, child-birthing methods, and cholera remedies suggest important facts about day-to-day living in Missouri that have been long forgotten.

This avenue of research represents a significant trend in the feminist analysis of women's historical personal documents as a way to give voice to women who were denied access to more public forms of communication. I believe that they also serve as examples of ways in which women's voices can be reintroduced into the collective memory of American culture. Letters, then, can be documents that complement, construct, and personalize cultural memories.

Letters are usually intended as everyday devices that serve a practical need of communication for individuals separated by distance. However, the mundane situation from which letters typically emerge need not preclude them from also being considered works of art. In his book *Art and Experience*, John Dewey notes that in Western cultures, art has traditionally been conceptualized as works of high culture that should be isolated from the general public and observed only with great reverence in halls designed for that function.[26] Dewey problematizes this definition of art as something "isolated from the human conditions under which it was brought into being" by noting that when it is defined in this limited way, it is removed from "the human consequences it engenders in actual life-experience."[27] Said differently, when art is conceptualized as merely a valuable specimen, any other function it may have is rendered opaque; in effect, art becomes a useless bauble. Dewey's definition of art opens up this category and explains how letters might be appreciated as artful in both style and form.

Stylistically, letters may be understood as a mode of literary expression. Thank-you letters, condolences, and invitations may be considered conventional because of the stock phrases and clichés of which they are composed, yet we may easily find the rendering of convention artistic when it reflects the human agent who created it. Importantly, this is not a skill limited to the great novelists or poets. Although we may look to the published letters of great writers for examples of artistic letters, we may just as easily find art rendered in the style of an anonymous everyday letter writer.

Similarly, letters may take on artistic appearance as material form. As noted earlier in this chapter, letters have a material presence, and letter writers often focus on the physical appearance of their letters as much as on content. As such, many personal letters appear as beautiful objects apart from their function as correspondence. During periods of history when letters were a more necessary form of communication, penmanship was frequently taught in schools and practiced by schoolchildren. During the Victorian period, for example, graceful flourishes of handwriting were favored, and this stylized form of writing has become valued for those who collect letters as objects of art. Likewise, different forms of stationery have been favored during different historical epochs. During the mid-nineteenth century, paper made of pressed flowers was favored, but by the turn of the century, monogrammed ivory stationery was considered more fashionable.[28] Further, different occasions for writing letters have sometimes necessitated special types of paper and envelopes. Perhaps the most striking example of this phenomenon is middle to late Victorian-era mourning stationery, which was paper and matching envelopes framed with a black border.[29] There are other possible adornments of letters as well. Some letter writers have decorated their envelopes with personalized wax seals; others have burned the edges of their paper to frame their words or have accentuated their verbal messages with hand-drawn pictures. Letters retain aesthetic appeal, often regardless of the communicative content or the personage who composed it.

There is a third possibility stemming from the materiality of the letter that I have yet to investigate, and that is the use of the letter as collectible. Individual collectors have increasingly been drawn to privately accumulating and studying letters and the objects associated with their production, much as academicians have accumulated these objects in public libraries and archives.[30] In fact, an entire industry has developed around the collecting of items related to letter writing and the post, including letters, stamps, pens, pencils, inkwells, blotting paper, vintage stationery, postcards, sealing wax, letter openers, letterboxes, and more.[31] Collectors of these items have self-organized into national and international groups or societies that allow them to gather information about their enamored items. For example, the American Philatelic Society is an organization dedicated to the collection of postal stamps; it was first formed in 1886 and boasts over 50,000 members in 110 countries.[32] Thus although the everyday nature of letters and letter-writing paraphernalia have made the epistolary arts somewhat transparent for some scholars of social communication, it has worked in exactly the opposite way for modern collectors.

What is it about objects like letters that attract a collector? Walter Benjamin has argued that collecting in the modern world is primarily a way of participating in culture. Benjamin himself was an avid collector of books and postcards, and he understood collecting as a phenomenological, passionate experience that occurs in the present while alluding to the past.[33] This allusion to the past is an important part of collecting for Benjamin and is closely related to his notion of insightful remembrance, which I address momentarily. Presently, however, I wish to focus on his concept of the act of collecting as passionate participation in contemporary culture.

For Benjamin, the unique relationship with the object of their desire defines the collectors' existence. He argues that collectors engage in a "mysterious" relationship with the objects they acquire, "a relationship to objects which does not emphasize their functional, utilitarian value—that is, their usefulness—but studies and loves them as the scene, the stage, of their fate."[34] Benjamin believes that ownership is the "most intimate relationship that one can have to objects" because the collector is able to live in and through the collected objects.[35] Further, he argues that given this relationship of collector to collection, the highest priority of a collector is the preservation and transmissibility of the collection. In this way, the collector participates in his or her culture as a steward, for even if the collection is housed in private, the goal of the collector is ultimately public in nature. In preserving objects, collectors recognize that their objects will remain after their death and will continue to be representations of cultural identity.

However, letters as objects differ somewhat from other items, like teapots or miniatures, which are frequently associated with collecting. Letters carry with them narrative content and the presence of an author and an audience, and they are not valued for their appearance or age alone. Although letters may be written on beautiful paper with handsome script, rarely is the collector of a letter attracted to its outward appearance as is the collector of many other objects. Instead, letters are given value based on their authorial presence, their content, the importance of the writer or receiver, their age and rarity, and only as a final consideration their condition and appearance. Letters as collectibles, then, differ dramatically from other objects that can be displayed in a curio cabinet or on a shelf; the whole of the collection is not what necessarily interests the letter collector so much as the content or author of each individual letter.

A useful way of negotiating the differences between the collection of letters versus the collection of other objects is through the Aristotelian notion of entelechy. Individual letters have something that most individual teapots in a collection do not; each letter in a collection is a thing complete in itself such that it may be read, valued, and admired alone. Individual letters do not necessarily depend on other letters in a collection for context. Such is not the case for the individual teapot in a collection of many teapots, which often rely on the context of other teapots in a collection for meaning. This is the very distinction between the souvenir and the collection noted by Susan Stewart in her work on desired objects.[36]

In her analysis of the collection, Stewart notes that "The spatial whole of the collection supersedes the individual narratives that 'lie behind it.'"[37] By this, Stewart means that collections depend on the principles of organizing and classifying object members into a unified whole. Further, she argues that "the collection represents the total aestheticization of use value. The collection is a form of art as play, a form involving the reframing of objects within a world of attention and manipulation of context."[38] Thus, according to Stewart, collections are classifications whose order obliterates the context of temporal and spatial history. Objects in the collection become merely a part of a larger whole, removed from their original place and use in human history. In this way, objects that populate a collection have a "formal" interest rather than a "real" interest in the mind of the collector.

However, letters do not quite fit the mold of a collection as defined by Stewart. Certainly, one may group and classify the letters that one has collected, and may keep a special box or desk drawer in which to house them. Yet when one reads an individual letter that is a part of her or his collection, the whole of the collection is dissolved into its parts, which is a characteristic of the letter unaccounted for by Stewart in her definition of a collectible. Stewart notes that this is a characteristic of the souvenir more than the collection. For Stewart, souvenirs such as memory albums, wedding rings, or memory quilts are traces of a life history that can be reduced into manageable physical dimensions. She argues that souvenirs are personal mementos that construct a narrative for the owner such that "The souvenir displaces the point of authenticity as it itself becomes the point of origin for narrative."[39] Further, unlike the collection, the souvenir cannot be extended to encompass the experience of anyone because "It pertains only to the possessor of the object."[40] For this reason, Stewart concludes, "We cannot be proud of someone else's souvenir unless the narrative is extended to include our relationship with the object's owner."[41]

Given Stewart's description of the souvenir, we might be inclined to infer that only the collector of letters addressed to oneself can truly appreciate them as a souvenir. Yet this is an oversimplification of Stewart's formulation in that it ignores one of the primary qualities of letters as a communicative genre. Recall that letters may carry with them the presence of their author. Because of this, letter collectors may read letters from authors who have long since died and still feel the presence of that individual through their words. In this way, the new owner of a letter develops a personal relationship, albeit a nonreciprocal one, with the original writer. Thus letters collected from unknown others can become a part of the collector's narrative, making the individual letter more of a souvenir than a collection in Stewart's terms.

Yet even this understanding of letters as a collection of individual souvenirs lacks the richness that one may experience in gathering and reading personal letters from the past. This is likely because of the somewhat artificial way that I have thus far discussed letters as *either* memory, art, *or* collectible. In reality, each letter that has been preserved from the past and collected in an archive or attic is *at once* memory, art, *and* collectible. Only when we consider the letter as the intersection of all these functions can the fullness of the letter be revealed.

Earlier in this chapter I alluded to a relationship between memory and the collection that was central to Walter Benjamin's writings on the act of collecting. The recognition of this relationship between memory and the collection is what he called the "magical" work of the collector—the ability to see the past through an object. He wrote, "Every passion borders on the chaotic, but the collector's passion borders on the chaos of memories."[42] As a collector of old books, Benjamin felt that he understood this passion and claimed that it was a process of renewal. "To renew the old world," claimed Benjamin, "is the collector's deepest desire when he is driven to acquire new things."[43] Absent from Benjamin's notion of collecting is the staleness suggested by the dusty collection housed in a museum because, for him, the past becomes intertwined with the present only in an *individual's* collection. Not surprisingly, then, Benjamin favored individual private collections over public ones, although the latter may be more socially and academically useful. He argued, "The phenomenon of collecting loses its meaning as it loses its personal owner."[44] Only through the interaction of collector and object does the past become revitalized in the personal narrative of the collector. The serendipitous element of this activity was not lost on Benjamin, who noted that "the chance, the fate, that suffuse the past before my eyes are conspicuously present" in his collection.[45] In sum, Benjamin understood collecting itself as a form of practical remembrance.

Like Benjamin, Stewart also recognized the function and value of memory in her conceptualization of the souvenir. She argues that souvenirs are markers of the transference of experience from event to memory. Stewart notes that when an object functions as souvenir, "it moves history into private time" and its function is to "envelop the present within the past."[46] In fact, it is the role of memory in the souvenir that separates it from the collection in Stewart's terms: "The collection does not displace attention to the past; rather, the past is at the service of the collection, for whereas the souvenir lends authenticity to the past, the past lends authenticity to the collection."[47] In this way, Stewart makes a distinction between the collection and the souvenir, based on the concept of memory, which is absent in the work of Benjamin. I believe that we can look to Benjamin's personal habits in collecting to begin to understand why he grants the function of memory to the collection in a way that Stewart does not.

Although Benjamin was interested in disparate types of cultural objects, many of his desired objects contained language in them. He was a collector of books and postcards, and he wrote about collections in the context of his own personal experiences with those items. In her introduction to *Illuminations*, Hannah Arendt notes that Benjamin was also a collector of quotations that he copied into "little notebooks with black covers which he always carried with him . . . [and] showed them around like items from a choice and precious collection."[48] If we consider that Benjamin focused on collecting as the accumulation of literary objects, the distinction that seems to exist between Benjamin's collection and Stewart's souvenir dissolves because both are centered on the relationship between the object and the owner in creating and preserving memory. What differs is merely the source of the memory. For Stewart, the souvenir object stirs personal memories from within the owner, but for Benjamin, the collection inspires and creates personal memories from the culture at large. And it is the living quality of language that allows literary collectibles to convey and create cultural memories in the private realm that most other objects cannot accomplish from Stewart's view.

To better understand the living, social function of language and its importance to the conveyance of memory to literary collections of books or letters, I would like to invoke M. M. Bakhtin's notion of the dialogic utterance.[49] Bakhtin reminds us that in prose (as in life) every utterance is an active participant in social dialogue. Every utterance is at its core social in such a way that each word is in constant interaction with other words in society, a phenomenon that he calls the "internal dialogism of the word."[50] Yet words are not only dialogic with other words in the present; they also invoke the past and suggest the future. He argues, "At any given moment of its historical existence, language . . . represents the co-existence of socio-ideological contradictions between the present and the past, between differing epochs of the past, between different socio-ideological groups in the present . . . all given a bodily form."[51] In this way, all language is drenched with the resonance of its culture. For Bakhtin, language as a social phenomenon "is becoming in history, socially stratified and weathered in this process of becoming."[52] Bakhtin's use of the word "becoming" in this quote is quite intentional because he sees language foremost as a living entity in flux as it is used and reused time and time again.

Thus when we encounter a letter written by an unknown author who lived in some distant past, we encounter the words as living representations of a particular time and place. Bakhtin uses the term "chronotope" to describe the connectedness of time and space as it is artistically expressed in language and that governs the spatial and temporal flow of the narrative. As Bakhtin explains, "Time, as it were, thickens, takes on flesh, becomes charged and responsive to the visible; likewise, space becomes charged and responsive to the movements of time, plot, and history."[53] As a representation of time and space, chronotopes define a concrete feeling of reality through the narrative, such that readers may feel that they are experiencing the words and events of the narrative they are reading.

As readers, we do not encounter the represented chronotopes as inanimate material. Instead, Bakhtin argues,

> This material of the work is not dead, it is speaking, signifying (it involves signs); we not only see and perceive it but in it we can always hear voices (even while reading silently to ourselves). . . . The text as such never appears as a dead thing; beginning with any text—and sometimes passing through a lengthy series of mediating links—we always arrive, in the final analysis, at the human voice, which is to say we come up against the human being.[54]

Precisely because these representational chronotopes are alive in this way, dialogue between a reader of a letter and its author is possible. Thus "the work and the world represented in it enter the real world and enrich it, and the real world enters the work and its world as part of the process of its creation."[55]

The notion of the dialogic word and its situatedness inside a chronotope explains clearly why letters cannot be understood as merely an object of collection in Stewart's terms. But they may be items that populate a collection when invoking Benjamin's more memory-driven definition. Collectors of letters, like Benjamin with his books, are able to enter into a dialogic relationship with the past through the act of reading the narrative of language contained within. Like Benjamin's books, letters are objects that are collected and preserved because they are composed of and are representations of the language of memory. Stewart's conceptualization of the collection, then, seems ill suited for logocentric collectibles like letters or books when compared with Benjamin's richer, more complex formulation.

Similarly, we cannot understand the letter as a collectible or as memory without also considering its aesthetic nature. As discussed earlier, the letter as a category of art stems from both its materiality and its literary qualities. As such, we cannot fully appreciate the social heteroglot nature of the language of any given letter without considering its literary style. According to Bakhtin, "Social dialogue reverberates in all aspects of discourse, in those relating to 'content' as well as the 'formal' aspects themselves."[56] The very style of the language, then, contributes not only to the value of the letter as a collectible but also to its ability to trigger memory.

Here, at the intersection of art, collectible, and memory, is where the concept of aura and auratic traces becomes invaluable to an understanding of the letter. Part of the value of any collection is the rarity of the individual objects that populate it. Letters are often incredibly rare in that each letter is generally different, with the possible exception of form letters. In this way, letters, like works of art, carry with them what Benjamin calls an aura.[57] There is only value in the authentic, original artwork, and the same is true of the letter. Copies of letters make one painfully aware that they do not posses the original because mechanical reproduction strips the original of its aura. As Benjamin notes, the aura of an authentic work of art involves its unique existence as a presence in time and space. When art is reproduced, it is stripped of this aura. Likewise, letters have auras and are valued because of them. This is why, argues Stewart, "When one wants to disparage the souvenir, one says that it is not authentic."[58] Thus the aura of the letter as collectible object is also inseparable from its artistic appearance and its memory-laden content.

Personal letters constitute a genre of communication that hints at a residual trace of the past. They arise from a condition of absence between two individuals, but remarkably they establish a sense of presence between the reader and the writer. They are mediated while at the same time intensely personal. They are simultaneously experience and the representation of experience. And, unfortunately, they are a genre of communication that has fallen by the wayside in our fast-paced, technology-oriented society. Yet letters from the past remain in the world as objects that suggest the trends, values, and needs of our ancestors. Ultimately, although the familiar letter may be understood as a communicative genre that functions as a repository of collective memory, as a work of art, or as an item of collectible worth, it is perhaps best understood as the sum of each of these qualities. The meaning of any particular letter is carried through the form and style of the artifact. Through its meaning, form, and style comes its value as a collectible. Yet only because of the letter's value as collectible (or souvenir) does the letter survive to become a repository of memory or a work of aesthetic merit. That the genre of letter writing serves these multiple, inseparable functions is suggested by the paradoxical qualities of epistolary itself. Yet it is precisely these diverse functions and elusive forms that makes letters so worthy of preservation and that makes researching them so potentially rewarding for modern scholars.

Notes

1. Ivar Ivask, "The Letter: A Dying Art?" *World Literature Today* 64 (Spring 1990): 213.
2. John L. Brown, "What Ever Happened to Mme. De Sevigne? Reflections on the Fate of the Epistolary Art in a Media Age," *World Literature Today* 64 (1990): 215.
3. Ibid., 216.
4. Ibid., 217.

5. Wayne E. Fuller, *The American Mail: Enlarger of the Common Life* (Chicago: University of Chicago Press, 1972), 2.

6. S. A. Taintor and K. M. Monro, *The Handbook of Social Correspondence* (New York: Macmillan, 1936), 45.

7. A. J. Malherbe, *Ancient Epistolary Theorists* (Atlanta: Scholars Press, 1988), 21.

8. Charles Bazerman, "Letters and the Social Grounding of Differentiated Genres," in *Letter Writing as a Social Practice*, eds. D. Barton and N. Hall (Amsterdam/Philadelphia: John Benjamins, 1999), 18.

9. Jacques Derrida, *The Postcard: From Socrates to Freud and Beyond* (Chicago: University of Chicago Press, 1987).

10. Charles A. Porter, "Forward," *Yale French Studies* 74 (1986): 2.

11. Martyn Lyons, "Love Letters and Writing Practices: On Ecritures Intimes in the Nineteenth Century," *Journal of Family History* 24 (1999): 234.

12. Joanne E. Copper, "Shaping Meaning: Women's Diaries, Journals, and Letters—the Old and the New," *Women's Studies International Forum* 10 (1987): 96.

13. David Barton and Nigel Hall, "Introduction," in *Letter Writing*, eds. D. Barton and N. Hall, 6.

14. Roland Barthes, *A Lover's Discourse: Fragments*, trans. R. Howard (New York: Hill and Wang, 1978), 158.

15. Ibid., 15.

16. Porter, "Forward," 11.

17. William Merrill Decker, *Epistolary Practices: Letter Writing in American before Telecommunications* (Chapel Hill: University of North Carolina Press, 1998), 40.

18. Carey Burkett, "The Joy of Letter Writing," *Sojourners* 26 (November/December 1997): 48.

19. Lyons, "Love Letters and Writing Practices," 236.

20. Lucille M. Schulz, "Letter-Writing Instruction in 19th Century Schools in the United States," in *Letter Writing*, eds. D. Barton and N. Hall, 109–25.

21. Barbie Zelizer, "Reading the Past against the Grain: The Shape of Memory Studies," *Review and Criticism* (June 1995): 214.

22. Tom Brokaw, *The Greatest Generation Speaks: Letters and Reflections* (New York: Random House, 1999).

23. Ibid., 3.

24. Margaret Baker Graham, "Stories of Everyday Living: The Life and Letters of Margaret Bruin Machette," *Missouri Historical Review* 93 (1999): 367–85.

25. Ibid., 385.

26. John Dewey, *Art as Experience* (New York: Pedigree, 1980).

27. Ibid., 3.

28. Nigel Hall, "The Materiality of Letter Writing: A Nineteenth Century Perspective," in *Letter Writing*, eds. D. Barton and N. Hall, 98.

29. Ibid., 99.

30. Maurice Rickards and Michael Twyman, *The Encyclopedia of Ephemera: A Guide to the Fragmentary Documents of Everyday Life for the Collector, Curator, and Historian* (New York: Routledge, 2000), 281–82.

31. Hall, "The Materiality of Letter Writing," 104.

32. American Philatelic Society, *APS Online* (1999), January 25, 2004, http://www.stamps.org/.

33. Walter Benjamin, "Unpacking My Library: A Talk about Book Collecting, " in *Illuminations*, ed. Hanah Arendt, trans. Harry Zohn (New York: Shocken Books, 1968), 59–67.

34. Ibid., 60.

35. Ibid., 66.

36. Susan Stewart, *On Longing: Narratives of the Miniature, the Gigantic, the Souvenir, the Collection* (Baltimore: John Hopkins University Press, 1984).

37. Ibid., 153.

38. Ibid., 151.

39. Ibid., 136.

40. Ibid.

41. Ibid., 137.

42. Benjamin, "Unpacking My Library," 60.

43. Ibid., 61.

44. Ibid., 67.

45. Ibid., 60.

46. Stewart, *On Longing*, 138, 158.
47. Ibid., 151.
48. Hannah Arendt, "Introduction," in *Illuminations*, 45.
49. M. M. Bakhtin, "Discourse in the Novel," in *The Dialogic Imagination: Four Essays*, trans. and ed. Michael Holquist (Austin: University of Texas, 1981).
50. Ibid., 278.
51. Ibid., 291.
52. Ibid., 326.
53. M. M. Bakhtin, "Forms of Time and the Chronotope in the Novel," in *The Dialogic Imagination*, 186.
54. Ibid., 252–53.
55. Ibid., 254.
56. Bakhtin, "Discourse in the Novel," 300.
57. Walter Benjamin, "The Work of Art in the Age of Mechanical Reproduction," in *Illuminations*, 217–51.
58. Stewart, *On Longing*, 159.

11 The Celebration of a "Proper Product": Exploring the Residual Collectible through the "Video Nasty"

Kate Egan

Video collecting as a consumption practice has only just begun to be explored by film and media academics, mainly since the advent of the laserdisc and other digital forms of home cinema. To date, the key work on video collecting has come from Barbara Klinger, whose 2001 essay on post-video collecting cultures remains a prominent intervention into the field. Klinger has not only attempted to explore film's transformation, through video, into a collectible object that can be aligned with books, records, comics, or stamps, and to explore some of the predilections that characterize film collecting as a specific practice, but, perhaps most crucially, she has also been the first to consider, systematically, how film collecting theories can be informed by theories based on comparative collecting practices, including work on collecting from Walter Benjamin, Jean Baudrillard, and Susan Stewart. In addition, she has sought to build on this work (much of which portrays collecting as "personal and idiosyncratic") by placing film collecting "within a cultural frame."[1] It is this focus on the relationship between collecting and culture that has allowed Klinger to address the way film's transformation, through video, into a mass-produced commodity *and* a possessable object has created "an intense link between private and public spheres," allowing video collecting to "emerge . . . as a complex activity situated suggestively at the fluid intersection between public and private."[2]

To explore the importance of cultural and external influences on film collecting, Klinger takes as her focus video collectors of contemporary home cinema formats (with particular emphasis placed on laserdisc), and, in the process, posits a number of observations about the key discourses and dispositions that characterize contemporary video collecting. Crucially, she achieves this by focusing on how public discourses generated by marketing and "the discourses of new media technologies" have not only "stimulated the growth of film collecting"³ but also constructed the concerns of the contemporary film collector.

First, as she argues at length, media industries and film distribution companies have sought to market contemporary collectibles through the creation of languages of scarcity or exclusivity (where videos or discs are marketed as limited collector's editions, collector's special editions, or classic collectibles) and through the inclusion of an array of extras and other forms of background information on the production and postproduction history of the film concerned. For Klinger, such discourses and forms of background information help imbue contemporary video collectibles with an aura of authenticity while, in the process, targeting collectors as "insiders," who are able to gain a heightened sense of "intimacy"⁴ with a particular film and reinforce their status as discerning collectors and serious cinephiles.

However, if this is one way in which contemporary video collectors are addressed as discerning cinephiles, then, for Klinger, the other, and perhaps defining, way in which this is achieved is through the creation of a key criterion of quality and value for such collectibles—namely, the extent to which the video version of a particular film reproduces the original theatrical conditions under which the film was viewed (achieved through the reconstitution of original theatrical aspect ratios and high-quality sound and image transfers). For Klinger, the ultimate consequence of this "preoccupation with sound and image" is that home cinema collectors are beginning to valorize filmic elements that emphasize a video version's technical quality (for instance, sound and cinematography) over the quality of other elements (its social and cultural themes, its narrative, etc.), a process that gradually aligns "a passion for cinema" with "a passion for hardware" and, ultimately, associates "cinephilia with technophilia."⁵

Through an exploration of some of the key concerns of contemporary video collecting, Klinger is therefore able to construct a persuasive "top-down" model, focusing on the discursive practices of media industries and hardware manufacturers (who label and construct serious collectors, and generate the key forms of knowledge and value that they bring to their collecting habits), and thus highlighting the extent to which public discourses shape supposedly private collecting identities. However, although such arguments make Klinger's intervention groundbreaking and important, it is also an account that, at points, posits some generalizations about the nature and character of contemporary video collecting practices.

Most prominently, although Klinger is specific, throughout the course of much of her essay, about the type of contemporary video collector that she is choosing to focus on (the "high-end" collector who has both a "passion for the cinema" and an engagement "with technological developments that mimic the conditions of the movie theatre within the home"),⁶ she frequently allows her observations on this type of collecting practice to speak for contemporary video collecting practices as a whole. This is most markedly the case when she begins to align herself with the arguments of Charles Tashiro, a laserdisc enthusiast who had previously published an account of his own personal collecting habits in a 1998 edition of *Film Quarterly*. Like Klinger, Tashiro focuses on the degree to which technical criteria reigns supreme in his collecting habits, subsuming the importance of plot and characterization to sound and image qualities of particular film texts. However, such subjective observations on his collecting dispositions allow Tashiro to make much grander claims about the characteristics of video collecting in general. For him, as for Klinger, because technically oriented collecting practices are so focused on the contemporary standards of digital technologies (on notions of constant improvement, progress, and an obsession with the technically "perfect copy" of a particular film title),[7] technological value is always seen to negate any sense of historical value in the contemporary video collection.

To demonstrate the extent of this negation, Tashiro chooses to compare video collecting with Walter Benjamin's account of his own book collecting habits, where historical value (the appreciation "of a book's individual history as an object") serves as Benjamin's key collecting criterion. On the basis of this direct comparison, Tashiro is able to argue that "if there is a consequence to the 'historical pleasure' of book collecting, and the absence of that pleasure from *video collecting* [my emphasis], it lies in the shape and feel of the books as objects," and that whereas a book may therefore be valued for its physical and material ability to "announce its history, its simultaneous existence in space and time," "physical markers of history . . . are present on the surface of a disc only to its detriment."[8] By harnessing this claim in her own account, Klinger is also able to make sweeping claims about contemporary video collecting, arguing that the centrality of the precious, scarce, and historically authentic "dusty, dog-eared volume" or "elusive first edition" is something that is "sorely lacking in this context."[9]

What therefore characterizes both accounts is the sense that forms of "historical pleasure" have been completely replaced in the contemporary video collecting world by an all-encompassing technophilia and, consequently, that there is always a distinct hierarchy among video collecting formats: where laserdiscs and DVDs are seen as the most important and valuable form of collectible, whereas VHS videotapes, which degrade and possess more potential to show their age, remain video collecting culture's "second-class citizens."[10] In this respect, Klinger and Tashiro can be seen to make grand claims about video collecting in general, painting a picture of a contemporary video collecting world where any sense that VHS videos can have their own, format-specific sense of authenticity is entirely negated, and without the acknowledgment that an appreciation of any collected object may vary according to the nature of the object *and* the reasons for its appreciation, its value, and its organization within the confines of the collection. As Susan Stewart has argued, "to ask which principles of organization are used in articulating the collection is to begin to discern what the collection is about,"[11] and it is the reasons and motivations that underpin a particular collection's organizing sensibility that can therefore frequently determine, hierarchically, whether old or new objects, versions, or formats are most highly valued and what key external discourses (whether historical or technological) come to inform a collection's structure of value.

To demonstrate this, I want to focus on the case study of the "video nasties," a particular historical incident that has produced a cluster of distinct, context-specific cultural meanings for a specific group of video collectors in the UK.

In the early 1980s, as video rental slowly started to establish itself as a potentially lucrative investment for small entrepreneurial companies, a ragtag group of films from a variety of post-1950s exploitation and horror film–making traditions began to be released, in many cases for the first time, onto the UK market. Because of the initial hostility of Hollywood studios to the encroachment of video, the majority of product that distributors had at their disposal were largely obscure titles that had previously been rejected by British film censors or shown only in marginal exhibition venues on the edges of London. These included a number of Italian and Spanish horror films from directors like Mario Bava, Dario Argento, and Jesus Franco, European Nazi and cannibal films, such as *SS Experiment Camp* and *Cannibal Holocaust,* and low-budget American titles that had emerged from the drive-in circuit (including *Blood Feast* and *The Last House on the Left*) or the early 1980s slasher tradition (including *The Burning* and *The Evil Dead*). At this time, there was no voluntary or compulsory regulation governing these videos, and they were therefore released without age certificates and without the need to be approved by any censorship body or trade organization.

As a result of a series of excessive, and now infamous, video sleeves and trade advertisements, such small-time video distributors ensured that the introduction of video into the British cultural climate would be both highly visible and highly controversial. And, as a direct consequence, articles condemning the availability of these films soon began to appear (from 1982 onward) in a variety of right-wing national newspapers, which increasingly called for curbs on the easy access to such films, which were labeled by press and moral campaigners as the "video nasties."

At first, the Police and Obscene Publications Squad seized horror titles from video shops and dealers through the use of the 1959 Obscene Publications Act (OPA), and, over time, the Home Office constructed a list of potentially prosecutable titles (which became known as the video nasties list) as a guide for the authorities. However, with a mounting press campaign gaining force from 1983 onward, the Thatcherite government of the time finally met calls to put a new system of state censorship into place, and thus, in 1984, passed the Video Recordings Act (VRA), legally requiring that all video releases be submitted to the British Board of Film Classification for approval and certification. Under the terms of the new act (which had stringent restrictions on the depiction of sex and violence), it was clear to distributors that a large number of the video nasties and other precertificate horror titles would fail to gain certificates, and they were therefore removed from the shelves of video shops and, effectively, became banned while the distributors who had bound themselves so publicly to these titles and endured endless prosecutions under the OPA, slowly, one by one, went bankrupt.[12]

However, while the VRA therefore sounded the commercial death knell for both these videos and their distributors, the consequences of this act would continue to be felt in the post-VRA world. First, the campaign against the video nasties had created, however inadvertently, a new video-specific context of origin for these film titles, collapsing their previous and distinct production and exhibition histories and reconfiguring their historical origin as an entirely video-based one in a contemporary UK cultural climate. Second, the availability and then sudden removal of these video titles had generated a taste for the forbidden and the banned among early horror video renters—a taste that, from 1984 onward, crystallized around an increased fan interest in seeking out and collecting the video nasties and the subsequent growth of a secondary market around these titles.

As Kerekes and Slater note in their account of the growth of the video nasty collecting culture, old ex-rental copies of video nasties (sold off by video dealers before the VRA came into force) began to reemerge and proliferate, throughout the mid-to-late 1980s and early 1990s, via two distinct sources—as items for sale in classified advertisements in British horror magazines like *Fear* and *The Dark Side*, and in car boot sales, film fairs, and market stalls throughout the UK.[13] Crucially, the VRA gave "nasty" collectors a "principle of organization"[14] and a key criterion of value for their collections. For not only did the existence of the video nasties list act as a "shopping guide"[15] for collectors who wanted to obtain a full set of video nasties, but it also established the primacy and importance of obtaining "original" prerecord versions of particular precertificate titles.

As Kerekes and Slater demonstrate, this initial concern with original versions in the early years of video nasty collecting was largely based around the need to obtain the best quality and most politically authentic version of a banned title. However, as the video nasty collecting culture expanded and developed throughout the 1990s, and newer, more uncut versions of the nasty titles began to proliferate, these concerns can be seen to have mutated into a focus on the historically authentic value of such videos, in both a cultural and more personal sense.

On a collective, more community-wide level, British fan-oriented publications and fan-produced Web sites, from around 1998 onward, have appeared to consistently foreground the artifact status of the original nasty videos, by focusing on their historically distinctive cover art and the minute historical details of their original distribution and censorship histories in the pre-VRA era.[16] What therefore, arguably, has come to be emphasized (at least within some of the most public sectors of the video nasty fan community) is not only the continued centrality of forms of specialist knowledge about pre-VRA covers, distributors, and labels, but also the now marked importance of the original nasty video's cultural history and its reconstituted status as "origin object" within British horror video collecting culture.

On a more personal level, however, original video nasties, for some, have also come to be valued not just as collective cultural artifacts, but also as personally meaningful reminders of a key moment in a collector's individual consumption history. This process can be clearly traced in *The Dark Side*, a publication whose letters page operates as a reliable biographer of the twists and turns of the nasties' changing relationship with those who collect them. From the late 1990s and onward, a number of letters from older veteran nasty collectors, who had experienced the pre-VRA era firsthand and who were generally male and around the age of thirty, were published in *The Dark Side,* and all heavily employed nostalgia when discussing their collecting habits and their relationship to the video nasties. In a 1998 issue, for instance, a letter from a reader named Dave Green thanks *The Dark Side* for its video nasties coverage, and then provides a lengthy account of his days at the video shop as a child, searching and renting nasties before they were banned. As the letter notes, "For all its sleaze, *The House of Video* was my spiritual home, and I miss it badly sometimes. It's another era now, and perversely there was more freedom when I was a kid than now when I'm an adult." Such an approach is common in the magazine at this time, with other readers also writing to *The Dark Side* to thank them for their coverage of the nasties, which has allowed them, in the words of one reader, to go back to "the best viewing time of my life."[17]

What such letters seem to convey, then, is the extent to which the currency of the shared pre-VRA experience of veteran video nasty collectors has been consolidated throughout the course of the 1990s, and how the "ardent" veteran video nasty collector has emerged as a distinct type in the contemporary world of horror and banned video collecting in Britain.[18] To further explore both the status and dispositions of this particular type of veteran collector, I want to focus, for the rest of this chapter, on the collecting practices of John,[19] a self-proclaimed video nasty collector whom I met with on two occasions in 2001.

From the accounts that John gave me of how he began to acquire his own nasty collection, it became clear that he was one of these veteran collectors. First, he had experienced the era of video nasty hysteria firsthand. From the age of around ten or eleven years onward, he had rented a large number of precertificate video nasties from video shops, watched them with friends, and spent a large amount of time at school discussing titles like *The Evil Dead* and *I Spit on Your Grave*. Second, the commencement of his video nasty collecting activities in the early 1990s involved the employment of some of the traditional methods of video nasty acquisition just outlined. Through a friend in work, John had learned how to read advertisements in *The Dark Side* that discreetly advertised "rare horror collections" and to learn which video labels to look for and which markets and fairs to attend to acquire old versions of video nasty titles.

There are perhaps two distinct reasons why I'm proposing that focusing on a "thick description"[20] of John's collecting history and habits might be productive within the realms of this chapter. Firstly, taking a bottom-up approach to video collecting practices and focusing on a specific *veteran* video collector's acquisition and classification rituals, can, arguably, allow for a tracing of the felt impact of an array of historical, and thus external, influences on the private collection (influences that frequently coalesce with some of the habits outlined in Benjamin's account of his book collecting practices). An exploration of how videos can accrue value over time can demonstrate how, just as with Benjamin's coveted book collectibles, residual layers of meaning (including "the realization" of a video's "individual history as an object" and the realization of a specific "collector's history of possession"[21] and consumption) can affect the value, and inform the video collector's appreciation, of his collected objects. In addition, a mapping of the historical formation of an individual's collecting dispositions can also highlight how forms of knowledge based around what Benjamin calls "tactical instinct" and "experience"[22] can slowly gain cultural currency in a collecting community and lead to the construction of "insider" identities generated not by contemporary marketing discourses, but in relation to the changing context of a secondary market, and to format-specific video distribution histories.

However, second, it is also important to note that although John's video nasty collecting and consumption history allows him to be typed as a veteran video nasty collector, it is also the case that he possesses particular collecting dispositions and practices that appear entirely peculiar to him. Not only is he resolutely antidigital (refusing, unlike many other veterans,[23] to include digital versions of nasty titles in his collection or succumb to any modern forms of video acquisition, via the Internet), but he also engages in a number of seemingly idiosyncratic categorization and modification rituals (the creation of pre-VRA label archives and the restoration of certain videos), which, at least on the basis of my own research into video collecting dispositions and discourses, do not appear to be widespread within the video nasty fan community. In this respect, John can also be perceived as a particularly committed example of a veteran nasty collector, in terms of his consistent attempt to move away from more central practices, construct entirely personally constituted rituals, and thus reconstitute himself as creator of particular, personal systems of meaning and value within the collection.

Although this may appear a methodological shortcoming, this focus on a partially divergent video nasty collector does allow for a deeper exploration of something glossed over in Klinger's account. For although Klinger does acknowledge, early in her essay, that the "zealotry" of the video collector is "undoubtedly characterised by the whims and obsessions of the individual,"[24] her fervent attempt to highlight the primacy of public discourses in collecting cultures means that she relies, throughout the majority of her account, on industry discourse and marketing materials at the expense of a more detailed, empirically based exploration of what the individual collector brings to their collecting practices.

By therefore focusing in depth on a collector who, although shaped by community-wide dispositions and economies of value, also strives to create his own personally meaningful rituals within the collection, the chapter explores and illustrates how complexly interweaved personal meaning and culturally determined meaning can *be* in each collection, and how, paradoxically, personally constituted collecting practices can succeed in more forcibly highlighting the impact of commercial and cultural meanings, categories, and discourses on individual collections.

In what follows, two key areas that characterize John's video collecting practices are touched on. First, to establish the reasons why certain videos are acquired and have value, the chapter looks at the importance of video labels, different video versions, and original or rare videos to the veteran video nasty collector. Second, to further explore the variety of complex intersections between public and private in a particular contemporary video nasty collection, it also looks at the particular secondary consumption spaces through which John has acquired his videos, and at how these videos have then been renovated and categorized to become part of his collection. To explore the extent to which John's collecting habits converge with wider dispositions and practices among veterans and other collectors in the video nasty fan community, my account is also informed by a wider discussion of UK nasty collecting from *The Dark Side* and from the UK-based Internet message board, *Dark Angel's Realm of Horror*.

While it should be noted that the majority of this research was conducted during 2001, and that the dynamics of the video nasty collecting community may have changed since this time, the following account is offered as a historical snapshot of the video nasty community (and of a specific video nasty collector's practices) at a key historical juncture, as DVD and the Internet were beginning to usurp traditional forms of uncertified or illicit video acquisition, and at a time parallel to the publication of Klinger's crucial essay on video collecting.

Discussions with a Nasty Collector: Issues of Value, Consumption, and Exchange

Since the banning of the original video nasties in 1984, an array of different video versions of particular nasty titles (some which are cut and others which are uncut) have been released in the UK and overseas.[25] Over the course of my conversations with John, it became apparent that there were a variety of insider ways to identify these different video versions of nasty titles, which could be picked up via horror magazines like *The Dark Side* or through accumulated knowledge and experience, word of mouth, or particular contacts. In some cases, this would simply be based around identifying particular scenes that would only appear in an uncut version of a particular nasty title. For instance, John gained an uncut copy of *The Burning* when, on borrowing a friend's copy, he realized the previously removed "shears scene" was intact; and on a 2001 *Dark Angel* message board posting, advice is given on how to find a particular scene in *I Spit on Your Grave*, which would identify whether a particular version of the film was uncut.[26] In other cases, telltale signs for collectors include serial numbers, dates, or (if a letter from *The Dark Side* is taken into consideration) particular artwork used on the video sleeve or on the video itself.[27]

However, perhaps the most effective way to identify particular versions of nasty titles (or at least the way that appears to have been used most frequently by John, *Dark Side* readers, and *Dark Angel* participants) is to have extensive knowledge about video labels and distributors. As John explained, "You can identify which video labels are cut films. You know which label to look for."[28] Although the key use of this label literacy is therefore to identify and find uncut video versions, it also appears to have encouraged many nasty collectors to use knowledge of labels and versions as a valuable research tool. Indeed, an illuminating example of this relationship between collectors and video companies occurs in a 2001 discussion of different versions of the nasty title *The Beyond*, on the *Dark Angel* message board. Here, when one Finnish collector realized, to his dismay, that his Cosa Nostra copy of *The Beyond* was cut, other British participants managed, through a comparison of copies of *The Beyond* from their own private collections, not only to provide background information on the video distributor but to expose them as producing cut bootleg videos and to even identify which video distributor's version of the film these bootlegs had been copied from.[29]

What emerges from such an example is an idea of the insider, which concurs with Klinger's definition (someone who has "obtained apparently special knowledge, possessed by relatively few others"),[30] but which manifests itself in a different way. Not through searching for the best transfer with the most deleted scenes and most behind-the-scenes extras (all supplied legitimately by a contemporary video, laserdisc, or DVD distributor), but through constructing a more active and interrogative relationship to distributors, both old and new. For, to identify the value and status of a version of a particular nasty title, collectors have to build up knowledge not only about the history of particular video distribution companies, but also their historical and commercial relationship to each other and to versions of particular nasty titles. As with Benjamin's book collecting practices, what this therefore seems to demonstrate is how a focus on a film title's history (its history of international distribution and the litany of video versions this history has generated) creates the potential for an array of different video versions to be sought out, and a rich, vast field of collecting knowledge and learning to be constructed around such versions. Allowing, as Benjamin notes, "the whole background" of each collected video ("the period, the region ... the former ownership") to turn the collection into "a magic encyclopaedia" of information and history about the video distribution of particular film titles.[31]

Although video nasty collecting therefore appears to be characterized, on the one hand, by the insatiable search for the most complete or uncut version of a film in existence, it is also characterized, on the other, by another fetish that is clearly informed by this conception of the collection as a historical information source—namely, to own every video version of a particular film title in existence. This tendency was identified by John as being central to his collecting habits on a number of occasions, including in terms of a discussion of video nasties, but perhaps the motivations that lay behind this tendency came across most clearly when he began to discuss the seven different versions of George Romero's *Dawn of the Dead* that exist in his collection. As he explained,

> The reason I've got so many [copies of] *Dawn of the Dead* ... is because different versions pop up all the time, and I thought "I wonder what's in that one? ... why's that one trimmed?" Different versions have been on TV and I got the 124-minute print off a mate. The one of the *Dawn of the Dead* I watch is the last video release. [The] retail release, which is one of the best quality copies I've got and it's pretty much complete. I've got rental ones from years ago. There's chunks of film missing out of them ... so, they're just now part of the collection, more than anything.... It's because it's my collection and it would seem now, if I removed any part of it, that it would be incomplete. I've got to keep it all.[32]

What John clearly conveys in this comment is the extent to which nasty collecting habits are frequently informed by two distinct notions of completeness—the need to possess the most complete, uncut video version of a film (the "retail release," which is "pretty much complete"), *and* the need to make *the collection* complete via the acquisition of all the different available video versions of a particular film title (which, if removed, would make the collection "incomplete"). Crucially, what this distinction highlights is the nasty collection's ability not only to function as a usable historical library and archive (where different video versions are compared to establish the identity of the most complete version) but also as a historical monument, where older, less complete items serve to strengthen and extend the historical worth of the collection as a whole.

This dual valuation and use of older, more incomplete video versions of nasty titles is also identified by horror video collectors in a number of *Dark Angel* postings, where, in 2001, discussions about the exchange and acquisition of particular horror videos frequently occured. In one *Dark Angel* posting, for instance, a collector justified his efforts to try to obtain an ex–rental, pre-VRA copy of *Cannibal Holocaust,* when uncut copies of the film were quite readily obtainable from abroad, by arguing that "I know it was pre cut by Go Video, but I want the pre cert to compare it with modern versions . . . cut or uncut it is a collectors item."[33]

In this respect, the valorization, by these nasty collectors, of older pre-VRA videos can be seen to revolve, on the one hand, around their use as comparative objects in the collection (which can be compared with so-called modern versions). On the other, they also operate as "collectors items," which function as necessary building blocks or components that extend both the quantitative mass and historical value of particular parts of the collection—something which, once again, appears to align these video nasty collectors with Benjamin and his fellow book collectors who, as Tashiro notes, could "only add to a collection" and "never" replace or "part with parts."[34]

However, although this is one justification for why video nasty collectors have "got to keep it all" and allow an uncut video version to coexist alongside versions that have "chunks missing," then it is also the case that original pre-VRA versions of the nasty titles (whether cut or uncut) *can* also be appreciated, in their own right, as meaningful and valuable "prize pieces" of the collection. What is suggested by the importance of these prize pieces is that videos in the collection are frequently valorized not in relation to standard, public forms of monetary value but to more personalized forms of value, and, based on John's discussions, it can be argued that two types of personalized value relate to these older defunct videos: aesthetic/historical value and the value of the rare or the singular.

As Susan Stewart notes, for Jean Baudrillard, "a 'formal' interest always replaces a 'real' interest in collected objects," and, as she argues, "this replacement holds to the extent that 'aesthetic value' can frequently 'replace . . . use value'" within the confines of the collection.[35] If the predominant use-value of a video is to be watched, and certainly John seems to watch some of his videos, then it is also the case that particular videos in his collection also seem to be appreciated on this purely formal or aesthetic level. When quizzed about why he continues to valorize VHS over DVD, when his VHS copies are often of bad quality and cut, John made the following comment: "I like the big boxes . . . I like the fact you feel as if you've got a proper product—a bulky big tape, especially with the old ones. You know how much a proper video nasty tape weighs. It's like a brick. . . . That's how all videos used to be."[36] In addition to this, John also waxed lyrical, throughout our discussions, about an old pre-VRA *Dirty Harry* video that he'd seen at a car boot fair on an old Warner Brothers label, noting that "I got excited about that *Dirty Harry* on the Warner Video . . . [with] all similar sleeve and spine designs, [a] picture, black border and Warner Brothers logo."[37] In addition, and as John later revealed, his old Warner Brothers tapes and his tapes from another defunct video nasty–era video company, VTC, had not only been purchased on the basis of their distinctive box, spine, and sleeve, but had been categorized together in, as he puts it, "a little collection on its own."[38]

What these comments seem to emphasize is the frequency with which the value of particular videos in John's collection are delineated in relation to their status as outmoded objects that, to him, represent the way "all videos used to be." For, in these examples, pre-VRA ex–rental videos appear to be appreciated not for their actual content or use value (i.e., the actual aesthetic, generic or cultural quality of the film *or* how complete and uncut it is), but for the meaningful historical value of their material form (the bulky bricklike tapes, the big battered old boxes, and the distinctive spines, logos, and designs on the video sleeve).[39]

This personally constituted link between aesthetic and historical value perhaps relates most clearly to the Benjamin-inspired idea that collecting "becomes a form of personal reverie, a means to re-experience the past though an event of acquisition," and indeed that a collected object, through its aesthetic form, can also "conjures . . . memories of its own past."[40] This idea, that the aesthetic form of a particular video can bring pleasure to the collector by conjuring up its past, and notably (and also in line with Benjamin's arguments) the collector's own past, is also clearly conveyed in two other accounts of video nasty collecting. In a *Dark Side* article on "The Death of the Video Nasty," journalist and video nasty collector Jay Slater notes that "seeing those large video boxes with fabulous artwork (remember VTC's vacuum-sealed monstrosities!) takes you back to a bygone age,"[41] whereas, for a veteran video nasty collector cited in Kerekes and Slater's book, it is the smell of pre-VRA video nasties, as well as their aesthetic form, that allows for this "personal reverie." As the collector explains, "There was a musty smell to the videos, too. Not an unpleasant odour—just what became an associative 'video nasty aroma.' . . . The smell lingers vaguely and a quick sniff transports me back to the halcyon days of seeking out banned movies."[42]

When taken as a whole, what these accounts suggest is that, for veteran collectors, the appreciation of original video nasty versions often has a clear relationship to nostalgia and to "personal autobiography,"[43] and that these collectors therefore frequently valorize these videos because of their relationship, through video artwork, labels, and shapes of boxes, to their adolescent past and prior consumption of these titles in the era of video nasty hysteria. Indeed, the fact that consistent reference is made, in *Dark Side* classified advertisements, to the fact that items for sale are "pre-1984 originals," "originally boxed films," or "pre-certificated originals"[44] indicates how frequently the value of original nasties is articulated in relation to their status as emblems of a prior consumption era that has key personal meanings to the veteran nasty collector.

In many respects, then, the nostalgic value of these old, ex–rental videos emphasizes the extent to which, in Stewart and Baudrillard's terms, collecting can be an entirely personal and "transcendent" domain, where the collector is able to "marshal" their "own discourse," which is always controlled and determined by the collector's own forms of value rather than those of the outside world.[45] Conversely, however, it is also the case that the determining influence of the cultural and external can still clearly be detected beneath the surface of John's appreciation and acquisition of these original pre-VRA videos.

For, as should be apparent, an articulation of the authenticity of the original nasties relates not just to memories of their prior consumption in the pre-VRA age, but to their distinctive material appearance (the materiality of their box and sleeves), and, as John's references to "proper products" and "similar sleeves and spine designs" seem to indicate, to past markets and past forms of marketing. In addition to this, although John justifies his acquisition of old pre-VRA versions from defunct video companies by explaining that these labels don't "exist anymore, so they're not going to be doing any more stuff," he also freely admits that, for him, one of his VTC tapes, *Hitchhike,* is a terrible film and that it has largely been acquired because of the need to include it in his VTC collection.[46]

Indeed, the fact that *Dirty Harry* and *Hitchhike* weren't on the video nasties list, but are connected to the video nasties era purely via their distribution companies and their material form, highlights the potential for John to be perceived as the creator of his own personal discourse, but a discourse which is still dependent on externally controlled forms of meaning and categorization. For here, in these examples, John appears to be purchasing certain videos not necessarily because he watched them in the pre-VRA era, but because of the need to complete a certain set of videos in his collection and acquire all the products from a defunct video company that doesn't "exist anymore." In cases like this, as Baudrillard argues, collectors can "get so carried away that they continue to acquire titles which hold no interest for them," allowing "the pure imperative of [their] association" to particular defunct video companies and their "distinctive position within the series" or archive of similar objects to be the only reasons needed for acquiring them.[47]

John's attempt to create a personal discourse for his old collected videos (and only acquire outmoded videos that have nostalgic value or meaning for him) is therefore complicated by the parallel need to imbue parts of the collection with wider cultural significance (in the sense that, through acquiring these videos, they can come to function as a complete archive of a defunct video company, rather than just being a set of videos that only have a relationship to *his* past video consumption experiences). However, if the need for John to classify his personal investment into meaningful cultural categories can sometimes lead to a tension between personal and wider cultural justifications for acquiring videos, John is still frequently able to imbue his collection with more personal and unique meanings via the acquisition of more singular (and thus uncategorizable) videos, which, through the process of their singularization, can be reconstituted as personally prized rarities in the collection.

Indeed, if the seeking out of the uncut is one means by which video collecting gains its sense of "bravado,"[48] then, as Tashiro argues, another clearly revolves around the seeking out and acquiring of a rare or unique title. Aside from the previously mentioned *Hitchhike*, the most valuable video in John's collection is a pre-VRA copy of the 1960s exploitation title *Mark of the Devil*. Although he will not lend out his original version of *Hitchhike* (only agreeing to make copies for others to borrow), the sacredness with which he treats his version of *Mark of the Devil* reaches new heights—to the extent that he has never watched it, will never watch it, but would never part with it. As he explained,

> I've got one that's in a slipcase that did actually go to rental, but it was a promo copy—*Mark of the Devil*. The tape's slightly damaged, so I've never watched it.... It's probably worth 75, 80 [pounds] . . . something like that. Dead rare. I think it came out on Redemption [a horror sell-through label], but it was heavily cut. Because I've got a promo copy, it's all there, the whole film.[49]

Although the value of this particular video is partly explained through its status as a complete uncut film, the fact that John can never truly appreciate this (i.e., that he can never watch it) suggests this is based more markedly around its rare status, with the irony of its extreme rarity being that, although the copy is seen as an heirloom, it is, in a practical sense, damaged and, in some ways, useless. The fact that the video is a promotional copy clearly has some bearing on this because acquiring a promotional copy is, arguably, a further indication of John's status as an insider (in the sense that a promotional copy of this video was never, strictly, supposed to be available for mass public consumption), and, indeed, a number of other videos in John's collection (including his previously mentioned copy of *The Burning*) were also promotional videos.

What seems significant about these promotional ex–rental videos is that, largely because of their status as videos that weren't originally intended for sale *or* for mass-market consumption of any kind, they seem to negate the basic idea of a commodity (as one of many copies that can be purchased and owned easily), and this can, potentially, succeed in giving the collector a sense of personal ownership over an object that seems to exist completely outside of the confines of consumer culture. Indeed, as Susan Stewart argues, "The . . . [unique] object has acquired a particular poignancy since the onset of mechanical reproduction; the aberrant or unique object signifies the flaw in the machine just as the machine once signified the flaws of handmade production."[50]

In this respect, then, the commercially aberrant and thus rare promotional video acts as proof that some objects can negate mass production and escape commodity status, something that allows collectors like John to reconstitute them as seemingly unique objects, whose "absolute singularity," as Baudrillard notes, then "depends entirely upon the fact that it is I [the collector] who possess[es]" them.[51] Indeed, John's ability to cut off his treasured copies of *Hitchhike* and *Mark of the Devil* from certain primary market-influenced forms of value is not only clear through the way that he freezes these videos in his collection (turning them into emblematic ornaments that can't be used, removed, or borrowed) but also the way that he reiterated, throughout the course of our discussions, their entirely personal and thus independent value and meaning by noting that "they're worth far more to me than they are to anybody else" and that he would only be able to sell his copy of *Hitchhike* at its true price, and thus value, "if I met another person like me."[52]

However, paradoxically, and as John's comment that his *Mark of the Devil* copy is "probably worth 75, 80" pounds indicates, collectors like John still need to reference these videos' specific secondary market value, to articulate and underline their status as singular rarities. Indeed, throughout the course of our conversations, *Mark of the Devil* and *Hitchhike* were the only videos that John valued via the naming of specific secondary market prices and moments where these videos were discussed represented the only points in our conversations where John admitted that particular pre-VRA labels have an effect on the economic (rather than just historical) value of defunct videos in the contemporary secondary video marketplace (something which, in the process, reinforces the sense that the valuation of such videos is frequently based on shared, community-wide valuations).

Once again, what this seems to demonstrate is the contradictory relationship collectors like John have to both commercial culture and to past and present market economies. For although the process of reconstituting an object as unique (rather than just promotional or 'aberrant') may allow collectors to disentangle themselves from a complex web of prior commercial uses and meanings, a fervent reiteration of a video's personal value as a rarity still frequently needs to be related back to shared systems of value in order for a particular video's worth to be articulated. As Igor Kopytoff notes, what this means is that collectors like John are frequently "caught" in an economic "paradox," where "as one makes . . . [objects] more singular and worthy of being collected, one makes them valuable; and, if they are valuable, they acquire a price . . . and their singularity is to that extent undermined."[53]

In many respects, then, such contradictions can, potentially, restrict collectors of defunct and secondhand videos, blunting their ability to entirely escape the culturally constituted meanings and changing forms of secondary market value that impact on particular secondhand objects. However, in situations where they have harnessed their knowledge of these changing forms of value within the networks and spaces that they use to acquire their collected objects, collectors like John can be seen to regain some personal control over their collection, and, in the process, reinforce their status as all-powerful insiders.

For, over the course of the two interviews, it became clear that, to acquire his videos, John had constructed networks based around particular contacts and particular consumption spaces, and that his relationship with these spaces and contacts appeared to be based around issues of autonomy and power (swinging both ways, between John and the spaces that he frequented). Like many other nasty collectors, the main sources that John uses, or has used, to obtain secondhand videos include a film fair, a car boot sale, and a weekly market, all based in the Nottingham area in the UK. What was particularly significant about his accounts of these places was that contacts and networks had to be built up over a long period of time and required skill and effort on John's part. Indeed, it was notable that many of John's descriptions of exchange practices at these markets and fairs tended to begin with the phrase "once you got to know them." However, as he also related, once this obstacle of getting to know them had been bypassed, John's efforts could be rewarded in kind, with perks and gifts that only added to his status as a particular type of secondary market insider. As John explained (about his primary Nottingham car boot contact),

> You have to get to know him first. You have to be interested in what he is and then he'll slowly show you. He's got certain videos laid out, but the majority is crap which you wouldn't even consider. But you'll start talking to him, and he'll go in his car and get an NTSC copy of *The Evil Dead* out, [and he'll say] "interested in this for £10?" He knew I wasn't going to run off to the police. He showed me his files, so [I've] got [a] lot of stuff from him and cheap prices as well. And covers, if he'd got them.[54]

What is clear from this account is that if uncertified video versions, whose initial value lies in their transgressive status as objects forbidden by British video classification laws, are to be acquired, then skill, patience, and knowledge need to be carefully utilized by video nasty collectors. However, although, on the one hand, this requires nasty collectors like John to win over traders who possess equivalent levels of knowledge about the value of particular videos, on the other, collectors can also use acquired skill and knowledge to gain control over the space of the boot fair and the means by which a desired video is acquired. For John, a primary part of gaining this control is to build up information about a particular trader, to detect flaws in their knowledge of video nasty versions and to then exploit these flaws for his own purposes. For instance, when discussing the Nottingham film fair that he used to frequent, John noted, "you'd get collectors there. Although it was nice to see it, . . . collectors knew the price of things, they knew the value of things. . . . They'd charge 25, 30 [pounds] . . . a film.'[55]

Although, for John, such potentially more challenging traders could be managed through slowly "getting to know them" (as was the case with the car boot contact just mentioned, and his primary contact at the film fair, whom he often telephoned in advance to secure titles), other traders were often easier to manage if, unlike these veteran collectors and traders, their knowledge was lacking.[56] As John explained,

> [sometimes] it was just walking round and seeing what you could find. Sometimes you'd find somebody who didn't realise what they'd got, and they'd have a table of videos and sometimes you'd find one rarity there. And it would only be three [pounds] . . . , so you'd grab it and think "I'll sell that for 25."[57]

What these examples once again reinforce is the extent to which prior cultural circumstances can affect the active and interrogative nature of exchange practices in these secondhand consumption spaces. For, in order for John to acquire the videos that he desires at the most reasonable prices (and considering that particular pre-VRA videos or other versions of the nasties are often illegal, outmoded, or secondhand), he is forced to interact with unpredictable local and temporary consumption spaces (where goods change from week to week and where traders possess varying degrees of knowledge about the items they sell). As a result, John has to hone the necessary skills needed to operate effectively within these spaces by acquiring, as he puts it "knowledge [that] you pick up along the way"[58] about the value of particular video versions and the depth of collecting interest and knowledge that characterize those who sell them.

Nicky Gregson and Louise Crewe have written a number of articles that attempt to consider and explore the car boot sale as a newly marked type of consumption space from the 1990s and onward, existing in clear opposition to the more predictable and rigid commercial spaces of the shopping center or mall. Although their work tends, on the whole, to concentrate on more traditional types of secondhand goods (e.g., bric-a-brac, books, and so on), a number of their observations appear to correlate closely with some of John's video acquisition practices.

For Gregson and Crewe, although the mall is more controlled, and mass-market based, the alternative consumption space of the car boot sale is temporary, unpredictable, and thus, for the people who frequent it, can be empowering and challenging. As they argue,

> here, clearly, is a form of exchange which re-creates earlier—in this case pre-capitalist—forms of exchange. Here is a form of exchange which . . . is dominated by cash, which looks back to an epoch without credit cards, store cards and cheque books. This is a consumption space in which we can be, at one and the same time, vendors and purchasers . . . where every price is negotiable, presenting the opportunity for haggling and, on occasion, barter.[59]

It is therefore this potential for the vendor and the purchaser to be interchangeable at the boot fair that seems to give such spaces, in Gregson and Crewe's terms, their alternative and "pre-capitalist" slant, in the sense that this view of the shopper seems in marked contrast to the idea of a restricted consumer at the mall or shopping center. Indeed, if this argument seems to correlate with John's video-acquiring activities, then the ideas that Gregson and Crewe go on to discuss—ideas of the "skilled shopper" (who knows how to get bargains) and the "small-time entrepreneur" (who acquires bargains to sell on to others)—are also terms that can be applied to the roles that John has adopted in the boot fairs and markets he has attended. Further, the adoption of malleable commercial identities also demonstrates how the public removal of the video nasties and other pre-VRA videos has enabled video nasty collectors like John not only to operate as archivists who collect, research, and preserve distributor output and the distribution histories of particular nasty titles, but also to ape these video distributors by reconstructing themselves as commercially minded entrepreneurs within the context of a secondary marketplace.[60]

The ability to flex forms of acquired knowledge and experience, to control price in relation to value, and to adopt a variety of primary market roles are therefore all factors that help to construct this idea of the active and empowered veteran video nasty collector. However, for John, it is the subsequent modification of some of his collected videos that enables him to yet further extend this malleable collecting identity. For Crewe and Gregson,

> What is important about purchases made at car boot sales, particularly when these are second-hand goods, is that people are often not buying the commodity we see but a particular attribute of it which will be realized only when they return home and either renovate, alter, transform or display it. Possession rituals are particularly important in the case of items which are unique, old, second-hand or rare.[61]

This focus on the value of an attribute can clearly be seen in John and other nasty collectors' acquisition of particular labeled or boxed pre-VRA videos, which are bought for the historical value of the video label and packaging, rather than for the video's content. However, another, perhaps more idiosyncratic,[62] example of this approach is John's occasional renovation of acquired pre-VRA videos. As he explained, because a large number of pre-VRA tapes had been destroyed after their public removal, and ex–rental boxes and sleeves had been lost or damaged, videos acquired at fairs and boot sales would often be incomplete objects—with tapes frequently being copied, rather than original prerecorded videos, covered only with cardboard slipcases or plain video boxes.

As a result, John's acquisition of some of his pre-VRA videos would, sometimes, only be part of the process of possessing them and collecting them, and, on a number of occasions, further work would need to be carried out on these videos before they truly became part of his collection. Although some sleeves could be acquired via the Internet (particularly in the case of some of the central video nasty titles), John frequently acquired others via local video shops, where old ex–rental sleeves could be purchased for twenty pence.

As John openly acknowledged, the consequence of this process is that a certain proportion of his collection is made up of videos with "original rental boxes and sleeves, and copied tapes."[63] Although, in some respects, this could appear as a disadvantage (in the sense that certain pre-VRA videos in his collection are only partly historically authentic), it does demonstrate the extent to which possession rituals can play a large part in placing a personal stamp on particular secondhand video collections, and how such rituals can, potentially, enable John to act as not only a collector of these videos, but also as a restorer and re-producer of their original commercial status and value as original pre-VRA videos.

In turn, what this, once again, emphasizes is the extent to which personally constituted rituals and the historical and commercial frameworks that inform the value of his collected videos frequently collide and mesh in John's collection. On the one hand, these forms of renovation can be seen to allow John to act as a shadow or echo of the original manufacturers of pre-VRA videos, allowing him, in the words of Benjamin, to take on "the attitude of an heir" to these videos and demonstrate "an owner's feeling of responsibility toward his property"[64] by restoring them to their original pre-VRA material standards. On the other hand, and in a more personal sense, the ability to change and alter the meanings of these videos also allows John to take even more active control of all aspects of a particular video's recirculation around a secondary market economy, and, in the process, to reinforce the Benjamin-inspired idea that "the most important fate of a copy is [always] its encounter with him" and "his own collection."[65]

When placed in context, this conception of John as a restorer and manufacturer lines up convincingly with the more widely shared community roles of archivist, researcher, skilled shopper, and entrepreneur to present a powerful picture of a collector as the creator of his or her own alternative economic and cultural sphere. And, arguably, these collecting roles and dispositions are alternative because they are not primarily generated or shaped by the discourses of contemporary marketing and distribution discourse, and specialist because they, instead, hinge so fervently around both situated and context-specific historical and cultural meanings, and the intimate knowledge and personal consumption and collecting history of the veteran video nasty collector.

If, as Klinger argues, "the phenomenon of collecting" can therefore help "us to explore . . . cinema's fate as private property,"[66] then perhaps what the collecting practices previously outlined suggest is that it is the residual and secondhand end of the video collecting spectrum where video collecting identities not only have the potential to be at their most active, complex, and malleable, but where videos remain at their most open to privatization. Often circulating outside the restrictions of contemporary commercial constraints and national regulation, and often existing as discarded or battered objects (almost like traces of videos that were previously buried by censorship or damaged by the ravages of time), the original, ex–rental pre-VRA video seems to cry out to be preserved and reconfigured within the confines of the British horror video collection.

Conclusion: The Death of the Nasty and the Threat of New Technologies

In a 2001 *Dark Side* article, Jay Slater heralded "the death" of the ex–rental video nasty, arguing that the "video nasty generation" had "grown up and moved on" and that the type of residual collecting enacted by John and other veterans was becoming distinctly outmoded.[67] For Slater, whereas the original video versions of the nasty titles were damaged goods, the new uncut Region One DVD versions of such titles as *I Spit on Your Grave*, freely available to order and import from the United States or Japan via the Internet, were the way forward for contemporary British horror video collectors. In a form of logic that mirrors Tashiro's comparison of the shiny, perfect laserdisc and the impermanent shoddy video, Slater notes,

> why should a film fan pay over-the-top costs for an inferior quality video with possible tape damage and a cassette encrusted with chocolate,

booze and engine oil that is often pan-and-scanned when they can buy a new and digitally improved 16.9 anamorphic widescreen print with plenty of mouth-watering extras such as audio commentary and additional scenes?[68]

Although it is true that the Internet and the concurrent availability of imported DVDs has revolutionized nasty collecting (something that post-1998 letters and articles in *The Dark Side* and 2001 discussions on *Dark Angel* clearly connote), what is, arguably, missing from this account is an acknowledgment of the pleasures and intricacies of VHS-centered video collecting, and its relationship to issues of authenticity, exclusivity, and the material life histories of objects.

Indeed, if Slater's arguments are considered in light of the issues I've discussed in this chapter, it could be argued that the availability of import DVDs and the accessibility of the Internet may not necessarily cause the death of the pre-VRA video nasty version. If the import DVD version of a nasty title is the most complete and uncut, will it not extend and become the most complete copy in the video version archives of many nasty collectors? If the DVD is smaller and differently packaged and shaped, does this suggest that it will be valued, by some, as an entirely different aesthetic object and hence form the basis of an entirely new collection, rather than replacing a VHS one?

Perhaps most significantly, as DVDs become more accessible, and forms of acquisition via the Internet become quicker and less risky, will the original video nasty versions become even rarer and more historically authentic video objects— the new vinyl to the DVD's compact disc, obtained in ever more intricate, time-consuming, and thus appealing and attractive ways? However the relationship between VHS and DVD develops over time, perhaps, in some ways, the discussions outlined here suggest that there is a need to not only consider video collections where new modern collectibles replace old, but also to consider the ways in which a proliferation of different forms of the video collectible extend the implications, and enrich the possibilities, of ascribing meanings to a film within the confines of the home.

Notes

1. Barbara Klinger, "The Contemporary Cinephile: Film Collecting in the Post-Video Era," in *Hollywood Spectatorship: Changing Perceptions of Cinema Audiences*, eds. Melvyn Stokes and Richard Maltby (London: British Film Institute, 2001), 133.
2. Ibid., 147.
3. Ibid., 139, 134.
4. Ibid., 147.
5. Ibid., 136.
6. Ibid.
7. Charles Tashiro, 'The Contradictions of Video Collecting," *Film Quarterly* 50, no. 2 (Winter 1996–97): 16.
8. Ibid., 15.
9. Klinger, "The Contemporary Cinephile," 144, 138.
10. Tashiro, "The Contradictions of Video Collecting," 12.
11. Susan Stewart, *On Longing: Narratives of the Miniature, the Gigantic, the Souvenir, the Collection* (Baltimore: John Hopkins University Press, 1984), 154.

12. For further background on the video nasties panic and the subsequent implementation of the Video Recordings Act, see Julian Petley, "A Nasty Story," *Screen* 25, no. 2 (March–April 1984): 68–74; Martin Barker, ed., *The Video Nasties: Freedom and Censorship in the Media* (London: Pluto Press, 1984); Kim Newman, "Journal of the Plague Years," in *Screen Violence*, ed. Karl French (London: Bloomsbury Publishing, 1996), 132–43; and David Kerekes and David Slater, *See No Evil: Banned Films and Video Controversy* (Manchester: Headpress, 2000).

13. Kerekes and Slater, *See No Evil*, 289.

14. Stewart, "On Longing," 155.

15. The portrayal of the nasty list as a "shopping guide" occurred frequently in *The Dark Side* during the course of the 1990s. See, for instance, an advertisement for the *Dark Side* book on the video nasties in *The Dark Side: The Magazine of the Macabre and the Fantastic* 73 (June–July 1998): 48–49.

16. This continued focus on covers and labels, to delineate the historical value of the original video nasties and other pre-certificate videos, can clearly be seen in the codes that continued to be used by collectors in classified advertisements in *The Dark Side*. With, for instance, a June–July 2000 advertisement noting that "original horror films" were available, including "many rare films" from such pre-VRA era labels as Go and Vampix, and a June–July 2001 classified including an "original Intervision release of *Dawn of the Dead*, boxed with sleeve" among its list of "wants." *The Dark Side* 85 (June–July 2000): 55; and *The Dark Side* 91 (June–July 2001): 23.

17. *The Dark Side* 71 (February–March 1998): 12; and *The Dark Side* 77 (February–March 1999): 21.

18. Kerekes and Slater delineate the "ardent" video nasty collector as someone who was still "prepared to pay the sometimes exorbitant price for original pre-certificated films," despite the fact that, from the late 1980s, it was much easier and cheaper to get hold of copied bootleg versions of particular banned titles. Kerekes and Slater, *See No Evil*, 293.

19. The name of the collector has been changed.

20. Clifford Geertz, *The Interpretation of Cultures* (London: Fontana, 1993), 3–30.

21. Tashiro, "The Contradictions of Video Collecting," 15.

22. Walter Benjamin, "Unpacking My Library: A Talk about Book Collecting," in *Illuminations*, ed. Hannah Arendt, trans. Harry Zohn (London: Fontana Press, 1992), 64.

23. Not only does the earlier cited Dave Green also collect laserdisc versions of video nasty titles, but a concurrent collecting of DVD versions of nasty titles is also common among evident nasty veterans on *The Dark Angel* message board, and also championed by the most public nasty veteran, Allan Bryce, editor of *The Dark Side*.

24. Klinger, "The Contemporary Cinephile," 137.

25. The re-release of cut sell-through versions of nasty titles in the UK began to occur in 1992, when distributors like Vipco, Redemption, Apex, and Elephant took a lead from Palace Pictures' success at getting a cut version of *The Evil Dead* through the British Board of Film Classification.

26. Dark Angel, "I Spit . . . ," online posting, 28 June 2001, The Dark Angel's Realm of Horror Discussion Forum, http://disc.server.com/discussion.cgi?id=126992&article=2365&date_query =994004884.

27. This letter actually refers to the UK release of Dario Argento's *The Stendhal Syndrome*. As the letter explains, although the distributor, Marquee Pictures, had withdrawn the uncut version of the film and replaced it with the BBFC-approved one, the uncut version could still be sought out and identified via the black-and-white design on the video itself. *The Dark Side* 85 (June–July 2000): 12.

28. Nasty collector, personal interview, September 27, 2001.

29. Dark Angel, "A Twist to the Tale," online posting, June 29, 2001, The Dark Angel's Realm of Horror Discussion Forum, http://disc.server.com/discussion.cgi?id=126992&article=2367&date_ query=994004884.

30. Klinger, "The Contemporary Cinephile," 139.

31. Benjamin, "Unpacking My Library," 62. As should be apparent, this focus on the importance of labels and the establishment of forms of connoisseurship around labels and versions is something that also frequently characterizes record collecting. For further discussion of this link, see Kate Egan, "The Amateur Historian and the Electronic Archive: Identity, Power and the Function of Lists, Facts and Memories on 'Video Nasty'-Themed Websites," *Intensities: The Journal of Cult Media* 3 (Spring 2003), http://www.cult-media.com/issue3/Aegan.htm.

32. Nasty collector, September 27, 2001.

33. Adey, "Re: bloodymoon," online posting, June 28, 2001, The Dark Angel's Realm of Horror Discussion Forum, http://disc.server.com/discussion.cgi?id=126992&article=2361&date_query= 994004884.

34. Tashiro, "The Contradictions of Film Collecting," 16.
35. Susan Stewart, "On Longing," 154.
36. Nasty collector, September 27, 2001.
37. Ibid.
38. Ibid.
39. This tendency to centralize historical markers that authenticate videos and also act as key nostalgic reminders is also pinpointed in Kim Bjarkman's recent article on collectors of television recordings (published since this chapter was drafted). In Bjarkman's account, television recording collectors are seen to valorize such elements as the era-specific television advertisements recorded on each tape, something that highlights, once again, how collecting cultures focused around historical valuations tend to focus on the most appropriate element to contextualize and effectively date the object (whether textual or material). Kim Bjarkman, "To Have and to Hold: The Video Collector's Relationship with an Ethereal Medium," *Television and New Media* 5, no. 3 (August 2004): 217–46.
40. Klinger, discussing Walter Benjamin's work, in "The Contemporary Cinephile," 137.
41. Jay Slater, "The Death of the Video Nasty," *The Dark Side: The Magazine of the Macabre and Fantastic* 90 (April–May 2001): 50–51.
42. Cited in Kerekes and Slater, *See No Evil*, 288.
43. Klinger, again discussing Benjamin, in "The Contemporary Cinephile," 137.
44. For examples of the use of such terms in classified advertisements, see *Fear* 17 (May 1990): 66; *The Dark Side* (May 1992): 66; and *The Dark Side* 47 (July 1995): 61.
45. Stewart, "On Longing," 165, and Jean Baudrillard, "The System of Collecting," in *The Cultures of Collecting*, eds. John Elsner and Roger Cardinal (London: Reaktion Books, 1994), 16.
46. Nasty collector, September 27, 2001.
47. Baudrillard, "The System of Collecting," 23. Indeed, it is notable that, although both Stewart and Baudrillard generally construct collecting as an entirely personal and "transcendent" domain, the determining influence of the cultural and external comes sharply into focus, for both critics, when they discuss the importance of categorization and classification to collectors (and particularly to book collectors).
48. Tashiro, "The Contradictions of Video Collecting," 15.
49. Nasty collector, September 27, 2001.
50. Stewart, "On Longing," 160.
51. Baudrillard, "The System of Collecting," 12.
52. Nasty collector, September 27, 2001; Nasty collector, personal interview, June 14, 2001.
53. Igor Kopytoff, "The Cultural Biography of Things: Commoditization as Process," in *The Social Life of Things: Commodities in Cultural Perspective*, ed. Arjun Appadurai (Cambridge: Cambridge University Press, 1986), 81.
54. Nasty collector, September 27, 2001.
55. Ibid.
56. Kerekes and Slater also acknowledge the existence of these two types of trader in the video nasty collecting world. As they note, although "many [market] traders unwittingly threw these tapes in together with legitimate releases, others understood their black market value and potential risks involved, only offering them to select clients." Kerekes and Slater, "See No Evil," 289.
57. Nasty collector, September 27, 2001.
58. Ibid.
59. Nicky Gregson and Louise Crewe, "Beyond the High Street and the Mall: Car Boot Fairs and the New Geographies of Consumption in the 1990s," *Area* 26, no. 3 (1994): 262. For further discussion of the cultural implications of such secondhand spaces, see Nicky Gregson and Louise Crewe, *Second-hand Cultures* (Oxford: Berg, 2003).
60. Again, the potential for video collectors to adopt malleable roles is also centrally raised and explored in Kim Bjarkman's article on television recording collectors. Bjarkman, "To Have and to Hold," 217–46.
61. Louise Crewe and Nicky Gregson, "Tales of the Unexpected: Exploring Car Boot Sales as Marginal Spaces of Contemporary Consumption," *Institute of British Geographers Transactions* 23 (1998): 48.
62. The only two pieces of evidence I've uncovered that may suggest that others indulge in such practices are a September 1995 *Dark Side* classified advertisement selling individual UK video covers, and a 2001 discussion on the *Dark Angel* message board, where a participant suggests, rather mysteriously and elliptically, that a particular Internet covers site might be useful for those who "need covers for some of your vids, due to obviously misplacing . . . them." *The Dark Side* 49

(September 1995): 44; and Bubblezzz, "Video Coverzzz,' online posting, August 7, 2001, The Dark Angel's Realm of Horror Discussion Forum, http://disc.server.com/discussion.cgi?id=126992&article=2881&date_query+997540380.

63. Nasty collector, September 27, 2001.

64. Benjamin, "Unpacking My Library," 68.

65. Ibid., 63. Crucially, in the following issue of *The Dark Side,* editor Allan Bryce confirmed that the article had "sparked a fair amount of controversy" among "most readers," with one printed letter arguing that the availability of DVD versions didn't "make the original video releases worthless" and that "the original versions" still sold "for big money to the 'true' collector." *The Dark Side* 91 (June–July 2001): 3, 10.

66. Klinger, "The Contemporary Cinephile," 133.

67. Slater, "The Death of the Video Nasty," 49.

68. Ibid., 48.

12 Going Analog: Vinylphiles and the Consumption of the "Obsolete" Vinyl Record

John Davis

Now that we are well into the age of digital media, when even compact discs are said to be failing,[1] it might be worthwhile to ask: what happened to vinyl records? Even though the greater weight of scholarly attention has been directed toward the study of diffusion and consumption of new forms of popular media, activities surrounding the decline of older technological forms, including analog media formats such as the vinyl record, should also stimulate scholarly analysis and discussion. The aging of the vinyl format, and its continued importance to a group of vinyl-oriented vinylphiles, offers a case study of what can happen when an established, once-popular medium becomes a residual medium.

Examining the case of the vinyl record format provides an interesting view of how forms of residual media—and their production and consumption—are shaped by social forces. Vinyl records, as discussed here, include two different analog formats for recorded sound: first, the 33⅓ RPM LP (Long Play) album, and, second, the 45 RPM (or 7-inch) single. These formats originated in the so-called format wars of the 1950s, as the recording industry sought to revivify a stagnated music marketplace with "high-fidelity" microgroove records.[2] The status of vinyl records remained remarkably stable until the late 1970s, when, affected by global economic recession, the worldwide marketplace for recorded music contracted significantly, dropping some 11 percent in the United States alone.[3] The collapse of disco was also blamed for a sharp decline of music sales in both record and audiocassette tape formats in the period.[4] Prerecorded audiocassette sales, however, began to improve the industry's prospects in the early 1980s, buoyed by the popularity of new portable cassette players such as the Sony Walkman Mark II and the boombox, as well as in-dash players for automobiles.[5] At the same time, both the LP and 45 RPM formats began to lose market share, which accelerated with the introduction of the compact disc (CD) format.

In the United States, CDs accompanied changes in the manufacturing and retailing of recorded music. Major record labels began to slow the release of new albums on vinyl, shifting production to cassettes and CDs.[6] Even though the National Association of Record Manufacturers (NARM) argued for the commercial viability of vinyl, major labels began to reduce their back catalog in the format while shifting newly produced catalog releases to CD, and, by 1985, sales of audiocassettes overtook vinyl.[7] The major turning point for the vinyl format came in 1989, when the seven major labels stopped accepting unsold records and began unloading their back catalog in vinyl at reduced prices.[8] This decision made it extraordinarily difficult for music retailers to continue stocking even new vinyl record releases.[9] The reduction in new vinyl titles, the halt in catalog releases on the format, coupled with the increased sales of the cassette and, later, the CD format, helped doom the vinyl record format in the mainstream marketplace for recorded music.

Even as the record was disappearing from the mainstream, however, a subculture of vinyl aficionados emerged who refused to abandon the format. Plasketes described this subculture as composed of white males who were traditional vinyl-record buyers and fans of genres such as alternative rock, the blues, R & B, and soul.[10] The predominance of men in this subculture led Straw to describe the practice of vinyl collecting as a "homosocial" activity that allowed heterosexual men to deal with the uncertainties of maintaining male-to-male nonsexual relationships.[11] Later, Shuker noted the vinyl record collectors he interviewed reported that acts of collecting were themselves the central motivation and source of pleasure in vinyl record collecting.[12] The creation of a collection not only gave collectors pleasure but displayed the collector's acquisition skills to other interested parties. Writing about the record collections of electronic dance music DJs, Thornton described the display of a record collection as the physical manifestation of "subcultural capital," a physical representation of the collector's taste and sophistication in their genres of interest.[13]

In the years since these articles, the vinyl format has receded into even greater cultural obscurity, even though it is arguably easier today to acquire secondhand vinyl records than it was in the early 1990s.[14] The vinyl collector, armed with a personal computer and access to the Web, can pursue the targets of their interest in a global marketplace drawing on an informal network of vinyl hunters who search for their prey in the dense jungles of secondhand pop culture artifacts. Interviews with vinylphiles indicate vinyl collecting continues among these aficionados even as the format's obsolescence becomes ever more pronounced.[15]

In this age of digital media, the vinyl record format *is* obsolete, in the technological sense. Obsolescence, as used here, reflects both a formal status of technical incompatibility with popular and widely used digital media such as CDs, iPods, and various PC-based media systems, but also a symbolic status. This chapter is less concerned with the issue of technological difference than with the ways in which the definition of obsolescence haunts the vinyl format. As Williams explains, technologies are cultural forms, both transmitting meaning in the form of content but also encoded with meaning in their design and function.[16] In terms of media formats, obsolescence is a best concieved as a symbolic status accorded certain outmoded technologiescontinue to survive. Media technologies that age and are no longer used for any purpose should perhaps be considered dead media, a concept originated by author Bruce Sterling.[17] Obsolescence is a discursive construction that involves technological aspects but is not necessarily determined by them; rather, whether a medium is obsolete depends largely on its status as shaped by structural forces both formal and informal.

Social shaping theory contends new technologies emerge from the interaction of groups involved not only in research and development, but also other social forces involved in manufacturing, regulation, and marketing.[18] The collective influence of these groups shapes the intended uses of a medium, which are then encoded in aspects of product design and commercial advertising.[19] Although these social forces exert direct control over most forms of new technology, media technologies produced as consumer products are also subject to the influence of another group, whether we call them audiences or users. Once a medium enters the consumer marketplace, its uses and even its meanings can change as people adapt it to their existing and emergent needs. In du Gay, Hall, Janes, Mackay, and Negus's study of the evolution of the Sony Walkman, the Sony Corporation was described as including dual headphone jacks on its original version, intending the unit to allow simultaneous music listening for couples.[20] Instead of using the Walkman together, however, early adopters simply bought their own individual units. This form of adaptation lead Sony to remove the second output jack from future versions of the technology, as well as influencing a shift in its advertising to promoting the Walkman as a portable medium for personal music listening.

The vinyl record format was itself shaped by similar forces during the course of the twentieth century. At the dawn of the era of the talking machine, acoustic recordings were considered "inauthentic" reproductions of live music performance.[21] Consumers of early recorded sound technologies, however, seem to have been less concerned about fidelity than with the novelty of recorded sound, and, later, of the pleasures of the listening experience.[22] The public grew to prefer the phonograph as an entertainment medium, rather than as the utilitarian office dictation device intended by the Edison Company.[23] Edison had also sold the phonograph for creating recordings at home, but phonograph salespeople and others developed "coin-in-the-slot" players (the forerunners of the later jukeboxes) for public consumption at entertainment arcades. The commercial success of these coin-in-the-slot players peaked in the first decade of the twentieth century, but the redefinition of the record as an entertainment medium shaped the emergence of an industrial system of production devoted to the manufacture of recorded music.[24] Edison's chief rival, the Victor Talking Machine Company, skillfully adapted to the situation, using advertising as a means not only to define recorded music as a product in itself but also to shape record buyers into a "discerning" audience with sophisticated new tastes.[25]

With the establishment of the recording industry, new developments in sound and music recording technologies were shaped by companies involved in making the records. As described by Andre Millard, this history was one of bitter rivalries, culminating in the "format wars" of the 1950s, from which emerged the vinyl medium. Vinyl records became the quintessential form for recorded music for most of the latter half of the twentieth century, and its decline was considered representative of the extensive social and cultural changes occurring in American life at the dawn of the twenty-first century.[26]

The *fin de siecle* accounts of the death of vinyl also revealed the existence of a diehard group of vinyl fans, unwilling to abandon the now-obsolete medium. In a larger sense, the existence of such a group reveals the significance of recorded music audiences as an influence on the establishment of meaning of residual media. The decline of vinyl records as a popular music format was shaped by music labels, retailers, and consumer electronics companies, but this activity did not render vinyl records obsolete for everyone. A group of vinyl aficionados evolved for whom the format is more than a medium for music consumption. These vinyl record vinylphiles are best understood as a group of medium-specific collectors for whom the format's evolving significance motivates ongoing acquisition and possession.[27] Moreover, the actions of vinylphiles—the collecting practices that keep this residual medium alive—continue to reshape the meaning of the vinyl record format.

In consumer behavior research, collecting is envisioned as a category for practices involving the acquisition and possession of "things removed from ordinary use and perceived as part of a set of non-identical objects or experiences."[28] Certain cultural artifacts deemed of personal and/or cultural significance—such as postage stamps, comic books, or Beanie Babies—are sought and added to existing repositories of similar (but individually exceptional) objects. Collectors can render even mundane objects meaningful by interpreting them as special and unique, or linking up those objects to others, creating a set that has its own significance as a "collection."[29] Research indicates that collections are often formed by a desire for security or are based on an aesthetic sense linked to the uniqueness of objects.[30] Belk has argued that mass-produced media forms are not unique, and thus not collectible, but vinylphiles engage in activities very similar to other forms of collecting.[31] Those who collect vinyl in the late stages of the format's decline often respond to it as a special, almost sacred artifact.

For vinylphiles, vinyl's sacred element is strongly associated with a sense of the format's authenticity. Belk, Wallendorf, and Sherry write about consumer products that come to be recognized as the quintessential form of a consumer good, such as Harley-Davidson motorcycles, Zippo lighters, and Levi's blue jeans.[32] Among vinylphiles, the vinyl record is the quintessential recorded-music format, providing a truly authentic listening experience superior to existing digital media. The listening experience includes, but is not limited to, the sound of the vinyl record, from its scratches and pops to the perception of its more extensive bass range. Vinylphiles also respond to the format in a textual sense, as a cultural form that itself invokes meaning. The quintessential qualities of the vinyl record include its size (7 or 12 inches), its cardboard sleeve, cover art, enclosed materials such as liner notes, and even the color and shape of the record itself. For vinylphiles, these elements provide an essential accompaniment to the listening experience.

Among vinylphiles, vinyl record consumption also extends to the act of playing the records. Taking the album off the shelf, removing the record from its sleeve, placing it on the turntable, cleaning it of dust, and even putting the tonearm on the record offers a sense of authenticity to the consumption process that CDs cannot evoke. As one vinyl aficionado put it,

> [Playing vinyl] is just more of an experience, the experience of playing and actually getting [out] the vinyl and putting it on. There is something kind of exciting about . . . owning something that was actually manufactured and produced in that time frame, [knowing it was] produced in 1965 and that is when it came out. It is an original issue, it is not a reproduction, it is not something that was recorded and dubbed over 25 years later. This is the real deal.

The physical act of playing the record invokes a holistic listening experience that involves past associations with the music and/or with the medium itself. By comparison, the CD experience for vinylphiles suffers in all of these areas, although the degree of acceptance to the CD format partially defines the social boundaries of divergent vinylphile groups.

The social boundaries of vinylphile groups are largely defined by affinities for certain musical genres. Vinylphiles should be first understood as members of larger music-consuming subcultural groups, including indie rock fans, analog audiophiles, and those who are vinyl collectors in the strictest sense. Vinylphiles are arguably on the fringes of these various musical subcultures: the popular "rock music" mainstream, classical music audiences, and collectors of other secondhand goods. The unifying distinction among vinylphiles is a maintained affinity for the vinyl record format and an ongoing engagement with the format through practices of acquisition and continued possession of a collection.

Of all vinylphile groups described here, fans of indie rock music navigate closest to the mainstream popular music marketplace, even though they are generally the furthest from that marketplace in terms of generic interests. Indie rock is a label for rock music produced by "financially sound independent distribution companies (for) independently owned retail record stores and college radio stations."[33] The indie rock scene in the United States emerged from the punk rock subculture described by Hebdige and Lull,[34] and the later postpunk indie labels have traditionally employed vinyl releases to differentiate themselves from major labels and artists.[35] Despite the shift to CDs in the mainstream market, indie rock artists, labels, and audiences employ the vinyl format as a material signifier of status in many of the same ways Thornton describes in her study of vinyl-wielding DJs.[36]

For indie rock fans, the production of vinyl records signifies a kind of subcultural legitimacy for artists as much as the consumption of vinyl does for their audiences. To release an album or single on vinyl is an indicator of authenticity marking the artist or audience member as a sophisticated consumer not beholden to the commodification of music practiced by the corporate recording industry—as well as that industry's emphasis on CDs. Although independent music scenes are in perpetual threat of cooptation by the mainstream, the importance of authenticity in the indie music subculture traditionally kept these artists within the independent realm, lest they lose appeal for the indie rock audience.[37] Although indie rock record labels more or less follow (on a much smaller scale) the models of production, promotion, and distribution applied by the major conglomerate-owned record labels that produce popular music, some labels have an orientation to the marketplace that offers a clear and specific critique of the mainstream system of popular music production. At least in the early 1990s, releasing indie rock on vinyl indicated allegiance to an alternative to the major labels' collective move to CDs and the reduction and eventual elimination of vinyl. Today, in the wake of the vinyl record's almost-total invisibility in the popular music marketplace, coupled with the increased cost for indie labels to release music on vinyl, what remains of the indie affinity for the format is almost entirely a strategy to reach a small, sophisticated, and exacting audience of indie rock aficionados.

A similarly sophisticated vinylphile group, the analog audiophiles, seek out the vinyl record format primarily for its aural qualities. Among these vinylphiles, vinyl records are considered a superior medium for recorded sound, at least in comparison to other media. Audiophiles in general are music listeners who express a preference for lifelike high-fidelity sound, with "a strong emotional investment in their stereo systems," which are constantly upgraded or "tweaked" to optimize their fidelity.[38] Analog audiophiles are a smaller group within the larger audiophile community, one whose members seek out vinyl records because their listening skills discern aural problems with other formats. To analog audiophiles, most CDs offer inferior sound reproduction when compared to the aural qualities of a well-recorded pristine vinyl album on an optimized stereo system. Analog audiophiles employ specially constructed and precisely maintained playback equipment, including vacuum tube amplifiers, and high-end turntable components (tonearm, cartridge, platform) to enjoy "high-fidelity" sound. For analog audiophiles, the vinyl record format is *the* essential component in this chain of recorded sound technology. Because surface flaws, such as dirt, dust, and scratches, can add unwanted sound to a vinyl record, analog audiophiles also prefer their records to be in pristine, or near-perfect condition. A record that is scratched or warped—even slightly—is undesirable chiefly because that record is technically unable to reproduce a high-fidelity sound.

The final group of vinylphiles discussed here should be considered purely vinyl collectors, with emphasis on the *collector* element. Unlike indie rock fans or analog audiophiles, vinyl collectors are motivated to seek out the format for its connections to the past, either in terms of an individual or collective experience. Vinyl collectors are a subgroup of the larger subculture of collectors of nonmusical pop culture artifacts, as well as collectors of music-related artifacts other than vinyl records. Vinyl collectors also resemble some popular music audiences. Like indie rock fans, vinyl collectors focus their acquisition efforts on certain specific genres, including jazz, hard rock, or classic country music, but they are also known to range across other music genres. Although vinyl collectors tend to prefer pristine or "clean" records to those with flaws (not unlike analog audiophiles), vinyl collectors often pick up records that are not in excellent condition, if a record is known to be rare or desireable. The ultimate goal of the vinyl collector is to acquire the desired album by a specific artist or work that is part of a preferred genre *on vinyl*. Without such an orientation to the format, the vinyl collector would be largely unremarkable in their similarity to most other audiences for recorded music.

Vinyl collectors, however, are different from other vinylphiles in terms of their unique orientation to the past. The various symbolic elements of the record, from the music to the form of the record to its packaging and the practices involved in playing it, can resurrect elements of the past that are either personally important, like an album of family photographs, or culturally significant, like a documentary film. Qualities such as the unique sound of old vinyl, including the scratches and pops of well-used records, and the ways in which records were marketed and purchased are often rooted in collectors' past experience. For other collectors, vinyl is associated with some larger historical sense as an authentic artifact of the times or as a symbol of a past not experienced directly. Old records, produced in the past, are considered by vinyl collectors to be more authentic than CDs, even though reissues can often be found more easily on CD. The vinyl record, for these collectors, offers a material symbol bridging the present and past.

For some vinyl collectors, certain records have even greater personal significance. In one sense, the vinyl record can refer to one's personal history, where it is a reminder of a directly experienced past. Belk envisions collectible objects as physical parts of "the extended self," and for vinyl collectors old records provide an extension of their interior life world.[39] For many of vinylphiles interviewed, the vinyl record format held a position of significance in their early life. Vinyl records are described as providing a very significant connection between experiences of listening to recorded music and to other major and minor life experiences. One vinylphile recalled a childhood experience of singing along with soft rock records on a peaceful Saturday morning in the 1970s, while another remembered painful memories of their alcoholic "drinking years," a phase they associated with their 1950s honky-tonk records. These responses illuminate the kind of "curatorial collecting" described by McCracken, with vinyl records, much like old photographs, providing an artifact manifesting memories of the past.[40] To the vinyl collector, maintaining a contemporary collection of records revives pleasant memories of childhood, youth, and/or young adulthood.

For some vinyl collectors, the influence of personal experience on the sense of the vinyl record as an "authentic" medium cannot be underestimated. Vinyl records were, in some cases, the first consumer products they remember wanting or the first ones they owned. For these vinylphiles, the vinyl format stands as the preeminent medium for music consumption. These elements of attached meaning are largely—but not always totally—absent from vinyl collectors' experiences with the CD format. Even for collectors, however, the recorded music on the record, rather than the format itself, is the primary text to which meanings are attached.

For some vinyl collectors, records have a cultural significance beyond the individual level. As a residual medium, secondhand vinyl records are documents of cultural history. Beyond offering a reminder of one's personal history, a vinyl record can be engaged as a referent to a past *not* directly experienced by the vinylphile. Vinyl records are like documents of specific times, places, and/or events that may be linked to a more generalized social or cultural experience. A well-known musical release such as the Beatles' *Sgt. Pepper's Lonely Hearts Club Band,* for example, can recall the Beatles' position in rock music history, the album's position within the developing musical genre of rock, public events occurring at the time of the record's release (i.e., the Vietnam War, recreational drug use, etc.), and so on.

Although each of the individual groups just described have their own unique motivations for collecting the vinyl record format, indie rock fans, analog audiophiles, and vinyl collectors should be considered as constituents of a medium-specific vinylphile subculture. Although these groups are fairly isolated from one another in terms of social memberships, they do share an affinity for this obsolete medium as well as a number of record-collecting practices. The practices of vinylphiles resemble other kinds of specialized collecting *rituals* at play among other pop culture consumers, especially in the acts of vinyl record *acquisition, possession,* and *use.* Each of these activities reflects, reveals, or reemphasizes the vinyl record's sacred nature for the vinylphile or for others who come into contact with the record collection.

The trajectory of vinylphile record collecting begins with *acquisition,* first of information about records and, later, of the records themselves. Acquisition rituals are specialized methods used to obtain artifacts of personal significance. Acquisition is the initial step in the transformation of an object from its profane, commodity state to one in which its sacred nature is revealed and celebrated.[41] Practices of acquisition assist in the *recontextualization* of a common object chiefly by its removal from its location of origin and placement in new contexts that display its special nature.[42] The acquirer's home or "life space" is a particularly important context in this respect, as it is a place where acquired artifacts can be juxtaposed with other special objects, including other collections. Moving the item into one's life space also helps further divest an artifact of past associations, including its status as a commodity or as something once owned by others.

Vinylphiles acquire secondhand records primarily for their own personal enjoyment, but some enterprising aficionados acquire records for resale to others. This distinction is common enough among nonvinylphile collectors that Belk describes it as a difference between *terminal* and *instrumental* collecting.[43] In instrumental collecting, artifacts are acquired to help accomplish other goals, either through sale or trading of the object of interest. By contrast, terminal collecting involves the intent to acquire objects only for the purposes of entering them into a permanent collection. Generally, vinylphiles engaged in terminal collecting are careful to distinguish their practices from those with instrumental motivations, but even the latter group actively separate their personal collections from those bought for resale. One jazz record collector who was interviewed kept his terminal collection in the living room near the stereo system, while the instrumental collection was kept further away inside a closet whose door was closed. Vinylphiles who otherwise engage in terminal collecting may also remove records from their private collections and sell them for a variety of reasons, including using the profits to buy records they desire for their own collection.

For many vinylphiles, acquiring vinyl records is described as a form of artifact *hunting*. Hunting involves the search for certain desired vinyl records among stores of less-desirable records and other commodities.[44] Among vinylphiles, it is more common to talk about hunting vinyl records than to use the more generic term "shopping," which Straw notes transforms the experience from a feminine behavior to a more masculine one.[45] The two major practices of hunting among vinylphiles include *learning* about what vinyl records are available and *salvaging* them from their isolation among other profane objects.

Learning involves the identification of significant objects. What is acquired in learning is information that assists in the identification of desired artifacts, whether a rare recording by Italian progressive rock group PFM or the release of a new vinyl LP by indie rock group Pavement. The importance given to the acquisition of information among vinylphiles indicates the "seriousness" of their pursuit of records and helps them individuate vinyl collecting from other forms of music consumption. Shuker describes how vinyl collectors he interviewed "constantly referred to both the effort and the pleasure involved in systematically gathering information from peers, older siblings, the music press . . . collector magazines, discographies, price guides, and the back catalogue."[46]

In a marketplace of independent record stores and a variety of secondhand retailers such as thrift stores, antique shops, yard sales, and online and offline auction services, acquiring information can identify a potential object of desire and/or orient the vinylphile to the "hunting grounds." For vinylphiles, learning about records, and about where to find them, requires engagement in interpersonal and mediated communication. Word-of-mouth communication is a key component of social action among vinylphiles, a practice that not only spreads knowledge about records and music but also enables acquisition of records. At a local level, vinylphiles gather information from peers when shopping for vinyl among various outlets. Members pass along what records are in stock, the quality of the selection, what genres a retailer specializes in, the condition of the records, prices, personalities of the store personnel, and, for "bricks-and-mortar" outlets, the store's atmosphere/environment.

On a wider scale, vinylphiles also acquire information from media sources such as vinyl record price guides, periodicals, and the Internet. Price guides, such as Tim Neely's *Goldmine Record Album Price Guide* (2001) and Les R. Docks's *American Premium Record Guide* (2001) have two uses, the first being as standardized sources for the cash value of secondhand records and the second as a reference for discographies of past releases by musical artists. Periodicals devoted to the topic of vinyl record acquisition are extremely genre specific, with the exception of *Goldmine* magazine, mentioned at least once by almost all vinylphiles who were interviewed. *Goldmine*, billed as the "world's largest marketplace for collectible records, tapes CDs, and memorabilia," is a biweekly periodical published by Krause Publications. *Goldmine* provides not only articles about both popular and obscure musical artists but also attracting a vast array of advertisers from across the United States selling secondhand vinyl records. Finally, an extremely important medium for information sharing among vinylphiles is the Internet. On the Internet, the most popular sources of vinyl-related information are Usenet newsgroups and Web sites. Among the tens of thousands of public newsgroups—covering topics both familiar and obscure—there are at least two groups specifically devoted to vinyl records: the first, rec.arts.music.collecting vinyl, that focuses on topics and issues related to the collecting of. vinyl records; and the second, rec.arts.music.marketplace. vinyl, which focuses on the actual buying and selling of records among individuals. Web sites also function as the electronic equivalent of the community bulletin board or public newspaper, and indeed, many of the significant print magazines related to vinylphile interests (*Goldmine*, *Discoveries*, etc.) also have electronic counterparts. Often these Web sites overlap with retailer sites, including those for record labels and vinyl distributors.

Beyond the discovery of information about records, vinylphiles also actively hunt records in these various venues and salvage them from isolation in the world of common, profane objects. According to Belk, Wallendorf, and Sherry, "Collectors often sacralize objects by finding and rescuing them from those who do not understand the objects' worth or value."[47] Salvaging "sacralizes" an object by removing it from locations where it is presented as common or unspectacular, then moving it into the collector's personal possession and thereby divesting it of its mundane status. Collectible objects can be salvaged from immediate physical threats, such as abandonment or being discarded, but for most vinylphiles the rituals of acquisition consist of purchasing a record directly from its owner or a retailer. Although this exchange has profane associations, buying a desired object can establish a sacred quality. For collectors who have no other practical route to acquiring a desired record, buying is an essential act that starts the separation of the object from its previous owner. Moreover, the cost to acquire the record can become one element of its sacred quality, as when one pays hundreds of dollars to acquire a much-desired LP. There can also be an inverse relationship between price and a vinyl record's sacred nature, as when an extremely rare album is found in a "Used Records" box at a flea market and purchased for a dollar while among aficionados it is worth thousands more.

Prior to the act of purchase, however, many vinylphiles engage in salvaging rituals best thought of auditioning. Auditioning rituals involve practices intended to verify that the vinyl record in question is still a viable medium for aural listening. A beat-up secondhand vinyl copy of the soundtrack to the 1967 motion picture *Casino Royale* has value to collectors regardless of its condition, but the playability of the record still means a great deal, even for vinylphiles who are not analog audiophiles. Prior to purchase, many vinylphiles attempt to gauge a record's condition by checking the record visually for dust and scratches or by previewing it on a nearby record player. Some vinyl record retailers offer turntable units for auditioning the record, recognizing the difficulty of estimating playability by sight. One vinylphile who was interviewed described bringing a portable, battery-powered record player to audition secondhand 7-inch singles at flea markets and antique shows. Because physical condition is a major determinant of the monetary value of a vinyl record (rarity is the other), auditioning is an important mechanism in the process of acquisition.

After acquiring their desired records, vinylphiles engage in activities of possession by which those acquisitions become part of an already-established collection. Whereas acquisition rituals help vinylphiles identify and gather desired records, transforming them into significant objects, practices of possession are those that both individualize those records and make them part of a more meaningful whole.[48] Once found and acquired, artifacts must be introduced into a symbolic and symbiotic relationship with others in the acquirer's life world. The status of a particular artifact reflected in its inclusion or exclusion from the space set aside for the collection of similar objects. In their placement among those other records, the individual record "take(s) on meaning beyond their individual existence."[49] Essentially, possession rituals construct an identity for an object that reflects both its individual existence and its place in a symbolic collective. Practices of possession are ways "to store, display, and conserve" the meaning(s) of each individual object as well as the collection as a whole.[50]

Among vinylphiles, rituals of possession include collectivizing and preserving vinyl records. Collectivizing rituals are activities that transform an individual object from an isolated piece into an artifact of significance within a collection of similar artifacts. A collection is a collectivity of objects that are "special, unique, and separate from the everyday items [collectors] have and use."[51] Making a collection also "transfer(s) to goods . . . the meaning of the collectivity" of all the individual objects.[52] Those individual objects do not lose their own unique status, however.[53] To adhere to a standard of "no two alike," an object must be different enough to require inclusion while being similar in some way that makes it worthy of addition to the collection. A vinylphile, for example, might place a rare early recording by country music artist Buck Owens among other recordings by Owens, including those that are fairly common. The same record might also, in another vinylphile's collection, be placed among recordings by contemporaries such as Conway Twitty or Ferlin Husky, or in another collection, among records of the country music genre, and so on. In totality, each individual object's own unique status both adds and reflects a collection's symbolic completeness or unity.

Even though not all vinylphiles are collectors in the traditional sense, vinylphiles regard and describe their accumulated vinyl records as their "record collection." The collection is defined as a whole or as a set, wherein all the vinyl records are symbolically unified as "my record collection" or the "core collection" in relation to other cultural materials, including other collections of cultural artifacts, from designer furniture to cartoon-character lunchboxes. A core collection is made up of the records that the vinylphile chooses to retain, for whatever reason, and generally will not part with. This collection receives the bulk of the vinylphile's attention, and maintaining it is concomitant with preserving the physical integrity of each individual record.

The other major set of possession rituals should be called *preserving* practices. Preservation of vinyl record collections involves efforts to sustain the sacred nature of the records in one's possession. Among vinylphiles, practices of preservation not only serve to maintain a pure state, true to their essential nature and unsullied by contact or incorporation with other profane objects in the collector's life world. One aspect of preservation is maintaining the collection itself as a unique artifact in itself within the vinylphile's home.[54] Preservation practices also seek to maintain the physical condition of special objects over time. Among vinylphiles, storage practices resemble those of comic book collectors who use plastic or Mylar sleeves and acid-free cardboard boxes to preserve their comics from the negative effects of direct sunlight and elevated humidity levels, changes in temperature, and damage through physical handling.[55] Other curatorial tools include acid-free paper sleeves to preserve the record itself inside the card stock cover and acid-free cardboard storage boxes with lids.

Where possession rituals allow vinylphiles to maintain a record collection and foster a sense of its uniqueness, rituals of *use*, the final set of collecting practices discussed here, involve actions that engage the functional attributes of vinyl records. Sacred objects are typically not for everyday use, but vinylphiles frequently develop specialized practices of use that help preserve the status of their records.[56] In some instances, rituals of use are required to refresh or reinvigorate the collector's interest in the object. Such practices are synonymous with the kinds of "grooming" rituals described by McCracken, which typically involve preservation activities, including cleaning, reorganizing, or "refreshing" those objects of significance.[57] These practices allow vinylphiles the means to enjoy the content recorded on the record while preserving the record's unique qualities and status. The physical properties of the vinyl record medium require both rudimentary and complex maintenance, from cleaning and preserving the turntable stylus to removing dust and grime from the record's surface. Failure to practice maintenance risks the chance of permanent damage to the record, and, by extension, a potential disruption to one's enjoyment of the recorded content and/or the listening experience.

Although enjoyment of music content is a central motivation common to most audiences, medium-specific rituals of use among vinylphiles make vinyl record listening different from other forms of recorded-music consumption. Maintenance practices help vinylphiles to sustain the uniqueness of their collected records, both in terms of form and sound. For most vinylphiles, listening to a vinyl record requires a systematic process, especially among those—such as the analog audiophiles—for whom the perfection of sound is of absolute importance. Playing a vinyl record in a way that optimally draws out the best sound can be a process more complex than using other recorded sound media. Procedures such as adjusting the turntable tonearm, replacing the turntable stylus, as well as cleaning activities intended to remove dust and grime from the record, are all intended to prepare the record for optimal sound reproduction.

Sometimes, the value or importance of a vinyl record's condition must be weighed against the desire to listen to the recording on the record. To preserve the special qualities of the most valuable records they own, some vinylphiles create *surrogates*, or copies of the album or song in question on another format. A surrogate, such as a copy on audiocassette tape, helps maintain the condition of the original record and can allow the vinylphile to listen to the music outside of the home. Vinylphiles also make surrogates for others to enjoy, reinforcing their own status while facilitating the aural experiences of others. Most vinylphiles would prefer to listen to their music from the actual vinyl record itself, but surrogates provide access to music when vinyl is not feasible, such as when jogging or driving an automobile.

All of the collecting rituals described here help vinylphiles sacralize the vinyl record format, making it a special and unique component of their life world. By engaging in these practices of acquisition, possession, and use, vinyl collectors not only respond to the meanings encoded in this obsolete format, they also actively shape those meanings. This situation provides a niche for the vinyl record that keeps the format from slipping further into obscurity and maintains its status as a residual medium.

Vinylphiles should be understood as a medium-specific subculture, one whose members cross the virtual boundaries of music genre. The existence of groups strongly oriented to specific residual media is a phenomenon that could stand further investigation. Vinyl record collecting emerged from a specific set of conditions—from the long successful history of vinyl to the shaping of its obsolescence—that have offered vinylphiles distinctive cultural opportunities. The larger decline of physical storage media from the vinyl record to the CD has also been predicated on new developments in how media content is produced, encoded or recorded, packaged, and distributed. The obsolescence (planned or otherwise) of contemporary media forms such as the CD and the DVD could foster their own unique collecting situations. On the other hand, new media forms, including online music streaming, personal video recorders (PVRs), video-on-demand (VOD) services, and portable devices such as the Apple iPod and iRiver's multimedia players, challenge our traditional sense of the media collection itself. These new technological forms have already inspired their own forms of consumer adaptation, from illicit digital media file exchange to multimedia mashups, and the degree to which these new technologies deliver on the long-promised "celestial jukebox"[58] will strongly influence how people collect recorded music, television shows, and motion pictures in the future. We need to be attentive not only to the ways emerging media forms are made obsolete but also to how groups of users and audiences adapt to these kinds of changes and perhaps contribute to the survival of future residual media.

Notes

1. Kevin Hunt, "Compact Disc Obsolescence Could Be Just a Decade Away," *(Louisville, KY) Courier-Journal*, August 20, 1999, E4; Wilson Rothman, "DVDs? I Don't Rent. I Own," *New York Times*, February 26, 2004, 1G; Ben Raynor, "At the Age of 20, the CD Is Crashing and Burning," *Toronto Star*, November 16, 2002, J10.

2. Andre Millard, *America on Record: A History of Recorded Sound* (New York: Cambridge University Press, 1995).

3. Simon Frith, "The Industrialization of Popular Music," in *Popular Music and Communication*, ed. James Lull (Newbury Park, Calif.: Sage Publications, 1992), 49–74.

4. Ken Terry, "1982–1992: Talkin' 'Bout a Revolution," *Billboard*, September 26, 1992, CD4.

5. Matthew Killmeier, "Theorizing Automotive Radio: Prosthesis, Technology, and Cultural Form," paper presented at the annual meeting of the Association for Education in Journalism and Mass Communication, Phoenix, Arizona, August 2000); Paul du Gay, Stuart Hall, Linda Janes, Hugh Mackay, and Keith Negus, *Doing Cultural Studies: The Story of the Sony Walkman* (Thousand Oaks, Calif.: Sage Publications/The Open University, 1997); George Plasketes, "Romancing the Record: The Vinyl De-evolution and Subcultural Evolution," *Journal of Popular Culture* 26, no. 1 (1992): 109–22.

6. Terry, "1982–1992: Talkin' 'Bout a Revolution."

7. Earl Paige and Geoff Mayfield, "NARM Meet Concludes LP Viable," *Billboard*, October 4, 1986, 39, 41; Ken Terry and Dave DiMartino, "Majors Accelerate Vinyl Phase-out," *Billboard*, September 17, 1988, 85.

8. Negativland, "Shiny, Aluminum, Plastic, and Digital" (2000), http://www.negativland.com/minidis.html.

9. Plasketes, "Romancing the Record."

10. Ibid.

11. Will Straw, "Sizing Up Record Collections: Gender and Connoisseurship in Rock Music Culture," in *Sexing the Groove: Popular Music and Gender*, ed. S. Whiteley (New York: Routledge, 1997), 3–16.

12. Roy Shuker, *Understanding Popular Music*, 2nd ed. (New York: Routledge, 2001), 202.

13. Sarah Thornton, *Club Cultures: Music, Media and Subcultural Capital* (Hanover and London: Wesleyan University Press, 1996).

14. This article deals exclusively with vinyl record collectors who are not electronic dance music or rap and/or hip-hop DJs. For a treatment of issues specific to residual media and the DJ, see chapter 6 (Rietveld) in this volume.

15. John D. Davis, "'Vinylphilia:' Consumption and Use of the 'Obsolete' Vinyl Record among the Vinylphiles," unpublished doctoral dissertation, University of Kentucky, Lexington, 2003).

16. Raymond Williams, *Television: Technology and Cultural Form* (New York: Schocken Books, 1974).

17. Bruce Sterling, "The Life and Death of Media," Electronic Frontier Foundation, (2003), http://www.eff.org/Publications/Bruce_Sterling/Dead_Media_Project/media_life_death_sterling.speech.

18. Donald MacKenzie and Judy Wacjman, eds., *The Social Shaping of Technology*, 2nd ed. (Philadelphia: Milton Keynes/Open University Press, 1999); Wiebe E. Bijker, Thomas P. Hughes, and Trevor Pinch, eds., *The Social Construction of Technological Systems: New Directions in the Sociology of History and Technology* (Cambridge, Mass.: MIT Press, 1989).

19. du Gay et al., *Doing Cultural Studies*.

20. Ibid.

21. Thornton, *Club Cultures*.

22. Millard, *America on Record*.

23. Lisa Gitelman, "How Users Define New Media: A History of the Amusement Phonograph," in *Rethinking Media Change: The Aesthetics of Transition*, ed. David Thorburn, Henry Jenkins, and Brad Seawell (Cambridge, Mass.: MIT Press, 2003), 61–80.

24. Jonathan Sterne, *The Audible Past: Cultural Origins of Sound Reproduction* (Durham, N.C.: Duke University Press, 2003).

25. Marsha Siefert, "The Audience at Home: The Early Recording Industry and the Marketing of Musical Taste," in *Audiencemaking: How the Media Create the Audience*, ed. James S. Ettema and D. Charles Whitney (Thousand Oaks, Calif.: Sage Publications, 1994), 186–214.

26. Millard, *America on Record*.

27. Davis, "'Vinylphilia.'" This chapter summarizes conclusions from Davis (2003), a two-year qualitative study of twenty vinyl aficionados from various locations in the midwestern and southern United States.

28. Russell W. Belk, *Collecting in a Consumer Society* (New York: Routledge, 1995), 67.

29. Grant McCracken, *Culture and Consumption: New Approaches to the Symbolic Character of Consumer Goods and Activities* (Bloomington: Indiana University Press, 1990).

30. Brenda Danet and T. Katriel, "No Two Alike: Play and Aesthetics in Collecting," in *Interpreting Objects and Collections,* ed. Susan Pearce (New York: Routledge, 1996), 220–39.

31. Laura Hilgers, "McStuff," *Attache* (December 2000), http://www.attachemag.com/stories/archives/12-00/story2/story2.htm.

32. Russell W. Belk, Melanie Wallendorf, and John Sherry, "The Sacred and the Profane in Consumer Behavior: Theodicy on the Odyssey," *Journal of Consumer Research* 16, no. 1 (1989): 1–38.

33. Matthew B. Smith-Lahrman, *Selling-Out: Constructing Authenticity and Success in Chicago's Indie Rock Scene,* unpublished doctoral dissertation, Northwestern University, Evanston/Chicago, 1996, 184.

34. Dick Hebdige, *Subculture: The Meaning of Style* (London: Methuen, 1979); James Lull, "Thrashing in the Pit: An Ethnography of San Francisco Punk Subculture," in *Natural Audiences: Qualitative Research of Media Uses and Effects,* ed. Thomas R. Lindlof (Norwood, N.J.: Ablex Publishing, 1987), 225–52.

35. Smith-Lahrman, *Selling-Out.*

36. Thornton, *Club Cultures.*

37. Will Straw, "Systems of Articulation, Logics of Change: Communities and Scenes in Popular Music," *Cultural Studies* 5, no. 3 (1991): 368–88; Thornton, *Club Cultures.*

38. J. O'Connell, "The Fine-Tuning of a Golden Ear: High-End Audio and the Evolutionary Model of Technology," *Technology and Culture* 33, no. 1 (1992): 6.

39. Susan Pearce, *On Collecting: An Investigation into Collecting in the European Tradition* (New York: Routledge, 1995).

40. McCracken, *Culture and Consumption.*

41. Ibid.

42. Danet and Katriel, "No Two Alike."

43. Belk, *Collecting in a Consumer Society.*

44. Formanek; quoted in Belk, *Collecting in a Consumer Society,* 92.

45. Straw, "Sizing Up Record Collections."

46. Shuker, *Understanding Popular Music.*

47. Belk et al., "The Sacred and the Profane," 20.

48. Belk, *Collecting in a Consumer Society,* 94.

49. Belk et al., "The Sacred and the Profane," 19.

50. McCracken, *Culture and Consumption,* 44.

51. Belk et al., "The Sacred and the Profane," 19.

52. McCracken, *Culture and Consumption,* 86.

53. Per Danet and Katriel: "For an assemblage of objects to be considered a collection, each item must be *different* from all others in some way discernable to the collector" ("No Two Alike," 225).

54. According to Belk et al., collections are "separated from other objects to reinforce their sacred, non-utilitarian status and to prevent their entrance into the profane world where they might be consumed or used" ("The Sacred and the Profane," 21).

55. Jonathan David Tankel and Keith Murphy, "Collecting Comic Books: A Study of the Fan and Curatorial Consumption," in *Theorizing Fandom: Fans, Subculture and Identity,* eds. Cheryl Harris and Alison Alexander (Cresskill, N.J.: Hampton Press, 1998), 55–68.

56. Belk, *Collecting in a Consumer Society*; Belk; quoted in Laurence Zuckerman, "Why Hunt and Gather a Trove of Stuff? Studying the Ageless Need to Amass Collections," *New York Times,* January 22, 2000, B9.

57. McCracken, *Culture and Consumption.*

58. Janelle Brown, "The Jukebox Manifesto," *Salon,* November 2000, http://archive.salon.com/tech/feature/2000/11/13/jukebox/.

PART IV

Media, Mediation, and Historiography

13 Neglected News: Women and Print Media, 1890–1928

Maria DiCenzo and Leila Ryan

The demand which *Votes for Women* has to meet is twofold. In the first place, there is a growing desire for knowledge on the part of the outside public to learn what it is women are really striving for.... In the second place, it has to supply to all those women who are at work within the ranks a bulletin of the doings of the Union which shall keep them in touch with all the ramifications of the movement.
—*Votes for Women*, October 1907

The Englishwoman is not addressed only to those who are already fully convinced of the justice of the Women's Movement.... It is intended for the general public.... The question of the Enfranchisement of Women is not one . . . that interests only a struggling minority, and we trust that we may add to the already increasing number of women who desire a more equitable distribution of political power and responsibility.
—*The Englishwoman*, February 1909

Given the extent to which Victorian and Edwardian women activists were both "making the news" in the press of the day as well as "making" their own news—producing and distributing their own papers—there has

---

The authors gratefully acknowledge the generous financial support of the Social Sciences and Humanities Research Council of Canada and Wilfrid Laurier University. Maria DiCenzo would also like to thank the Women's and Gender Studies and the Department of Societies & Cultures at the University of Waikato, New Zealand, for their support while on research leave.

been remarkably little attention paid to the wide range of women's political newspapers and periodicals produced in Britain between 1890 and 1928.[1] These print media are crucial to gaining an understanding of the scope and activities of a women's public sphere at the turn of the twentieth century because they were instrumental in shaping opinion and establishing and mobilizing large- and small-scale activist networks and reform campaigns. This chapter provides a snapshot of the types of official organs and journals that proliferated in these years—with a particular focus on suffrage/feminist publications—and it explores the functions that these papers served for specific campaigns and organizations, but also, crucially, for the general public. The suffrage/feminist press in this period constitutes a residual media form at a variety of levels. In historiographic terms, these periodicals, overlooked by most media scholars, challenge dominant narratives of press and media history, suggesting a longer and more continuous history to the nineteenth-century radical print tradition, as well as earlier roots and precedents to postwar developments in feminist and alternative media. In historical terms, they are residual in two further senses. First, they are the literal residue or traces of pivotal developments in the history of the emancipation of women—the artifacts and evidence of an elaborate and complex social movement. Second, they relied on what were by this point residual forms because feminists drew on established practices and formats of earlier reform movements, such as the Chartist, radical, and socialist presses, using print media to articulate and circulate ideas. But by inflecting existing (and increasingly dated) forms in new and unconventional ways, the early feminist press offers a case in point for Williams's argument that the "residual" and the "emergent" may be effectively linked through their alternative or oppositional relation to the dominant culture. This press was overtly politically partisan at a time when the mainstream press, according to most media historians, was being depoliticized under the pressure of commercialization. These publications adapted long-established forms but generated new meanings and effects in the process of waging a political struggle for a disenfranchised constituency that was gaining a louder and stronger public voice.

Media Studies, History, and Methodology

We begin by relating current theoretical debates in media studies to a discussion of the suffrage press to demonstrate how recent frameworks can illuminate the complexities and dynamics of earlier media that have been either ignored or dismissed. The aim is to expand the historical dimensions of issues often assumed to be relevant only to contemporary media and social movements—highlighting the continuities, rather than the ruptures. Taking our cue from James Curran's criticism that much media history fails to draw the necessary links between media forms and larger social trends, the chapter argues that there is a great deal that the late nineteenth- and early twentieth-century feminist press as a residual form can tell us now, particularly in terms of disrupting the generalizations that are too often made about the scope and diversity of early women's movements and particularly how the articulation of feminist ideas through a range of print

FIGURE 13.1 *Miss Kelly, "a champion* Votes for Women *seller," was captain of the Charing Cross pitch. Courtesy of the Museum of London.*

media came to influence attitudes toward women's roles in public life.[2] There has been an active process of reflection in recent years on the theoretical and methodological issues related to media history. On one level, critics have noted the disproportionate emphasis in media studies on contemporary forms and developments. The criticisms are often raised in specific contexts, but they point to the implications for media studies more generally and in different national contexts. For instance, Neils Brügger and Søren Kolstrup lament the waning of the important role that historical studies played in the 1970s, and their collection of essays grows out of the need to address why historical scholarship became "the neglected child of media studies."[3] In some cases, the concerns are more directly related to particular genres or media forms. Martin Conboy offers what he argues is a "much needed critical history" of the popular press because only a historical approach can provide the kind of long-term perspective necessary to contextualize the relationship between the popular and the press today.[4]

John Downing, in his examination of alternative media, also stresses the need for historical approaches that illuminate the development of cultural forms and processes over time. He argues,

> A recurring and insidious temptation in media studies is to assess media from the singular vantage point of the contemporary moment. Both the impact and the origins of media become extremely foggy as a result. This is not least true of radical alternative media and oppositional cultures, which are already vulnerable to premature dismissal as ephemeral and therefore irrelevant.[5]

Similarly, in his attempt to expand the definitions of alternative media through an analysis of recent forms, Chris Atton stresses the importance of history and bases his argument "at least on historical 'congruence'" and regards his study as "grounded in the histories of alternative media from the past two centuries."[6]

On another level, those within the field of media history have posed fundamental questions about what constitute appropriate objects of study and approaches. Hans Fredrik Dahl interrogates the disciplinary boundaries, as well as the meaning and implications of the very term "media history"; the fact that its objects range from economics to culture and its methods from quantitative to qualitative leaves it without "a clear thematic identity and great depth in time."[7] Tom O'Malley takes the problem of the relationship among media studies, communication studies, and history further, and he offers a more systematic and chronological account of the emergence and development of media history as theorized and practiced in the UK.[8] These tendencies toward retrospective and meta-analyses of larger trends can also be found in Curran's recent work in which he identifies and distinguishes the main interpretations of media history that have emerged in explaining the historical role of media in Britain.[9]

These overviews raise fundamental issues—ranging from whether or not media history should deal with institutions, media content, or cultural effects, to the tensions between theoretical and empirical approaches—but they do not draw any clear conclusions except to point to the inevitably interdisciplinary nature of the field and to call for increasingly comprehensive and contexualized methodologies. They clearly all agree that history matters and that, as Curran (like Brügger and Kolstrup) puts it, media history is "the neglected grandparent of media studies: isolated, ignored, rarely visited by her offspring."[10] Nevertheless, this compelling case for the value of reinvigorating media history raises other, namely historiographical, problems. In this turn to history, we might justifiably ask, whose history or whose versions of history will influence the structure and concerns of the history of media in a given period? The work of historians of radical alternative and feminist media suggests that this self-conscious analysis of the field has not been entirely reliable in terms of recovering or addressing particular traditions of work that have gone missing in the scholarship. For instance, Curran acknowledges the extent to which feminist narratives have been "usually entirely ignored in conventional media history" and attempts to give the area extended consideration in his overview. But, in reconstructing women's history, he ends up reinforcing accounts (i.e., with an emphasis on domestic ideology, minor uprisings, followed by more containment, until the women's liberation movements post 1960s) that have been rendered problematic, even discounted, by feminist historians.[11] As a result, the scope and profile of feminist activism, from roughly the mid-nineteenth century through the interwar period and the 1950s, and particularly the media of these women's movements, are not accounted for. These omissions and oversimplified historical narratives are by no means confined to media studies, but they remind us that where media historians look and the questions they ask have important implications for what they find and how they interpret the findings. As Downing argues, "What might abstractly seem a bland and low-key instance could, in a given context, be wielding a hammer blow at some orthodoxy. . . . So context and consequences must be our primary guides to what are or are not definable as radical alternative media." We argue that the early feminist and suffrage press offers a case in point and provides a particularly compelling example of why and how residual media warrant further consideration.[12]

Feminist/Suffrage Media History: The Scholarship to Date

There is a rich body of work devoted to Victorian periodicals of all kinds, and women's domestic or commercial magazines before and after World War II, so the dearth of studies about women's political periodicals from the late 1800s to the 1920s becomes all the more surprising. Official organs of suffrage organizations, for instance, are part of a long tradition that includes the Chartist, socialist, labor, and radical presses of the nineteenth and early twentieth centuries. Like these other movements, the women's suffrage campaign was committed to electoral as well as more general social and political reforms, and many of its leagues organized and distributed publications nationally. Yet suffrage and other feminist print media are conspicuously absent in the narratives of British press history—the rise and fall of the radical/popular/political presses—chiefly because these accounts focus on class politics, rather than gender politics. They take for granted the decline of a politicized press due to the commercialization or "capitalization" of the press by the end of the nineteenth century, without accounting for the fact that increasing numbers of women were intensifying a long-standing public battle to gain citizenship rights.[13]

Additional reasons for this neglect range from the position of early twentieth-century periodicals, journalism, and activist literature in the academy—particularly how these forms slipped between the disciplinary cracks of history and literary studies—to the labor-intensive nature of the empirical research that often relies on materials difficult to obtain. Historians have tended to draw on them as sources but have not focused on them in terms of "genre" or as objects in their own right. In literary studies, genre as well as periodization complicated the recognition and inclusion of periodicals that appeared between 1900 and 1918. The field of Victorian studies has produced an impressive and extensive body of scholarship on periodicals, but even those who use the "long nineteenth century" often end at the turn of the century. It is modernism—with its emphasis on an avant-garde and formalist aesthetic—that constitutes the major narrative of early twentieth-century literature and little magazines.[14] At the same time, as noted earlier, the explosion in media studies that has tended to focus on the period after 1945 has only recently begun to address the issue of historical scope in relation to political media.

The impact of new historicist and cultural materialist challenges to the field of literary studies has been important, in combination with the growth of feminist approaches to literature and history and the commitment of women's studies to recovery and revisionist projects. Related to these developments is the increasingly interdisciplinary nature of research in the humanities and social sciences, particularly the literary, historical, and sociological discourses and perspectives that are themselves the product of debates within and between disciplines concerning the "cultural" (or linguistic) turn, on one hand, and the "historical" turn, on the other. In these ways, new critical perspectives and approaches have been instrumental in the recovery of residual practices and artifacts. These factors and developments inform the methodological issues related to, and the overall position of, feminist periodical research on the late Victorian and Edwardian years. The early feminist and suffrage press is finally gaining some of the critical attention it deserves. These print media represent the traces of networks of individuals and organizations who worked to effect social and political change for women of all classes. They provided the vehicles through which allies and opponents of women's emancipation tried to negotiate different and sometimes conflicting definitions about women's roles in the public sphere.

Margaret Beetham describes the advent of journals advocating women's rights as "a thin trickle which became a stream, if not a torrent, of words between 1900 and 1914 when, for the first time ever, there was a lively and diverse periodical press which spoke to and for women in terms of their rights."[15] There is no question that radical and progressive papers were outnumbered by the even greater growth in domestic magazines for women at the end of the nineteenth century, most of which were characterized by either a complete unawareness of "women's rights-ism" or a conscious rejection of these developments, defining femininity in opposition to the New Woman.[16] They are the kinds of publications that have had the highest profile in studies of early women's print media. But the radical and progressive papers were numerous and diversified enough to indicate a broader range of interests, reading habits, and practices for women in the period.

The mid and late years of the Victorian period saw the development of progressive journals that covered a wide variety of issues, ranging from work, education, and law to reports of specific reform campaigns such as antivivisection, anti–Contagious Diseases Act, and the suffrage movement. They are recognized as papers that were produced by women, for women, but most of them stressed again and again that they were addressing a wide readership that included men. Some of the notable titles include *The English Woman's Journal* (1858–64), which David Doughan describes as "the early British feminist movement working out its theoretical base in public"; *Woman's Opinion* (1874); *Women's Penny Paper* (1888–90), subtitled "the only paper in the world conducted, written [printed and published] by women"; *Shafts* (1892–1900), geared in part to the working classes; and *Woman's Signal* (1894–99), which was linked to the temperance movement.[17] Also emerging, particularly in the years during and after the First World War, were a series of trade/work-related papers that often referred to themselves as "organs" for these groups and constituencies. Even though the features could vary from the serious (the abuses of employers) to the more frivolous (what to wear), they assumed and promoted women's rights to independence and the need for collective effort to ensure their rights as employees and to improve their status and working conditions. The titles suggest different areas of employment for women in these years such as *The Woman Teacher* (1911–61); *The Business Girl* (1912); *Humanity* (1913–14), subtitled "Devoted to the Emancipation of Sweated Female Workers; *The Woman Clerk* (1919–31); and *The Woman Engineer* (1919–present).[18] It is often assumed that the women's movement died out by the interwar years, but surviving feminist periodicals tell a different story.[19] Some of the major suffrage journals continued to publish into these years, often under other new titles: *The Vote* ran until 1933, and *The Common Cause* became *Woman's Leader* from 1920 to 1932. Also, the well-known and long-running feminist political weekly, *Time and Tide*, was founded in 1920.[20]

Although it is important to indicate the spectrum of progressive periodicals in the period, we turn our attention to a selection of overtly feminist journals that developed out of and in reaction to the Edwardian suffrage movement. Narrowing the focus allows us to offer a more detailed discussion of particular genres, generated by what remains one of the most visible and familiar campaigns. The emphasis is on two types of publications. The first type is what were referred to at the time as "official organs" and what scholars have variously termed pressure group periodicals, publications of special interest groups, campaign journals, and suffrage newspapers. In this sense, suffrage organs such as *Votes for Women* and *The Common Cause* are part of a long tradition of publications growing out of movements devoted to electoral as well as other social and political reforms. The second type, the nonaffiliated/independent feminist reviews, appeared in both mainstream (*The Englishwoman*) and more avant-garde forms (*The Freewoman*), modeling themselves on the radical and literary weekly and monthly reviews of the period. All of these labels distinguish these publications from commercial and domestic magazines and the public press of the day, even though they referred to themselves variously as papers, journals, periodicals, and newspapers.

The Role of the Suffrage/Feminist Press: Strategies and Functions

Not only did the suffrage campaign generate an extensive and lively press of its own, monitoring and challenging the mainstream dailies at every turn, but as a movement, it was highly diverse, with internal divisions that affected its

FIGURE 13.2 *Inside* The Woman's Press, *156 Charing Cross Road, 1910. Courtesy of the Museum of London.*

relationship with authorities and the general public. The papers often acknowledged these issues directly:

> The Press and their Public: There is no question but that the attitude of the majority of the daily Press is a gross breach of faith with their readers. The public expects news from them, not misrepresentation. Men buy a daily paper to learn what is going on. To conceal or misrepresent what is going on is to obtain money under false pretences. We give in our news columns an account of what has been done by the *Women's Social and Political Union* during the past weeks, that our readers may see for themselves the extent to which facts have been suppressed. (*Women & Progress*, November 2, 1906)

There is a tendency to dismiss suffrage newspapers as mere newsletters or as serving a solely propagandist function, but this overly reductive thinking obscures the variety of approaches to form and content, and—when read in relation to one another—the provocative and often conflicting messages they conveyed. These papers were part of a conscious campaign of counterinformation designed to influence public opinion and to build and mobilize support for the cause, as well as a broader agenda of social and political change.[21] In virtually all of the senses in which Downing and Atton define them, these papers constitute early examples of radical alternative media, serving the same functions then as those we attribute to the media of social movements now. Atton stresses "the alternative press's responses [to the social construction of mass media news] as demonstrated not simply by critiques of those media but by their own construction of news, based on alternative values and frameworks . . . alternative media provide information about and interpretations of the world which we might not otherwise see and information about the world that we simply will not find anywhere else."[22] Downing attributes two main purposes to radical alternative media: "(a) to express opposition vertically from subordinate quarters directly at the power structure and against its behaviour; (b) to build support, solidarity, and networking laterally against policies or even against the very survival of the power structure."[23] It is through the latter of the two that we see how social movements come to constitute "alternative public spheres." Although Downing's use of the concept is a reformulation of Oskar Negt and Alexander Kluge's notion of a proletarian "counter public realm," it echoes the use of terms such as "feminist public sphere" and "counterpublics" posited by feminist critics of Habermas.[24] What the terms all point to are oppositional groups/cultures/discourses that Nancy Fraser describes as "parallel discursive arenas where members of subordinated social groups invent and circulate counterdiscourses to formulate oppositional interpretations of their identities, interests, and needs."[25] Print media played a crucial role in the formation of a feminist public sphere at the turn of the twentieth century, part of a larger context that saw the increasing differentiation and pluralization of the public sphere. But it is important to note that this process of segmentation did not necessarily imply segregation. In this respect, the relationship between the suffrage press and the public sphere once again illuminates the dynamic relationship between residual and emergent forms. Although it was emergent in the sense that it contributed to the differentiation of the public sphere, it was residual in the sense that it did not aim to isolate or ghettoize itself. Rather, as we elaborate later, it addressed a general public that included men and as well as women. So although it drew new attention to women's issues, it did not regard itself in specialized terms and consciously engaged with a wider political arena.

The "lateral" function of building solidarity that Downing identifies was complicated for suffrage activists because although they expressed open opposition, vertically, vis-à-vis the government, they had to capture the attention of and convince a male-dominated public (and Parliament) to grant them citizenship rights. So in building and mobilizing support, the papers addressed themselves internally to existing memberships and sympathizers but also externally to the general public. They tried to fill the gaps in existing press accounts and to counter negative coverage. Balancing these different functions was a continual challenge. As Helena Swanwick notes in one of her editorials for *The Common Cause,*

> We would like the paper to meet the needs and wishes of many sorts of people. There is the old, convinced Suffragist, who is sick and tired of "arguments" and who wants to have news to be kept abreast of the movement. There is the new convert, who is hungering for fresh reasons wherewith she may defeat the enemy in dialectics. There is the educated man or woman who wants special articles, and there is the illiterate, for whom we would like to cater. There is the secretary of the small society who wants the names of the local people and their speeches recorded, and there is the large body of the frivolous or the tired, who want "something readable."[26]

This problem of negotiating internal and external needs also presents a tension in terms of how we situate and discuss these media now, between understanding alternative media as "ghettos" or as part of more complex alternative forums/spheres that exist and must be taken seriously within and in relation to the wider public.[27]

On one level, the women's/suffrage movement had developed into its own sphere, complete with office headquarters for various organizations, women's presses, literature shops, holiday resorts, and private clubs that provided rental accommodation and libraries. Some organizations had sophisticated promotional strategies, embracing commercialization and exploiting fully the possibilities of niche marketing. For instance, the Women's Social and Political Union ensured that members could have their rooms decorated or buy bicycles, jewelry, china, and playing cards in the colors of the Union. Although these practices complicate, they do not contradict the political aims of such groups and their publications, suggesting instead a dynamic engagement with commercial culture that was not incompatible with political activism. At the levels of production and consumption, periodicals were markers of involvement in the cause, and the suffrage newspaper as "object" came to be linked to the image of the "modern" or "political" woman. They were crucial to establishing institutions and networks, offering a forum for participation (through articles or correspondence) and bringing news of the activities of the leadership and branches to members nationally, thus connecting them across geographic and even social lines. It is this unifying role that is often stressed in studies focusing on the potential of print media to influence "collective identity formation" and to create what Kate Flint calls "reading communities."[28] Feminist scholars have noted how the papers represented new discursive spaces for readers and provided what Martha Solomon suggests were "replacements for old social stereotypes about the nature and roles of women . . . encouraging readers to envision new roles and activities."[29] In her discussion of the role of print in the politics of gender formation, Beetham underscores the genre of the periodical as the "crucial site for the debate around the meaning of gender, sexuality and their relationship," but she also points out that "women were actively engaged, not only in defining themselves, but in shaping the press, as it went through its own late Victorian crisis."[30] Beetham's insistence on relating these practices and their effects to a broader context is important. There is a tendency to ghettoize feminist media—like other alternative media—limiting the scope and impact of these discursive arenas and obscuring their contribution to larger political discourses.[31]

A New Take on an Old World: April 1912

Gaye Tuchman has argued,

> News is a window on the world. . . . The news aims to tell us what we want to know, need to know, and should know. . . . But, like any frame that delineates a world, the news frame may be considered problematic.

> The view through a window depends upon whether the window is large or small, has many panes or few, whether the glass is opaque or clear, whether the window faces a street or backyard. The unfolding scene also depends upon where one stands.[32]

Her analogy remains a simple, but productive way to consider how we position and interpret news media. In the case of suffrage organs alone, there was a proliferation of papers after 1907, each with a different mandate, each claiming to offer an approach or perspective not currently addressed by the press or other women's/suffrage papers. The growing number of suffrage papers ranged in terms of the political spectrum from *Woman's Dreadnought* (1914–18), organ of the East London Federation of Suffragettes (Sylvia Pankhurst's socialist paper geared to poor and working women) to the *Conservative & Unionist Women's Franchise Review* (1909–16), which restricted membership along party lines, was opposed to universal suffrage in any form, and clearly drew on the ladies' magazines of the period. The spectrum included the high-profile organizations (Women's Social and Political Union, National Union of Women's Suffrage Societies [NUWSS], and the Women's Freedom League) whose papers are discussed later, as well as denominationally based groups, men's groups, regional associations, the international alliance, and of course the antisuffragists who launched their own review in 1908.[33] The only two serious precursors were Lydia Becker's *Women's Suffrage Journal* (1870–90) and *Women's Franchise* (1907–11), both designed to bring news from different suffrage societies to the movement as a whole. Two significant journals to appear during the Edwardian campaign were *The Englishwoman*, an independent monthly review, clearly connected with and sympathetic to the NUWSS, and *The Freewoman*, perhaps the most notorious, albeit short-lived magazine of the period, edited by the former militant suffragette Dora Marsden.[34] The paper announced itself as a feminist review, claiming,

> The publication of THE FREEWOMAN marks an epoch. It marks a point at which Feminism in England ceases to be impulsive and unaware of its own features, and becomes definitely self-conscious and introspective. For the first time feminists themselves make the attempt to reflect the feminist movement in the mirror of thought." (November 23, 1911)

What surprised and angered some was that its editorial stance was openly critical of the emphasis on the suffrage campaign as the means to advancing women's rights. The proliferation of suffrage papers reflected not only the expansion of the movement but also the growing diversity of opinion about goals and tactics—and, in turn, the need to make them visible. Although each might be seen to fulfill similar functions, they necessarily distinguished themselves and created the need for others to define their own positions. Like the associations they represented, these periodicals varied considerably in terms of form, content, and style.

Ragnild Nessheim, in her study of the mainstream press and the suffrage campaign, notes, "Almost without exception, authors of books about press and politics quote very sparingly from the letterpress of newspapers. . . . The reader learns what individual papers said and stood for, not from the horse's mouth (i.e., the letterpress of the papers themselves) but from the press historian's indirect rendering of editorial content."[35] One might go even further in the case of this material to suggest that the closer we look at the content and format of suffrage papers, the harder it becomes to generalize about them. Although each established its own particular template—using standard features ranging from editorials, feature articles, and interviews, to book/theater reviews, short stories/poetry/drama, correspondence sections, branch reports, advertisements, and classifieds—they also differed in terms of what they deemed newsworthy and allotted space accordingly. For papers ostensibly covering the same movement, in addition to events nationally and internationally, at times there was almost no overlap in content. Complicating the process of reading these papers now is that the conclusions one might draw from any given sample of issues are easily undermined by those in subsequent months/years. For example, *Votes for Women*, official organ of the WSPU, with its popular journalistic style, had a tendency to be quite narcissistic (promoting a cult of celebrity around high-profile figures like Emmeline and Christabel Pankhurst, and the Pethick Lawrences) while at other points providing more substantial and informative features. The shifts in style are especially evident in long-running periodicals that were subject to major changes in personnel and editorial policy.

In this final section we offer some general points of comparison between leading periodicals on opposite sides of the militant/constitutionalist divide, as well as those claiming to be independent. Taking April 1912, a period of turmoil for the suffrage movement, we provide a brief case study to demonstrate the different and conflicting ways in which these periodicals interpreted major events for their readerships.[36] Although these papers claimed to speak on behalf of the women's movement, they did so in different ways. The defeat of the Conciliation Bill on March 28, 1912, (by a vote of 222 to 208 on second reading) was a flash point for politicians and suffrage activists alike. This was a development fraught with intrigue, accusations of betrayal, and attributions of blame. It involved the role of the press, namely the appearance in *The Times* of the famous missive from self-styled scientific observer, Sir Almroth Wright, describing the militant suffragists as "warped," "immoral," "sex-embittered," and "incomplete" women. Not only did most of the feminist papers denounce and ridicule this diatribe, they also interpreted the decision by *The Times* to publish it on the day the bill was debated in Parliament as a deliberate attempt to sabotage its passage. For example, *The Englishwoman* noted that *The Times*, although it later modified its support of Wright, did not "scruple to make use of these exaggerated statements against women when they could be turned into a weapon to defeat women's suffrage."[37]

Virtually all of the periodicals foregrounded the extent to which parties, particularly the Liberals and Conservatives, were willing to exploit the suffrage cause to their political advantage and equally willing to abandon it in the interests of expediency. This was evident in the significant number of MPs who had pledged support but voted against the second reading. *The Freewoman* reminded its readers of the "overwhelming majority by which the Bill was carried twelve months ago, and also the various majorities by which the principle of Woman's Suffrage has been affirmed over and over again by Parliament."[38] Even though the WSPU had long discounted politicians' reliability and honesty, for those who were not yet convinced, the defeat of the Conciliation Bill proved they could not rely on promises of party support. *The Vote* stated,

> The House of Commons . . . has its own peculiar brand of humour; it is not so long since honourable gentlemen greeted with hilarious laughter the mention of the stomach-pump and the forcible feeding of political women prisoners. The laughter, the cheers, and the hysterical outburst which greeted on March 28 the successful smashing of pledged words is another instance of this sense of humour as well as a fitting comment on their sense of public honour.[39]

Reinforcing a similar message, both *Votes for Women* and *The Common Cause* published division lists of the names of all those who voted in favor, against, and who were absent and unpaired—a tactic they frequently employed after major parliamentary decisions. Several papers also reproduced extracts from the speeches made in the House of Commons during the debate.

Especially damning were the views expressed across the board about the role of Irish Nationalists, the overwhelming majority of whom voted against the bill, in spite of their previous support. It seemed that all were agreed on the motives of the Irish Nationalists. *The Freewoman*, for example, stated clearly that the Nationalists were "using their power simply to keep the Government in for the purpose of obtaining Home Rule."[40] Even the more sympathetic *Common Cause* lamented that "The fact is Nationalists are in mortal terror lest anything should interfere with the passage of their Bill this year. They should take longer views, and remember that they may need the help of women for two more years. They surely do not propose to fight the women for that length of time?"[41] Also interesting is the way in which the Labour Party emerged in this crisis as the only consistent supporter (the only party for which all members present voted in favor of passage). In calculating the small margin of votes by which the bill was defeated, *The Vote* regretted that "thirteen Labour Party representatives of the miners were obliged to be absent unpaired on account of the coal crisis."[42] This would prove a pivotal point in the party's relationship with the NUWSS; the constitutionalists, traditionally Liberal supporters, felt particularly betrayed by the government's handling of the situation and soon after began to form an alliance with Labour.

But underlying all these factors in the defeat of the bill and central to the coverage was the issue of WSPU militancy. The WSPU had intensified its window-smashing campaign in the weeks leading up to the debate, and the other suffrage organizations were not only openly critical of their actions but blamed the militants for encouraging defeat of the bill. The statements were unequivocal. *The Common Cause* claimed on its front page (which regularly included "Non-Militant" in bold letters): "we regard its defeat as a distinct piece of success for the W.S.P.U. They are in the position of having predicted the death of the Bill and then having ensured the success of their prediction by administering poison."[43] Regarding the organization of women workers as a more urgent priority, *The Freewoman* expressed impatience with "the childish obsession of grown women as to the value and potentialities of the vote" but saved its most biting sarcasm for the WSPU, accusing it of operating only on the basis of self-interest:

> That society, however, had its organization to save. . . . If the W.S.P.U. get their way, we shan't have votes for women this ten years. It is not to their interest to get it. With their policy, with its extraneous thrills; their society—a jolly club . . . they can get along for another decade very comfortably. It is the people with work to do who feel the pinch.[44]

It becomes easier to see why *The Freewoman*'s radical take was seen to verge at times on the side of antisuffragism. Even the other militant society, the Women's Freedom League, seemed compelled to reiterate its support of the bill and was willing to suggest that WSPU violence had played a role in its defeat:

> We cannot and do not share the view that it is a good thing the Bill is killed; on the contrary, we consider nothing more unfortunate could have happened at this juncture . . . The W.S.P.U. has been quite candid . . . they have said repeatedly in their paper and on platforms that they were not interested in the Bill . . . and indeed, it seems likely that the militancy which took place only three weeks before this critical Second Reading was deliberately designed to wreck the Bill.[45]

The story was far from front and center in the WSPU's *Votes for Women* for the same week, which featured on its cover a political cartoon depicting women firefighters breaking windows to save lives. Instead, the story took second billing to news of Mrs. Pankhurst's release from prison, the conspiracy charges against the Pethick Lawrences, and a meeting in Albert Hall. In addressing the defeat of the bill, the paper stated simply that "The result of the division on the Conciliation Bill simply illustrates afresh the futility of the attempt to legislate on a matter of this importance without a lead from the Government and pressure by the Government" and denied that militancy was a reason for its defeat, arguing that "the far more serious militancy of the miners, with all its attendant suffering and financial disaster, did not prevent but actually brought about the passing of the Minimum Wage Bill."[46]

The reactions to the Conciliation Bill provide a brief glimpse into the ways in which the periodicals tried to hold the authorities to account, just as the Almroth Wright story indicates how they responded to such displays of misogyny, reinforcing how public these battles indeed were. Equally, however, they reveal both the solidarities (all were united in their satirical treatment of antisuffragists) and the divisions within the movement and the conflicts over specific demands (i.e., the wording of bills defining who would get the vote) and the tactics/means by which they should try to secure their demands. These stories were only two among many in the first week of April 1912. It is also necessary to situate them among the other regular features that suggested business as usual—branch reports, announcements for upcoming meetings, recent pamphlets and books, and ads for everything from "ready-to-wear frocks" to vegetarian room and board. Here we see how these networks functioned and how organizations tried to encourage continued efforts in the face of major setbacks. These print media made movement-related news their primary concern, but they were also vehicles for "news" generally, offering their perspectives on the events of the day. The sinking of the *Titanic* was a major story at the end of April 1912 and provided an occasion for some to explore the policy of "women and children first" and for others to lament the death of the newspaper figure and champion of women's causes W. T. Stead.

In conclusion, these periodicals provided a feminist window on the world for readers then and offer a window onto earlier women's movements for readers now. For all their differences, what they shared—activists and opponents alike—was a fundamental belief in the power of reasoned argument to influence public opinion at time when print media were the most effective means of circulating ideas. Through them, early feminists were able to generate the public discussion of issues that were formerly excluded, performing a central function Fraser attributes to counterpublics, namely that they "expand discursive space," forcing "assumptions that were previously exempt from contestation . . . to be publicly argued out."[47] At the same time, they worked out their diverse and sometimes conflicting definitions of what women's roles in public life should or could be. The diversity of this early women's press reveals the range of political positions held by individuals and organizations; women's struggles were internally divided, and these media reflected those internal divisions. Published weekly and monthly for many years, the papers provided commentary on a wide variety of topics ranging from women in local government, to the rights of married women, prostitution, labor issues, and the struggles of women in other parts of the world.

These are the residual traces of women's active engagement in the political process, but at the same time they also demonstrate the ways in which women's voices were implicated in broader political struggles. In other words, they represent the attempts to articulate a particular set of concerns in the context of the public political arena. These papers reported on political developments nationally and internationally. On another level, even this brief case study of April 1912 offers evidence of how these publications were vehicles for the intersection of other social and political movements, ranging from the labor movement to Irish nationalism. Although the process of recovering these documents for media history has stressed their status as organs within a specific movement, we must recognize the risk of such an exclusive focus. As Patricia Gibbs and James Hamilton have noted, there are advantages to using an umbrella term like "alternative media" to describe what have otherwise been identified as specific efforts such as the "labor press" or the "feminist press," arguing "it is extremely useful to see them together because such a move emphasizes their collective resistance to increasingly monolithic commercialized media systems and products."[48] Defining alternative media solely in terms of targeting specific publics, interests, and constituencies—in their emergent capacity—has served to contain and marginalize their significance. Whether or not a broader conceptual strategy of the kind posited by Gibbs and Hamilton would militate against the reductive, homogenizing, and even dismissive accounts of these media is not clear. Attention to the "temporal lags" certainly reminds and forces us to consider the inextricable links between residual and emergent forms at particular points in time, revealing the ways in which the former give rise to, or act as catalysts, for the latter, in their challenges to the dominant—whether consciously or not. One might argue that postwar feminism saw itself in revolutionary terms because the neglected history of earlier movements obscured the sense of what it was "continuous" with. In the process of forging new identities and discourses, it was, however, also instrumental in the recovery of this history as feminists went in search of their precursors, just as Victorian and Edwardian activists had done before them. In similar ways, historians of alternative media are contributing to a growing body of theoretical and empirical studies of media that have remained obscure until now and, in the process, they are "expanding the discursive space" of media history research and offering new insights into the continuities of media forms.

Notes

1. The generic terms "press" and "public press" were used widely in the period by a variety of publications to refer to the mainstream press (mainly the daily newspapers).

2. James Curran, "Media and the Making of British Society, c. 1700–2000," *Media History* 8, no. 2 (2002): 135.

3. Neils Brügger and Søren Kolstrup, eds., *Media History: Theories, Methods, Analysis* (Aarhus, Denmark: Aarhus University Press, 2002), 7.

4. Martin Conboy, *The Press and Popular Culture* (London: Sage Publications, 2002), 2.

5. John Downing, with Tamara Villarreal Ford, Genève Gil, and Laura Stein, *Radical Media: Rebellious Communication and Social Movements* (Thousand Oaks, Calif.: Sage Publications, 2001), 6.

6. Chris Atton, *Alternative Media* (London: Sage Publications, 2002), 2.

7. Hans Fredrik Dahl, "The Pursuit of Media History," *Media, Culture & Society* 16, no. 4 (1994): 551, 553.

8. Tom O'Malley, "Media History and Media Studies: Aspects of the Development of the Study of Media History in the UK 1945–2000," *Media History* 8, no. 2 (2002): 155–73.

9. Curran, "Media and the Making of British Society," 135–54. See also the longer version of the same paper in James Curran, *Media and Power* (London: Routledge, 2002).

10. Curran, *Media and Power*, 3.

11. Ibid., 8. For a more detailed critique of Curran's account of feminist interpretations in media history research, see Maria DiCenzo, "Feminist Media and History: A Response to James Curran," *Media History* 10, no. 1 (2004): 43–49.

12. We use the terms "suffrage," "feminist," and "women's movement" in deliberate ways to distinguish between specific campaigns and more general attitudes and activities related to the advocacy of women's rights in the period. In the limited space available, note that we use "feminism and/or feminist" to refer broadly to the conscious attempts to change and improve the social, economic, and political conditions and status of women vis-à-vis men. We use "the women's movement" to encompass "feminism" but also to point to the wider scope of women's participation in the public sphere, including women who lobbied for social change but did not regard themselves as breaching the boundaries of conventional gender roles, as well as those who lobbied to preserve the status quo for women (e.g., antifeminists and antisuffragists). The suffrage movement constitutes a particular campaign, geared toward gaining enfranchisement for women. Although we try to

maintain consistent usage, all three terms (in addition to "the Woman Question") were employed in overlapping and conflicting ways in the period in question.

13. For a more detailed account of this problem, see Maria DiCenzo, "Militant Distribution: *Votes for Women* and the public sphere," *Media History* 6, no. 2 (2000): 116.

14. There is perhaps no better example of how influential narratives shape and distort our understanding of decades and centuries than the disproportionate emphasis on modernism over all other forms of writing in early twentieth-century literary studies. See Ann Ardis, *New Women, New Novels: Feminism and Early Modernism* (New Brunswick: Rutgers University Press, 1990) for a discussion of these issues in relation to women's fiction of the period.

15. Margaret Beetham, *A Magazine of Her Own? Domesticity and Desire in the Woman's Magazine, 1800–1914* (London: Routledge, 1996), 174.

16. Beetham, *A Magazine of Her Own?* 174.

17. David Doughan and Denise Sanchez, *Feminist Periodicals 1855–1984: An Annotated Critical Bibliography of British, Irish, Commonwealth and International Titles* (Brighton: Harvester Press, 1987). This reference guide remains an indispensable source for identifying and locating women's periodicals.

18. The titles are too numerous to list here, but these and many others can be found in the Harvester Microform collection entitled *Social and Political Status of Women: Radical and Reforming Periodicals for and by Women.*

19. See also Cheryl Law, *Suffrage and Power: The Women's Movement, 1918–1928* (London: I. B. Tauris Publishers, 1997), and Johanna Alberti, *Beyond Suffrage: Feminists in War and Peace, 1914–1928* (Houndmills: Macmillan, 1989).

20. For a more detailed discussion, see Dale Spender, *Time and Tide Wait for No Man* (London: Pandora Press, 1984).

21. It is important to stress that the campaign for the enfranchisement of women had broader goals and implications than simply securing a parliamentary vote. It was for many suffragists at the time a means to an end, rather than an end itself; citizenship rights were a first step in securing an official voice in matters relating to their lives generally. As the first editorial in *Common Cause* made clear, "Women's homes, their houses and children, their food and drink and work and sickness, the attendance upon them in labour, every minute matter of their daily life, from the registering of their birth, to their final old-age pension and death certificate, is bound round, hedged in, prescribed by law, and the laws are not always what the women approve—they are by no means what they would be if the women's voices were heard" (*Common Cause*, April 15, 1909).

22. Atton, *Alternative Media*, 10, 12.

23. Downing, *Radical Media*, xi.

24. See also John Downing, "The Alternative Public Realm: The Organization of the 1980s Anti-Nuclear Press in West Germany and Britain," *Media, Culture & Society* 10, 163–81 no. 2 (1988): The terms he uses to describe "the alternative scene," which encompassed the antinuclear press in Germany in the 1980s, could be applied directly to the suffrage movement: "It was a zone of multiple disagreements, not a little self-delusion and potential careerism, but also of alertness to a variety of issues not permitted on to the plateau . . . of the official public realm . . . also a zone of interaction for a considerable variety of purposes. Its foibles often satirized . . . In the 1980s, however, it was a way of life, partial for most, total for some" (171). For discussions by feminist critics, see Rita Felski, *Beyond Feminist Aesthetics: Feminist Literature and Social Change* (Cambridge, Mass.: Harvard University Press, 1989), and Nancy Fraser, "Rethinking the Public Sphere: A Contribution to the Critique of Actually Existing Democracy," in *Habermas and the Public Sphere*, ed. Craig Calhoun (Cambridge, Mass.: MIT Press, 1992), 109–42.

25. Fraser, "Rethinking the Public Sphere," 123.

26. *Common Cause*, April 14, 1910.

27. Atton, *Alternative Media*, 37.

28. See Mary P. Ryan, "Gender and Public Access: Women's Politics in Nineteenth-Century America," in *Habermas and the Public Sphere*, and Kate Flint, *The Woman Reader 1837–1914* (Oxford: Clarendon, 1993).

29. Martha Solomon, *A Voice of Their Own: The Woman Suffrage Press, 1840–1910* (Tuscaloosa: University of Alabama Press, 1991).

30. Beetham, *A Magazine of Her Own?* 118.

31. For a more detailed discussion of the internal and external orientations of these papers in relation to the feminist public sphere as formulated by Nancy Fraser and Rita Felski, see DiCenzo, "Militant Distribution," 117–18.

32. Gaye Tuchman, *Making News: A Study in the Construction of Reality* (New York: Free Press, 1978), 1.

33. Many of these groups published their own papers, such as *Church League for Women's Suffrage*, *Catholic Suffragist*, *Free Church Suffrage Times*, *The Altruist*, and *Jus Suffragii*. Some of these continued after the franchise bill of 1918 under new names. Other titles reflect the divisions and splinter groups within the movement. For instance, the founders of the Women's Freedom League defected from the Women's Social and Political Union to form a separate militant, but democratically run, league, with their own paper, *The Vote*. Emmeline and Christabel Pankhurst founded *The Suffragette* after their split from the Pethick Lawrences, who continued to edit *Votes for Women*, and eventually Sylvia Pankhurst, too, was ousted and founded *Woman's Dreadnought*.

34. Its notoriety stemmed from its explicit discussion of sexual mores and for advocating free love. Ironically, given its short run, this periodical is perhaps one of the best known of the period. It reappeared as *The New Freewoman* and then became the well-known modernist journal *The Egoist*, edited by Ezra Pound.

35. Ragnild Nessheim, *Press, Politics and Votes for Women, 1910–1918* (Oslo: Scandinavian University Press, 1997), 17.

36. The process is even more interesting when we take into account that people subscribed to and read more than one paper at a time, which is clear from personal papers, diaries, scrapbooks, and membership lists. Space is too limited to deal with the evidence of the close monitoring of daily and specialist papers, but the papers routinely reproduce lengthy passages from a wide range of daily newspapers and other publications. Even the editors of the *Anti-Suffrage Review* clearly read all the suffrage organs and cite them regularly.

37. *The Englishwoman*, May 1912.

38. *The Freewoman*, April 4, 1912.

39. *The Vote*, April 6, 1912.

40. *The Freewoman*, April 4, 1912.

41. *The Common Cause*, April 4, 1912.

42. *The Vote*, April 6, 1912.

43. *The Common Cause*, April 4, 1912.

44. *The Freewoman*, April 11, 1912.

45. *The Vote*, April 6, 1912.

46. *Votes for Women*, April 5, 1912.

47. Fraser, "Rethinking the Public Sphere," 124.

48. Patricia L. Gibbs and James Hamilton, "Alternative Media in Media History," *Media History* 7, no. 2 (2001): 117.

14 The New Techno-Communitarianism and the Residual Logic of Mediation

James Hay

Enduring Freedom in Iraq: Acting upon and Governing (through) "Communities of Practice"
As we gain experience developing business-based communities, it will become easier to apply such methods *outside the firm and even beyond networks of firms*. . . . Firms that understand how to translate the power of communities into successful knowledge organizations will be the architects of tomorrow—not only because they will be more successful in the market-place, but also because they will serve as a learning laboratory for exploring how to *design the world as a learning system* (emphasis added).
—Etienne Wenger, Richard McDermott, and
  William M. Snyder, *Cultivating Communities of Practice*

Since the early 1990s there has been increased recognition of and support for the informal networks of colleagues that operate within the bounds of our corporate and government structures. It is within these voluntary groups that novices are mentored, . . . experts are identified, knowledge is shared, and answers are rapidly provided to an ever-widening array of inquiries. . . . Along with the recognition of the value of such groups, has come the provision of tools and support structures [e.g., databases and information systems] to facilitate the launching and sustaining of [these informal] communities. *Outside of the formal bureaucracies, but aligned with their strategies, communities are increasingly recognized as*

> *successful mechanisms for the management and sharing of organizational knowledge"* (emphasis added).
> —"The Development of Communities of Practice at USAID," projects document, U.S. Agency for International Development, May 27, 2003

My intervention in a book about "residual media" is purposefully indirect, in part because understanding media and media power involves considering how communication technology matters within environments and arrangements that cannot be reduced to media—or to "root causes" such as economy, politics, or culture (conceived as separate forces). This objective necessitates two brief introductory points of clarification. First, this chapter is interested in communication technology as technology of community and in media as technology of mediation. On this point, my project acknowledges the relation between technologies of communication and community but also is interested in how communication technology and networks are imagined and applied as a civic good within communitarian programs. By "technology" of community, I refer to the practical devices, skills, and knowledge, and to the scientific rationality, for engineering community as a mediated state, but also to the instrumentality of community (to how community is deployed and made to matter politically) as a mediated state. Second, I emphasize "environments" and "arrangements" because my interest in residual or emerging media—the technology of mediation—is geographic and spatial. My intervention into a book about residual media in these terms begins by considering the intersection of the two epigraphs to this chapter—statements that comprise what Foucault would describe as a discursive formation but that also constitute interdependent programs of government in which community has been instrumentalized (engineered, mobilized, and acted upon), in part through communication technologies. To discuss the historic and geographic relation of these programs, I turn to an event, and a third program of government, where they converged. Their convergence offers a useful way of considering how current forms of Western domination have rationalized a new regime of communication technology (a regime of the "new" and of "emergence") through communitarian programs that bespeak a residual logic about mediation and governance, even as these programs and technologies are being linked to the "reinvention" and "advancement" of liberal government (in various parts of the world).

For the Bush and Blair administrations, the success of Iraq's so-called liberation and the introduction of a new liberal government to the Middle East through Iraq depended on a military operation comprised of an unprecedented number of privately contracted ("outsourced") workers, who (by force and example) were supposed to lay the groundwork for privatizing Iraq's state-governed institutions—a transformation that would replace despotic rule with the latest, most "modern" principles of management, including those valorized in the U.S. Agency for International Development (USAID) document above.¹ In the wake of this military action (titled Operation Freedom, and subsequently Operation Enduring Freedom), the USAID began sponsoring an initiative that it called the Iraq Community Action Program (ICAP). In April 2003, the USAID's first Web page outlining the objectives and purpose of the ICAP stated that it

> will promote diverse and representative citizen participation in and among impoverished communities throughout Iraq. . . . The program will create community committees responsible for identifying and prioritizing community needs, mobilizing community and other resources, and monitoring project implementation. Broader cluster committees will be formed with representatives from various community committees to increase participation and cooperation on issues of regional concern, and to promote ethnic, religious, and tribal cooperation on the basis of shared interests.²

Whose "shared interest" indeed? In these times, such is community's connotative robustness and goodness, not to mention instrumentality. Today, community makes for good government. Community is instrumental not only because it implies the opposite of self-interest and alienation. Certainly for the Bush administration, introducing a program of community action/activism—identifying, prioritizing, and mobilizing community in Iraq (tantamount to cooperation and shared interest)—was a strategic and political interest. In part this was because Iraq's status as nation-state had developed out of a long history of Western-imposed boundaries and Western-imposed institutions for managing, in the USAID's terms, longer-standing ethnic, religious, and tribal differences (such that "regime change" either required redrawing boundaries, perhaps devising three new, smaller nation-states, or a federalism with new institutions for assessing and administering to/through these differences). The strategic/political value of community also had to do with the Bush administration's having arrived in office in 2000 eschewing the Clinton administration's penchant for "nation-building" (on which a "regime change" through "community mobilization" presumably would improve).³ In addition to these objectives, however, *acting upon* community activism—fostering a program of "community action," activating/energizing community, indeed the mantra of community in the USAID's document—has become the *reason* (the pretext and the rationality) of good, wise, fair government, not only as a formula for a new "democratic" Iraq (previously the epitome of "bad government") but also as central to a reasoning in the United States about liberal governance, which the Bush administration and U.S. administrations over the last twenty years have sought to advance and defend, abroad and in the United States.

In the terms adopted by this policy directive, implementing a program of "community action," that is, *acting upon* community, was envisaged by the United States as a basic and (in conjunction with military occupation) a primary technology for governing Iraq (for making Iraq governable)—a fundamental object/objective of the new reasoning/rationality of liberal-democratic government that was supposed to improve Iraq (to make it more modern) and to make the world (and the U.S. vision for the world) more secure and civil. It is not only that making Iraq governable involved translating something called community in Iraq but that introducing liberal-democratic government in Iraq involved *liberating* (making active, giving agency to) and acting on resources that were considered as natural and plentiful as oil. The political and policy rationale for community action, both as natural potentiality waiting to be freed and as a way to know, think about, and economically organize/calculate resources for governing, was basic to good government's rationality (which the Bush administration frequently claimed had been lacking under the maniacal/irrational and repressive dictatorship of Saddam Hussein). Reinventing an Iraqi citizen(-ry) in these times and on these terms not only linked the institution(-s) of liberal democracy in Iraq to rationality, but it made citizenship in Iraq subject to a particular reasoning—a regime of knowledge (identifying, prioritizing, monitoring "community needs")—and in this way made this citizenry recognizable and responsible (and thus a "reasonable" people) with respect to a technical reasoning about government in the United States and Britain. Through community action, Enduring Freedom thus was cast not only as a liberation of an "oppressed" people but the installation of a new governmental rationality: liberalism as an advancement/improvement over previous forms of government there (demonized as irrational). "Community development" represented nothing short of community's universality and of the centrality of community in a managerial rationale that aspired (to use Wenger's expression above) to "design the world as a learning system." Community development was to affirm a Western rationale about rights and freedoms and about the path to a "civil" society in Iraq, and through Iraq toward the maintenance of a "global civil society."

Community is arguably a Western ambition, complexly entangled with Western conceptions of modernity (or the West's conception of its own modernity), but as the ICAP demonstrates, community also is increasingly inscribed in U.S. policy and in state-sponsored programs of government. As indicated by both epigraphs at the opening of this chapter, community is so deeply accepted in the United States as a key term and technology of corporate management as well as political government as now to seem a quite unremarkable and virtuous means of improving Iraq, a fundamental term/condition of fashioning Iraq as a liberal democracy, wherein "cooperation" is tantamount to transforming Iraqis into active participants in their own governance. Political agency (Iraqi citizenship) thus involved the institution of a "new" and decidedly U.S./Western-centric way of reasoning about liberal government and its resources. If, as Nikolas Rose has argued, community has become an important object/objective of neoliberal government, of "governing at a distance," then a program of community action in Iraq was not only a recent experiment by the United States in fashioning civil society and *economic government* but, as such, a transnational (transhemispheric) form of governing at a distance—of "sharing" interests with the United States and acting upon the cooperation of a new Iraqi citizenry through community.[4]

Whereas the early/primary task for U.S.-led regime change in Iraq involved identifying, categorizing, prioritizing (making rational), and mobilizing community committees, as the molecules of a new political science and economic management for Iraq, by December 2003 the USAID Web site modified/refined its initial statement of purpose about the ICAP, linking infrastructural reparation and investment, community activism, and good/effective government:

> The Community Action Program (CAP) enables Iraqis to address critical needs of their communities. Through the program, Iraqi community associations identify and prioritize their needs, and subsequently develop and implement projects that address those needs. This process gives Iraqi citizens a voice in the decisions that affect their communities. CAP projects include repair to local sewerage systems and roads, rehabilitation of schools, and renovation of health clinics. Women, youth, and minority groups participate in the program.
>
> Under the repressive rule of Saddam Hussein the Iraqi people suffered economically and socially for over a decade. A lack of investment led to decreases in productivity, wages, and jobs. In addition, social services also eroded significantly. Iraqi citizens were excluded from the local decision-making process. CAP encourages citizens to become involved in addressing the issues that affect their communities.[5]

Unlike the earlier statement of purpose, this one de-emphasizes the role of the USAID in the path to self-sufficiency and self-governance, attributing responsibility and an active citizenship (rather than dependency) to "Iraqi community associations," under the *auspices* of the USAID and various nongovernmental organizations (NGOs). Progress is measured (literally calculated) in terms of communities that mature and become more responsible, rational (assessing their own needs), and self-sufficient within programs of community development and mostly private patronage (rather than directly through military/police actions).[6] Linking the ICAP with infrastructural accomplishments, carried out exclusively by U.S. businesses, the USAID rationale valorized privatization in two interdependent ways: as a condition and outcome of liberal government (of engineering/enabling a political state and a "civil society" wherein citizens become involved in their own government through community and wherein the state recognizes/instrumentalizes community as a resource for governing) and as an outcome and condition of economic liberalization (the prominent role of privatization in achieving economic government and a state that reflects on its own limits and values outsourcing as a means to the economic welfare of its citizenry). The USAID's list of accomplishments above articulates welfare, health, and quality of life in both of these terms, gauging benefits and success for a new Iraq and for the United States in these terms.

The USAID's concern with community mobilization and community activism, as a fundamental problem and solution for maximizing self-sufficiency, self-governance, and good (liberal-democratic) government, thus found its key objective and key concept in "communities of practice." Communities of practice allowed the agency to rationalize (to make intelligible to itself) the fashioning of a liberal democracy in Iraq, and to rationalize (make reasonable) its role in coordinating and overseeing the development/improvement of a civil society in Iraq through privatized initiatives, which had as much to do with "cooperative" business ventures (e.g., imagining that oil revenues would defray reparation costs) as with the "sharing of knowledge" and "designing the world as a learning system" (bringing civil society to Iraq as a step toward building/maintaining a global civil society). Within this political and managerial rationality, communities of practice operate as technologies of participation and thus of a citizenship in a liberal-democratic Iraq. Communities of practice are not just passive, dependent objects merely to be studied (even though the term conjures certain veins of anthropology and sociology); they are potentialities for mobilization, organization, and management (i.e., ways of knowing, governing, and securing Iraq across existing activities and ways of life (hence the emphasis on "practice"). Communities of practice are sites/technologies that need to be mobilized to make management more productive and efficient—more *economic*.

As both of the passages cited at the outset of this section affirm, "communities of practice" has become part of the technical vocabulary, knowledge, and reasoning in the United States about business management and organization, and as the USAID's proposal affirms, the idea/term is as prominent in the private sector as in contemporary U.S. government policy (or rather in the current intersection— "cooperation"—between corporate and state programs of management).[7] In this respect, communities of practice serve as a mediating rationale, model, and technology between public and private, state and corporate governance. As Wenger notes, "Business organizations [that organize and manage themselves as communities of practice] are . . . becoming a catalyst for civic communities of practice."[8] Mediating multiple spheres of governance requires the reinvention of community as "knowledge organizations" that act and are acted upon through "smart" tele-technologies. For Wenger, "it is not communities of practice themselves that are new, but the need for organizations to become more intentional and systematic about 'managing' knowledge, and therefore to give these age-old structures a new, central role in business."[9] What makes corporations organized through communities of practice "architects of tomorrow" and "learning laboratories for how to design the world as learning system" is not their exclusive access to information technologies but their rationalization of themselves (their sense of their new role in the world) as domains of technical knowledge networked as a flexible governmental technology.

As the USAID's literature on communities of practice points out, communities of practice has become a salient idea as organizational communication/knowledge has been increasingly practiced through Internet and Web-based technologies. These days, good government and corporate management both valorize "intelligent" communities of practice—spheres of life, civic activity, productivity that depend on knowledge and information distribution/coordination—that can be known and thus implicated (made productive and "smart") within a new managerial/governmental rationality.[10] Organizations of management such as the USAID have rationalized their own programs in these terms, through this literature, describing their purview as "our community of communities of practice" and organizing their Web site about the ICAP through hyperlinks with their other communities of practice. Communities of practice, as the fashioning of community through "informatics" (the science, networks, and technologies of information and knowledge), mediate the inside and outside of corporate and state management, that is, the relation between private and public sector governance and the linking of "communities" into networks of "practice" beyond specific corporate or state organizations. Within this reasoning about management and government, communities of practice are valued for maximizing the productivity of knowledge/information (and thus the activity and self-reliance) among groups ("communities") of workers (or corporate citizens) whose knowledge, activity, productivity, and volunteerism/cooperation—whose relative autonomy and freedom as producers—is suited to and dependent on informatics and the technologies of mediation that informatics makes rational. Communities of practice are thus sites of management whose simultaneous dispersion and coordination (through informatics) is believed to encourage productivity and freedom, collectivities that within the network are entrepreneurial and self-directing. Ideally, communities of practice operate through and sustain (are made to endure/last through) a managerial reflexivity made possible through informatics and the ongoing implementation/experimentation ("practice") of this reflexivity as network—chains of proliferating and coordinated nuclei expected to secure, surveil, and regulate various domains and productivity of knowledge, with the goal of making organizations less dependent on instructions from above and more efficient and economical. In all of these senses, communities of practice are best understood in terms of the program's felicitous acronym, CoPs, which casts communities as "voluntary groups . . . [wherein] novices are mentored, expertise is nurtured, experts are identified, knowledge is shared, and answers are rapidly provided to an ever-widening array of inquiries" (passage at beginning of this section). Introducing CaPs in Iraq was thus intertwined with CoPs as a program of "enduring freedom": casting agencies such as the USAID as "mentors," as the largest community (a center of technical expertise) mobilizing and acting upon other communities of practice that are not only waiting to be recognized/liberated as productive of a new Iraq but that, in their activism/activity, freedom, self-reliance, and productivity, become models for the latest experiment in liberal-democratic government/management—an advancing and more "advanced" liberalism.

Although community-action initiatives were listed last on the USAID's list of infrastructural accomplishments in early 2004, this prioritizing (part of the rationality of contemporary government in the United States) not only implies a relation between "liberalized" economic development and liberal government (good government as *economic* government), it also suggests the importance of community mobilization in establishing and representing the *moral economy* of an Iraq that is being politically and economically organized through a liberal-democratic governmental rationality. Good government requires a moral/cultural economy, a form of self-governance that transforms "ethnic, religious, and tribal" attachments into community and civic institutions (by extension, a *civil society*), into forms of political agency (citizenship through communities of practice), and hence into the fundamental resources and technologies of good government. And while a military occupation and the management of infrastructural tasks by U.S. businesses may be cast in policy as the primary rationale for calculating/representing success, progress, and reform in Iraq, the USAID's identification/recognition of community as a political and economic resource plays an important role in representing economic and political reform in Iraq as not managed by the United States, and in this sense as moral.

The rest of this chapter discusses how the naturalness of a program/policy of community action in the United States (its interest in "shared interest" through knowledge sharing and information networks), and how and why community mobilization became such a fundamental reason of and for liberal-democratic government in Iraq, developed out of a set of "problems" and "solutions" pertaining to community, media/technology, and governance that emerged over the 1980s and 1990s. How, I ask, did the governmental discourse and rationale of community develop in the United States over the 1980s and 1990s, and as such an accepted and basic form of intervention and of linking a lasting/"enduring" freedom (economic and political liberalization) and government? Accomplishing this in one chapter will involve the strategic consideration of a couple of examples—strategic to emphasize the relation between neoliberal governance and the changing conception and place of community in an environment/geography organized through emerging media/communication/information technologies. Thinking about media in this way also allows me not only to address the contradiction of (what is oxymoronic about) a *new* communitarianism but to consider an emerging communitarianism's negotiation of a modern reasoning about mediation. My chapter is therefore less about residual *media* per se than it is about a logic of mediation that has been integral to modernity and its conception of community (and indirectly communication/media) as problem and solution. Addressing this issue leads me to consider the significance of place/space in understanding both the legacy of a modern (and postmodern) reasoning about community, technology, and power as well as in suggesting an alternative perspective. I therefore seek to understand emerging programs/technologies for organizing social space in terms of residual and emerging conceptions of community, particularly as community pertains to the "advancement" (the new spaces and spatial productivity) of liberal governance in the United States. As a way of working back around to considering the implications of the ICAP in these times, and of reconsidering the rationale's interest in securing a framework for participation and cooperation through community (fashioning/refashioning community in Iraq), I consider the new communitarianism's relation to recent technologies for making social space productive, cooperative, and governable as community.

Governmental Rationalities in the Age of "Intelligent Machines"

To discuss neocommunitarian discourses and programs, the interdependence of these discourses and programs, their preoccupation with media and communication technologies, and their negotiation of modern understandings of community, this section begins by considering briefly why it is important to recognize the relation of these discourses and programs to modernity, a starting point for developing a countermodern analytic media/power. In accounts of modernity, *community* often figures paradoxically: its loss or obfuscation within the formation of modern societies becomes the basis for rationalizing its reinstitution. Community is thus both a problematization of the present and an objective in the present. Its premodernity is seen as a problem, embedded within modern conceptions of sociality ("society" as a modern development that displaces community), and as a solution that lies temporally just beyond the conditions on which modern societies are conceived and rationalized (by science, commerce, and government).

Ferdinand Tonnies's famous distinction between gemeinschaft and gesellschaft thus attributed the organicity of relations of community to community's pre-modernity and extra-urban-ness: "Community [gemeinschaft] is old, Society [gesellschaft] is new, both as an entity and as a term. . . . Wherever urban culture flourishes, 'Society' also appears as its indispensable medium. Country people know little of it."[11] Through formulations such as Tonnies's, relations in modern societies are thus posed negatively to (i.e., as the absence or fabrication of) community's organicity. For Tonnies, social relations (unlike relations of community) are a mental abstraction ("a purely mechanical construction existing in the mind") whose immateriality (and, in the extreme, inauthenticity) he attributes to the *rationalization*, calculative ordering, and functional division occurring within and across all spheres of modern societies, in which individuals are reduced to "elementary units of labor, like atoms."[12] The wholesale rationalization and functional division of modern societies thus replaces the wholeness of community (as one aspect of its organicity, its natural and direct bonds) with a more abstract and (politically and economically) instrumental totality of individual parts. Community thus remains an unattainable ideal in modern societies, problematizing the climate of self-interest (what Kant had described as "unsociable sociability") fostered in modern societies.[13] In that community has been central to modern conceptions of the social, it has figured prominently in the formation of sociology, whose most functionalist projects have affirmed the immateriality/negativity of relations—the "social construction" of relations—in modernity and whose scientific rationality is supposed to make that atomization and immateriality intelligible/rational and thus available for analysis as a coherent, bounded entity/system.

There are indeed important ways that modern conceptions of community often have been interwoven with modern conceptions of culture and communication—or with a "techno-culture," once associated with "mass media" but more recently with "informatics." In the first half of the twentieth century, the culturology of "critical" German social theory (e.g., Kracauer's "The Mass Ornament," Benjamin's "The Work of Art in the Age of Mechanical Reproduction," Adorno and Horkeimer's "The Culture Industry") and the "mass culture" thesis in the United States (most famously advanced by Dwight Macdonald) all emphasized the new relation of culture to the rationality of modern industrial production and commerce and contrasted this development to the organicity of premodern and rural culture. The remnants of this organicity often were linked on the Left to the urban working class, whereas European fascism aspired to a much broader, classless conception of organicity in its conception of the "popular culture." In the 1960s and 1970s, the writing of Raymond Williams and James Carey articulated culture and community through the term "communication." Whereas Williams connected culture to communication to rethink nineteenth-century hierarchical conceptions of culture and to develop an expansive conception of culture as social practice that included modern forms of communication, Carey connected culture to communication to rethink the objectives of North American communication research that had developed out of (arguably as a branch of) the social and behavioral sciences. In both these latter formulations, community is the historically specific *commonality* of culture as communication. In Carey's terms, thinking about communication as culture "exploits the . . . common roots of terms 'commonness,' 'communion,' 'community,' and 'communication.'"[14] Although cultural studies since the 1960s often have understood communication or signification to be central to culture (as a "whole way of life" and as a sphere/formation of political struggle), they have tended to consider community (and occasionally communication and culture, by extension) as a term that, as Williams has noted, "seems never to be used unfavorably, and never to be given any positive opposing or distinguishing term."[15]

I prefer to disarticulate "community" from these two terms, not only to avoid conflating culture, community, and communication, and to emphasize instead the meaning and mattering of community contextually and conjuncturally (through these and other terms), but more importantly to consider how community has become a specific object(-ive) of a discursive formation ("communitarianism") and a technology of a new ("neoliberal") political rationality and activism in the late twentieth and early twenty-first centuries. Thinking about community as the object of communitarian discourses and programs (and as a technology of liberal government) also allows me to consider how media and communication technology are conceived and deployed in the service of these programs, rather than assuming that culture, or modern forms of community, are primarily or only forms of communication and representation.

Although the invention of "cyber-community" is rooted in a modern problematization of community, over the 1980s it became an object of intersecting discourses and programs about techno-culture and communitarianism. The most renowned spokesperson/homesteader of cyber-community was Howard Rheingold, whose *The Virtual Community: Homesteading on the Electric Frontier* (1993) was praised by a wide variety of sources as having brought a humanistic perspective to the possibilities of the emerging regime of information technology.[16] As the *New York Times* review on the book's jacket asserted, "Rheingold is concerned with the question of whether computer communications will end up centralized, in the hands of big business and government, or decentralized, in the hands of ordinary people," a statement that valorizes and not only equates communication with community but also the civic virtue (and humanity) of "decentralization." Rheingold's ruminations on the formation of the WELL (the Whole Earth 'Lectronic Link, "a computer conferencing system that enables people around the world to carry on public conversations") described his conversion from an initial uncomfortableness with the "coldness" of an electronically linked community to his warming up to the "coziness" of its world (to the world as a cozy place, "hidden within the walls of my house").[17] The modern virtue of "computer-mediated communication" (CMC) was its capacity to enable virtual community. And the civic virtue of virtual community was, for Rheingold, its solution to the modern problem of isolation/alienation, a condition that presumably resulted from an earlier reorganization of social space through media/communication technology (e.g., feeling cut off from the world at home) but that could be redressed through an emerging regime of computer communication located within a new domestic sphere. As Rheingold described it, the WELL "felt like an authentic community . . . because it was grounded in the everyday physical world,"[18] a cyberspace that paradoxically restored a relation to physical place even as it overcame one's location in physical places.

Rheingold's "humanistic," neocommunitarian representation of emerging digital communication systems located the coming virtual community as much in terms of the Whole Earth projects from the 1970s as in terms consonant with literature that derided the postsuburban landscape for its inhospitableness to community and that imagined the techniques for rediscovering community through a renewed "sense of place." *The Virtual Frontier* thus cites Ray Oldenburg's *The Great Good Place* as a template for recognizing the potential of CMC to shape "third places," those spheres of sociality (once authentic communality) that had been typical of towns and the premodern city but that were eradicated ("shred") by the "agorae of modern life": automobility, suburbia, fast food, shopping malls. The term "agora," or marketplace, not only rearticulates the modern lament about commercialization (not restricted to Marxisms) but also about a series of developments since the Second World War (developments not attributed primarily to capitalism) that collectively have eroded the spheres of life on which community once was predicated: "American life-styles, for all their material acquisition . . . are plagued by boredom, loneliness, alienation." Invoking his own 1950s analogy, Rheingold adds that cyberspace might be "one of the informal public places where people can rebuild the aspects of community that were lost when the malt shop became a mall."[19]

Although the civic virtue that the Rheingoldian account of "computer-mediated communication" associated with community was a long-standing modern ideal (represented through nineteenth-century Saint-Simoneanist rationales about international peace through communication, and more decidedly through the Cold War invention of "outer space" and the U.S. "space program" as integral to programs of "global communication" as a path to global peace and citizenship),[20] Rheingold's statements about the civic virtue of CMC also pertained to a rethinking of liberal government and citizenship in the United States during the 1980s and 1990s. Not only did Rheingold describe communication technology as a solution to the problems located in a modern society, but he viewed a Net community's civic virtue as its naturalization/universalization of certain rules among *Netizens*. This perspective is evident in the inscribed frontispiece of *The Virtual Community* where Rheingold inserted a statement from communitarian M. Scott Peck's *The Different Drum: Community-Making and Peace* that envisages a new evolutionary stage from being "social creatures" to "community creatures," achieved through the *innately* understood rules of community that can have a "healing effect upon our world." Aside from begging the question about whose world is being healed, for Peck and Rheingold these rules (innate but forgotten within the modern conditions of sociality) need to be rediscovered/reinvented to achieve a "genuine community" and the "freedoms and openness" that accrue from community's presocial forms of regulation. Whereas Peck's quote casts the modern idea of freedom as a virtue once found in community, Rheingold translates Peck's communitarianism as a call to rediscovering lost civic virtues through *natural forms of regulation*, articulated by Rheingold to the potential of CMCs. The modern idea of freedom as a virtue once found in community is, in a new age of information networks, a *stage* (a new experiment) for reinventing liberal government.

As a counterpoint to his view of the civic virtue of CMC/community, Rheingold invokes Foucault's *Discipline and Punish*, rehearsing a common criticism (particularly in the United States) that Foucault offered a dark and cynical account of government control by a big brother (and wrongly stating that *Discipline and Punish* was about how "the machinery of the worldwide communications networks constitutes a kind of camouflaged Panopticon").[21] Not withstanding that Foucault's writing on governmentality and liberalism was a response during his late career to this kind of criticism about his most widely read (and arguably most misunderstood) book, Rheingold's very successful account of a coming virtual community required adopting some very modern binarisms that Foucault's account of power had complicated or rejected: about freedom as the opposite of government/power, about resistance as the opposite of control, about surveillance as something performed by others ("totalitarian rule") rather than the self, and about community as a form of regulation anathema to the programs and policy advancing liberal government. Indeed, on this last point, Rheingold's communitarian account of the "digital age" made him an attractive expert/consultant to the U.S. Congress Office of Technology Assessment and was quite consonant with policy development and efforts to "reinvent government" during the Clinton-Gore administration, particularly the discourse of "the Information Highway" associated with Gore and that continued to be integral to Gore's campaign for the presidency in the late 1990s.

Rheingold's account of virtual community in the early 1990s is a somewhat arbitrary starting point from which to explain how a new communitarianism found its way into research and initiatives for advancing communication/information technology, and how Net communitarianism (the current regime of techno-communitarianism) has become relevant to experiments in reinventing liberal government in the United States since the 1980s. It is nevertheless a useful starting point for describing the historical convergence between a reasoning about government and a reasoning about "new information technology" and CMC through which community came to matter and, in its emplacement, was made productive of (instrumental in *advancing*) a new sociospatial arrangement, one of whose central problems/concerns is connectivity at a distance.

Performing similar discursive labor was Robert Putnam's account of the loss of "social capital" in America, a condition most famously represented by his figure of the solitary bowler. Relying on "indicators of civic engagement," Putnam charted the decline of social connectedness, the decline of trusting and joining, and a deepening inability to participate in older (and, for Putnam, better) forms of civic engagement, all of which he linked to the disappearance of community. Rejecting a considerable body of writing that understood social, political, and economic citizenship in terms of the unequal distribution of resources among different classes and populations (e.g., gendered citizenship, racialized citizenship, suburban citizenship) and relying on the most modern empirical procedures for calculating indicators of civic engagement, Putnam argued that the "decline in social capital" since the 1950s has affected everyone, everywhere (even though his account was decidedly about the United States). For Putnam, the pervasiveness of this decline since the 1950s could have only one culprit: television and its prominent place in an "electronic revolution in communications technology" that was displacing earlier forms of civic engagement and community.[22] In 1995, in his address for the Ithiel de Sola Pool Award and Lectureship, Putnam linked his epochalist narrative of electronic revolution and bowling alone to the problem of liberal conceptions of freedom and government: "As a classic liberal, he [Pool] welcomed the benefits of technological change for individual freedom, and, in part, I share his enthusiasm.... On the other hand, some of the same freedom-friendly technologies whose rise Pool predicted may indeed be undermining our connections with one another and our communities."[23] Noting that Pool, as a "classic liberal," probably would have agreed, he concluded with a statement by Pool corroborating their affinity in thinking that a deepening privatization (which they associated with communication/media technologies) had become the current problem for liberal government: "We may suspect that [the technological trends that we can anticipate] will promote individualism and will make it harder, not easier, to govern and organize a coherent society."[24] That is, of course, were community (including the techno-community envisaged by Rheingold) not also a technology of liberal government. That Putnam referred to neighborhood watch groups as an "artificial replacement for the vanished social capital of traditional neighborhoods, a kind of sociological Astroturf, suitable only where you can't grow the real thing"[25] failed to acknowledge the instrumentality of new or old neighborhood associations as programs adapted, in the past and present, to various objectives of liberal government, which relies on the ongoing development of private forms of governing, the resources for governing at a distance.

Within the historical conjuncture mobilizing the two positions represented by Rheingold and Putnam—each with different understandings of communication technology's relation to civic involvement/activism, each with diverging understandings of the *place* of community within a techno-culture, yet both with a deeply modern conception of community—there emerged a body of research that took as its object the community as network. This research is worth mentioning in part because, although rejecting the conception of community that Putnam uses to gauge a decline of social capital, it accepts (as a calculative project) Putnam's conception of social capital as a measurement of citizenship and community. As in Putnam's writing, measurement perpetuates the modern mode of constructing, locating, and/or verifying community, of making community rational, scientifically intelligible in the age of intelligent machines (employed to assess community). In this latter research, the Net is central to linking communication and community; it is both the object of and the tool for measuring/predicting something called community. And although this research recognizes the "glocalization" of community in the age of computer-mediated communication/networks (the neighborhood/village that is globally connected), the research perpetuates the modern assumption that community is a universal ideal (and not an objective of modern, Western, liberal government): the strategic New Frontier conception of a global village/community through the new global communication network organized by a U.S. space program during the 1960s.

One of the consistent claims/assumptions of the diverse and proliferating research about community as network is that the Internet became just as viable a basis for connectivity (to family and beyond) as traditional forms of connectivity; however, unlike Rheingold's view that community has been lost but can be reinvented through CMC, this research accepts that modern forms of community have occurred at a distance. Thus Barry Wellman's *Networks in the Global Village* rejected the modern sociological conception of community as place bound (as *gemeinschaft*) and argued instead for accommodating the postmodern idea that community is a network unrestricted to places: "All this time communities have continued to thrive around the world, if only people knew how to look for them—and how to look at them. The traditional approach of looking at community as existing in localities—urban neighborhoods and rural towns—made the mistake of looking for *community*, a preeminently *social* phenomenon, in *places*, an inherently *spatial* phenomenon. . . . The principle defining criterion for community is what people do for each other and not where they live."[26] Approaching a somewhat countermodern analytic that understands Net activities as part of everyday life, Wellman has contended that social capital, civic activism, and community have had as much to do with what people do offline as online. However, his research (like most of the research on community as network) locates Net activity at the center of its explanation of citizenship and community formation. He thus argues, unlike Putnam's view of TV, though in as technologically determinist an inflection as Putnam's, that the Internet has the potential to reverse the "decline of social capital,"[27] particularly for those who devote a great deal of time online. He posits that although individuals who are online for long periods have "no greater general sense of community, they do have a greater sense of on-line community."[28]

Because Wellman's account lacks a narrative of community lost, he views his own analysis as corroborating neither "utopian" pronouncements of enhanced overall community nor "dystopian" views of alienation. Neither, however, does this rationale understand power in any terms other than the neoliberal lexicon of "capitalizing" (enhancing social capital) on the Net, as privatized activity and civic activism. Like Rheingold's and Putnam's rationales, with which this research is directly engaged, community is an unquestioned ideal, problematized mostly in terms of how well it is achieved rather than how well it has been instrumentalized as a technology of liberal government and through liberal rationales about citizenship.

Not only do the programs attest to a new rationality of community, predicated on cyber know-how and networking, and calculated through the scientific discourse of social-capital indicators of Internet activity, they have articulated a Net communitarianism to new programs of local government (as models for the reinvention of liberal government and citizenship). Over the late 1990s, these programs represented themselves as experiments in "community informatics." Community informatics emerged in the United States out of perceived failures and limitations of 1970s-era public broadcasting, and subsequently out of the experiments in "community access TV" accompanying the proliferation of cable/satellite TV distribution during the 1980s, and it became the latest, most advanced, kind of initiative for improving the delivery of social welfare ("*enabling* communities with web-based information")[29] and for facilitating civic activism through the new governmental rationality of community as network. After Rheingold and the community-as-network research, community informatics rationalized its projects as a solution to the modern problem of community.[30] "The old or 'traditional' community was often exclusive, inflexible, isolated, immutable, monolithic, and homogeneous . . . [with] geography as the sole orienting factor in a 'community,'" according to Doug Schuler in *Community Informatics*.[31] More than the previously mentioned communitarian discourses and programs, however, a community informatics articulates itself to the rationality of a *government* reorganized as an "information technology network." This process is a new form of social welfare in that it is conceived and practiced as a process of improvement and enlightenment (of making communities and citizens "smarter," more "intelligent"), which is facilitated by Web designers as professional guides and by the "interactive" features attributed to information technology. Citizens do not wait for government services to arrive; they seek out resources for helping themselves via Web sites. Thus community informatics rationalizes its projects in terms of "civic intelligence" and "citizen technology." Interactivity is a key component to rationalizing government this way—the key to a new civic entrepreneurialism and activism, and to a new sense by a mobilized and activated citizenry of itself as an empowered community: "A new community . . . needs to be fashioned from the remnants of the old community. . . . A new community is marked by several features . . . [its] *consciousness* [a high degree of self-awareness]—both of itself (notable its capacities and needs) and of the milieu in which it exists). . . . The consciousness of the new community must be intelligent and creative."[32]

Many of the community informatics initiatives have involved representing the community network as a mechanism for improving and advancing citizenship through a virtual geography or atlas of the traditional institutional organs and sites of local government. These "free-Net" systems vary among cities, but they commonly provide links to post offices, arts buildings, schools, science and technology centers, libraries, and public squares. As such, the initiatives formulate a map of governmental sites (physical and virtual) where, as interactive Web user, one can practice a new, enhanced (more active and entrepreneurial) form of civic participation. The Seattle Community Network (www.scn.org), however, was organized less in geographic categories than in degrees of civic activism:

Web sites where individuals can find informational resources consonant with their commitment to certain endeavors, related, for instance, to health, finance, recreation, the arts, neighborhoods, science and technology, and spiritual matters. Although the Seattle Community Network represents itself as a multiuse organ (a search engine, a resource for Web mail), it links these individual practices to a program of mobilization, where encouragement for volunteerism and activism frame its primary and ancillary Web sites.

As a virtual framework for performing the new/restored virtues of civic life, community informatics represents the moral imperative (what Rose has called the "ethico-politics") of good government through community. Whereas "informatics" speaks to the new rationality/economy of liberal government in the age of intelligent machines, "community" refers to the principledness of a government that conceives of (inter-)active citizens looking after their own welfare. Community informatics' idea of community development thus offers itself as a path to a moral economy (intelligent, expeditious, user friendly, and thus "good" government) for the digital age. Echoing the Rheingoldian humanism of cyber-community (whose rules Rheingold contended were innately recognized by Netizens), Schuler claims that "Rebuilding—and redefining—the community is not optional, nor is it a luxury. It is at the core of our humanity; rebuilding it is our most pressing concern. . . . The new community has both *principles* . . . a set of core values that maintain its 'web of unity.'"[33] As such, community informatics understands its primary objective to be the enhancement of *civil society*, the ethico-political terrain of liberal government. As a governmentality adapted for localities/communities, a community informatics rightly sees itself as solving one of the problems of a neoliberalism: governing through new programs and procedures (the cyber-technologies/networks) of citizenship. If, as Gurstein contends, community informatics is warranted by "the cutting back . . . of the 'social safety net,' and the imposition of the 'cost discipline' on the range of public services,"[34] then it is incumbent on community informatics to acknowledge how it also operates as a new and localized resource for neoliberal government—that is, as a practice that fulfills the objectives of government interested in developing localized, interactive mechanisms for restoring trust in government and for making self-government (as a basis for a civil society) more effective.[35] Volunteerism, after all, has become integral to the needs/objectives of a government increasingly committed to outsourcing and privatized forms of welfare and social security. Furthermore, although it is one thing to emphasize (as do proponents of community informatics) the inequalities of the resources/technologies of citizenship, it is another to assume that those technologies (and community network as a technology) is the primary resource of citizenship within the daily lives of different populations or that a "community network" is inherently democratic. Rather than seeing good government as an objective of community, as has community informatics, how might one think about community as an objective of liberal government? Following the latter path would emphasize a different set of questions about how the Net became a more effective mechanism than broadcasting for dispersing and coordinating spheres of government (as community) on different scales and at different sites—how the current stage of liberal government has come to require and to depend on privatized and localized mechanisms of government that can also be trusted/accepted for mobilizing volunteers, for coordinating/managing its penchant for dispersion, and for governing at a distance.

Considering how "new information technologies" (NITs) have come to be viewed by communitarians as useful and as more effective than forms of broadcasting (at a time when broadcasting has been rearticulated to/through NITs) therefore would lead to a consideration of how communication networks (and community as network) become the objectives and resources for programs of government through the production of space. This leads me to consider briefly the historical and spatial conjunction of the programs discussed above with the initiatives of the New Urbanism. Community informatics, by understanding social inequality in terms of both physical space and cyberspace, implies that citizenship and community are not simply questions of expertise and possessing the cultural capital (know-how) required of a cultural citizenship through communication technology and networks, but they also are matters of physical access to spheres of citizenship, even as they presume that community networks (interactivity online or at home) improve access to institutional sites of civic activity, which long have involved issues of physical access and mobility. The New Urbanism, in contrast, continues to understand citizenship and the path to good government more in terms of physical access and mobility, although their planning, advocacy, and mobilization initiatives rely on NITs. Through associations, initiatives, colloquia, books, pamphlets, and Web sites over the 1990s, the New Urbanism has become a banner for organizing architects, city planners, and policymakers around a common purpose: redesigning the urban and suburban environment into "neighborhoods" as a means of restoring "community."[36] Vincent Scully, an advocate of the new urbanism, thus has described it as the "new architecture of community." In certain respects, the aims of the new urbanism are rationalized in terms similar to Rheingold's *The Virtual Community*. Philip Langdon, for instance, contended that "the building block" of community is neighborhood, and that "lessons about how to build better communities can be extracted from the nation's successful older communities," as represented in the design of older houses and residential spaces. The New Urbanism's communitarianism, often described as a *neotraditionalism*, focused on the *recovery* of a spatial practice and sociability lost to suburbanization and automobility in the United States, but whose model is the premodern, European city structure organized around piazzas, propinquity, and pedestrian paths (even as its implementation—making the late twentieth-century U.S. city more rational and healthy—follows the general objectives of modern Western planning that seemed to have become anathema with the wave of mass suburbanization after the 1940s).[37]

Central to the New Urbanist rationale is its discourse of so-called smart growth. In one sense, smart growth casts the New Urbanism as a solution to hyperliberalized programs of growth (urban sprawl perceived as unfettered) that has occurred most intensely in the United States since World War II. "Smart," in this sense, implied restoring rationality and restraint, by private citizens/

technicians (architects and planners as guides acting in consort with "community activists"). "Smart growth" also referred to the health and well-being ensuing from wise, rational management of cities and sub-/urban populations transformed into neighborhoods and communities. Like the Net-community discourse and the Putnam-esque account of bowling alone, the policy and programs representing smart growth justified themselves as a remedy to a decline of neighborhood and community accompanying this period of mass suburbanization. And "smart" arguably implied a style or look: the "cleaning up" of blighted inner cities ("infills") and the introduction of presuburban design motifs into newly built suburbs. In all these respects, the new urbanism and Net communitarianism found a commonality of purpose in Ray Oldenburgh's view that the "third spaces" of social and civic life (beyond the sites of work and home life) had disappeared, and rediscovering them was tantamount to restoring community.[38]

As I have argued elsewhere, New Urbanist community restoration (the city rethought/reconfigured around neighborhood and "third spaces") was one model, program, and site of efforts to reinvent liberal government during the 1990s.[39] New Urbanist planners, for instance, reasoned that enhancing civic space required facilitating access to third spaces by a more diverse citizenry than mass suburbanization had allowed (and whose diversity represented the communitarian spirit of urban planning). So New Urbanism envisaged planned residential areas of a more mixed demographic, as well as more mixed uses, than were found in U.S. suburbs. This aim of New Urbanist projects perpetuated a liberal-democratic ideal of inclusiveness (not always achieved in practice) while being consonant with neoliberal programs of localized self-management through neighborhood and community.

The New Urbanist contribution to reinventing government and to neoliberal programs of privatization, local autonomy, and self-management also developed through its role as consultant to political government. This always occurred locally, as part of programs sponsored through municipal government, as in Austin, Texas, where new urbanist programs became integral to city government's initiatives to manage traffic and to enhance the city's old neighborhood centers as a response to suburbanization. However, it also occurred in conjunction with the Clinton-Gore administration's efforts to reinvent government by redesigning forms of social welfare, as occurred when the new urbanists became consultants in 1996 and thereafter to the Hope VI program, an initiative by the Department of Housing and Urban Development. Hope VI was a program that (in the spirit of the Clintonian "third way") not only demonstrated a path to "reinventing" liberal government (i.e., literally remaking public housing as a remnant of New Deal–era policies of social welfare) but also required the active participation by public housing residents (guided by new urbanist planners) to remake themselves as/into a community.

Beyond these instances where New Urbanist design and planning conjoined with the efforts of government policy to localize and privatize the mechanisms of social welfare, the New Urbanism also was (as Andrew Ross's account of Disney's Celebration affirms) easily articulated to the proliferation of "gated communities" as a suburban administrative mentality that valued privatized means of achieving the security/safety, welfare/healthiness, responsibilization, and self-sufficiency of community as neighborhood.[40] And it was no small coincidence that Gore campaigned in 2000 in part on his commitment to smart growth, an agenda represented in a policy report (describing itself as a "guide" and "toolbox") that was attributed to Gore, posted on a government Web site, and titled, "Building Livable Communities: Sustaining Prosperity, Improving the Quality of Life, Building a Sense of Community" (June 2000). Smart growth (consonant in all these respects with Clinton's third way) thus implied reasoned and measured economic/commercial freedoms: "soft," privatized, and localized forms of restraint, and simultaneously a form of economic liberalization/privatization that was necessary and reasonable, particularly if it was guided by professional, rather than bureaucratic, planners. Similar to third way reform, the New Urbanism eschewed the distinction (a long-standing argument in the United States during the twentieth century) between urban planning or municipal zoning and privatized suburban development, adopting instead the term "smart code" to represent their effort to make the objectives of communitarian "civic-minded" planners attractive to private developers.[41] Smart growth and smart codes had as much to do with the domestic *political* salience of the New Urbanist regime for building healthy communities in suburban America as with a distinctly American way of justifying and enacting neoliberal programs.

The New Urbanist conception of smart growth, however, bears a noteworthy relation—discursively and programmatically—to "new" or emerging media/communication technologies, to the networking capacity of "intelligent" (self-communicating and self-monitoring) machinery facilitating virtual forms of community. Although the New Urbanist rationale for smart growth was only indirectly about engineering community in the age of "intelligent" machinery, emphasizing this dimension of the New Urbanism is one way of thinking about its linkage to Net communitarianism and their joint value within a neoliberal reasoning about citizenship and government, and within an economic rationality (the "information economy") that sought to advance the liberal ideal of "economic government." In 1993 (the year that Clinton became president), the Congress of the New Urbanism was founded and Rheingold's *The Virtual Community* was published. The New Urbanist conception of smart growth, which rejected the insularity and "disconnectedness" of mass suburban enclaves in favor of more "integrated systems" (neighborhoods "networked" into the urban fabric), reiterated the Rheingoldian terminology (if not the ideal) of community through networks.[42] However, whereas the Rheingoldian idea of virtual community was about the transcendence of physical/corporeal impediments,[43] viewed as having contributed to various modern problems in sociability and the unhealthiness of urban societies (i.e., their loss of community), the New Urbanism (after Putnam, in some respects) rationalized its programs as a response to the unhealthy legacy of residual media, particularly television, as a narcotizing influence and thus as responsible for passive, immobile bodies and populations. The *Charter of the New Urbanism* devotes a page to a photograph of an overweight boy in front of a TV set (the iconic "couch potato") and to the warnings by clinical psychologists about the ill effects of too much TV. For the new urbanism, television was a symptom of the malaise of hyperinteriorized life under mass suburbanization and a factor in the erosion of the third places—the collective public sphere of town squares and front porches—that they attributed to life in the United States before World War II. New Urbanists' moral dichotomy between nurturing public spheres and immobilizing domesticity, however, mostly was silent about Web technologies that, over the 1990s, became tools for mobilizing various networks of New Urbanist civic activism, such as New Urbanist literature, and they also never acknowledged the array of programmable technologies (including media technologies) that had made the household "smart"—having the capacity to govern itself, and thus hypothetically to free up subjects and enhance mobility at/from home. Moreover, in its condemnation of old media, the deleterious legacy of television in suburban life, the New Urbanism also shared the dream of interactivity (of "active citizenship") attributed to virtual community (and, for that matter, the economy and sociality of new digital and Web-based media).[44] Thus, for the physical emplacement of community imagined/practiced by the New Urbanism, as well as for the virtual emplacement of community imagined/advocated by Rheingold, community was a basis for the new (neoliberal) models of (inter-)active citizenship facilitated by smart growth and smart homes, respectively.

The relation between smart growth and smart homes is not a play on semantics but a congruence—a historical conjuncture—between technologies of citizenship and governance whose primary objective has been instrumentalizing community. Smart growth and smart homes both involved localizing and privatizing mechanisms of connectivity, trust, and active citizenship on which liberal government increasingly relied. Both sought to advance a new governmental rationality of social spaces (homes, cities, and nation-states) believed to have developed through broadcast media and to have produced inactive citizens. Furthermore, each program was imagined to solve a set of problems about emplacing community that the other program was not equipped to address. So, even though Net communitarians and new urbanists both mobilized around the loss of community,[45] each program's enactment of community involved particular limitations (in physical space and cyberspace, respectively) and thus operated as separate but corresponding/interdependent technologies for a new (albeit retro-oriented) model of citizenship.

Here it is useful to return briefly to the discussion of "communities of practice" introduced in the opening section of the chapter. To understand and map the correspondence between Net communitarianism and the New Urbanism within an emerging diagram of power (and, through these specific programs, to understand the value of community within a neoliberal reasoning about government and citizenship), it is worth recognizing the value of communitarian programs for a new managerial rationality among businesses increasingly organized through information technologies, and for economies oriented toward producing and distributing these technologies. Communities of practice refers to the clustering of workers into specialized fields of knowledge, mobilized (made interactive) through micro-networks of communication and information sharing that in turn are linked upward (as fonts of knowledge production and efficiency) to a corporation's centers of management. As the American Productivity and Quality Center stated, "Communities of Practice are the next step in the evolution of the modern, knowledge-based organization." Over the 1990s, communities of practice became one of the dominant managerial models, instrumentalizing community in the "knowledge-based" and "service" industries, but more generally in corporate organizations that understood themselves as a system of communication.[46] For businesses relying on digitalized networks of communication, community of practice represented a new model of corporate citizenship and productivity, a new basis for discerning and achieving value in the new workplace, and thus a new rationality of business management. Within the increasingly global scale of networks to which (virtual) corporations were connected over the 1990s, furthermore, communities of practice became a strategy for expanding a scale of operation—extending the boundaries of the workplace—without losing micro-collaborations. As a *managerial* discourse/model, communities of practice thus represented the governmental value of community (communicating micro-organizations and corporate Net communities) for the global and virtual corporation.

Introducing communities of practice to an analytic of neoliberal government is useful because it underscores a changing relation between government and economy, intended here not as a political economy of emerging/residual media but to demonstrate how recent liberal governmentalities found their way into the corporate economy of NITs, and conversely how new models of active/productive corporate citizenship, and of management at a distance within corporations, found their way into a neoliberal reasoning about political government. (Contrary to the usage of some writing on globalization, "neoliberalism" does not refer to an economic formation, in a narrow sense, but to programs of government/management, some of which are directed to reorganizing the workplace and the territorialization of production/distribution.) Bringing communities of practice into an analysis of the New Urbanism and Net communitarianism thus underscores what neither of these programs acknowledges about themselves: the instrumentality of community within an emerging relation between government and economy. If community is imagined to be lost and in need of recovery, what is the value of recovering community for particular economies, as well as for diverse programs of government that make these economies knowable, productive, and manageable?

In conclusion, and as a way of returning to my introductory discussion of the ICAP, I ask what has changed in the new millennium and specifically in the shift from the Clinton-Gore to the Bush-Cheney administrations. Should "neoliberal" refer in the United States to a governmental rationality supported as much by recent Democrat as Republican administrations? If liberal government has always been involved in assessing and instrumentalizing "civil society"—liberal government's resources for governing "at a distance"—then might it be better to describe the current stage of liberal government in part as having supported and instrumentalized community and communitarianism (and as having reinvented itself this way)? Certainly the new communitarianism, and its perpetuation of a deeply modern problematization of community and the technologies of mediation, has not been more endemic to one or the other of these administrations, although the Bush-Cheney administration could instrumentalize community and communitarianism because of its value to liberal government over the preceding Clinton-Gore administration. The Bush-Cheney administration's first major policy directive (and the most prominent strategy for representing its "compassionate conservatism") was the "Faith-based and Community Initiative (FBCI)" for privatizing forms of social welfare through church and civic associations. In Bush's address to the U.S. Senate commending them for support of his "faith-based" legislation in 2001, he stated that the initiative will "encourage more charitable giving and rally the armies of compassion that exist in communities all across America." As of 2003, the initiative had been located in no less than seven government agencies, all coordinated by an Office of Faith-based and Community Initiatives and supported by the Compassion Capital Fund.

Although the program was criticized by many Democrats as blurring the relation between church and state, the program developed out of and rearticulated the value of community in the "reinvention" of liberal government. According to the Web site for the Office of Faith-based and Community Initiatives in 2003,

> [T]he FBCI represents a fresh start and bold new approach to government's role in helping those in need. Too often the government has ignored and impeded the efforts of faith-based and community organizations. Their compassionate efforts to improve their communities have been needlessly and inproperly inhibited by bureaucratic red-tape and restrictions.

Therefore, if the FBCI seemed reasonable and valuable, it was because it built on the instrumentality of community/communitarianism within a governmental and economic rationality from the previous decade. Yet in the Bush administration, communitarianism became a framework not only for reinventing government (a deepening of the Clinton administration's "reinvention of government" through a "third way") but also for coordinating/administering numerous government programs. FBCI put on display the importance of "partnerships" between political government and local, private, communitarian initiatives (the value of "looking after" and "taking care of" oneself). It also articulated the church and communitarianism, making social welfare moral and redefining the "virtues" of social welfare as a new relationship among government, the church, and communitarianism. Whereas the Clinton-era Hope VI program (mentored by New Urbanists) was about "restoring" community activism in the reinvention of public housing (and, by example, social welfare), the HUD arm of the Bush administration's FBCI explicitly tied FBCI to HUD's new mission: linking home ownership to good citizenship and assuring that HUD's programs of rehabilitation were accomplished less directly through "government intervention" and more through community development corporations. A July 2002 HUD report that explains the objectives of CDCs even asks how public housing can be made "smarter."[47]

To say that the Bush-Cheney administration instrumentalizes community as a technology of government may be too general, too similar to the communitarianism that emerged over the 1990s. As a political rationalization of public welfare, of a new relation among government agencies and services, and of a new relation between public and private administrative institutions (reinventing liberal government in all of these ways), the FBCI represents a deeper commitment to the model/rationality of business management for instrumentalizing community. ("Building the Organizations That Build Communities: Strengthening the Capacity of Faith- and Community-Based Development Organizations," the collaborative publication by the FBCI and HUD cited above, is a vivid example of this.) And in this way, the FBCI, CoPs (as a model of corporate citizenship/management), and community informatics (as a model of Net-based, interactive citizenship) become relevant to one another. If community is how Western societies have problematized modernity and modern, liberal government, the FBCI, CoPs, and Net communitarianism are about the unlimitedness, reasonableness, and naturalness of community—about governing not just in the United States but wherever experiments of liberal government are enacted around the world. If community is required and needed, then for whom? Answering that question makes community matter politically and governmentally, as part of a diagram of power. Community is tantamount to bringing freedom, to maximizing liberalization in an orderly, rational, disciplined, and civil fashion. It is no small coincidence, therefore, that CoPs—and, behind it, the FBCI—became the rationale/justification for the invention of community in Iraq through the ICAP.[48] Communitarianism is also about (inter-)active citizenship and self-enterprise. In this sense, the New Urbanism and community informatics have become templates for a kind of "nation-building" that sees itself less as nation-building than as community activism, an activism of self-enterprising citizens. Bush's description of the FBCI as mobilizing "the armies of compassion that exist in communities all across America" applies as much to the administration of public welfare in the United States as to the welfare of Iraq because of the reasonableness and naturalness of community as a universal objective of "free societies." Mobilizing the "armies of compassion" is not just about mobilizing all of America but of directing that mobilization globally, of reinventing government ("regime change"?) through rebuilding community. The outcome of this experiment in Iraq will undoubtedly demonstrate, among other things, the limits of this rationale: that community is not an essential truth or referent but a historically and geographically deployed technology for advancing liberal government. In this respect, it is just as important to recognize how the ICAP developed through the communitarianisms of the 1990s as it is to rethink the historical and geographic claims of these communitarianisms by charting their relation to an ICAP, and the global reach (the universalist aspiration) of an FBCI.

Postscript: The "Power-Geography" of Enduring Freedom's Residual Media

Considering the ICAP's relation to communitarian programs in the United States is useful for underscoring that an analysis of emerging programs of community in their relation to modern/residual logics of mediation involves recognizing the *power-geography* of community as a governmental technology. An analytics of government, particularly one that is informed by a spatial materialism, are countermodern—ways of thinking about the limits and situated-ness (and thus the strategic-ness) of modern conceptions/deployments of community (e.g., "community development"), about the different scales and global unevenness of modernization and modernization's preoccupation with "community development."

A few months following the occupation of Iraq under the banner of Enduring Freedom, the United States appointed an "interim government" (the Coalition Provisional Authority), which officially sponsored the introduction of Al-Hurra, "The Free One," or "Freedom TV" (or, "TV Freedom"). For a national territory that had only one TV network and in which citizens were forbidden to own satellite dishes during the regime of Saddam Hussein, Al-Hurra was represented by its U.S.-government sponsors and producers as one indication of—one way of exhibiting—the coming of a new liberal democracy and as one program for mobilizing a newly "liberated" citizenry, now *free* to watch TV programs that they had been deprived of watching. Al-Hurra also affirmed the shifting power-geography of TV (the globally situated and dispersed *futures of residual media and residual logics of mediation*) in the age of an emerging pan–Middle Eastern broadcasting (e.g., Al-Jazeera). Because a majority of Iraqis relied more on TV than on other technologies (the Internet or radio), in regions of a country where a population was mobilized daily just as often through speakers mounted on mosques, Al-Hurra's invention (as an experiment in enacting liberal government, alongside the ICAP) calls attention as much to the current political instrumentality as to the limits of residual media and residual logics of mediation.

Notes

1. The number of private outsourced U.S. workers directly involved in the military operation exceeded the number of British soldiers, the second largest national participants in the military "coalition."

2. U.S. Agency for International Development, April 2003, www.usaid.gov/iraq.

3. This distinction was implied in Donald Rumsfeld's statement just after the U.S. military action in Afghanistan and just prior to the Iraq campaign: "The objective is not for us to engage in 'nation building'—it is to help Afghans, so they can build their own nation. That is an important distinction. In some 'nation building' exercises, well-intentioned foreigners arrive on the scene, look at the problems, and say, 'Let's go fix it for them.' This can be a disservice. Because when foreigners come in with international solutions to local problems, it can create a dependency" ("Against Nation Building," Secretary of Defense Donald Rumsfeld at the 11th Annual "Salute to Freedom," Intrepid Sea-Air-Space Museum, New York City, February 14, 2003). Particularly for the Bush-Cheney administration, community was integral to a governmental and political rationality whose primary objective was to avoid linking the welfare of nations with forms of dependency.

4. The term "economic government" refers to the liberal ideal that good and wise government is little government and to the modern, physiocratic ideal that economy is a self-correcting system requiring little interference by the state. As Graham Burchell, and others who have written about governmentality have noted, economic government relies on unofficial programs of government in civil society. Graham Burchell, "Liberal Government and Techniques of the Self," in *Foucault and Political Reason*, eds. Andrew Barry, Thomas Osbourne, and Nikolas Rose (Chicago: University of Chicago Press, 1996).

5. In an attached report (identified on the Web site as the "twelfth report"), the USAID lists a set of *cooperative* initiatives that have "established 383 Community Action Groups of 400 targeted in 16 governates," all "part of a campaign targeting grassroots democratic development." And in this

attachment, the USAID Web site lists the material contributions of the CAP ($32.8 million for 1,114 projects, with "457 projects . . . already completed") that have combined with the contributions of NGOs (CHF, ADCI/VOCA, IRD, Mercy Corp, Save-the-Children Fund), noting the "Iraqi communities have contributed $9.8 million—one quarter of the total project funding—to community projects" (www.usaid.gov/iraq, December 2003).

6. This is not to trivialize the enormous financial expenditure that links infrastructural improvement with the engineering of a civil society comprised of interfacing communities. Although the USAID calculates progress occurring through NGOs *and* "grassroot" efforts by Iraqi "communities," the accounting rationale casts U.S. financial contributions as a considerable financial contribution by "Iraqi communities." In this respect, the current program of leveraging and mobilizing community remains mostly an outline and set of objectives for political reform *proposed* for Iraq by the Bush administration.

7. "The Development of Communities of Practice at USAID" cites, among other literature, Wenger, McDermott, and Snyder, *Cultivating Communities of Practice: A Guide to Managing Knowledge*, 2002, and the APQC Best Practices Report, *Building and Sustaining Communities of Practice: Continuing Success in Knowledge Management*, 2001.

8. Etienne Wenger, Richard McDermott, and William Snyder, *Cultivating Communities of Practice: A Guide to Managing* (Boston: Harvard Business School Press, 2002), 232.

9. Ibid., 6.

10. The term "governmental rationality/reasoning" is central to a Foucaultian analytic of government (a study of governmentality) with which my own project is engaged. See, for instance, *Foucault and Political Reason: Liberalism, Neo-liberalism, and Rationalities of Government*, eds. Andrew Barry, Thomas Osborne, and Nikolas Rose (Chicago: University of Chicago Press, 1996).

11. Ferdinand Tonnies, *Community and Civil Society*, ed. Jose Harris, trans. Jose Harris and Margaret Hollis (Cambridge, UK: Cambridge University Press, 2001), 19.

12. Ibid., 57.

13. Tonnies's thesis is, in this respect, consonant not only with many of the assumptions of early German social theory (e.g., Weber, Kracauer, Benjamin, and Adorno) but also with the projects of Robert Park and the Chicago School of social research in the early decades of the twentieth century.

14. James Carey, *Communication as Culture* (Boston: Unwin-Hyman, 1989), 18.

15. Raymond Williams, *Keywords: A Vocabulary of Culture and Society*, rev. ed. (New York: Oxford, 1976/1983), 76.

16. By its sixth edition, a cover page of *The Virtual Community* by HarperCollins included praise from Silicon Valley (the *San Jose Mercury News*), Mitch Kapor (founder of the Lotus Development Corporation and the Electronic Frontier Foundation), the *New York Times*, *Business Week*, and the fledgling *Wired* magazine.

17. Howard Rheingold, *The Virtual Community: Homesteading on the Electric Frontier* (Reading, Mass.: Addison-Wesley, 1993), 1.

18. Ibid.

19. Ibid., 26.

20. See Armand Matterlart, *Networking the World: 1794–2000*, trans. Liz Carey-Libbrecht and James A. Cohen (Minneapolis: University of Minnesota Press, 2000), as well as James Hay, *Governing the Free-Range City: Houston's Way of the Future* (forthcoming).

21. Rheingold, *The Virtual Community*, 15.

22. Robert Putnam, "Tuning In, Tuning Out: The Strange Disappearance of Social Capital in America," The 1995 Ithiel de Sola Pool Lecture, *Political Science and Politics* (December 1995): 680.

23. Ibid.

24. Ibid.

25. Ibid.

26. Barry Wellman, *Networks in the Global Village* (Boulder, Colo.: Westview, 1999), xiv.

27. Ibid., 295.

28. Ibid., 313.

29. This expression is the subtitle of Michael Gurstein, ed., *Community Informatics: Enabling Communities and Communications Technologies* (Hershey, Pa.: Idea Group Publishing, 2000).

30. "Community Informatics is an emerging 'discipline,' but it is also a creative and practical response on the part of communities to the difficulties which they and their members are facing in the modern world." Gurstein, *Community Informatics*, 19.

31. Doug Schuler, "New Communities and New Community Networks," in Gurstein, *Community Informatics*, 175.

32. Ibid.

33. Ibid., 174.

34. Gurstein, *Community Informatics*, 19.

35. "Community Informatics is concerned with enhancing civil society and strengthening local communities for self-management and for environmental and economically sustaining development" (Gurstein, *Community Informatics*, 2).

36. See the *Charter of the New Urbanism*, eds. Michael Leccese and Katherine McCormick (New York: McGraw-Hill, 1999). The official Web site for the new urbanism is www.newurbanism.org.

37. The new urbanism frequently valorizes the importance of restoring "civic buildings" and "civic space" to a postsuburban landscape (Andres Duany, "Civic Buildings and Public Gathering Places Require Important Sites to Reinforce Community Identity and the Culture of Democracy," in Leccese and McCormick, *Charter of the New Urbanism*). In this respect, the new urbanism claims to update the objectives of the city beautiful movement (one dimension of the first wave of suburbanization in the late nineteenth century) for a postsuburban United States.

38. Ray Oldenburgh's conception of third spaces in *The Great Good Place* (St. Paul, Minn.: Paragon House, 1991) became the template for Rheingold's conception of the WELL and of computer-mediated networks/community more generally, and it figures prominently in numerous rationales by new urbanists and as suggested reading in *The Charter of the New Urbanism* (New York: McGraw-Hill, 2000).

39. I have written at greater length about the new urbanism's relation to a new communitarianism and to a recent stage of liberal government in "Unaided Virtues: The (Neo-Liberalization of the Domestic Sphere and the New Architecture of Community," in *Foucault, Governmentality, and Cultural Studies*, eds. Jack Bratich, Jeremy Packer, and Cameron McCarthy (Albany: SUNY Press, 2003).

40. Andrew Ross, *The Celebration Chronicles: Life, Liberty, and the Pursuit of Property Value in Disney's New Town* (New York: Ballantine Books, 1999).

41. Andres Duany discussed the new urbanist emphasis on smart codes at a public address at the University of Illinois–Champaign-Urbana, March 2004.

42. That the new urbanism and Rheingold's WELL both developed out of or have affinities with post-1960s holistic environmental (green) movements in the United States is yet another point of convergence, yet another facet of the two initiatives' aspiration to be both cosmopolitan (a "whole earth" network) and restorative of community (networking as the path to third spaces).

43. As Jody Berland has noted, the idealization of "virtual" communication (and arguably of virtual community) fancied the ongoing achievement of purer states: the transcendence of physical place and situated bodies.

44. See for instance, Mark Andrejevic, *Reality TV: The Work of Being Watched* (Lanham, Md.: Rowman & Littlefield, 2003).

45. Charles Brewer, the founder of Mindspring, confessed his conversion to the new urbanism in 1996, noting that the public square was, hands down, a more attractive sphere of sociality than the Net could ever be.

46. The first use of the expression "communities of practice" began to appear in the literature of corporate management and organization communication in the early 1990s, roughly the same period as the communitarian programs/discourses of the new urbanism and Net communitarianism.

47. In June 2003, HUD's Center for Faith-Based and Community Initiatives and the Office of Policy Development and Research convened a symposium: "Building the Capacity of Faith- and Community-Based Organizations Summit." From this symposium HUD produced a 300-page collection of papers, "Building the Organizations That Build Communities: Strengthening the Capacity of Faith and Community-Based Development Organizations," available through the HUD Web site.

48. It is worth noting that although the FBCI was the Bush administration's first policy initiative, it lost some of its prominence following the attacks of September 11, 2001, when the problem of which "faith-based" organizations were the appropriate ones to receive government funding. The FBCI did continue to be a rationale for coordinating/realigning government branches and services after September 11, 2001, and the ICAP became one solution to the governmental problem that September 11 posed for the future of the FBCI: If money could not go to fund non-Christian organizations in the United States, bringing programs of "community action" to Iraq could be cast as part of the rehabilitation—the civilizing—of Iraq.

15 Unearthing Broadcasting in the Anglophone World

James Hamilton

The use of "broadcasting" to describe both the sowing of seeds by hand and the electromagnetic dissemination of visual/audio programming seems initially to be a curious but ultimately irrelevant coincidence. The former refers to a single person sowing seeds using individual muscle power, and the latter is popularly seen as the "sending [of] a wide range of sounds . . . from powerful transmitters to thousands of individual receivers."[1] (See Figure 15.1.)

Yet if one understands broadcasting as more than a technology and a kind of media institution, this "coincidence" defines its cultural and historical core. As Raymond Williams remarked more than forty years ago regarding the study of media and communication, reliance on political and economic (one could add technological) description alone "can leave out too much. How people speak to each other, what conventions they have as to what is important and what is not, how they express these in institutions by which they keep in touch: these things are central."[2] Recent work by Peters and Scannell relies on similar rationales to suggest that broadcasting can be seen as a complex of social relationships, not only technologies and institutions.[3]

This chapter works in the wake of these and similar assertions. It argues that the articulation of agricultural and media practice through "broadcasting" indicates more than an archaic, metaphorical equivalence between the spreading of seeds and the spreading of messages. More generally, it asserts that the application of "broadcasting" from eighteenth-century English debates about farming technologies to early twentieth-century American debates about the nature of media systems in modern societies charts the emergence of new ideas about the role of communication media and the nature and constitution of publics in the context of accelerating capitalism and consumerism.

To explore this formative development, this chapter investigates expressions regarding broadcasting from the contexts of agriculture to the early days of mass media, recovers the incorporation of the first usage into the second, describes the implicit, conventional frameworks of understanding that made its incorporation meaningful, and interprets the significance of such frameworks for democratic

FIGURE 15.1 *Broadcast sowing "prevailed in its essentials until well into the nineteenth-century, when the seed drill at last came to be widely used." From G. E. Mingay,* A Social History of the English Countryside, *reprint ed. (London and New York: Routledge, 2002), 31. The woodcut image is from William Cullen Bryant,* The Song of the Sower *(New York: D. Appleton, 1871), 23.*

communication.[4] Such a purpose rests on the claim that expressions not only reflect commonly held beliefs but actively produce them, ourselves, and our societies. In addition, it proposes that such a process—linked as it is to rationales for political decisions and actions of all kinds and at all levels—is an exercise of power.[5]

"Unearthing broadcasting" refers not only to the expansion of broadcasting over time from agriculture to media industries. It underscores the extent to which broadcasting, despite its modern sheen, is a residual medium based in and continually resuscitated from hundreds of years of accumulated materials, practices, and ideas, yet done so in continually emergent conditions. The recognition of its residual status also underscores the necessity of historical recovery as a means of understanding media and society.⁶ "Unearthing broadcasting" also encapsulates the primary argument to be made here, that the trajectory of "broadcasting" can be understood as a complex historical process of mystification and naturalization—the transmogrification of a human muscle-powered activity into a nonhuman, invisible force of nature. Evangelical, fundamentalist Christian social activism during the nineteenth century forged the initial articulation of "broadcasting" with media in the context of vanguard mass politics. In the context of technological innovation, the rationalization of the state, and the beginning of a consumer society in the late nineteenth and early twentieth centuries, "broadcasting" emerged not only as a means of harnessing an electromagnetic property for centralized delivery of information and entertainment to the general public but, as will be seen, in some inflections as the physical energy flow itself that takes place throughout the natural world. After describing this gradual articulation, the chapter reflects on the implications of this heritage for "broadcasting" as a form of democratic communication.

The Roots of Broadcasting in Early Modern England

Paradoxically, "broadcasting" first appeared in the English language in the mid-1700s as part of arguments in favor of its disappearance. Literally speaking, "broad-casting" meant walking through a field, scooping a palmful of seeds from a seed bag slung over one's shoulder, then throwing them through the air to the ground in the hopes they would germinate and grow into crops to be later harvested. Arguments against broadcasting appeared commonly in English farming books, a hybrid genre part self-help manual, empirical study, promotional advertisement, and political treatise. Fussell traced the genre to 1523, noting that "they were the classics, part of every gentleman's education of the day."⁷ By contrast to this labor-intensive, imprecise, and inefficient method of sowing, proponents argued that seed drilling (the use of animal-drawn machines to place seeds at a precise location and depth) made it possible to cultivate fields continuously without requiring fields in intervening years to lie fallow, thus enabling farmers to extract greater economic value from the land they owned or rented.⁸ This benefited not only the individual farmer, but the nation, recalling a linkage articulated by classical political economists of the day. By the early 1800s, such a linkage had the status of an axiom for one writer, who asserted that "every step that a man makes for his own good [also] promotes that of his country."⁹ Working for the good of the country maintained an abstract nationalist pride, but it also sought to avoid poverty, hunger, and resulting widespread rioting, a practical policy objective not only for England, but for "all European governments in the eighteenth century."¹⁰ To help the country as well as oneself, one had to farm using the most productive, efficient methods possible.

When seen more broadly, critiques of broadcasting were responses to the increasing subjection of agriculture by logics of accelerating expansion and profitability. Gentry farmers aimed to become more productive, in part, by minimizing their dependence on and the cost of peasant labor. A writer at the end of the eighteenth century noted that, compared to "hand-hoeing [that] requires a great many laborers," mechanized cultivation "requires but a few labourers . . . and therefore may be practiced in many places where the other cannot, for want of labourers."[11] And, as Fussell notes, another inventor "designed his seed-drill to combat the intransigence of his workers . . . after they had refused to do the planting in rows that he required."[12]

By contrast to gentry proponents of seed drilling, agricultural laborers preferred broadcasting for practical reasons. Not only were seed drills notoriously unreliable and difficult to fix, the mechanization of agriculture was gradually depriving rural laborers of a means of making a living.[13] Seen in a broader scope, the technical rationale for seed drilling was part of the accelerating economic and political rationalization of the national economy throughout the eighteenth century, with local customary law becoming increasingly superseded by a national policy that embodied classical political economy and its particularly orthodox brand of market economics.[14] Rationalization had its own costs, however. Decreasing the need for laborers meant increasing the burdens on local communities to take care of them when unable to find work, a trend contrary to the longstanding "social code of the countryside and the unwritten law that the onus of finding work should be spread among all employers in the parish."[15] Such a trend was not accepted lightly. The examples of Luddism and the agricultural riots of the eighteenth and nineteenth centuries suggest the depth of the working-class response to attempts to "control time, the wage, and levels of skill . . . both [through] the law and new machines."[16] What the context of its emergence indicates is that "broadcasting" from the beginning was not a simple description of a technique. It marked a terrain constituted by politics, technology, and class, emerging just as it was seen as a thing of the past.

Sowing Reform in Nineteenth-Century America

The term "broadcasting" does not appear in the Bible. Yet the practice to which broadcasting refers operates as a central trope to embody a claim about the process of finding Christ. Rather than privilege the power of the text by making a mechanical equivalence between messages and seeds, the parable of the sower argues that the effect of The Word is determined by who encounters it.[17] As Peters argues using contemporary terms, the parable "offers a receiver-oriented model in which the sender has no control over the harvest." In Peters's summary of "the synoptic Gospels of Matthew, Mark and Luke . . . The Word is scattered uniformly, addressed to no one in particular, and open to its destiny."[18]

By contrast, this indeterminate conception was remade by Christian fundamentalist activism in nineteenth-century America, particularly in the emergent mass-publicity practices of evangelical abolitionists during the nineteenth century in the United States. The mystification of broadcasting can be traced to the remaking of "broadcasting" within evangelical, millennial conceptions of the spread of The Word and its relation to salvation of the masses. Bible and tract societies played a central role in this redefinition, which in many ways created a key template for strategies, organizations, and technologies of modern mass media.[19] Nord argued that it was "evangelical Christian publicists in the Bible and tract societies ... [who] first dreamed the dream of a genuinely mass medium—that is, they proposed to deliver the same printed message to *everyone* in America" [emphasis in original].[20] American evangelicals owed much to older tract societies in England, which "inspired the Americans with their vision of what religious mass media should be."[21]

For the nineteenth-century tract societies, "broadcasting" was a means of message dissemination seen as eminently suitable for reaching lower classes, from whom emanated "the rising tide of democracy and irreligion."[22] It is here that the class indexing performed by broadcasting increases in complexity, illustrating how meaning, significance, and class operations depended on historical context. Broadcasting in agriculture was disparaged by gentry and cultural elites as something suitable only for lower classes. Yet with the ideological merger with evangelical activism and publishing, "broadcasting" was not a popular practice in opposition to elite preferences but a popularizing practice to aid elite purposes. The shift is based in earlier agricultural and biblical usages, which were remade to fit the new context, thus emerging from but also severed from the past.

Correspondingly, evangelical leaders drew from the traditional past as well as dreams about the future. Nord characterized these evangelical ideologues as "staunch conservatives, believers in tradition, deference, and hierarchy," a blend of archaic elitism and forward-looking faith in industrial development. As "17th-century men, mis-born into the 18th century, trying to come to terms with the 19th," they sought to merge a fundamentalist Calvinism and an "ordered religious communitarianism" with a Hamiltonian Federalist program of industrialization and centralization.[23]

Although their expressly ideological aim was to stem the tide of irreligion and democracy and thus restore the past, their practical contribution consisted of innovations in media reproduction and organization. They helped develop and refine stereotyping, steam-powered printing presses, and mechanical papermaking. They also pioneered mass means of financing and distribution in such ways as "the building of a genuine national organization" composed of "elaborate systems of local auxiliary societies and branch distributors" that were "subsidiary to a centralized, national organization."[24] The emergent media strategy, first put into practice by the New York City Tract Society in 1829 (fully four years prior to the *New York Sun* of Benjamin Day, typically identified in the United States as the initiator of mass media in the form of the penny press) was to "place in the hands of every city resident at least one tract, the same tract, each month." Such a strategy and practice exemplified for Nord "the ideology of modern mass media: to have everyone reading and talking about the same thing at the same time."[25] (See Figure 15.2.)

As seen in a number of abolition and popular newspapers and magazines, an increasing confluence appears between broadcasting in the agricultural sense and the then-new practice of mass media in the service of mass political organizing.[26] Despite its great variability, this confluence can be mapped along some key

FIGURE 15.2 *The millennial, messianic power of broadcasting typical of nineteenth-century Christian fundamentalist conceptions is suggested by this woodcut that appeared in William Cullen Bryant,* The Song of the Sower, *22.*

dimensions. The first concerns the concreteness of the practice. On one hand, a metaphorical usage expressed the natural widespread distribution of nonmaterial qualities or ideas. On the other hand, a practical usage named the intentional human action of distributing media products, such as newspapers, books, pamphlets, and broadsides. Each position tended to place human agency differently. To the degree that the metaphorical usage removed human agency except as implied recipients of this process, it mystified and reified the process and its effects. Thus authors wrote of "the growth of those [desirable] affections whose seeds are sown at broadcast in the natural relations of life."[27] Horace Mann stated in a speech that "the way is open here in Kentucky for sowing the seeds of freedom broadcast."[28] And others argued that "the seeds of oppression had been scattered broadcast over this land."[29] By contrast, the practical usage placed people as the instigators and concrete means of distribution. Writers advocated to their followers that "such notices [for abolition] should be thrown into the tract form, and sown broadcast over the land."[30] The facts of "the brutal character of American slavery" are available for all to read because they "are sent broadcast over the British empire, and are published in nine-tenths of all the British newspapers."[31]

Yet variations existed that render simplistic an attempt to rigidly equate the metaphorical with the mystified. Some items referred to the active distribution of media products, but as a natural process unaided by human intention or intervention, thus also mystifying their production and use. One statement noted how documents seemingly under their own power "would go broadcast all over the free States in this country."[32] On the contrary, other variations referred to an intentional human activity, but of the spreading of ideas that lacked any material embodiment. People would "go sowing broadcast the seeds of rich blessing," "sow also broadcast and with a liberal hand, the seed of truth," or have "sown broadcast the seeds of bitterness."[33]

Specifying this exploration was the increasing frequent discussion throughout the 1850s and 1860s regarding the role of media and broadcasting in fomenting social change. Such items express an increasingly vanguard relationship between the activist group and its target publics. The pace of such mentions accelerates throughout the 1850s to the point where "broadcasting" describes the political work done by media employed by vanguard groups. Whether metaphorical or practical, "broadcasting" in the service of the movement increasingly meant, for the instigators, an intentional, concrete practice and, for the audiences, reception and acceptance. A commentary in *The North Star* remarked on the "mighty and overwhelming" ability of the press to spread "truths broadcast in the world."[34] In a promotion for a tract, the "friends of Freedom will, it is hoped, send their orders immediately for this seasonable pamphlet, and cause it to be spread, broadcast, over the whole country."[35] Similarly, "anti-slavery books, tracts, and newspapers should be scattered broadcast over the land; [and] the question of abolition should be brought home to every man's hearth-stone."[36]

As time went on, the certainty of effect increased. As one writer put it in 1856, no one can "doubt for a moment the utility of scattering broadcast over our land, such documents and speeches as will have a tendency to enlighten the public mind."[37] When distributed, manuscripts, books, and tracts are seen to enable the spread, "broadcast, [of] the very truths which are there contained."[38] Others note that media distribute "the seeds of anti-slavery truth, which shall thus through your instrumentality be scattered broadcast over the land."[39] Readers were increasingly urged to "inform the masses. . . . Circulate the documents broadcast through the land, until every honest farmer and mechanic had read and understood the principles of our party."[40] One organization suggested that "our publications should be sown broadcast in all our advances, and keep pace with our advancing missionary operations, as a sure means of permanent success."[41]

By the early 1860s, this complex of relationships and practices was even more explicitly and uniformly expressed. "Broadcasting" had come to mean a vanguard relation between advocacy groups and their targets, and a reified spread of sentiment as well as a practical tactic of political agitation. The key tactic of mass social activism was broadcasting, whatever the technology.[42] Although some items admonished readers to promote abolition "in all practical ways" including "spreading broadcast the issues of that mighty engine, the public press . . . ," others preferred to privilege the means by advocating that "every true friend of liberty and humanity" should do everything possible to "scatter broadcast the seeds of Anti-Slavery truth"[43] By 1859, the irrelevance of the specific technology in favor of a recognition of the power of broadcasting was increasingly axiomatic. "Scatter information broadcast among the people," urged a writer in 1859. "The harvest will be sure."[44]

However, the power of media to broadcast new ideas was not always seen as beneficial. Those who opposed abolition felt the press "contributed to bring about a ferocious discontent, which needed only the insidious and inflammatory articles spread broadcast over the land by designing men to fan into an insurrection."[45] This class-bound association of "broadcasting" with nonelites emerged within broader debates about the moral effects of mass commercialized popular culture. On one hand, "broadcasting" expressed an authentic popular culture as framed within and linked to the Protestant tradition of direct access to God and the popularization of religious practice. As one Protestant evangelical interpretation of medieval society remarked, "The Word of Truth [at that time] was read only by the learned men and cloistered monks, and its benign influence was not shed broadcast over the earth as now."[46] However, the moral effect of other kinds of materials and sentiments broadcast were less laudable, noted one commentator: "For if these masses are not soon educated and Christianized, they will sow broadcast the worst forms of infidelity imported from the Old World, and by and by they will overturn the very foundations of our free institutions."[47] Growing concerns were expressed about, for example, the popular distribution of French novels "sown broadcast through our land [that] corrupt the imaginations of our young men and maidens, and . . . waste the time of more mature readers."[48] Similar claims appeared regarding how "the mails are extensively prostituted to immoral and vicious purposes, and that through this channel obscene books, circulars, &c. are sown broadcast throughout the country."[49]

The evangelical articulation not only extended "broadcasting" conceptually into media and politics. It newly defined its social nature and class status. "Broadcasting" described not only a way of distributing ideas but one uniquely suited to vanguard intentions and uses. It—and not any specific media technology—was increasingly seen as the determinant, fundamental agent and cause of moral and political transformation. Unlike its agricultural iteration, evangelical broadcasting was not a practice of nonelites but a practice of elites in political programs directed at the rest of society.

Yet existing in the midst of this dominant articulation of "broadcasting" was an emergent articulation with democracy and, implicitly, with consumerism. Within it, the power of broadcasting was seen as something other than audience responses subsumed within the wishes of a vanguard. Some articles expressed an active shaping response by the audience consistent with orthodox classical-liberal conceptions of the marketplace, such as in an account of a broadside author who had "sent so many thousands of 'lower law sermons' broadcast throughout the free States." He must, argued the commentator, "meet the penalties of the popular will" should his argument not be received approvingly.[50] The fit of this political statement about the ideal workings of a democracy as the sovereignty of the people with an implicit parallel claim about the sovereignty of consumers as conceived in classical economic thought became only more apparent as the century closed and a new one dawned.

The Nature of Broadcasting at the Turn of the Century

In the contexts of the increasing pace of consolidation, centralization, and rationalization of the state, and the popular celebration of the creative capacities of Americans to exert control over their world (seen perhaps most broadly in the geopolitical consolidation of the continent under the banner of Manifest Destiny), the vanguard social relations underlying "broadcasting" that were pioneered by Bible tract societies became increasingly articulated through forms and organization of commercial media.[51] This emergent articulation became increasingly important to the point where, in 1907, inventor and entrepreneur Lee de Forest could conceive the general outlines of a broadcasting system in terms not of centralized political activism directed at the masses, but as equally centralized popular entertainment delivered to the masses for their individual pleasure. In a diary entry of that year, the evangelical aim of creating a social movement had become a secular wish to extend the political community beyond its geographic borders. In the entry, de Forest explained that "My present task is to distribute sweet melody broadcast over the city and sea so that in time even the mariner far out across the silent waves may hear the music of his homeland, sung from unseen sources."[52]

The conception of a broadcasting system as a means of achieving social solidarity via the centralized distribution of entertainment derives from long-standing millennial hopes and dreams regarding electronic communication and its almost magical power by which communities could be (re)constituted.[53] Furthering the mystification and naturalization of "broadcasting" was its articulation with the popular fascination with electronic technologies and the seeming disappearance of the materiality of the message.[54] Such a fascination was in part a response to sectional strife, significant labor activism, social and cultural dislocation related to urbanization and the reconstitution of the United States as an industrial rather than agricultural society, and large-scale immigration and resulting debates regarding its impact on American society and identity.[55] Such a view appeared in Britain as well, which was experiencing many of the same pressures. A British writer in a popular science magazine of 1884 concurred that "everything which can knit a community together and which can cause a rapid interchange of sentiment and ideas, annihilate distance, isolation and prejudice, is of the greatest happiness to the greatest number [of people]."[56]

An overlap with its evangelical articulation expresses this community building in the form of broadcasting religious services. A laudatory item promoted one such effort in Old St. Patrick's Church in Pittsburgh during two weeks late in 1921, whose author estimated (perhaps optimistically) that one million people "every night [were] 'listening in' to these messages picked out of the air." The result was that "the whole world is brought to our very ears while we lounge at ease amid the comforts of our own drawing-rooms, libraries, or bed-rooms." By this means, the author hoped that "the seed of further conversions could be sown and scattered wherever human beings congregate."[57] (See Figures 15.3 and 15.4.)

The emergent conception of human involvement in broadcasting also continued to change. Rather than being a human practice or spiritual manifestation, broadcasting was becoming not only a nonhuman force of nature to be tamed by human ingenuity but, for some commentators, the infinite, unimaginable energy flow itself in the natural world. Some early twentieth-century writers regarded broadcasting as the harnessing of the "æther" [sic], which was seen to constitute the enabling medium of wireless (described in a 1924 popularized account of broadcasting as a kind of material denser than the finest steel but undetectable by human senses).[58] Another British writer in 1923 remarked on the wonder of harnessing this force. Even more amazing than how "the electrical influence from the aerial emerges out of the unknown and manifests itself at some place, perhaps hundreds or even thousands of miles away" is the discovery that this force "is actually under control and can be made to carry messages."[59] Yet other contemporary commentators elevated broadcasting from the human harnessing of this force, to the force itself, such as one in 1924, which asserted that "nature has been 'broadcasting' since the earliest thunderstorm. With the first lightning flash, wireless waves were sent rippling across space, penetrating primæval forests, rocky caverns and the haunts of such animal life as then existed." But, by contrast, people have been broadcasting for little more than twenty-five years, "using harnessed forces of nature."[60]

This ambiguity regarding control (had humans tamed this force, or were they simply riding a tiger that could turn on them at any time?) found expression, alongside approval of such technologies, in concerns about electronic media as a demagogic means of subverting or erasing traditional social hierarchies.[61] Concerns about the social and political consequences of the telephone had appeared less than a year after Alexander Graham Bell's initial public demonstration. A March 15, 1877, cartoon in the *New York Daily Graphic* "illustrated 'the terrors of the telephone' by showing an announcer at a telephone-like device mesmerizing masses of people listening simultaneously throughout the world."[62]

Within this overall transformation from human to heavenly to natural force, a parallel reifying trajectory can also be identified. A preoccupation with the wonder of the technology and a celebration of individuals and their technical achievements soon became the broader celebration of the teleological emergence of consumerist "'media' for the millions."[63] As explained in a promotional book published by the Radio Corporation of America (RCA), the problem to tackle after point-to-point telephony was how to devise "a system whereby one voice,

FIGURE 15.3 *Despite differences in apparatuses, the conception of media broadcasting as a particular social relationship remained quite similar, as seen in this pair of images separated by twenty-four years. This first example is from "The Radiophone at the Electrical Exhibition,"* Scientific American *27 (May 1899): 347.*

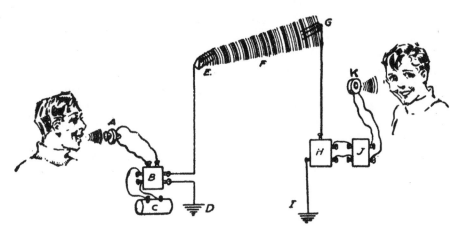

FIGURE 15.4 *A later, very similar diagram can be found in Ellison Hawkes,* The Romance and Reality of Radio, *(1923) 170: "A radio-telephone is exactly similar in principle to an ordinary line-telephone. a, Transmitter; b, Modulator; c, Oscillation generator; d, Earth; e, Aerial; f, Radio waves; g, Aerial of receiving station; h, Tuning apparatus; i, Earth; j, Detector; k, Telephone receiver."*

one speech, one concert of music might be heard instantaneously and simultaneously in millions of homes," with the answer being broadcasting, seen as the next evolutionary stage in the development of communication technology, with the need met by the newly formed National Broadcasting Company.[64]

Yet the emergence of broadcasting systems was not the natural fruition of intrinsic properties of communications technology, as RCA asserted. By contrast, its emergence was directed by conceptions of what "broadcasting" should be, thus picking up on the vanguard service pioneered by evangelicals and print. As Briggs notes (corresponding to Nord's claim about evangelical publishing), the idea of "everyone everywhere sharing the same 'programmes' at the same time preceded the introduction of the required technology."[65] In addition to evangelical statements, conceptions of "broadcasting" were publicly worked out in such forms as "cartoons, articles, and even popular songs [that] lauded the potential wonders of the telephone, including the distance dissemination to mass audiences of music, information, drama, and education."[66] News and information had been broadcast to ships telegraphically for almost two decades before the first commercial "broadcast," and propaganda by the Germans and others during World War I were also noted as early instances of broadcasting, albeit with telegraphy rather than telephony.[67]

Other attempts to realize broadcasting using prewireless point-to-point electronic technologies took place in 1877, when Edison stumbled on a means of using the telephone as a public receiver. After further experimentation and refinement, "by placing the [modified] telephone in the centre of the room[,] persons seated around could hear [transmitted music] with perfect distinctness."[68] Conceptual experimentation included French engineer Clément Ader's patent application for "Improvement of Telephone Equipment in Theaters," in which he proposed to put telephones on the stage so subscribers could listen to the performance at home.[69] De Forest's own contribution occurred in 1906 through hanging a microphone on the stage of the Metropolitan Opera House in New York City and sending the signal to several receiving sets installed in several public locations throughout the city for people to encounter and listen to.[70] Yet another example occurred later in 1906, with Canadian Reginald Fessenden's Christmas Eve transmission to ships of a program consisting of "a person speaking, then a woman singing, then someone reading a poem, followed by a violin solo, and verses [read] from the Bible."[71]

In this complex of the agricultural, the vanguard, and the consumerist contexts, "broadcasting" emerged as a synthesis during the 1910s in the activities of physics and engineering faculty and their students at midwestern U.S. universities. As part of learning more about electronic theory, transmission facilities set up in such departments distributed to farmers "reports from time to time on weather conditions, crops, producer prices, Department of Agriculture advisories, and other things that an isolated farmer might otherwise have to wait days to learn."[72] A parallel effort took place in the then-rural Santa Clara Valley of California, where Charles Herrold developed a wireless information-distribution service during the 1910s "programming music and talk" for "amateur men" in the area. Hilliard and Keith note that this effort "evolved out of his role as the headmaster of a wireless trade school"—with the school clearly providing the institutional basis, rationale, and technical facilities, but the social uses and relations deriving from that of a headmaster with his students.[73] (See Figure 15.5)

FIGURE 15.5 *By the 1920s, representations of broadcasting focus on the originator and the apparatus, such as in this example of the set used in January 1920 for early broadcasts from Chelmsford. From H. M. Dowsett,* Wireless Telephony and Broadcasting, *vol. 1 (London: Gresham Publishing, 1924), 54.*

Increasingly in the United States, however, the template for organizing broadcasting was commercial, with a broadcasting system becoming a primarily consumerist pastime that fit the basic redefinition of social relations in an accelerating capitalist economy.[74] A consumerist, often paternal vanguardism as opposed to political vanguardism drew on formulas developed by mass-circulation newspapers, popular magazines, popular books, and live popular entertainment. A broadcaster "must be both showman and concert manager, newspaper editor and music-hall director"—all of which derive from earlier commercial forms of popular entertainment.[75] A contemporaneous assessment asserted that "broadcasting can only reach its ultimate goal through centralization of efforts."[76] A non-Anglo example noted in U.S. press accounts was the "telephone newspaper," *Telephon Hírmondó,* whose customers "receive[d] the news as they would ordinary telephone messages." By applying long-standing ideas of organizing live commercial entertainment (such as scheduling the day and time of performances as akin to programming) and commercial forms of media production (rationalizing, professionalizing, and centralizing its practice), the telephone newspaper also helped pioneer what became the media practice and organization of broadcasting.[77] Even amateurs' early experimentation worked through a commercialized intention of pleasing an audience. As one author put it, early amateurs' means of assessing their success consisted of "timid [inferred] requests" of their audience: "'Did you like it? Would you like more of it?'"[78]

The emergent mass-consumerist articulation of broadcasting pushed the mystification of "broadcasting" further than the evangelical, in that it appeared to be a form of direct delivery to individual households and a direct connection between performance and its experience. This was perhaps the most powerful social and cultural effect, which helped set terms of participation, although not without detractors.[79] Unlike newspapers and magazines, which in their presentation and narrative structure were overtly professionally gathered and produced, broadcasting relied on the aesthetic evaporation of the organizational means by which the message was propagated, prompting the then-current term to describe audiences' use as a direct, unmediated "listening-in."[80]

In these ways, the move from evangelical to consumerist turned on its head the power relationship implicit in vanguard forms. Instead of placing the public in the service of elites, the public was ostensibly being served by a coterie of scientists, inventors, visionaries, and, later, entrepreneurs and national corporations. Yet the very separation of makers from listener-consumers was inherited from the evangelical form, aided by rationalizing organization and the general shift from distributed, pragmatic, individual use to centralized, entertainment, "mass" use. Such a development was institutionalized with the merging of governmental and industrial interests and the creation in 1919 of the Radio Corporation of America (RCA), which, in Hilmes's estimation, constituted the conditions by which "American broadcasting would never again be as accessible to individuals."[81]

Broadcasting and Democratic Communication

By the early twentieth century, "broadcasting" had become an industrialized process that not only partitioned makers from audiences but that institutionalized a form of sociality paradoxically collective in its pervasiveness and individualized in its experience. A seemingly unrelated 1921 news item suggests this by noting that a seaboard wireless station "is used partly for broadcasting Press and other messages to ships, that is, sending out messages without receiving replies."[82] Reply and dialogue are neither expected nor technically possible. Whereas an agricultural usage deems a nonresponse to be a bad crop, this usage suggests that response is irrelevant. Ships receiving the broadcast message can act on it or not, depending on their specific situation. Thus, by the early twentieth century, a conception of "broadcasting" even more fundamental than a wireless means of disseminating commercial entertainment is lack of reply.

To the extent that "broadcasting" in the sense unearthed here guides the media practice of progressive groups and forms of legitimate participation, hopes for a fully democratic means of communication are shortchanged by the social and cultural fragmentation of the process. Yet such an unearthing also helps suggest directions for building and enabling quite new kinds of social collectivities, transnational, disjointed in time and space, fluid and hybrid as they increasingly are.[83] Even as early as 1930, and while admitting that "the technique for all such projects has still to be developed," Brecht noted that a democratic media system should "be directed towards the prime task of ensuring that the public is not only taught but must also itself teach."[84] Only by enabling self-renewing, general participation can there be a way of, in Williams's words, "get[ting] rid of the idea that communication is the business of a minority talking to, instructing, leading on, the majority."[85] It is by recovering and critiquing the class-bound, vanguard, and consumerist tradition of broadcasting that such a rethinking might start.

Notes

1. John Cain, *The BBC: 70 Years of Broadcasting* (London: BBC, 1992), 7.
2. Raymond Williams, "Communications and Community," in *Resources of Hope* (London: Verso, 1989), 23.
3. John Durham Peters, *Speaking into the Air: A History of the Idea of Communication* (Chicago: University of Chicago Press, 1999); Paddy Scannell, "Introduction: The Relevance of Talk," in *Broadcast Talk*, ed. Paddy Scannell (Newbury Park, Calif.: Sage, 1991), 1–13. See also Thomas Streeter, *Selling the Air: A Critique of the Policy of Commercial Broadcasting in the United States* (Chicago: University of Chicago Press, 1996).
4. Key works include Erik Barnouw, *A History of Broadcasting in the United States*, 3 vols. (New York: Oxford University Press, 1966–70); Raymond Williams, *Television: Technology and Cultural Form* (New York: Schocken, 1974); Susan Douglas, *Inventing American Broadcasting, 1899–1922* (Baltimore: Johns Hopkins University Press, 1987); Robert Britt Horwitz, *The Irony of Regulatory Reform: The Deregulation of American Telecommunications* (New York and Oxford: Oxford University Press, 1989); Robert W. McChesney, *Telecommunications, Mass Media, and Democracy: The Battle for the Control of U.S. Broadcasting, 1928–1935* (New York: Oxford University Press, 1993); and Susan Smulyan, *Selling Radio: The Commercialization of American Broadcasting, 1920–1934* (Washington, D.C.: Smithsonian Institution Press, 1994).
5. Raymond Williams, *Marxism and Literature* (Oxford: Oxford University Press, 1977); Andrew Milner, *Re-Imagining Cultural Studies: The Promise of Cultural Materialism* (London: Sage, 2002).
6. Perry Anderson, *In the Tracks of Historical Materialism: The Wellek Library Lectures* (Chicago: University of Chicago Press, 1984), 11; Hanno Hardt and Bonnie Brennen, "Introduction: Communication and the Question of History," *Communication Theory* 3, no. 2 (May 1993): 130–36.
7. G. E. Fussell, *The Old English Farming Books: From Fitzherbert to Tull, 1523 to 1730* (London: Crosby Lockwood & Son, 1947), 1.
8. Asa Briggs, *A Social History of England*, 3rd ed. (London: Penguin, 1999), 192.
9. Adam Smith, *An Inquiry into the Nature and Causes of the Wealth of Nations*, 6th ed., ed. Edwin Cannan (London: Methuen, 1961); William Amos, *The Theory and Practice of the Drill Husbandry* (1802), v–vi.
10. Douglas Hay and Nicholas Rogers, *Eighteenth-Century English Society: Shuttles and Swords* (Oxford: Oxford University Press, 1997), 71.
11. John Anstruther, *Remarks on the Drill Husbandry* (London: T. Egerton, 1796), 133.
12. Fussell, *The Old English*, 20; G. E. Fussell, *The Farmer's Tools, 1500–1900: The History of British Farm Implements, Tools, and Machinery before the Tractor Came* (London: A. Melrose, 1952).
13. G. E. Mingay, *Land and Society in England 1750–1980* (London and New York: Longman, 1994), 41; G. E. Mingay, *A Social History of the English Countryside*, reprint ed. (London and New York: Routledge, 2002), 46–50; E. J. T. Collins, "The 'Machinery Question' in English Agriculture in the Nineteenth Century," in *Research in Economic History*, Suppl. 5, Pt. A: *Agrarian Organization in the Century of Industrialization: Europe, Russia and North America*, eds. George Grantham and Carol Leonard (Greenwich and London: JAI Press, 1989), 213.
14. Hay and Rogers, *Eighteenth Century*, 98–102.
15. Collins, "The 'Machinery Question,'" 206.
16. Ibid., 203; Hay and Rogers, *Eighteenth Century*, 124. See also Eric Hobsbawm and George Rudé, *Captain Swing* (London: Lawrence and Wishart, 1969); Barry Reay, *The Last Rising of the Agricultural Laborers: Rural Life and Protest in Nineteenth-Century England* (Oxford: Clarendon, 1990); and Lee Davison, Tim Hitchcock, Tim Keirn, and Robert B. Shoemaker, eds., *Stilling the Grumbling Hive: The Response to Social and Economic Problems in England, 1689–1750* (New York: St. Martin's Press, 1992).
17. See also V. George Shillington, *Jesus and His Parables: Interpreting the Parables of Jesus Today* (Edinburgh: T&T Clark, 1997).

18. Peters, *Speaking into the Air*, 35.

19. David Paul Nord, "The Evangelical Origins of Mass Media in America, 1815–1835," *Journalism Monographs* 88 (May 1984): 23. Two recently published books that describe in detail these developments are David Paul Nord, *Faith in Reading; Religious Publishing and the Birth of Mass Media in America* (New York: Oxford University Press, 2004), and Candy Gunther Brown, *The Word in the World; Evangelical Writing, Publishing, and Reading in America, 1789–1880* (Chapel Hill: University of North Carolina Press, 2004).

20. Ibid., 2.

21. Ibid., 4.

22. Ibid., 3.

23. Ibid., 4.

24. Ibid., 13–15.

25. Ibid., 20.

26. The body of materials consulted for this section includes publications available through Accessible Archives (http://www.accessible.com): *Godey's Lady's Book* (1830–80), *The Pennsylvania Gazette* (1728–1800), *Freedom's Journal* (1827–29), *The Colored American (Weekly Advocate)* (1837–41), *The North Star* (1847–51), *The National Era* (1847–60), *Provincial Freeman* (1854–57), *Frederick Douglass Paper* (1851–59), and *The Christian Recorder* (1861–1902).

27. C. M. Sedgwick, "Scenes from Life in Town," *The Lady's Book*, April 1843, 159.

28. "Speech of the Hon. Horace Mann," *The National Era*, 21 September 1848, 150.

29. "Selections: Address of H. W. Johnson," *The North Star*, 21 August 1848.

30. *The National Era*, 15 April 1847, 4.

31. "Monthly Illustrations of Slavery," *The North Star*, 25 August 1848.

32. "Counter Statement," *The Colored American*, 16 May 1840.

33. "Review of The Young Lady's Friend," *The Lady's Book*, April 1838, 167 "Spring!," *The North Star*, 14 April 1848 (Item 10109); "Michigan Politics," *The National Era*, 25 October 1849, 169.

34. "The Press," *The North Star*, 30 June 1848.

35. "Fugitive Slave Bill—Third Edition," *The National Era*, 21 November 1850, 186.

36. "Resolutions Adopted by the late Anti-Slavery Convention in Cincinnati," *The National Era*, 15 May 1851, 79.

37. "A Circular to the Friends of the Republican Movement," *The National Era*, 17 January 1856, 12 (Item 65776).

38. "'Facts for the People'—A Want Supplied," *The National Era*, 21 April 1853, 62.

39. "To the Members of the 'Rochester Anti-Slavery Sewing Society," *Frederick Douglass Paper*, 27 August 1852.

40. "Extracts from Our Correspondence," *The National Era*, 5 January 1854, 1.

41. "Have We Made Any Advancement?" *The Christian Recorder*, 4 June 1864.

42. "An Address to the Anti-Slavery Christians of the United States," *The National Era*, 24 June 1852, 101 "The Annual Report of the New York State Temperance Society," *Frederick Douglass Paper*, 2 September 1853 "Facts for the People," *The National Era*, 18 August 1853, 130 "The Present and the Future," *The National Era*, 31 March 1853, 50.

43. "The Present and the Future" "Extracts from Our Correspondence," *The National Era*, 11 August 1853, 128.

44. "Speech of Gov. Chase at Cincinnati," *The National Era*, 10 November 1859, 671.

45. Charles E. Blumenthal, "Develour; A Sequel to 'The Niebelungen,'" *Godey's Lady's Book*, January 1851, 51.

46. "Clerical Humanity," *The National Era*, 10 July 1851, 110.

47. R. D. Harper, "The Church—Her Danger and Duty," *The Christian Recorder*, 7 June 1862.

48. "Editors' Table; Novels in French and English," *Godey's Lady's Book*, March 1863, 304.

49. "Domestic Items," *The Christian Recorder*, 30 May 1863.

50. "Congressional Proceedings," *The National Era*, 8 January 1852, 5.

51. Stephen Skowronek, *Building a New American State: The Expansion of National Administrative Capacities, 1877–1920* (Cambridge and New York: Cambridge University Press, 1982); Richard L. McCormick, *The Party Period and Public Policy: American Politics from the Age of Jackson to the Progressive Era* (New York: Oxford University Press, 1986); Daniel Schiller, *Theorizing Communication: A History* (New York: Oxford University Press, 1996), 3–38.

52. Quoted in Lee de Forest, *Father of Radio: The Autobiography of Lee de Forest* (Chicago: Wilcox and Follett, 1950), 225–26. Courtesy of Hargrett Rare Book & Manuscript Library/University of Georgia Libraries.

53. Carolyn Marvin, *When Old Technologies Were New: Thinking about Communications in the Late Nineteenth Century* (New York: Oxford University Press, 1987), 63–108; Daniel Czitrom, *Media and the American Mind: From Morse to McLuhan* (Chapel Hill: University of North Carolina Press, 1982), 91–121.

54. In addition to Marvin and Czitrom, see Catherine L. Covert, "'We May Hear Too Much': American Sensibility and the Response to Radio, 1919-1924," in *Mass Media between the Wars: Perceptions of Cultural Tension, 1918–1941*, eds. Catherine L. Covert and John D. Stevens (Syracuse: Syracuse University Press, 1984), 199–220; Mary Mander, "The Public Debate about Broadcasting in the Twenties: An Interpretive History," *Journal of Broadcasting* 28, no. 2 (Spring 1984): 167–85; Hugh Aitken, *The Continuous Wave: Technology and American Radio, 1900–1932* (Princeton: Princeton University Press, 1985); James Carey and John J. Quirk, "The Mythos of the Electronic Revolution," in James W. Carey, *Communication as Culture: Essays on Media and Society* (Boston: Unwin Hyman, 1989), 113–41; Lisa Gitelman and Geoffrey B. Pingree, eds., *New Media, 1740–1915* (Cambridge and London: MIT Press, 2003); David Thorburn and Henry Jenkins, eds., *Rethinking Media Change; The Aesthetics of Transition* (Cambridge and London: MIT Press, 2003); and Martin Spinelli, "Democratic Rhetoric and Emergent Media: The Marketing of Participatory Community on Radio and the Internet," *International Journal of Cultural Studies* 3, no. 2 (August 2000): 268–78.

55. Raymond Wilson, *In Quest of Community: Social Philosophy in the United States, 1860–1920* (New York: John Wiley, 1968); Jean Quandt, *From the Small Town to the Great Community: The Social Thought of Progressive Intellectuals* (New Brunswick: Rutgers University Press, 1970); Fred Matthews, *Quest for an American Sociology: Robert E. Park and the Chicago School* (Montreal: McGill-Queens University Press, 1977); Martin Bulmer, *The Chicago School of Sociology* (Chicago: University of Chicago Press, 1984).

56. Asa Briggs, *The BBC: The First Fifty Years* (Oxford: Oxford University Press, 1985), 9.

57. Thomas F. Coakley, "Preaching the Gospel by Wireless," *Catholic World*, January 1922, 517–18. See also "The Gospel by Wireless," *The Literary Digest*, 4 February 1922, 32; "Preaching by Wireless," *Current Opinion*, May 1922, 649–50; and "Radio Sermons for Pastorless Churches," *The Literary Digest*, 17 June 1922, 27.

58. Arthur Richard Burrows, *The Story of Broadcasting* (London: Cassell and Company, 1924), ix. For an account of theories of the æther, see also H. M. Dowsett, *Wireless Telephony and Broadcasting*, vol. 1 (London: Gresham Publishing, 1924), 179–201.

59. Ellison Hawkes, *The Romance and Reality of Radio* (London: T. C. & E. C. Jack, 1923), 1.

60. Burrows, *The Story of Broadcasting*, ix.

61. Example cited in Robert L. Hilliard and Michael Keith, *The Broadcast Century and Beyond: A Biography of American Broadcasting*, 3rd ed. (Boston: Focal Press, 2001), 4.

62. Quoted in ibid., 4.

63. Briggs, *The BBC*, 9.

64. Radio Corporation of America, *The Place of Radio in World Communications* (New York: Radio Corporation of America, 1928).

65. Briggs, *The BBC*, 10.

66. Hilliard and Keith, *The Broadcast Century*, 4.

67. Burrows, *The Story of Broadcasting*, 17.

68. Hawkes, *The Romance*, 175.

69. Hilliard and Keith, *The Broadcast Century*, 4.

70. Ibid., 12.

71. Ibid., 7, 9.

72. Ibid., 16.

73. Ibid., 16, 20.

74. Raymond Williams, *Keywords: A Vocabulary of Culture and Society* (New York: Oxford University Press, 1976), 69.

75. Waldemar Kaempffert, "The Progress of Radio Broadcasting," *Review of Reviews*, September 1922, 305.

76. "About the Radio Round-Table: Opinions of Radio Leaders Regarding the Past, Present and Future of Broadcasting," *Scientific American*, December 1922, 379.

77. Briggs, *The BBC*, 8; "The Telephone Newspaper," *Scientific American*, October 1895, 267.

78. Kaempffert, "The Progress of Radio," 303.

79. For example, see "The Coming of the Wireless Telephone," *The Spectator*, 13 May 1922, 582–83.

80. Pierre Boucheron, "News and Music from the Air," *Scientific American*, December 1921, 104–5; "'Listening In,' Our New National Pastime," *Review of Reviews*, January 1923, 52; and Rex

Palmer, "Future of Broadcasting as a Social Force," *English Review*, August 1923, 210–14. Claims that equated the fidelity of signal and skillfulness of the production with the degree of audience pleasure can be seen in "Organs of Sound That Do and Do Not Broadcast," *Current Opinion*, July 1922, 68–69.

81. Michele Hilmes, *Hollywood and Broadcasting; From Radio to Cable* (Urbana and Chicago: University of Illinois Press, 1990), 13–14.

82. C. G. Crawley, "Maritime Wireless," *Discovery: A Monthly Popular Journal of Knowledge*, April 1921, 92.

83. Andrew Opel and Donnalyn Pompper, eds., *Representing Resistance: Media, Civil Disobedience, and the Global Justice Movement* (Westport, Ct.: Praeger, 2003).

84. Bertolt Brecht, "Radio as a Means of Communication: A Talk on the Function of Radio," in *Communication and Class Struggle*, vol. 2,: *Liberation, Socialism*, eds. Armand Mattelart and Seth Siegelaub (New York: International General, 1983), 171.

85. Williams, "Communication and Community," 29.

PART V

Training, Technology, and Modern Subjectivity

16 The Musicking Machine

Jody Berland

> While environments as such have a strange power to elude perception, the preceding ones acquire an almost nostalgic fascination when surrounded by the new.
> —Marshall McLuhan, *The Essential McLuhan*

What do musicians with traditional training—in my case, eleven years of piano lessons—have to offer the new musical culture of the twenty-first century? It is not hard to believe we are completely obsolete. A host of new digital and Internet services provide studio time and sound resources (I am not sure whether to call them instruments), online services, and live or digital collaborators to people interested in making music on their computers. To some musicians and music theorists, and to me when I'm gloomy, the digitization of musical production signals the end of music. But nothing can bring the end of music, and "obsolescence" is a more complex process than we ever imagined. "Postmusic" production is attached to long-standing critical debates surrounding sound reproduction, musical authenticity, authorship, and transformations of listening. My more modest purpose here is to revisit the history of the keyboard in the context of changing dynamics of space and time, automation, and (dis)embodiment dominating sound reproduction in the last century.

The piano was a central player in the transformation from Victorian domestic culture to the practices and ideals of modern consumerism, and the mechanization of the piano placed speed and convenience at the heart of early twentieth-century consumer culture. Through technical and commercial innovations designed to enhance musical reproducibility, modes of performance and reproduction have been continuously challenged and altered by newer forms. Through this process the piano has played a role in the elevation and containment of creative work performed by women, the dissemination and adaptation of bourgeois cultural practices and values across the new world, the symbolic enhancement and commodification of entertainment, and the efficacy of mechanization in fostering new ideologies of convenience and efficiency in leisure time as well as in work. Although these same dynamics have characterized a broad range of human–machine interactions, the piano evokes questions about time, embodied memory, and the "technologies" of creative expression that would be hard to explore in an examination of washing machines, typewriters, or computers. Every history of technology invites questions about progress, agency, technological determination, and power, but musical instruments also articulate motion and emotion, individual subject and social space, feeling and doing within the flow of material history and symbolic expression. The term "musicking," which I draw from Christopher Small's *Musicking: The Meanings of Performing and Listening* (1998), encourages us to focus on music as cultural practice rather than as musical text. Looking at musical practice today means confronting changing relations between musicking and machines. The history of the piano exemplifies the increasing interdependency of musical practice, mechanization of sound production, and altered technologies of the self that has defined musicking over the last century.

The tension between convenience and duration, and how this tension is fought out in the temporality and motion of the human body, is a central theme in the piano's history. Music exists in and makes meaning from time and duration. Music's essential finitude—the fact that a sound falls into silence—has been profoundly challenged by the technology of sound reproduction. Musical mechanization has been obsessed with "saving" time in two respects: by making musical performance reproducible, so it is never lost to time, so it can be endlessly reheard; and by making the production of such musical performance more efficient, so that millions of players are freed from the necessity to rehearse. A second and not unrelated theme taken up here is McLuhan's concept of media "obsolescence." McLuhan argues that new technologies create changes in the media environment much as the emergence of a new species, according to evolutionary biology, alters the interactions of species within its environment. Although I am increasingly skeptical about casual analogies between biology and techno-history I accept McLuhan's challenge to situate media innovation in the context of a material ecology of the media environment. As it establishes its place in the cultural marketplace, each new medium transforms this material ecology, redefining meanings and practices in older, now "residual" (in Williams's terminology) or "obsolescent" (as McLuhan worded it) technologies. We see this clearly in the design of sound technologies, which are continuously reengineered to simulate and displace recognized instruments, sounds, and musical practices. Pianos were embraced by middle-class families in part because of their ability to replace the more costly (and spatially demanding) chamber ensembles patronized by the aristocracy. The piano could sound multiple notes simultaneously and replicate the effect of many instruments playing together. Later, in an extension and reversal of this process, some mechanical pianos were designed to contain multiple instruments that could be automatically sounded within its frame. Today the piano may be selected as one sonic choice among many in the digital archive of sampled sounds. Sampling is a stage in a long process of technological alteration that has blurred conventional distinctions between the consumption and production of music,[1] and transformed instruments, performance practice, and the larger ecology of musical production.

A Brief Geography of the Keyboard

The upright piano appeared in England in 1795, in Austria in 1835, and by 1900, it had replaced the square piano as the favored product of American and Canadian piano manufacturers.[2] The first upright was produced in Ontario in 1853, but uprights did not replace the more awkward square pianos, which were particularly susceptible to damage from Canadian climate changes, until around 1875.[3] Pianos were then primarily an import business, an arrangement clearly evoked in the 1816 advertisement cited earlier.[4] It was well established as part of the civilizing process bringing culture to the New World. For many immigrants, possession of the piano was defined as a crucial, even "mandatory" part of the colonial settlement process.[5] Many settlers of European descent would not enter the Canadian wilds without one, and popular histories are replete with stories of pianos being hauled over mountains into mining towns, brothels, and remote northern communities. The piano's popularity was not limited to salon rituals of genteel society; more casual communities of the frontier, enlarged by the gold rush areas in British Columbia and the Yukon, found a valuable use for pianos in the dance halls and bars. The first transcontinental train contained a Heintzman piano (an older relative of my own upright).[6] Wayne Kelly claims that the piano "evolved" from a status symbol in the 1880s to "an almost mandatory household appliance by 1910." "The piano was the first and only form of creative expression most households would know," he concludes.[7]

By the turn of the century, following inventive technical adaptations to the Canadian climate (specifically the wood that was used, for Canada's extensive old-

FIGURE 16.1 *Illustration from the* Nordheimer Music Book, *Toronto, 1904. Nordheimer was the longest lasting brand in Canadian music history, operating from 1844 to 1964. From Wayne Kelly,* Downright Upright: A History of the Canadian Piano Industry *(Toronto: Natural Heritage/Natural History, 1991), 82.*

growth forests provided exquisite sounding boards), the piano manufacturing industry was a major contributor to the economies of a number of Canadian cities. Piano manufacture became one of the most profitable industries in central Canada, and Canadian pianos were being exported worldwide. Although individual craftspeople had been making (and teaching on) pianos since 1816, the first Canadian piano company opened in 1844, offering instruments, sheet music, and sewing supplies.[8] The congregation of musical and sewing supplies was not unusual in the nineteenth-century marketplace; together musical and sewing supplies provided the tools for women seeking to create a comfortable, respectable life in a frontier society. Such shops supported the important association between gentility, femininity, and domestic well-being that dominated Victorian culture and formed a bridge with its modern successors.[9] Then, as now, amateur pianists were predominantly women. In settler communities, music education was gradually established for "the daughters of the well-to-do," and wives-to-be of pioneers in the West insisted that they be provided with this "civilizing instrument before tying the knot."

Through music the "lady of the house" was supposed to provide solace, prestige, and aesthetic refinement. It was her task to provide the cultural and moral foundation for a household whose success depended on the occupation of the husband and, in the colonial context, on tangible ties to the European homeland. This assignment played a substantial role in the emergence of gendered consumerism, whereby "the lady's function in a capitalist society was to appropriate and preserve both the values and commodities which her competitive husband, father, and son had little time to honor or enjoy; she was to provide an antidote and a purpose for their labour."[10] The piano-player-salon assemblage accomplished this perfectly, constituting a domestic household that was simultaneously "private" (the domain of moral strengthening and parental discipline) and, through display to the "outer" world, a crucially visible site for the practice of conspicuous consumption. The social and aesthetic uses of the instrument helped materially and rhetorically to constitute the modern categories of public and private life. The piano mediated and sustained conventional heterosexual power relations within the domestic sphere and played an important role in the modernization of domestic space in capitalist societies of the West.

As they became essential amenities, pianos nurtured a growing business in manufacturing, teaching, and sales. The meanings of these activities were linked to (and contained by) their association with social and cultural conventions of European middle-class life and the obligation of women to ensure the endurance of these in the new homeland. These responsibilities were part of an explicit acculturation of women's work, as a consequence of which the social importance of the piano often prevailed over the creative horizons open to its interpreters. A good life housed elegant deportment, sensitivity, and discipline: in short, a piano and a good woman to play it.[11] Buying one for his wife or daughter provided the husband with a solid representation of public success and private felicity. If women took their music seriously or sought different purposes than this for their art, then the usual barriers to public space came into play. Professional concert pianists were almost exclusively men, amateurs mainly women.[12] The piano offered itself as a vehicle for women's personal expression and helped both spatially and symbolically to keep them in their place. The piano-woman-home assemblage was thus fundamentally unstable, notwithstanding the sizable solidity of the instrument itself. This ambivalence would help shape its ensuing history.

In some cases, music's association with gentility was stronger than its capacity as a medium of expression, as illustrated in an 1884 short story by the then-popular writer Isabella Valancy Crawford, "Five O'Clock Tea." "Musical girls," she notes,

> generally with gold eye-glasses on chill, aesthetic noses, play grim classical preparations, which have as cheerful an effect on the gay crowd as the perfect, irreproachable skeleton of a bygone beauty might have, or articulate, with cultivation and no voices to speak of, areas which would almost sap the life of a true child of song to render as the maestro intended.[13]

This caustic narrator, finely attuned to the cultural dynamics of colonial society, already associates the piano with a "bygone" era and counterpoises that "grim" sound to the imaginary songs of a child of nature (presumably a person uncorrupted by lessons). In this instance the piano manifests a geography of colonization as well as a normative gender performance, and at least some female observers were not captivated.

The piano may have been a reassuring possession for women in this environment, and for their husbands and neighbors. But the prospect of a musical culture in the possession of women was not reassuring to visionaries of the new society. Crawford's opinions were not unique in this respect. In the context of a colonial society dominated by European settlers, a musical culture comprised predominantly of women pianists and itinerant fiddlers, devoid of external aesthetic and professional standards, was an insufficient vehicle for the articulation of a solid, respectable aesthetic culture. The accomplished playing of the piano by wives and daughters may have enhanced the gentility of the family and select parts of the local community, but it was not in itself an adequate vehicle for the constitution of a cultural elite. Canadian schools of music presented little similarity to European conservatories; they were not professional schools but "taught all comers, anyone who would pay a fee."[14] "Music . . . was considered a frivolous pastime and considered important only as part of the education of young ladies," notes music historian Clifford Ford. "Should a talented young man or woman manage to survive both the prevalent social attitude towards music and the jungle of charlatans and frauds in the music teaching profession, he [sic] would still have to seek advanced training in the musical centres of Europe."[15] If piano playing finally lacked prestige outside the parlor, it was because the affinity among class, femininity, and piano performance was so well inscribed in the social meanings of the instrument. This association permeated popular fiction, social commentary, and commercial advertising through the nineteenth and early twentieth century. Musical performance helped constitute a self that was simultaneously a refuge from and the product of new regimes of everyday life being established through capitalism, colonialism, and the bourgeois family. The feminine gendering of the instrument and performance on it helped constitute its history as an instrument of modern culture that was simultaneously sacred and disposable.

The growing public for instrumental music in Canada through the second half of the nineteenth century was drawn by powerful concepts and feelings about music's importance to the cultivation of competent individuals in a flourishing modern society. By nurturing players and audiences for instrumental music, the settler community melded together the dynamic search for aesthetic experience and social status. The musical culture of Europe was favored not only because so many immigrants came from there, but also because it seemed the most appropriate means to mobilize artistic practices and conventions through which fully realized individualities could evolve. In this endeavor, the piano was an important instrument, for despite (or, really, because of) its complex mechanical nature, it favors fluctuating chromaticisms and intensities and rewards players with the sense of a fully realized musical voice—if they maintain the self-discipline necessary to actualize it.

These attributes appealed to middle-class women who could learn to express themselves in a relatively intimate form without disrupting their social situation. The attributes of the piano—its size, its solidity, its interdependency with the active self-discipline of practicing—lay the foundation for a new approach to the culture of the home. Through the symbolic and material effects of the piano, the home was redefined not just as a site for family life, with the wife at the center, but also as a site of leisure. If "home" was still only secondarily a site of leisure (particularly for women), the piano established the home as a site of cultural practice and musicking as something that occurred in this private sphere. Listening to music became something one did "privately" in conditions of individual preference and (later) ease, rather than something that occurred in the public domain.[16] The home was thus constituted simultaneously as cultural space and as "free" from the constraints of public space, a socio-topographic premise that carried with it the threat of unreliable artistic standards.

The piano was indispensable to and indexical of a social process through which the middle-class home was defined as a site of leisure, a space-time set apart from work and commerce. Piano-playing women were central to the affective images and material practices of domestic space. But "practicing" sounded too much like work, and failing to practice produced dispiriting results. Mechanical players were marketed in terms of the same semiotics of domesticity and refinement but linked to an explicit rejection of the discipline of practicing. As the market was inundated with new status-enhancing instruments for the home, it was not only the piano being displaced but also an accumulated history of feminine—and carefully feminized—culture and cultivation. The increased accessibility of electronic music reproduction went hand in hand with the loss of listeners for their performances. Stan Godlovitch suggests that the expectation of an audience is integral to musical performance. "Lectures and performances alike look for ears to fill," he writes. "Lacking the ears, nothing remains to fill save the room, scarcely a surrogate." A player without an audience is not a performer but a demented person, "almost pathological"; proceeding to play without listeners "smacks of desperation."[17] Godlovitch imagines this "defiance" in an empty hall, not a living room. But amateur pianists, attached through training, disposition and embodied memory to obsolete skills and archaic instruments, so many of us women, might feel some resonance with his description.

Mechanical Pianos

Changes in musical production seem so rapid today that they seem to turn the knowledge gap into microfractions of biological time. Microfractions will not serve us well in tracing this history, however, for experiments in mechanizing keyboards began well before the piano had been embraced as the dominant instrument for domestic entertainment and musical education. As early as the fourteenth century, when organs were the exclusive property of monasteries, builders were seeking technical improvements to them, not only in terms of sonority and temperament, but also in terms of their human–machine interface.

"The organ is an instrument strongly bearing the character of a machine," notes Max Weber in his study of music and industrialization. "In the middle ages its manipulation required a number of persons, particularly bellows treaders. Machine-like contrivances increasingly substituted for this physical work."[18] The desire for sublime registers of sound had apparently begun to conflict with the limited capacities of the human body. Once the split between body and spirit was inscribed in these instruments, technological improvement became the preferred means of resolution, and pianos and organs were subject to continuous renovation. The lighter clavecin, the piano's other precursor, was played by amateurs, "particularly and quite naturally all folk circles tied to home life," including women.[19] The greater accessibility of this instrument helped to accompany and advance new harmonies in popular music, but it seems not to have been subject to a comparable evolutionary trajectory in terms of mechanical augmentation.

This search for mechanical progress was not purely derived from spiritual vocation, of course. Henry VIII is known to have left behind "an instrument that goethe with a whele without playing upon" at his death in 1547.[20] The mechanical ingenuity of these instruments was itself a source of pleasure, indubitably enhanced by the exclusiveness of possessing them. By the early seventeenth century, however, such mechanical devices were no longer the exclusive property of monasteries and kings. A builder of "musical automata" in Germany made spinets that could be played by a drum or barrel, and mechanical instruments were heard in early performances of Shakespeare.[21] By the mid-eighteenth century, the piano had emerged as the favorite instrument of the northern European middle classes, reinforcing their greater focus on indoor life and, as Weber put it, the "culture of the bourgeois-like home,"[22] with its growing preoccupation with effort and discipline. Clockwork and self-playing pianos remained the subject of widespread technological innovation. By the mid-nineteenth century, mechanical pianos were popular in bars and music halls as well as some homes.

FIGURE 16.2 *The Tel-Electric Piano Player boasted the advantage that the action could be worked remotely, with an electric cable joining the bank of playing solenoids under the keyboard to a special cassette containing a thin strip of perforated brass. Country Life in America, November 1912, Paul Ottenheimer collection. From Arthur W. J. G. Ord-Hume,* Pianola: The History of the Self-Playing Piano *(London and Boston: George Allen and Unwin, 1984), 102.*

A substantial market was emerging for musical commodities—sheet music, instruments, and public performances by artistic celebrities—that were no longer confined to aristocratic or religious settings.

We should be cautious about imposing a retrospective evolutionary narrative on the life of these instruments; they had, and helped to make, a life of the time. Retroactively, however, we can identify one general dynamic shaping this process of change. With the emergence of mechanical instruments, musical notation and the mass printing of musical scores, music was increasingly mobile; it belonged less and less to a particular place or time. Obviously the piano itself was never a very mobile instrument. Rather, it served to incorporate simultaneously the aesthetic mobility of the middle classes and the domestic stasis of their homes. The compositions, the sheet music, the fashions, and, eventually, the players were more easily moved, and their mobility served to spread musical practices laterally through the "better" parts of society. Its appearance in so many homes constituted a solid link between culture as a specialized and privileged domain and the specific achievements of the middle-class domestic habitus, in happy contrast to the earlier, more limited privileges of monastery and court. In an increasingly secular, market-oriented milieu, the piano became the preferred symbol of middle-class ideals, achieving a mythological status that resonates nostalgically in film and literature today.

Such achievement was not popular with everyone. To some contemporary minds it was appropriate that mechanical experiments would thrive in this context. With the prospect of new musical compositions being purchased by "nameless grocers' daughters," as one writer put it, "composing for the market seemed unworthy of a skilled, deliberate effort; it seemed almost like something that could be equally well done by a machine."[23] It was not the technological reproduction of sound that irritated such writers, but the automatic-sounding live performances produced by multiple purchases of instruments, lessons, and popular compositions in print. The synchronized physical training of so many human hands was perceived as blurring the boundary between bodies and machines. Innovations in sound reproduction would henceforth be understood as further interventions and diverse delineations of that fluid boundary.

Sheet music, the first music-related mass medium, was a rapidly circulating commodity by the middle of the nineteenth century. In this situation "art is made mechanical," a disenchanted music teacher concluded, by "fast and furious amateurs."[24] The spread of amateurism invited the scorn of sensitive critics like this one who opposed culture to mechanization or massification of any kind. But the mass production of musical scores and the pressures to cultivate domestic performance were established cultural technologies of the Victorian middle class, whose combined embrace of self-discipline and fashion put thousands of piano students to work practicing away, "one eye on the music . . . the other on the mantelshelf clock."[25] If musicicking had been liberated from the privileged spaces of prebourgeois institutions, note how deeply it has become imbricated with a specific measurement and embodiment of time. The student is required to practice a half hour, an hour, two hours, depending on escalating levels of discipline, accomplishment, and commitment. From these descriptions, it doesn't sound as though these fledgling impresarios were having much fun. We need to be wary of this conclusion, to some degree, for the semiotically feminized and domesticated context of such practicing was bound to earn contempt from some quarters.

The changing technology of musical production incorporated and went beyond the simultaneous repetitive discipline of practicing identical pieces, having undertaken to automate those same movements through mechanical reproduction. Entrepreneurs in the music industry were learning to profit from the experiential "downside" of the established practices they themselves promoted. Experiments in designing machines for composing and performing were flourishing by the nineteenth century, not just because it was technically possible but also because a hospitable social context was emerging for them. Responding to the increased middle-class focus on interiority and domesticity and the growing willingness of households to invest in these qualities, inventors explored ways to enhance instruments as playback mechanisms for the home. They worked assiduously on shrinking supplementary sound sources to fit within the frame of the mechanized piano and on mechanizing the playing of them. Early mechanical pianos could render the music of an orchestra accessible to a family home.[26] More advanced player pianos offered greater verisimilitude through recordings of professional pianists while providing a regular keyboard as well. By the beginning of the twentieth century, retailers for player pianos and acoustic pianos were competing for the same market.

McLuhan suggests that when a medium is displaced by a newer medium, it becomes a work of art and acquires the aura associated with art in the modern era. In other words, its transparency as a medium—a consequence of the symbiotic relationship between hardware and software, among other things, which forefronts the "content"—disappears behind its newly foregrounded materiality, which, reframed by the newer media, acquires the "aura" previously denied to it as a mass-produced object. Nowhere is this process more clear than in the history of the keyboard. With pianos competing with their own technological successors, advertisers retreated from the conventional claims of competing technological prowess formerly used to promote the acoustic (nonmechanized) piano and emphasized rather the status of the older instrument as a work of art. Turn-of-the-century promotion for Steinways exemplifies the manner in which high-end pianos were being sold as "great works of art" in contrast to their mechanical competitors.[27] Some advertisers commissioned new graphic art to accompany depictions of the instruments, so that pianos appeared simultaneously as unique technological achievements and artifacts of an age not yet taken over by mechanical reproduction.[28] Shifting their emphasis from the instrument as a marvelously intricate manufacturing achievement, advertisers now associated the piano with fantasies of exquisite taste and unique individual expressiveness (what Roell terms "the basic *uselessness* that allowed the Steinway to thrive in the consumer's imagination").[29] It was the physical instrument, not the aesthetic comportment associated with it, that now occupied the center of the frame. The piano was offered not as something one did, but as something one had. Meanwhile "doing" was being reattached to a very different set of activities and skills.

The first decades of the twentieth century were crucial in cementing ties among fashion, consumption, and the ideology of the modern. Emphasizing the enhanced role of consumption in domestic management, the growing retail sector of department stores, advertisers, and magazines drew on new design strategies forefronting feminine taste and discrimination at the center of production and consumption. Celebrating what had been described paternalistically as "female accomplishments," this campaign depended on a heightened appreciation of distinctive items of domestic use and display, which women were now encouraged to buy rather than make.[30] Aesthetic aspirations for the home environment were intensified through the deliberate targeting of women as consumers of beautiful things. The newly coined phrase "the democratization of luxury" described the process through which products representing traditional feminine craftsmanship were entering department stores and graphic arts and being continuously updated by the more rapid temporal rituals of contemporary fashion. Feminine craftsmanship was promoted as a decorative resource just as it was being displaced from the process of production. Art deco in particular offered "a distinctively feminine version of the modern" that was richly decorative, openly commercial, and closely tied to domestic consumption. "It appealed to consumers internationally," Penny Sparke notes, "and was central to the transformation of the modern from an elite to a democratized base, a shift which helped to construct a bridge between masculine and feminine cultures."[31] The emphasis of this new social engineering was not on effort or accomplishment, but on taste and aesthetic judgment; here, as in other contexts, "the liberal subject was reformulated from an individual moral being into a subject of need."[32]

This new taxonomy of doing and having was equally influential in musical culture. Practicing was discursively redefined by being attached to negative connotations of social ambition, drudgery, and kitsch. Piano design and marketing sought to displace the need to practice with purchasable substitutes that were doubly valorized as conduits to pleasure and tools of democratization. Pianos began to register two temporalities at once, representing both the knowing sophistication of the up-to-date twentieth-century consumer, armed with a good eye for design and innovation but lacking the time to spend on unproductive leisure activity, and the harmonious stability of the nineteenth-century bourgeois home, where individuality was an expressive act and musical achievement the result of socially approved effort. Time and the benefits of saving it, carefully articulated to compelling rhetorics of pleasure, status, and need, formed central themes in the early twentieth-century marketing of consumer technologies. It only seems ironic that it was the piano that led the way.

The mechanized piano represented America's most fervent ideals in the early part of the century: the opportunity to participate in something important without needing to understand it; the pleasure of creating without the disciplines of work, practice, or the taking of time; and a fascination with exceptional talent (the early stirrings of celebrity). The idea of enhancing the personal power of consumers through mechanization recurred and intensified in this milieu. Mechanization also offered an enhanced sense of power to recording performers, who could correct their mistakes and perfect the recording before it was released, thus reversing the qualities of spontaneity and "natural" inspiration that justified the performance, the talent, and the technical preservation of it in the first place. Not surprisingly, the popularity of these techniques aroused a nascent anxiety about mechanization in the culture. This mood is evident not only in the advertisements for traditional pianos, where heritage and prestige are now emphasized, but also in ads for the more up-to-date mechanical instruments. It was not necessary to invest time, these texts implied, to reap personal rewards from music; at the same time, music could maintain, even enhance its abilities to inspire authentic and profound listening experiences.

Implicit anxiety about the authenticity of technologically mediated performance inspired copywriters to produce rhapsodic prose on the expressive possibilities of the new machines. The following instance appeared in 1921:

> The tel-electric piano player. It is the *one* player which you, *yourself*, whether an expert musician or not, can quickly and easily learn to play with all the individuality of a master pianist. It permits you to interpret perfectly world-famous compositions with all the original feeling, all the technique, and with all the various shades and depths of expression as intended by the composer.
>
> In using electricity as the motive force of the tel-electric we not only eliminate the tiresome foot-pumping and noisy bellows of the pneumatic player, but we place the instrument under your absolute control—ready to answer, instantly, your slightest musical whim.[33]

The constellation of pleasure, self-expression, ease, and heightened individual power invoked here clearly encapsulates the attraction of the new mechanical instruments. Like the conventional piano, the mechanical players were celebrated for their ability to enhance self-expression and social status. Both were advertised in terms of their ability to create clearly individuated sounds, their technical superiority to their competitors, and their indispensability for the creation of enviable domestic space.[34] But the mechanical piano offered a greater sense of personal power with a fraction of the effort. Music was no longer dependent on the "drudgery" of practice, with its now negatively connoted emphasis on "digital dexterity alone" and its capacity to make "capable and obedient machines of the fingers."[35] The joint purpose of instrument and owner was to entertain the guests and keep the offspring busy at home without making them feel like dexterity machines, advancing what Roell describes as "the gradual replacement of the Victorian work ethic with the leisure-oriented consumer ethic."[36] By sidestepping the need to practice, the pianola was able to satisfy "growing daughters and sons with a natural craving for some form of entertainment . . . who will seek it outside it if is not provided within the home."[37] Musical skill, formerly enshrined as the favored symbol of the Victorian work ethic and family togetherness combined, was now subjected to a systematic downgrading through promotional rhetoric. The piano was still a crucial item for "small or family parties," but the embodied skill, discipline, and time it took musicians (still mainly women, judging from illustrated advertisements of the time) to perform it had been made dispensable.[38] Women could continue to perform traditional feminine roles like playing the piano even if they had taken a job, for they no longer needed to practice; thus anxieties about gender roles were neatly transferred to (and presumably solved by) the consumption of new instruments.

Player pianos were marketed in terms of their promise to reproduce faithfully the brilliance and nuance of the most accomplished performances, without the user having to acquire such accomplishment herself. By playing on a special piano, professional pianists produced piano rolls that created markings to exactly correspond to the performer's strokes on the keys; these rolls were then played back on a reproducing piano powered by an electric motor. The pianists recorded in these rolls also worked as "part-time merchants," promoting piano manufacturers in concerts and testimonials.[39] In fact, the piano was the first major consumer commodity to be attached to brand loyalties. Musical celebrities endorsed Steinways and Chickerings in live performances and magazine advertisements. Through player pianos the cult of virtuosity was extended into the home, where music-loving consumers could listen to these performers on their mechanical pianos.

By the late nineteenth century, patent offices in Canada and the United States were deluged with applications by inventors seeking to improve the mechanisms of the piano and the sound of pianists playing them. The first self-contained "pneumatic piano" appeared in 1880.[40] The first commercially successful "reproducing pianos" for the home appeared in Europe and North America in 1901—just before the phonograph, another technology intensifying the move toward a more separated geography of family and domestic life. By World War 1, with sales of player pianos and piano rolls booming, the "silent" piano seemed on the verge of obsolescence; it seemed inevitable that, as Roell recalls it, "the benefits of music would soon be enjoyed by all."[41] In this democratic sonictopia, musical events displayed not the homely skills of the domestic performer but rather the recorded skills of the virtuoso, which could be purchased and brought home to enjoy in the family salon.

In his history of recorded sound, Michael Chanan describes the phonograph as the first sound technology to produce both hardware and software (record players and records) as interlinked commodities.[42] But that initiative preceded the phonograph with player pianos and the marketing of piano rolls by famous performers. As Lisa Gitelman shows, "The terms 'hardware' and 'software' might not have been applied at the time, but something of the relation that later emerged between them was already recognizable, working in the construction of 'self-playing,' as it would in the construction of machine-readability and of digital 'wares.'"[43] The sale of piano rolls depended on inciting a desire to purchase multiple recordings in consumers who already possessed player pianos. Whereas sheet music sales continued to depend on the ownership of instruments, "the newer makers of music rolls suffered the hardware obsession of the music trades."[44] If pianos had been marketed on the basis of womanly skills and middle-class domesticity, player pianos and piano rolls were advertised on the basis of new innovations in faithful sound reproduction, enhanced by the claim that through such innovations world-class musical performance could be appreciated on a more democratic basis. Here democracy and consumerism were adamantly joined together with the promise that individuals could flourish musically while bypassing the acquisition of musical skills. Rather than practicing, or submitting to someone else practicing in the home, one should simply develop one's musical discrimination and learn to appreciate the fine points of diverse recorded performances.

Today, looking back at the enthusiastic testimonials that pianists offered to the faithful reproduction of their artistry on player pianos, it must be remembered that they were contrasting these mechanical reproductions to the still elementary technique of acoustic recording, which made the sound of the piano "thin and tinny, liked the plucked string of a banjo or guitar." Rachmaninoff claimed that the early recordings of his playing made the piano sound like a Russian balalaika.[45] More successful recordings were made of the singing voice, most famously Caruso, whose voice, recorded in 1902, was so rich it drowned out the surface noise and made the inadequate apparatus sound "rich and vibrant."[46] Attempts to modify the piano to make it record more clearly with these early techniques did not improve the effect. The more sonically effective player pianos conquered the home market and made their way into the movie theaters by the early 1920s. Reproducing the work of professional musicians while enabling projectionists to switch in a timely manner from one musical-dramatic genre to another, they put thousands of pianists (typified in one account as "a jaded female beating a scarred upright whose duty it was to scare away the ghostly silence that otherwise enveloped the screen"[47]) out of work. Between 1925 and 1932, with the growth of synch-sound conversions, the number of musicians employed in motion picture theaters declined from 19,000 to 3,000 in the United States.[48] The transformation of music performance was even more dramatic for amateurs, who by the 1920s were being encouraged to imagine themselves as consumers, rather than producers, of music for the home.[49]

By the end of that decade, however, pianolas were being stockpiled, having fallen victim to the attractions of radio. With the advancement of mechanical sound reproduction, the piano, embraced and promoted as the center of middle-class domestic and musical life across North America, both expanded and lost that role. This paradoxical change transformed the economies of domestic musical culture. The larger scale of factory production necessary to produce such complex instruments in large numbers led to greater corporate concentration and the disappearance of multigenerational family businesses who did not survive the Depression. Like the nineteenth-century music halls later overtaken by cinemas, the piano as a "cultural technology" helped accomplish its own obsolescence.

We cannot see or hear the transformation of music production or domestic space in the sounds of an individual piano. A well-made piano, constructed as mine is from good Canadian wood, can sound as harmonious now as when it was made eighty years ago. It is not through the individual instrument but through the instrument's changing relationship to other technologies and practices that we must reconstruct this transformative process. The piano's agency as a medium or instrument of interaction between composers and performers, performers and listeners, teachers and students, people and machines, has altered over time in interaction with other social and technological changes. Although the materiality of the residual instrument has remained more or less the same (aside from the wood, for old-growth forests are now largely depleted; only the most expensive pianos could use such wood today), the piano has produced diverse effects, or acted with various kinds of agency, in different moments and geopolitical spaces. Today we can best understand the piano as the bearer of heterogeneous and sometimes incompatible registers of temporality and social memory.

The Mechanical Ideal

Traditionally, music has been the creation of human performers interacting with one another and with their instruments. The modernization of instruments contributed to an inventive search to create inanimate objects that could duplicate the work of performers. The emergence of virtual performance concerns not just whether we as audiences witness a performance "live" or mediated, but how the performance itself is produced.[50] The piano offers a suggestive case study of this transformation.

The mechanization of musical performance involved the transfer of compulsion from individual effort to technological innovation, with a new emphasis on the saving of time and the happy avoidance of unnecessary discipline. This transition was marked by the new capacity for performers to correct details in their performance before they reached their audiences, creating new standards for

FIGURE 16.3 *Social gatherings among the affluent rippled with debates about the status of artistry and interpretation in mechanical performance. In this advertisement ca. 1905, the performer uses the magical ambiguity of mimesis to promote the Welte-Mignon Autograph Piano, which is reproducing her own playing. From Arthur W. J. G. Ord-Hume,* Pianola, *262.*

performance perfection. More generally, the production of "virtual fingers and hands," as Gitelman describes it, produced new senses of "the self" that are complicatedly referenced by the concept of "self-playing." Automated playing enabled one to embrace the pleasures of sound separate from the requirements of producing such sound, and helped enact a new kind of subjectivity that was further supported by the rhetoric of democratic access attached to it. These experiences made innovation in musical machines even more attractive, and they altered the formulation and reformulation of musical taste tied to a quickened temporality of playing and listening pleasures. It was no longer the pressure of parents or teachers, but the novelty of player pianos and the currency of music "software" commodities available to program them, that were expected to inspire young people. In students' competitive efforts to memorize their music pieces, reproducing pianos came to replace their mothers as memory guides.[51]

Through musical mechanization, the modern idea of self-fulfillment was detached from skill and moral development and rearticulated to newer ideas of pleasure and convenience. With mechanized pianos, one could perform beautiful pieces of music for friends and family without having to spend the time learning to play them. All you had to do was push a pedal or lever. The idea that mechanization of instruments made music accessible to the unskilled efforts of toddlers and children became a central motif in their promotion, and it quickly became the chosen strategy for advertising automatic pianos. Most well known was the image of the Gulbransen baby, accompanied by the slogan "easy to play": "all the family will quickly become expert.... Without long practice! All the joy without hard work!"[52] Of course, as Roell notes, musical performance was thereby challenged as a vehicle for domestic virtue and harmony; if playing does not need to be nurtured and cultivated, if good performance doesn't demonstrate the sensitivity and self-discipline of the player, what is its value? And what is the moral worth of those who have achieved it? From being a key symbol of feminine virtue and eligibility, playing a keyboard was converted to something that, with the right prostheses, anyone can do. Such capabilities were thus disconnected from embodied or gendered constraints of time, learning, sensitivity, or skill. Consequently, "poor sister has hardly touched the keys," as a critic noted in 1919, recording a hearer's ambivalent response to the "horseless pianoforte."[53] A similar gendering of the dilemma also appears in John Philip Sousa's 1906 contestation for authorial rights over music rolls: "Printed notes, spots, and marks were no more equal to his musical thought or 'living theme,' he was sure, 'than the description of a beautiful woman is the woman herself.'"[54]

A 1903 article published in the *Chicago Indicator* assessed the instrument as follows:

> The increasing vogue of the piano player ... is now regarded—and rightly regarded—as one of the most significant phases in the life and advancement of this mechanical age. It is heralded by the enthusiastic as a portent of the dawning of a new epoch, when machinery will still be the motive power of civilisation, but will be applied to uses hitherto deemed sacred from its invading banners ... if this vision is to be realised, the piano player will certainly be the most prominent factor in

its accomplishment. The public in general will be pleased, and the piano trade will certainly not lag behind the rest of the world in similar feelings; for it is plain that, when the day of the player [piano] arrives, the field of piano enterprise will be greatly enlarged.[55]

Note that this ode to mechanization (and its conquest of "hitherto sacred" activities) appeared in the same decade that biomechanical research entered the sphere of factory production. Frederick Taylor was busily conducting detailed research on workers' physical movements to calculate ways to maximize their efficiency and the productivity of the assembly line. The idea of reducing to its essentials the motion required to produce a specific mechanical action was evidently not restricted to the factory but was being promoted with similar zeal in the voluntary domains of cultural and leisure activity.

This author asks what caused the vogue of the player piano to reach such heights: first, he suggests, the popularity of the piano as an instrument of the home; second, the "American habits of economising time."

> Our citizen loves music, but he has no time to spend in studying a complex technique. His daughter, perhaps, who would be the proper person to fill this want in his household, is busy working in a store or factory, or goes to high school and must study her Caesar or geometry when she gets home. And then there are many people who have no daughter, or none of the proper age. These matters seem trivial, but they nevertheless have a potent influence.
>
> The third fact is this: with a piano player in your house you can give a friend musical entertainment and discourse with him [sic] at the same time. Or if you have nothing to talk about—and this is a contingency that happens with remarkable frequency at social gatherings, especially small ones—you may set your piano at work.
>
> Still another cause lies in the admiration of the public for anything which acts, talks or plays automatically. They wonder at the thing. A wonder is a good thing to subdue and make your own. It pleases you; it will please others. Perhaps it will make them envy you; and what so sweet as envy to the envied? Again, the piano player is a novelty. In all ages novelties have been eagerly sought for, but never has there existed such a craving for them as now.[56]

Appearing in 1906, the same year Sousa attacked the adequacy of automatic playback, these comments are strikingly prescient. The author gives expression to themes that are now familiar in rhetorics of technological advocacy. We see the fascination with mechanization both for its own sake and for what it "portends" in terms of future progress; the promise of greater freedom from the constraints of embodiment; the pragmatic advantages of economizing time, following on the ambivalently viewed move of women into the workforce; the advantages of musical reproduction in allowing listeners to enjoy music in a state of genial distraction, without the need to converse with one another; the importance of up-to-date technologies for effective conspicuous consumption; the implied civic duty of contributing to industry through such consumption; and, perhaps most poignantly, yet with so much calculation, the magical qualities of sound reproduction with their ability to produce a childlike sense of wonder. If mechanization and rationalization were responsible for a growing "disenchantment" of the world, as Sousa implies, then paradoxically sound reproduction, with its magical disembodied siren song, promised to counter that process.

By the end of the 1920s, however, these effects were all being celebrated in and accomplished by radio. The dramatic drop in piano production was absorbed financially (for those protected by corporate mergers) through the dramatic rise in the sales of radios. This theme also remains a defining motif in discussions of the technological transformation of musical reproduction. The evocation of magical moments evoking wonder and desire occurs within the framework of a larger institutional process in which technical invention and obsolescence accumulate and recur. As soon as a technology has captivated and conquered its market, its interconnected network of machines, performers, performances, and related commodities crumbles in the wake of a newer device. Between 1900 and 1930, 2.5 million player pianos were sold in the United States,[57] while only 650,000 pianos a year were produced worldwide.[58] As soon as player pianos supplanted the sheet music industry, they were in turn supplanted by the phonograph (1907) and radio (1928). The market for pianos partially regrouped in Canada and the United States in the 1930s and 1940s, but it would never again reach levels from the First World War period, when there were seventeen piano firms in Toronto alone.[59] The Canadian piano industry reached its peak in 1966, and by the 1970s, it had given way to foreign subsidiaries. By the early 1970s, with Japan now the major producer of pianos, worldwide production was only slightly higher than it had been half a century earlier. By the early 1980s, it was estimated that 40,000 old pianos were being discarded—junked—annually in the United States.[60] The stereo sound system (and later its more mobile extensions, boom boxes, Walkmans, MP3 players, and Web radio) had replaced the piano as the center of home and family entertainment.

The rise of the piano coincided exactly with the rise of modern industry and modern culture. We can understand both its moment of dominance, and its later replacement by more mechanized means of sound reproduction, in the context of industrial modernism and postindustrial postmodernism as social and cultural formations. The history of the piano's technological evolution/displacement offers a cogent instance of McLuhan's idea of obsolescence, and it provides an exemplary case study in the paradoxical "progress" of cultural technologies in the twentieth century.[61] By advancing the required virtues of Victorian hardworking domestic morality, and articulating these values with the more casually compulsory imperatives of currency and innovation encouraged by modern consumerism, the piano contributed to both the creation and obsolescence of modern culture.

Obsolescence

McLuhan diagnosed the media as an environment, and he argued that this environment framed the conditions for our perception in general and our understanding of individual media like books, cameras, or radios in particular. When a new medium succeeds as a popular commodity, he maintained, the larger media ecology shifts and the role of each medium within that ecology is altered. As the new medium represents the forefront of commercial and technical development, previous media become "obsolete." That doesn't mean the older media disappear or even that they lose their social or material efficacy. In McLuhan's language, it means the new technology embraces the older one as its content; having as yet no medium-specific content to fill it, the new technology embraces the contents of its predecessor. It is this process that translates the earlier medium into art. McLuhan's account of technological change resonates with what Winner calls the process of "reverse adaptation": with a complex technological system in place, Winner suggests, investors in a new technology search for applications, and thus reverse the conventional relationship of means and ends.[62]

It may have been the invention of radio that first elicited this idea of reversed ends and means. "It was not the public that waited for the radio," argued Brecht, "but radio that waited for the public; to define the situation of radio more accurately, raw material was not waiting for methods of production based on social needs but means of production were looking anxiously for raw material. It was suddenly possible to say everything to everybody but, thinking about it, there was nothing to say."[63] Pursuing the idea of uneven development between a new medium and its available contents, media historian Friedrich Kittler echoes McLuhan's hypothesis; he describes "the housing of one form in another, creating a kind of piggyback from one technology to the next, bypassing the exploration of any salient features present in each new development."[64] Radio was quickly solidified socially and commercially as a "means of distribution," to borrow Brecht's phrase, only to be overcome by television; but radio artists have continued to explore the unique possibilities of radio as a medium.

McLuhan and Winner both emphasize the importance of the media environment as a whole, the interdependency of human and technological actors within that system, and the complex cultural reversals (as distinct from ruptures) brought about by technological change. In McLuhan's account, speech becomes the "content" of writing, drama the "content" of film, film the "content" of television. Critics lose their grasp of the specific effects of television, to take one notable instance, because they rely on critical methods derived from drama and film, rather than searching for what is specific to television as a medium. In this context, *pianoforte* performance was the "content" of the player piano, in that live performances were notated and reproduced by mechanical means. This simulation elevated the competence and expressivity required of musical performance, and reduced amateur practice to a marginal place in the ecology of musical production. Subjectivity and expressiveness were still important reference points for marketing mechanical instruments and music rolls, but these were machines for reproducing, not for creating sound. Ordinary people were encouraged to appreciate (and to act as vehicles for) the veracity and skill heard in the sonic representation of expressiveness in others.

Yet mechanical pianos encouraged consumers to attach themselves physically to the music-making process. They could sit at the piano in the accustomed

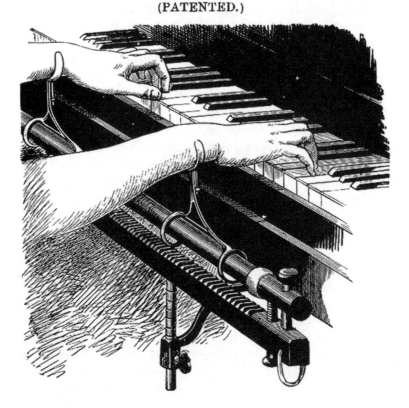

FIGURE 16.4 *Bohrer's "Automatic" Hand Guide. This ingenious Canadian invention is one of many transient techniques in the development of mechanized performance, reminding us of the long historical process through which automating mechanisms became the star performers. From Arthur W. J. G. Ord-Hume,* Pianola, *8.*

position and inflect these mechanical recordings with their own personalities by means of the manipulation of pedals, buttons, and levers that could alter volume, intensity, and speed. No musical training was necessary. It would take eighty years and the mergence of DJ practices before this idea actually succeeded as musical practice. Then, in another technical reversal, digital sampling made the DJ's recycling of records, or scratching, the "content" of a new music source that removed the necessity to work with turntables, reviving issues of musical memory and presence and generating new practices in the musical landscape.

With newer media taking up older "forms" as their content, McLuhan argues, the profound changes occurring with these technological shifts are sensorially and culturally obscured. We struggle to understand their significance in an environment in which our modes of understanding are themselves shaped and altered by the technological extension and suppression of our senses. Today the reconstruction of sound occurs at multiple levels of technological remove, with samplers and drum machines deconstructing and reprocessing musical performance that has been broken down to digital bits and reorganized via the language of the computer. These digital capacities suggest a radically new category in the process of obsolescence. It is not only specific media, like pianos or turntables or specific genres either, that provide the "content" of new digital media, but music as we know it, with its conventional modes of performance and their embeddedness in a particular mode of time. The sounds produced through this process of radical abstraction could challenge our understanding of "music," although so far they have not widely done so. Rather than the player completing the instrument through her performance, now it is the instrument (with all its digital resources, including previous performances) completing the on-site performer. Although both involve symbiotic relations between body and machine, the hinge of dependency has thus far shifted further toward the machine and its accumulated resources. Now, as in the time of the bellows organ, possibilities for innovative collaboration in musicking are being colonized by the commerce in new techniques. Now, as then, it is the mastery of new technologies that dominates the horizon.

Composer/producer/engineers stretch the conventional boundaries of music to encompass techniques like scratching, looping, condensing, or otherwise manipulating sampled sounds. These techniques ostentatiously confuse boundaries between music and machine, human and other sounds. It was this very confusion of boundaries that first inspired the earlier modernist art of noise. Is this use of musical language to set outside it what it means to imagine music as the "content" of the newer media? Most commentary has obscured rather than illuminated this potential transformation. When we think of the Internet, we think of downloading recordings that are sonically indistinguishable from what we could buy at the music store, and most discussion revolves around copyright issues arising from this practice. By conceptualizing the Internet as a kind of player piano, that is, as provider of prerecorded musical content, and by focusing on the economic and legal controversies that arise from this use, both users and critical commentaries have obscured the potential significance of the new media.

McLuhan was never one to underestimate the difficulties of understanding these shifts. To make sense of new media developments, Marshall and Eric McLuhan developed a tetrad of the "laws" of media to apply to an emergent medium. They pose these as questions: What does it enhance or intensify? What does it render obsolete or displace? What does it retrieve that was previously obsolesced? What does it produce or become when pressed to an extreme?[65] In their own responses, television enhances multisensuous experience, displaces radio, film, and point-of-view perspective; retrieves the occult, and reproduces the "inner trip."[66] The car enhances privacy, displaces the horse and buggy, retrieves the knight in shining armor, and in extreme form produces corporate piracy and traffic jams.[67] Just as television foregrounds and then displaces sensuality, the car emblemizes and then displaces speed.

Can new digital resources for music making be "read" in terms of these laws? What is enhanced? The spatial reach of sound technologies, interaction between digital and musical technologies, and the accessibility of musical sounds and "tools." What is obsolesced? Conventional interactions between musicians, audiences, skills, and instruments: everything that once constituted what we call "music," and the time it took to bring these together. Music education. Musical performance is no longer a representation of previous playing ("practicing"); and it is no longer a live (and therefore finite) moment in time. Spatially, digital resources render obsolete the physical boundaries and characteristics of the home, the studio, the concert hall, and the CD. What is retrieved? Interaction between musicians who might have retreated to the isolation of their studios or lost their access to expression in the highly commodified worlds of classical or commercial music. Sounds of the natural and technological "environment." The historical archive of sounds.

And what does it produce in extreme forms? Soundtracks for every space? Musical theft from afar? The end of record companies? The dissolution of music from property rights, and the disappearance of signature styles? Copyright wars between friends? The use of computers for music education and a renewal of elite access to learning musical skills? Tunes played out in brain-implanted microchips; no extensions necessary? Such speculations have throughout the history of modern culture shaped our dreams. These dreams, in interaction with social and material forces, continue to shape the contours of our technologies and our selves.

You Push the Button; We Do the Rest

Changes introduced by digital music making have transformed not only the instruments and the role and scope of playing, but the wider matrix or "ecology" of performance, reception, and meaning in cultural production.[68] We need to look at this in the context of everyday life. Today you can turn on lights without hitting a switch; search a library or museum without going to it; shop without entering a store; move up or down and sometimes across a building without using your legs; turn up the radio, change the channel, or flip to another CD without leaving your chair; cook without chopping or stirring; and then, utterly bereft of physical motion, sculpt your body shape "without strenuous time-consuming exercise!" as a recent TV commercial enthuses. There are few bodily functions that haven't been enhanced/displaced by technologies of convenience, whose advantages are promoted daily across the media. As a widely promoted value in contemporary culture, convenience translates as two interrelated advances: something that reduces the time it takes to perform a task, and something that permits us not to use our bodies, to substitute what McLuhan calls "extensions" or "amplifications" of the body in their place. Driven by the allure and escalating social necessity of convenience (for time is money), North Americans increasingly find physical endeavor too inconvenient to undertake—unless done *for its own sake*, in pursuit of expression or a better body, that is to say, with expressive intent, like art. Transcending the body through technological improvement redefines its achievements as limits, and it offers endlessly renewed marks of privilege and power. Seeking power, we empower our extensions. The paradoxes of this translation return inevitably to haunt its subjects.

Although instruments have changed over the centuries, the fascination with them has remained attached to a continuous rhetoric of improvement and delight. Technological innovations in musical performance technologies promise to endow their owners with the marvels of up-to-date knowledge and technical prowess. They deliver the convenience of simulated performance combined with an enhanced experience of individual expression. They offer an ever-enhanced blurring between human and mechanical agency; the enduring magic of sound reproduction; and the social status arising from being in possession of these wonders in their newest form. In sum, they promise to enhance the musician's sense of personal power through a constant renovation of sound technologies, and aim thereby to counteract the rather less empowering frustration of having to cope with their rapid obsolescence and the imperative of buying new ones.

The influential campaign to enhance pianos with electronic playback mechanisms was part of the construction of a consumer culture whose import goes well beyond the animation of consumerism as an economic practice. For it is embedded in, and serves to persuade its readers and listeners, of these twin ideas: saving time is a universal good, and physical effort is best avoided. Happiness is redefined as access to abstraction from time and the human body.

In "A Plea for Time," Innis describes this as the "modern obsession with present-mindedness."[69] If something *took your time* (even to think about), this was simply a problem in search of a solution. This characterization of time is an apt context for looking at the mechanization of piano performance. It was the struggle for time that fueled the social embrace of technological innovation. It is the detritus of this quest that inspires the increasingly popular preoccupation with so-called slow food, craft, yoga, and piano lessons, by means of which many people are trying to get in touch with a sense of embodied time. This is one contemporary purpose of "residual media," whose lasting popularity may help us think through and balance the achievements of the new.

Conclusion

Although instruments have changed over the centuries, the fascination with them has remained attached to a continuous rhetoric of improvement and delight. Technological innovations in musical performance technologies promise to endow their owners with the marvels of up-to-date knowledge and technical prowess. They deliver the convenience of simulated performance combined with an enhanced experience of individual expression. They offer the enduring magic of sound reproduction, an ever-enhanced blurring and reciprocal mimicry between human and mechanical agency, and the social status arising from being in possession of these wonders in their newest form. They promise to enhance the musician's sense of personal power through a constant renovation of sound technologies, and thereby to counteract the frustration of having to cope with rapid obsolescence and the imperative of having to research and invest in new hardware, new software, and the challenges of learning new techniques. The time once demanded by traditional musical skill acquisition has not disappeared, in other words; it has been transferred to another kind of *dispositif*, as we see with our predominantly male friends locked in front of the computer screen in fierce engagement with new programs and old ambitions, momentarily oblivious to time.

The question of the acceleration of time takes us to the boundaries of what we know about the "natural" rhythms of the human organism. As Richard Beardsworth notes in the context of a debate with Derrida about humans and other animals, "contemporary technoscience . . . constitutes an unprecedented speeding up of the dynamic relation between the human and the technical that 'risks reducing the difference of time, or the aporia of time' . . . to an experience of time that forgets time" (cited in Wolfe 2003, 75). This is an apt description of the musical-digital interface exacerbating what Cary Wolfe describes, in his discussion of this debate, as a radical asynchronicity of time among humans and other species.

If art really can, as McLuhan suggests, teach us about our environment through which we swim like oblivious fish, then new musical practices will draw our attention to these processes in some way. They will illuminate the tension between old and new techniques and the new possibilities that arise from them. They will remind us of music's ability to evoke history and memory, changing configurations of the human body, and the social production of time and space. Such technological changes have created not only alienation but also freedom; the freedom, as Steve Talbott eloquently suggests, "to form new hypotheses, to see things from fresh, previously unimagined perspectives, and even to consider how we ourselves might contribute to the future evolution of things."[70]

Notes

1. See Paul Theberge, *Any Sound You Can Imagine: Making Music/Consuming Technology* (Hanover and London: Wesleyan University Press, 1997).
2. Wayne Kelly, *Downright Upright: A History of the Canadian Piano Industry* (Toronto: Natural Heritage/Natural History, 1991), 22–23.
3. Ibid., 21.
4. Ibid., 24–25, 27.
5. Ibid., 7, 14.
6. Clifford Ford, *Canada's Music: A Historical Survey* (Agincourt: GLC, 1982), 59.
7. Kelly, *Downright Upright*, 14.
8. Ford, *Canada's Music*, 82.
9. Arthur Loesser, *Men, Women and Pianos: A Social History* (New York: Dover Publications, 1954).
10. Linda McDowell, *Gender, Identity and Place: Understanding Feminist Geographies* (Minneapolis: University of Minnesota Press, 1999), 161.
11. Ford, *Canada's Music*, 59.
12. Ibid., 55.
13. Isabella Crawford, *Selected Stories of Isabella Valancy Crawford*, ed. Penny Petrone (Ottawa: University of Ottawa Press, 1975), 22.
14. Arnold Walter, *Aspects of Music in Canada* (Toronto: University of Toronto Press 1969), 251. Cited in Clifford Ford, *Canada's Music*, 10.
15. Ibid., 35.

16. Keith Negus, *Producing Pop: Culture and Conflict in the Popular Music Industry* (London: E. Arnold, 1992). See also Paul du Gay, ed., *The Production of Culture/Cultures of Production* (London: Sage/Open University, 1997), 114.

17. Stan Godlovitch, *Musical Performance: A Philosophical Study* (London and New York: Routledge, 1998), 43.

18. Max Weber, *The Rational and Social Foundations of Music*, trans. Don Martindale, Johannes Riedel, and Gertrude Neuwerth (Southern Illinois University Press, 1958), 120.

19. Ibid., 119.

20. Arthur W.J.G. Ord-Hume, *Pianola: the History of the Self-Playing Piano* (London and Boston: George Allen and Unwin, 1984), 9.

21. Ibid., 10.

22. Weber, *Rational and Social Foundations*, 120.

23. Loesser, *Men, Women and Pianos*, 425.

24. Craig H. Roell, *The Piano in America, 1890–1940* (Chapel Hill: University of North Carolina Press, 1989), 57.

25. Ibid.

26. Weber notes that "Orchestra works were made accessible for home use only in the form of piano transcriptions. . . . The unshakable modern position of the piano rests upon the universality of its usefulness for domestic appropriation of almost all treasures of music literature, upon the immeasurable fullness of its own literature and finally on its quality as a universal accompanying and schooling instrument" (*Rational and Social Foundations*, 122–23).

27. Roell, *The Piano in America*, 177.

28. Ibid., 178–80.

29. Ibid., 177; emphasis in original.

30. Penny Sparke, *As Long as It's Pink: The Sexual Politics of Taste* (London: Pandora, 1995), 122–23.

31. Ibid., 123, 125.

32. Nicholas Rose, "Governing "Advanced" Liberal Democracies," in *Foucault and Political Reason: Liberalism, Neoliberalism and Rationalities of Government*, eds. A. Barry, T. Osborne, and N. Rose (University of Chicago Press, 1996), 40.

33. Ord-Hume, *Pianola*, 102.

34. "It is the peculiar nature of the piano to be a middle-class home instrument," Weber states (*Rational and Social Foundations*, 124).

35. From the 1901 advertisement "The Pianola: Its Mission," reprinted in Roell, *The Piano in America*, 111.

36. Roell, *The Piano in America*, 155.

37. From the 1904 advertisement "The Present of a Pianola Is a Present to *Every* Member of the Family." Reprinted in Roell, *The Piano in America*, 110.

38. A Dr. Busby, writing of the new "Cylindrichord" as early as 1825, explained that, "In small or family parties, where dancing to the music of the pianoforte is practised, a person totally unacquainted with music, a child or a servant, may perform, in the very best and most correct style, quadrilles, waltzes, minuets . . . or indeed any piece of music, however difficult" (Ord-Hume, *Pianola*, 18).

39. Theberge, *Any Sound*, 22.

40. Roell, *The Piano in America*, 41.

41. Ibid., 52.

42. Michael Chanan, *Repeated Takes: A Short History of Recording and its Effects on Music* (London, New York: Verso, 1995), 32.

43. Lisa Gitelman, "Music, Materiality and the Measure of the Digital; or, The Case of Sheet Music and the Problem of Piano Rolls," in *Memory Bytes: History, Technology, and Digital Culture*, eds. Lauren Rabinovitz and Abraham Geil (Durham, N.C.: Duke University Press, 2004), 204.

44. Ibid.

45. Timothy Day, *A Century of Recorded Music: Listening to Musical History* (New Haven and London: Yale University Press, 2000), 10.

46. Chanan, *Repeated Takes*, 30.

47. Loesser, *Men, Women and Pianos*, 548.

48. Ord-Hume, *Pianola*, 46.

49. See Theberge's study of popular musicians and musicians' magazines in *Any Sound You Can Imagine*.

50. Peter Auslander, *Liveness: Performance in a Mediatized Culture* (London and New York: Routledge, 1999), addresses whether the performances we see are live or mediatized, and he critiques the ontological preference for live performance. Auslander addresses the simulation of liveness from the vantage point of the audience, arguing rightly that rock music is mainly recorded music and should not be judged through the optics of live performance. But he does not address the changing situation of the performer or producer in the recording context. As Gitelman observes, "The specifics of materiality continue to *matter* much more to authors, to publishers, to 'labels'— that is, to potential owners—than they ever can, could, or will to listeners" ("Media, Materiality and the Measure of the Digital," 214). And to musicians, too.

51. Roell, *The Piano in America*, 194–95.

52. Ibid., 157. The Gulbransen Player-Piano ads with the famous Gulbransen Baby trademark are illustrated, 116–17 (dated 1921).

53. Roell, *The Piano in America*, 58. The Gulbransen ads are reproduced, 116–17, discussed, 156–57: "An eager public was bombarded with advertisements that emphasized how little effort it took to play automatic pianos.... Gulbransen advertising poignantly illustrates the contradictory ideology of promoting ease of play while espousing the individual creativity traditionally associated with the producer ethic."

54. John Philip Sousa, "The Menace of Mechanical Music," 1906, cited in Gitelman, "Media, Materiality," 202.

55. Ord-Hume, *Pianola*, 28–29.

56. Ibid.

57. Ibid.

58. Kelly, *Downright Upright*, 129, 126.

59. Ford, *Canada's Music*, 107, 183.

60. These numbers are drawn from Kelly, *Downright Upright*, 129, 126.

61. See J. Berland, *North of Empire* (Duke University Press, forthcoming) for a fuller discussion of cultural technologies and social change.

62. Langdon Winner, *Autonomous Technology: Technics Out-of-Control as a Theme in Political Thought* (Cambridge, Mass.: MIT Press, 1986). See Jody Berland, "Mapping Space: Imaging Technologies and the Planetary Body," in *Technoscience and Cyberculture*, eds. Stanley Aronowitz, Barbara Martinson, and Michael Menser (New York: Routledge, 1996), for a discussion of the "reverse adaptation" of technology in a different context.

63. Bertolt Brecht, "Radio as a Means of Communication: A Talk on the Function of Radio," trans. Stuart Hood, *Screen* 20, nos. 3/4 (1979–80): 24. Cited in Dan Lander, "Radiocastings: Musings on Radio and Art," in *Radio Rethink: Art, Sound and Transmission*, eds. Daina Augaitis and Dan Lander (Banff, Alberta: Banff Centre for the Arts, 1994), 19.

64. Lander, "Radiocastings," 11–12, also cites Friedrich Kittler, "The History of Communication Media," *On Line; Kunst im Netz/ Art in the Network,* Graz 1992, 76–77.

65. Marshall and Eric McLuhan, *Laws of Media: The New Science* (Toronto: University of Toronto Press, 1988), 7.

66. Ibid., 158–59.

67. Ibid., 148.

68. "You push the button, we do the rest" This famous Kodak advertisement first appeared in 1888. For discussions of its relevance to the history of music recording, see Michael Chanan, *Repeated Takes: A Short History of Recording and its Effects on Music* (London and New York: Verso, 1995), 27, and Patrice Flichy, *Dynamics of Modern Communication: The Shaping and Impact of New Communications Technologies* (London: Sage, 1995).

69. Harold Innis, "A Plea for Time," in Harold Innis, *Staples, Markets, and Cultural Change, Selected Essays*, ed. Daniel Drache (Montreal and Kingston: McGill Queen's University Press, 1995), 366, 368.

70. Steve Talbot, "Technology, Alienation and Freedom: On the Virtues of Abstraction," *Netfuture: Technology and Human Responsibility* 134 (July 18, 2002), http://www.netfuture.org/.

17 Mississippi MSS: Twain, Typing, and the Moving Panorama of Literary Production

Lisa Gitelman

I.

In a late sketch entitled "The First Writing-Machines," Mark Twain claimed that he had been "the first person in the world to apply the type-machine to literature."[1] He meant that he had been the first literary author to deliver typescript copy to an editor and publisher. Although Twain himself remembered the title in question as *The Adventures of Tom Sawyer*, it was actually *Life on the Mississippi* (1883) that earned him the distinction. Part travelogue, part memoir, part fiction, and part kitchen sink, *Life on the Mississippi* is partly about the residual media of authorship. Its kernel is a work called "Old Times on the Mississippi," which William Dean Howells published in installments in the *Atlantic Monthly* in 1875. The "Old Times" articles follow the progress of a "cub pilot," who has to study the river for what are to him "utterly invisible marks," until he knows them "just like ABC." When it came time to submit manuscript (MS) for *Life on the Mississippi*, Twain sent corrected tear sheets of his *Atlantic* articles as Chapters 4 through 17, and for everything else he appears to have sent eight "batches" of typescript, now lost.[2]

Genetic criticism is common coin in Twain studies, and along their way scholars have explained the meaning of Twain's missing typescript in some curious fashions. One critic has conjectured that Twain had the MS typed "in order that his own clean [holograph] manuscript might go to a friend and collector" who wanted it. Another offers the explanation that having the MS typed allowed Twain "to hide its blemishes, namely the use of source materials, . . . frequently [included] without proper acknowledgement."[3] Friedrich Kittler, who apparently considered only the "First Writing-Machines" sketch, suggests that Twain adopted typing "for entertainment purposes."[4] None of these explanations seems plausible to me, and this chapter pursues the issue further, more broadly, in readings of Twain's work against and within what I hope to characterize as the visual culture of textual production from which his work emerged. *Life on the Mississippi* and the contexts of its authorship and publication help show that typescript may be described as one element within the total visual economy of Twain's day, a visual economy importantly characterized by new media as well as by old familiar forms.

"Visual culture" or "visual economy" and literary criticism may seem like unlikely partners, but the Mississippi MSS suggest ways to connect American authorship and the practices of authoring to the concerns of visual critics like Leo Charney and Vanessa R. Schwartz or like Stephan Oettermann. Charney and Schwartz argue in their collected volume *Cinema and the Invention of Modern Life* (1995) that cinema, by the time it arrived a decade or so after *Life on the Mississippi*, was "both inevitable and redundant." Likewise Oettermann, in his history of the panorama—invented a century before—argues that the panorama emerged as a perfect reflection of its times, becoming "obsolete with its first appearance" and concealing its "anachronism by reproducing itself in countless and seemingly new variations" over the course of the nineteenth century.[5] For scholars such as these, visual media are fully produced by as well as productive of modernity, each medium emerging as an exact expression of its time, whether the liberalism of 1787, in the case of the panorama, or the "active synthesis" of the technological, cultural, and economic conditions that created the potent "epistemological pressure" of the 1890s, in the case of cinema. Added together, panoramas and motion pictures emerged as partial and varied embodiments of a modern subjectivity. They are the commodity visions of a new "unstable attentive subject," as Jonathan Crary puts it—a subject who, in the course of the nineteenth century, became increasingly "competent to be both a consumer and an agent in the synthesis of a proliferating diversity of 'reality effects,'" and who would become, in the twentieth century, "the object of all of the proliferating demands and enticements of technological culture."[6]

This narrative might well suggest that *Life on the Mississippi* partakes of a similar modernity, written and published as it was amid the long shadows of the panorama and on the eve of cinema. Yet it is striking how few instances there are in which modern texts, whether literary or vernacular, have been attended by their critics as elements of visual culture. We may look at the pictures and judge a book by its cover, but we rarely consider its linguistic content as having the fixed, visible form it happens to have in the edition we encounter. The visual culture of modern texts, to the extent there is one, tends to remain the narrow and arcane purview of descriptive bibliographers, the erstwhile publishers of facsimile editions, and professional type and book designers.[7] Rare indeed are textual critics who forward an argument like Charney, Schwartz, or Oettermann, connecting the visual form of a modern text with the epistemic climate of its production. (The fields of book history and media studies offer wonderful exceptions to the rule: I think of Megan Benton's work on fine printing in the 1920s and 1930s.[8]) Even critics who set their sights on authors like Dante Rossetti, with oeuvres "executed in two different media," graphic and linguistic, have been hard-pressed to cope with or even—given the economics of publishing—to encounter the two together, with the probable result that the poems and pictures remain relatively distinct objects of analysis.[9]

The great puzzle of this is that although texts are so seldom attended as visual forms, the work of authoring them—like editing, printing, and publishing as well as reading them—is resolutely (although of course not exclusively) visual. Consider authorship today. Whether you are the author of a novel, an office memo, an academic article, or an architectural drawing, the work of authoring that text is more than likely a screen practice, something you do at a keyboard while looking at a monitor. And you probably expect to check, preserve, annotate, recheck, and possibly circulate your authored work by somehow applying toner particles to paper, even if your ultimate publication will take a different, fixed form. What is more, variations within screen practices give us a varied experience of text as a visual form. Is it a raw .txt file? Is it a word-processed .doc file? An Adobe PDF file? A page in HTML? Does it have a scalable font? Is it searchable? Is it read-only, or can it be altered? Are its intertextual links hot or cold? In each case, the screen image of the text being authored resembles to a greater or a lesser extent—in appearance, in plasticity—the toner-and-office-paper version as well as any final, fixed form. Even the most rudimentary familiarity with these different electronic file types, like any familiarity with scanners, offers us an experience that helps locate text as an element of visual culture, texts that are somehow more or less like pictures of texts, where picturing arrives incumbent with the sorts of questions about subjectivity that visual critics ponder.

What has any of this to do with *Life on the Mississippi*? Twain can hardly be thought of as a representative author or a representative American of his day. Nonetheless, Twain and his work can help index the visual culture of textual production at a particularly important moment, the moment when typewriters were about to become widely used in America and Europe and, coincidentally, when halftones were about to begin their displacement of older, more expensive forms of printed illustration. A new scene of writing "as keystrokes, spacing, and the automatics of discrete block letters" soon dominated the authorship of texts both literary and vernacular;[10] at the same time that halftones proved "a major turning point in the visual status of the fashion commodity."[11] My point is simply stated. Careful readings of Twain's work and the associated conditions of its production, broadly construed, offer a glimpse of a long neglected element of visual culture ca. 1883, the typescript. Neither Twain nor his book register much about the impending triumph of halftones (the more than three hundred illustrations in the first edition of *Life on the Mississippi* are simple cuts supplied by its publisher), but Twain's typing, particularly in light of the way his work thematizes flow, sequence, letters, and legibility, gestures toward an awareness that resembles, in kind if not degree, the awareness that attends the screen practices of authorship today. Put another way, readers of certain ages and occupations will recall what their first encounter with a PDF was like ("My cursor just turned into a hand!"). Typescript was something like Twain's PDF.

II.

There are no typewriters and no typists in *Life on the Mississippi*, but there are plenty of authors trying to represent the river. As he had struggled to finish the final installments of "Old Times on the Mississippi" in 1875, Twain had complained to William Dean Howells that the world he was trying to write about did not lend itself easily to segmentation: "There is a world of river stuff to write about, but I find it won't cut up into chapters worth a cent. It needs to run right along, with no breaks but imaginary ones."[12] The finished work would have to have "imaginary" breaks because the actual river was so plentiful and continuous. Twain's complaint is mirrored by the problem facing the cub pilot in installments already published. Instructed by his senior to "get a little memorandum-book" and take note of what the experienced pilot tells him, the cub soon has a notebook full of details about the river. Although elsewhere quick to enjoy a sense of self-importance, the cub pilot here realizes with humility that even a full book and a completed one-way journey have left him with "a long four-hour gap" in his notes for every four hours when he was off watch sleeping.[13] Twain's imaginary breaks in the text are echoed by the pilot's literal ones; pauses in his labor create gaps—not of imagination but of attention—in his pages.

Later, when Twain began to extend "Old Times" into a book, his imaginary breaks had to be reimagined. In 1882, he turned seven installments into thirteen chapters, building a new whole out of newly imagined parts and adding forty-seven chapters and an appendix in the process. He created two more authorial figures to inhabit the book version, the first a Twain persona, the traveler and narrator who works from his own notes taken on the river, and the second a lecturer who has toured with a moving panorama of the river and who is on hiatus while he helps write a tourist's guidebook of the Mississippi. The unnamed lecturer can describe the riparian scenery in such florid detail that the Twain narrator is able, jokingly, "to imagine such of it as we lost by the intrusion of night."[14]

Many readers have noted that the panorama lecturer allows Twain to revisit one of his favorite themes, the fulsome fakery of American culture, with its premium on salesmanship and sham. But the lecturer and the moving panorama are more important than that. Critic Curtis Dahl was probably too ambitious when he suggested long ago that the form of the moving panorama had seeped into Twain's psyche at an early age, so that it "influenced what he saw," and later "unconsciously" helped determine the "loose, inclusive, free, [and] variegated" structure of his published works.[15] But Dahl was onto something. The balance struck in *Life on the Mississippi* between the cub pilot's nighttime attention gaps at one end, and the lecturer's imaginary filling in at the other end, help suggest the relevance of the moving panorama to that work. The moving panorama in *Life* forms part of Twain's response to and revision of "Old Times"; it introduces a conventional element of nineteenth-century visual culture and connects it by juxtaposition to the subjective vision of the cub pilot toiling over his little memorandum book. The moving panorama has less to do with the shape of the text (*pace* Dahl) than it does with the shape of the river, not the actual river either, but rather the "eluding and ungraspable" alphabetic object that the cub must learn.[16]

The moving panorama was a distinctively American obsolescence—to follow Oettermann—of the panorama form. The original panorama had been a sort of amusement panopticon, a 360-degree landscape painting invented by Englishman Robert Barker in the same year that Bentham conceived his 360-degree model prison.[17] Observers viewed the painting from a central platform and were encouraged to feel immersed or "entrapped in the real," an effect enhanced by music and lighting as well as by three-dimensional scenic elements that helped tie the painting to the platform.[18] But in the United States the most successful and widely viewed panoramas were moving panoramas, giant scrolls of painted canvas that were wound or unwound across an audience's field of vision as an evening's or an afternoon's entertainment. Unlike the 360-degree panoramas in their specially constructed sheds, the moving panorama had a traditional proscenium logic and could be toured around to any venue large enough for both painting and audience. Their reality effect was twofold. Audiences saw scenery pass before their eyes, giving them the effect of movement, and audiences saw realistic scenery for which vociferous claims of verisimilitude were made by the lecturer whose monologue accompanied the progress of the canvas.

Among the most widely known moving panoramas were five virtual trips along the Mississippi River that competed for audiences beginning in the late 1840s. Viewers headed up or down river, depending on which way the painting was wound during that exhibition. In either case, what they witnessed was supposed to be a view of the scenery along one bank of the river at a time. To the extent that the reality effect worked, then, travelers sailed up or down river with their heads cocked to one side, and they witnessed many miles of riverbank on hundreds of yards of canvas. The actual length of the paintings is difficult to determine because "All 'moving panoramas' boasted implausible, by all accounts fraudulent dimensions."[19] Banvard's famous "Mississippi from the Mouth of the Missouri to New Orleans" (1846) could not possibly have been 3 miles long, as its creator and promoter claimed. Yet he boasted that "Ten captains, ten pilots, and more than a hundred others" familiar with the Mississippi "testified before the mayor of Louisville that the picture was notable for its correctness to nature and truthfulness."[20] Something like a reduced Shakespeare play, the moving panorama succeed partly to the extent that its audience could take pleasure in experiencing the painting as both accurate and seamless, although any seeming seamlessness the moving panorama might possess must have jeopardized its verisimilitude: Even a three-mile painting could have little "correctness to nature" if the nature it represented were thousands of miles long.

The moving panorama worked by means of a dialectic established between imaginary gaps and attentive ones, between its sense of plenitude and continuity and the punctuated or rationalized labor that produced it. However modestly, the illusion of movement—in which seated audiences contradictorily enjoyed a sense of moving through space—suggested a continuous, mobile point of view. Yet in presenting recognizable features of the riparian landscape, the paintings had to rely on a sequence of separate views, each with its own stationary perspective attending a prominent geologic feature, a notable settlement, or a typical scene. A sense of continuity emerged partly to the degree that individual views could be integrated by dovetailing their margins and partly to the degree that the lecturer and his script might exert a kind of "rigid narrative" control.[21] But continuity emerged as well according to the suggestibility of the audience, particularly the ways in which its members might take pleasure in rejecting sequential points of view in favor of the suggested mobile one. No one was fooled; that wasn't the point. The varied pleasures of this reality effect must have arisen in both an appreciation of the "illusionistic virtuosity" of the painting and its mode of display,[22] and in the corresponding and potential enactment of a *willed* persistence of vision, the melting of one scene into another. These pleasures helped make moving panoramas into a popular form of entertainment and, arguably, an early "mass medium."

As such, the moving panorama became one ingredient of the emerging "logic of circulation" that would soon be applied to and adapted by cinema. Together with picture postcards, stereograph sets, illustrated travel lectures, and the still photography of exploring expeditions, the moving panoramas helped put images of circulation into circulation as commodities.[23] After the panorama had thus broached the question of representing mobile points of view, it was in one respect a small step from the "moving" of moving panoramas to the "motion" of motion pictures, to the extent that panoramas proved one popular subgenre of the early actuality films. There were river panoramas like the two Edison films of 1903 that together offer, "with only a minor break in continuity," an entire "sweep around the southern tip of Manhattan." And other gimmicks abounded. In films like *Panorama of the Golden Gate* (Edison, 1902) the camera plunges through space on the front of moving train. In *Panorama of the Moving Boardwalk* (Edison, 1900), taken at the Pan-American Exposition, the still camera is itself panned in the foreground by a popular midway attraction. Indeed, cinema might have facilitated a more complex, more heterogeneous and immersive sense of mobile perspective, but its early technical and formal vocabulary of frames, cuts, and continuities variously drew on a collapsed distinction between imaginary breaks and attention gaps, the same collapse toward which the moving panorama had already gestured and one that helped signal what Mary Ann Doane identifies and explores as "the emergence of cinematic time."[24]

By juxtaposing the cub pilot's mission in "Old Times" with the figure of the moving panorama in *Life on the Mississippi*, Twain distinguishes the complexity of the former while raising the problem of representing the cub's point of view. The pilot's threefold task must take him well beyond the precinematic complexity of the panorama. The cub must learn the shape of the river, he must learn the surface or "the face of the water," and he must learn its soundings. He has to learn the river as well as he knows his hallway at home, in the dark, and to learn it so absolutely, his senior tells him, that he can "always steer by the shape that's *in your head*, never mind the one that's before your eyes."[25] He must forsake reality effects at the same time that he must reject any positivist model of correspondence between visual stimuli and optical response in favor of a whole new phenomenology. For him, unlike the observers of the moving panoramas, "Nothing ever had the same shape when I was coming down-stream that it had borne when I went up." Worse, "No prominent hill would stick to its shape long enough for me to make up my mind what its form really was," he observes, "but it was as dissolving and changeful as if it had been a mountain of butter in the hottest corner of the tropics." He notices that he can fix on some prominent feature, "laboriously photographing its shape upon my brain," but just as this manufactured reality effect begins to succeed, he "would draw up toward it and the exasperating thing would begin to melt away and fold back into the bank!"[26]

It would be anachronistic to call this mobile perspective "cinematic." Instead, in "Old Times" and then in *Life*, the pilot's new phenomenology repeatedly involves a spatial alphabet. The cub's senior explains that if the shapes of the banks, "didn't change every three seconds they wouldn't be of any use":

> Take this place where we are now, for instance. As long as that hill over yonder is only one hill, I can boom right along the way I'm going; but the moment it splits at the top an forms a V, I know I've got to scratch to starboard in a hurry, or I'll bang this boat's brains out against a rock; and then the moment one of the prongs of the V swings behind the other, I've got to waltz to larboard again, or I'll have a misunderstanding with a snag.[27]

Soon the cub pilot comes to know the shores of the river "as familiarly as I knew the letters of the alphabet." He can read the face of the water for any dimple that becomes to him "an *italicized* passage" and "a legend of the largest capitals." The spatial alphabet is dynamic as well as aural. The water's surface becomes a book, "which told its mind to me without reserve, delivering its most cherished secrets as clearly as if it had uttered them with a voice."[28] Its soundings are accumulated by the pilot like an arbitrary sequence of letters sounded out. The narrator challenges his reader,

> If you were walking and talking with a friend, and another friend at your side kept up a monotonous repetition of the vowel sound A, for a couple of blocks, and then in the midst interjected an R, thus, A, A, A, A, A, R, A, A, A, etc., and gave the R no emphasis, you would not be able to state, two or three weeks afterward, that the R had been put in, nor be able to tell what objects you were passing at the moment it was done.

By contrast, the talented pilot has effortlessly and *"unconsciously"* "photographed" his bearings.[29] Twain's distinctive *flânerie* depends on the pilot's movement through and absorption in a changeable landscape of letters, letters in random—rather than alphabetic or word-forming—sequences, letters that move and monotonously repeat, letters with shapes, sounds, names, and spatial relations that exist at the limits of consciousness, intentionally learned but viscerally known.

After the pleasures of youthful employ involving the special expertise required of the pilot, the Twain narrator of *Life on the Mississippi* remarks wistfully,

"I became a scribbler of books, and an immovable fixture among the other rocks of New England."[30] Instead of moving through scenery, he became scenery itself, and soon the rock hounds—the autograph hounds—started to seek him out for his celebrity, not his spatial literacy or expertise. As if in comic self-critique, the older Twain heard no R's at all amid the A's when he returned to the river in 1882: "The educated Southerner has no use for an *r*, except at the beginning of a word," he notes. Expressions like "befo' the waw," "may lack charm to the eye, in print, but they have it to the ear."[31] The spatial literacies required by authors and printers are much different than those required of a riverboat pilot. In the course of becoming a celebrity, Twain learned first the spatial organization of letters in the California case. (That is, he learned to set type.) Later he learned the spatial organization of letters on the Remington typewriter keyboard.

III.

Before Twain began to write *Life on the Mississippi* in 1882, he traveled back to the river for a month to collect material. He took with him a notebook, already partly filled, plus an insurance company stenographer named Roswell Phelps. Joining them was James Ripley Osgood, Twain's close friend and current publisher.

Osgood was an affable companion and a practiced traveler. As a young man he had briefly been a partner in the esteemed New England publishing house of Ticknor and Fields. In that capacity he had managed Charles Dickens's American lecture tour of 1867–68 and had even traveled part of the time with the celebrated Victorian novelist. Never much of a success in business, Osgood now tied his fortunes to Twain's. The two had just published *The Prince and the Pauper* (1881) with disappointing results, but their hopes ran high for *Life*. Osgood also shared many interests with Twain that were not financial. For instance, in February 1880, Twain purchased four fifths of the patent for the "kaolotype" process. It was an engraving process that Twain hoped to have adapted into a process for stamping the designs onto book bindings. He was swindled. Likewise, Osgood had purchased the American rights to the "heliotype" process in September 1872. It was a photochemical process whereby a gelatin print was made into a printing matrix, so that more than a thousand impressions of any photograph could be struck. Unlike the kaolotype venture, Osgood's American Heliotype Company actually enjoyed some promise and even, for a short while, cash flow. It offered the first practical—if cumbersome—means of using "photography in the printing-press," according to its inventor, Ernest Edwards.[32] As such, the heliotype process helped identify a niche that it didn't or couldn't quite fill, one which halftones and later offset processes would enter into and make their own.

As they stood on the deck of a Mississippi riverboat in 1882, the disappointed American owners of kaolotypy and heliotypy shared a certain naivete in business, they shared a wide-eyed enthusiasm for new technology, and they shared a shifting sense of the distance between world and page. That is, they shared a necessary and professional absorption in the processes of textual production as well as an awareness that those processes were shifting and changing around them as certainly as the scenery. Heliotypes briefly called into question the representational qualities of book illustrations, and Twain raised a related question when he drew on his notes in *Life on the Mississippi* and when he reflected on typing and typescripts.

In *The Heliotype Process* by Ernest Edwards, James R. Osgood & Company published an explanation of the heliotype and offered twenty-eight lavish examples of the uses to which the process might be put. These included a variety of illustrations "copied" by heliotype, "printed" from different formats, and taken "from nature," "from life," and "from the objects." Each descriptive caption seems to shift the mimetic claims being made according to the subject being reproduced. If the heliotype was of an engraving, a woodcut, drawing, or sketch, then it "copied" its subject. If the heliotype was of a lithograph or an electrotype, then it "printed" its subject. And if the heliotype presented a portrait, an idyllic scene, or a couple of boots, then it did so directly from life, from nature, or "from the objects" (Figure 17.1). Yet in every case the heliotype process was the same: A photograph was taken and the resulting negative used to create a hard gelatin "printing film," which was then mounted on a metal support and used in a printing press, moistened and inked between impressions. However the captions to heliotype illustrations might present it, the pictures all stood in precisely the same relationship to "the objects."[33] Only the objects themselves varied in their

FIGURE 17.1 *"From the Objects." This is a halftone (printed in a book) of a photograph (courtesy of the Library of Congress) of a heliotype (James R. Osgood & Company).*

relationships to prior (in the double sense of previously available and profilmic) processes of reproduction.

Edwards noted that "the Heliotype is not an *originating* process," that "it cannot *originate*, it cannot *idealise* [sic],—it can only *realise* [sic]."[34] The very same assumptions lay behind Twain's engagement of Roswell Phelps the stenographer. The author would originate and then the stenographer would realize his original thoughts on paper. This was the simple instrumental logic behind shorthand reporting, which was then widely used in business and government. At different times throughout his writing career, Twain embraced and rejected dictation as a means of composition. Somewhat like Edwards making distinctions between heliotypes of one thing and heliotypes of another, Twain found some kinds of composition were better to dictate than others. The distinctions he drew between forms—autobiography versus fiction, for instance—tended to belie the simple instrumental logic he and others so obviously assumed for stenography and shorthand reports. The instrumentality of shorthand could hardly be simple, in other words, if it worked better for business correspondence and the self-representations of autobiography than for other things. For accumulating the occasional thoughts of a traveler returning to old haunts, it was apparently just the ticket.

Chapter 22 of *Life on the Mississippi* ("I Return to My Muttons") jumps twenty-one years forward in time, leaving the Mississippi of the narrator's youth and picking up the account of his return trip in 1882. As a transitional device, in this chapter the narrator draws self-consciously on his notes. But the notes Twain adapted for inclusion are the narrator's or "his" in the sense that Twain must have spoken them to Phelps, who must have taken them down in shorthand and then later transcribed them into a notebook, still extant, to which Twain referred. Twain's words in Phelps's hand offer, for instance, "Down here we are in the region of boots again," and the final version—written by Twain, typed for him, corrected by him, printed for and published by Osgood—runs, "Next, boots began to appear. Not in strong force, however. Later—away down the Mississippi—they became the rule."[35] The process of composition, like the larger process of textual production of which it was part, involved Twain in what John Matson has called "the specular economy" within which Twain's words, *as* Twain's, were experienced by him and became present to his readers.[36]

Twain's literal ownership of his words was hardly a banal point: *Life on the Mississippi* is probably unique to literature as a work bearing both author's copyright and trademark notices behind its title page.[37] But of equal moment was Twain's ownership in the sense that his writing might offer visible, material evidence of himself, his originality, his character or personality, with his unique imprint or signature. Chirography proved the point: His handwriting was his, an index of him, almost as certain as a fingerprint. Transcribed shorthand notes were less sure. The notes were his but in a different and much more complicated sense, one that clearly fascinated him as he dabbled with shorthand himself and perennially strove to represent "talk" on the printed page. And typescripts turned the problem over again.

In his "First Writing-Machines" sketch, Twain recalls that the first letter he ever dictated to a typist "was to Edward Bok, who was a boy at the time." Bok became famous as the entrepreneurial editor of the *Ladies' Home Journal*, and he had early in life gained notice as a successful collector of autographs. As Bok himself remembered it, he "was always interested in the manner in which personality was expressed in letters," so he sought "what were really personality letters," not autographs, from "the most famous men and women of the day."[38] Bok was making a distinction between one kind of letter and another. Autograph seekers looked for celebrity identity in the unique form of the individual characters that spelled a signature on the page. "Personality letters" allowed him to find celebrity identity in the characteristic style and content of an entire epistle. The distinction is face saving for Bok because Twain sent his letter typed, "*signature and all*," a fact that Bok neglects to mention. In "The First Writing-Machines" Twain believes that he had denied Bok what he wanted by sending a typescript, whereas in *The Americanization of Edward Bok*, Bok believes that he had succeeded in obtaining Twain's "personality letter."

Although Twain did not see his own hand in the typescript characters he applied to paper, typing nonetheless became an important and self-affirming step in his authorial process. Accustomed to handwrite manuscript, revise it, have it copied, correct the copy, have it typeset, and correct the proofs, Twain now had his copy typed, remaking it as his:

> After one has read a chapter or two of his literature in the type-writer character, the pages of the sheets begin to look natural, and rational, and as void of offence to his eye as do his own written pages; therefore he can alter and amend them with comfort and facility; but this is never the case with a book copied by pen. The pen pages [done by a copyist] have a foreign and unsympathetic look, and this they never lose. One cannot recognize himself in them. The emending and revamping of one's literature in this form is as barren of interest, and indeed as repell[e]nt, as if it were the literature of a stranger and an enemy.[39]

Because the hand of the copyist is effaced or eradicated by the hands of the typist and the typewriter, Twain's copy—the *thing* that all authors submit to press—becomes newly his own, a site for self-recognition, or, as Matson says, "an epistemologically stable space in which Twain can be seen and verified as himself."[40] The typescript copy begins "to look natural, and rational" because it looks more like—and looks mechanically produced like—the printed sheets of a finished book. Yet it remains plastic inasmuch as it forms an intermediary, a set of cleanly rendered instructions for the printer that can still be altered and amended with ease. Very like the screen practices of authorship today.

IV.

Any allusion to the screen practices of authoring today can offer only partial assistance in the task of locating the visual culture that produced and was in turn produced by typescripts in Twain's day. If, as visual critics assert, visual media emerge as quintessential expressions of their times, only the screen practices of the 1880s will do. One suggestive contemporary context for Twain's typescript was the work of American psychologist James McKeen Cattell. Cattell is best known for importing German experimental techniques to the United States. As a graduate student, Cattell studied with the important pioneer in experimental psychology, Wilhelm Wundt, at his lab in Leipzig. Cattell brought a Remington No. 4 with him to Leipzig, and he used the machine to write letters, to prepare copy for the press, and to conduct experiments. Before he began his research in October 1884, he typed home to his parents:

> This is my first letter on the new machine. It seems to work nicely, I have only practiced two or three hours and can already write quite rapidly, and with tolerable accuracy. I think it is easier than writing with the hand, and I shall soon be able to write about twice as fast, . . . One does not use the eyes at all, which of cours [sic] is a special advantage to me. It is further a great comfort to hav [sic] what one writes so thoroughly legible, especially if he is preparing a manuscript for the pres [sic].[41]

Like the philosopher Nietzsche a year or two before him, Cattell adopted his typewriter partly to reduce eyestrain, so he learned to type without looking.

My purpose in this brief conclusion is to hint that Cattell and Twain in some respects shared similar interests, despite the great disparity in their occupations. Suggesting any complementarity between them does nothing to forward a reading of *Life on the Mississippi*, of course, but it does gesture toward an underlying visual economy in which techniques of observation and problems of attention were applied to reading practices and, more importantly, in which reading practices helped in part consolidate the modern subject.[42] What it meant to *look* or to *view* became defined in part by what it meant to *read*, and did so specifically at a time when typescripts were calling new attention to processes of writing and aspects of readability.

In some of his experiments, Cattell used the typewriter to produce "numbers, letters, words, and sentences"; in others he relied on a set of "test-types," a pamphlet of type samples published explicitly for testing vision.[43] He cut up a homemade typescript or his test-types pamphlet and pasted the characters he needed into his experimental apparatus. In one version, the characters went onto a cylinder that rotated behind a screen. A slit in the screen revealed the letters as they passed, and the rate of rotation of the cylinder controlled the time they were exposed to view. In another version, the characters were put onto a card that stood behind a screen. In this setup the characters remained still while a slit in the screen fell across them like a guillotine blade, with the height of the slit controlling the exposure time. Subjects (usually Cattell and his laboratory assistants) were tested to determine the minimum exposure times required to read (that is, to identify) the characters they saw.

There were several questions at stake.[44] The first concerned the time it took for visual stimuli to produce an optical response. Unlike a simple vision test with test types or eye charts, a "no" or null answer for Cattell had a psychometric value. If subjects fail to see the big E on an eye chart, they do so because of physical, physiological impairment. But if subjects failed to see a letter flashed at them by a kymograph or tachistoscope, Cattell knew their eyes had seen it but that the nerve impulse produced was "too weak" to "excite sensation" in the brain. He could continue testing in increments, to address "the difficult subject of the threshold of sensation," or "what may be called the threshold of consciousness." A second question arose when it became apparent that subjects could recognize different letters, groups of letters, and groups of words at different rates. Groups of letters were more readily recognized when they comprised a word, and groups of words were more readily recognized when they made sense as a phrase or sentence, although neither remained the case if the words and phrases were in a language foreign to the subject. Moreover, it appeared that some fonts and some letters were more readable than others. Using lowercase letters, Cattell discovered a new alphabetical order, that of "distinctness": d k m q h b p w n l j t v z r o f n a x y e i g c s. And he lamented "the probable time wasted each day through a single letter as E being needlessly illegible."[45]

Cattell's lament about E is none too far from the cub pilot's anguish over his own "eluding and ungraspable" alphabet. Indeed, the psychologist's alphabet of distinctness, the pilot's dynamic spatial alphabet, and the typist's functional alphabet—QWERTYUIOP, and so on, on the Remington keyboard—each propose a new version of the ABCs. Each alphabet has a new alphabetical order, an organizing principle that works both in contrast to mere or habitual sequence—A, B, C,—as well as in concert with a profoundly new experience of text. Cattell's alphabet is based on the psychometrics of optical response. The pilot's is based on the changeable shape, surface, and depth of the river. And the typist's is based on a machinist's optimum, putting the keys in places where the corresponding type bars won't get jammed as they move in quick succession. Cattell's subjects can't trust their eyes because their nerves don't work fast enough. The pilot can't trust his eyes because the river incessantly plays tricks on them. And the typist doesn't look at all, except at the resulting typescript, where the authorial subject appears newly complicated and differently legible than in handwriting or in print. What these very different instances suggest together is a varied phenomenology of text, ca. 1883, an importantly new sense of letters and of texts as differently pictured in different states and conditions.

Notes

1. Mark Twain, "The First Writing-Machines," *The $30,000 Bequest and Other Stories* (New York: Harper & Brothers, 1906). The author would like to thank Charles Acland, Pat Crain, Matt Kirschenbaum, and John Matson for their helpful comments on earlier versions of this chapter.

2. Details on the genesis of *Life on the Mississippi* are drawn from Horst H. Kruse, *Mark Twain and "Life on the Mississippi"* (Amherst: University of Massachusetts Press, 1981), as well as from Judith Hale Crossett, *A Critical Edition of Mark Twain's Life on the Mississippi*, Ph.D. dissertation, University of Iowa, 1977. For criticism reading Twain's work as self-consciously concerned with writing, see Edgar J. Burde, "Mark Twain: The Writer as Pilot," *Publications of the Modern Language Association* 93 (1978): 878–92; and Edgar M. Branch, "Mark Twain: The Pilot and the Writer," *Mark Twain Journal* 23 (1985): 28–43; more recently, see Leland Krauth, *Proper Mark Twain*

(Athens: University of Georgia Press, 1999), chapter 4. In "Textual Apparatus," *A Connecticut Yankee in King Arthur's Court* (Berkeley: University of California Press, 1979), 569–827, Bernard L. Stein notes that *A Connecticut Yankee in King Arthur's Court* (1889) was the first MS Twain had typed from beginning to end.

3. Caroline Tucker and Horst H. Kruse, both from Kruse, *Mark Twain*, 137, note 7.

4. Friedrich Kittler, *Gramophone, Film, Typewriter*, trans. Geoffrey Winthrop-Young and Michael Wutz (Stanford, Calif.: Stanford University Press, 1999), 205.

5. Leo Charney and Vanessa R. Schwartz, eds., *Cinema and the Invention of Modern Life* (Berkeley: University of California Press, 1995), 10; and Stephan Oettermann, *The Panorama: History of a Mass Medium*, trans. Deborah Lucas Schneider (New York: Zone Books, 1997), 47.

6. Jonathan Crary, *Suspensions of Perception: Attention, Spectacle, and Modern Culture* (Cambridge, Mass.: MIT Press, 1999), 148.

7. Johanna Drucker makes this point neatly and has influenced my thinking here: "Poets, critics, scholars, teachers—all regularly and frequently overlook the bibliographical, graphical, and materially semiotic codes of the media of the printed texts on which they depend" (686); "Theory as Praxis: The Poetics of Electronic Textuality," *Modernism/Modernity* 9 (2002): 683–91. These codes may be overlooked but are everywhere effective, as Marshall McLuhan was among the first to recognize, in *The Gutenberg Galaxy: The Making of Typographic Man* (1962) and subsequent works; see *Essential McLuhan*, eds. Eric McLuhan and Frank Zingrone (New York: Basic Books, 1995).

8. Megan L. Benton, *Beauty and the Book: Fine Editions and Cultural Distinction in America* (New Haven: Yale University Press, 2000).

9. Jerome McGann, *Radiant Textuality: Literature after the World Wide Web* (New York: Palgrave, 2001), 13.

10. Kittler, *Gramophone*, 193.

11. Crary, *Suspensions of Perception*, 117.

12. Twain's letter is quoted in Crossett, "A Critical Edition," 718.

13. Mark Twain, *Life*, 87, 88–89. For the sake of convenience, my page references to both "Old Times" and *Life on the Mississippi* are from the electronic edition *Life on the Mississippi* (Chapel Hill: University of North Carolina, 1999). The electronic edition is a transcribed edition that recreates the pagination of the first edition of Twain's work (Boston: James R. Osgood & Company, 1883) and includes scans of its illustrations; the text was encoded by Jill Kuhn as part of *Documenting the American South*, a digitization project at the University of North Carolina, http://docsouth.unc.edu/twainlife/menu.html (2000).

14. Ibid., 579.

15. Curtis Dahl, "Mark Twain and the Moving Panoramas," *American Quarterly* 13 (1961): 20–32.

16. Twain, *Life*, 108.

17. The Bentham connection appears in Oettermann's *The Panorama*; Jonathan Crary notes that the panorama offered an inversion of the panopticon because it helped turn Bentham's subject-as-object "of attention and surveillance" into the modern, "attentive subject [who] is part of an *internalization* of disciplinary imperatives" (Crary, *Suspensions of Perception*, 73). The varieties of modern attention consolidated by such a subject may be gleaned from Walter Benjamin's catalog of panoramas: "There were panoramas, dioramas, cosmoramas, diaphanoramas, navaloramas, pleoramas (*pleō*, 'I sail,' 'I go by water'), fantoscope[s], fantasma-parastases, phantasmagorical and fantasmaparastatic *expériences*, picturesque journeys in a room, georamas; optical picturesques, cinéoramas, phanoramas, stereoramas, cycloramas, *panorama dramatique*"; Walter Benjamin, *The Arcades Project*, trans. Howard Eiland and Kevin McLaughlin (Cambridge, Mass.: Harvard University Press, 1999), 527.

18. Oliver Grau, *Virtual Art: From Illusion to Immersion*, trans. Gloria Custance (Cambridge, Mass.: MIT Press, 2003), 70.

19. Bernard Comment, *The Painted Panorama*, trans. Anne-Marie Glasheen (New York: Harry N. Abrams, 2000), 63.

20. Quoted in John Francis McDermott, *The Lost Panoramas of the Mississippi* (Chicago: University of Chicago Press, 1958), 39.

21. Martha A. Sandweiss, *Print the Legend: Photography and the American West* (New Haven: Yale University Press, 2002), 52.

22. P. Sternberger, quoted in Mary Ann Doane, *The Emergence of Cinematic Time: Modernity, Contingency, and the Archive* (Cambridge, Mass.: Harvard University Press, 2002), 154.

23. See Tom Gunning, "Tracing the Individual Body: Photography, Detectives, and the Early

Cinema," in Charney and Schwartz, *Cinema and the Invention of Modern Life*, 15–45, especially 16–17; see also Sandweiss, *Print the Legend*. Sandweiss explains the extent to which early daguerreotypists and photographers needed to adapt the narrative logic of moving panoramas to find audiences for their work. Before they succeeded, still photographs were meaningful primarily as a means to some more narrative end, and there was at least one moving panorama that was painted from a series of daguerreotypes. On the nineteenth-century experience of travel as itself panoramic, see Wolfgang Schivelbusch, *The Railway Journey: The Industrialization of Time and Space in the Nineteenth Century* (Berkeley: University of California Press, 1986), chapter 4.

24. These titles, catalog copy, and films are available through the *American Memory* Web site of the Library of Congress (1999), http://memory.loc.gov/ammem/edhtml/edhome.html. See Doane, *The Emergence of Cinematic Time*, especially chapter 5.

25. Twain, *Life*, 104.

26. Ibid., 108.

27. Ibid., 109.

28. Ibid., 118, 119.

29. Ibid., 154.

30. Ibid., 246.

31. Ibid., 449.

32. Ernest Edwards, *The Heliotype Process* (Boston: James R. Osgood & Company, 1876), 5; On Osgood, see Carl J. Weber, *The Rise and Fall of James Ripley Osgood* (Waterville, Me.: Colby College Press, 1959); on the heliotype, see Sandweiss, *Print the Legend*, 292–320.

33. See Sandweiss, *Print the Legend*, 311.

34. Edwards, *The Heliotype Process*, 11.

35. Twain, *Life*, 248–49. Phelps's notes in shorthand do not survive, but his transcription does. See *Mark Twain's Notebooks and Journals, Vol. III (1877–1883)*, eds. Frederick Anderson, Lin Salamo, and Bernard L. Stein (Berkeley: University of California Press, 1875), 522, 540. Notebook 20 (kept by Twain) and 21 (by Phelps) are described, transcribed, and annotated by the editors.

36. John Matson, "The Text That Wrote Itself: Identifying the Automated Subject in *Pudd'nhead Wilson and Those Extraordinary Twins*," unpublished MS, 2002, 24.

37. While writing *Life*, Twain was engaged in litigation (*Clemens v. Belford, Clark & Co.*, also known as the Mark Twain Case, 14 F. 728; 1883 U.S. App.) testing the efficacy of trademark law as a means of protecting the use of his pseudonym applied to his literary works.

38. Edward Bok, *The Americanization of Edward Bok*, 26th ed. (New York: Charles Scribner's Sons, 1923), 204.

39. Twain to H. M. Clark, April 24, 1883; quoted in Stein, "Textual Apparatus," 609.

40. Matson, "The Text That Wrote Itself," 16.

41. James McKeen Cattell, *An Education in Psychology: James McKeen Cattell's Journal and Letters from Germany and England, 1880–1888*, ed. Michael M. Sokal (Cambridge, Mass.: MIT Press, 1981), 124–25.

42. Jonathan Crary and Friedrich Kittler have already suggested as much, although Crary's interest in reading as a special form of observation is never explicit, and Kittler's is far from clear. For Kittler, Nietzsche's use of an index typewriter for a few weeks in 1882 "was a turning point in the organization of discourse." Friedrich Kittler, *Discourse Networks 1800/1900*, trans. Michael Metteer with Chris Cullens (Stanford: Stanford University Press, 1990), 193. I hope it is clear that I'm making no such overarching claims for Twain or Cattell.

43. Cattell, *An Education in Psychology*, 134–35, and "The Inertia of the Eye and Brain," *Brain* (1885): 295–312.

44. For an overview of Cattell's research program (only a bit of which had to do with reading), see his *An Education in Psychology*, 132–35, and "The Psychological Laboratory at Leipsic," *Mind* (1888): 37–51.

45. Cattell, "The Inertia of the Eye and Brain," 308, 309.

18 Streamlining the Eye: Speed Reading and the Revolution of Words, 1870–1940

Sue Currell

This chapter examines the development of new technologies that aimed to measure, and then enhance, the performance of the reader in response to a proliferation of mass media in the nineteenth and early twentieth centuries. In this sense, then, what is "residual" is not the media itself but the traces—or echoes—that are imprinted on the human body or within ideological conceptions of the human, even when the technology or media itself has disappeared or diminished. Despite this, these residual traces are constantly threatening to reappear in everyday culture, and even today they become emergent with the appearance of new technologies and media. Knowing the origins of these residual discourses concerning the human allows us to see their appearance less as a natural occurrence, one that emerges out of technological transitions, and more as a continuation of the historically determined past, revisited and recycled in the present. One example of this is clearly visible in recent concerns over the future of the book and the reader in a digital age, a concern that replicates and recycles earlier concerns that are examined in more detail in the following. In this way this study traces the residual in the present by exposing the concerns of the dominant in the past.

Although hardly a new technology, new perceptions about books began to emerge in the late nineteenth century as new forms of communication evolved in an environment of rapid industrialization. By 1924, educators, writers, and literary critics had begun to adopt an increasingly mechanistic view of books, illustrated by the claim made by I. A. Richards that "a book is a machine to think with."[1] To critics and linguists like Richards, gaining maximum efficiency from this machine—eliciting the optimum meaning, significance, and usefulness from the text—could ultimately solve the problem of universal communication that linguists and literary critics pursued with scientific ardor in the first half of the twentieth century.[2] Alone, however, the machine/book could not produce effective thought, and the instructions to operate it (that is, literary criticism) would ensure that the machine ran smoothly. Underlying this new perception of books was thus the implication that the reader was an inefficient operator of the machine. To put it another way, without the trained operator the machine was unusable/illegible.

The success of books as a form of communication for the twentieth century thus appeared to rely on the improvement of readers' abilities, skills that would enable them to acquire and retain the information that books contained. By the mid-twentieth century, the "mechanization" of reading had evolved into a huge business that focused on training to increase reading pace, often with the aid of machines. This chapter looks at how this discourse of reading efficiency came about and how it led to a new type of reading advice that was completely at odds with theories of effective reading a century earlier, making faster reading a requisite and symbol of the modern American into the twenty-first century.

Prior to the industrial revolution in America, instruction books on reading and self-improvement consisted of religious primers or spellers. Improvement in reading mostly involved monitoring the moral and intellectual quality of what was read, along with painstaking study that would make "difficult" reading more easily understood. While these books addressed personal improvement, others attempted to reform language itself in order to standardize communication and enable readers to benefit more fully from what they read. In a multilingual society this improvement was fundamental to national success and effective business exchange. Thus, by the late eighteenth century, a "debabelization" process was already under way in America to standardize the various spellings and pronunciations of the newly independent nation, aiming to reduce "a dozen local dialects to one harmonious language."[3]

Reading speed, however, was not part of this standardization. In fact, until the early twentieth century, rapid reading "as an exercise in skipping" was associated with vulgarity and a trivialization of literature that had resulted from what Henry James termed in 1884 as the "superabundance" of fiction writing.[4] Despite lamenting the impossibility of reading all the literary productions of the age, English writer Thomas de Quincey wrote that "rapid reading . . . belongs to the vulgar interest of the novel." It is a delusion of the mind, he claimed, to think "that it is reading to cram himself with words, the bare sense of which can hardly have time to glance, like the lamps of a mail coach, upon his hurried and bewildered understanding."[5] Incorporating the same metaphor of the eye as a lamp that hurried over the page, author and educator William Chauncey Fowler also counseled against rapid reading in 1876: "In rapid reading, [the mind] is nearly in the same state as yours when you are whirled through a country in a post-coach or a railroad-car. How much do you know of that country in the one case? How much do you know of the book in the other?"[6] Reading too much and thinking too little, readers were "incapacitated for high achievements" because "addicted to mental gluttony, [the reader was] thus suffering from mental repletion."[7] The "living lexicon" or "walking encyclopedia," like the overstuffed Victorian parlor, was encumbered by culture in the face of acceleration.[8] Despite this, Fowler could only suggest slow reading and better selection of texts (the assumed role of the critic) as the way to "transfer the views of your author to your own mind." New designs for the mind were emerging, however, that suggested ways of increasing speed and acquisition of knowledge. These changes in perceptions over reading were central to a modern formation of the reading/viewing subject as well as the emergence of new cultural forms of expression and entertainment.[9]

By the early twentieth century, concerns with the overstuffed Victorian mind transformed with the introduction of new technologies. Industrial expansion and technological invention after the Civil War in the United States created two new circumstances that profoundly affected reading: a huge proliferation of printed materials and a massive expansion and change in public readership. Unprecedented immigration created a new working class, many of whom were not native English speakers, and industrial technology appeared to speed up the pace of everyday life. Understanding reception and controlling communication was fundamental to the success of this new national formation.

Along with this material change in social organization, philosophical changes in visual knowledge coming out of experimental psychology dramatically altered ideas about the observer/reader. As visual historian Jonathan Crary has detailed, new discoveries about vision from the early 1800s dramatically changed ideas about the transmission and reception of reality, creating "a condition of possibility both for the artistic experimentation of modernism and for new forms of domination."[10] By the second half of the nineteenth century, studies in language, perception, and mental functioning took on an urgency that paralleled the imperialistic drive toward a unitary national identity that characterized American foreign and domestic policy at this time.[11] Experimental psychology quickly emerged in the United States "with a vigor unmatched elsewhere."[12] The impact of industrial methods on the study of philosophy and psychology led to a rationalization of the production of knowledge that required standardized methods and instruments, along with a standardization "of the experimental human subject as an introspecting instrument."[13] Epitomized by his book *Psychology and Industrial Efficiency* in 1913, American psychologists such as Hugo Münsterberg, president of the American Psychological Association, gained legitimacy and funding for experimental psychology, promoting it as a useful tool in the new industrial order.

At the outset, however, experimental psychologists were more concerned with understanding, recording, and measuring mental processes than with working with industry. In the early 1800s, reaction-time experiments that extended philosophic enquiries began to connect the human subject to mechanical apparatus; using equipment such as chronoscopes, stereoscopes, tachistoscopes, polygraphs, kymographs, sphygmometers, dynamometers, and telegraph keys, psychologists attempted to measure human responses to stimuli scientifically.[14] For optical measurements, new machinery was rapidly developed to determine how fast the eye responded to stimuli such as color, letters, words, or objects. These attempts to measure reaction time had a huge impact on ideas about the transmission of language and meaning through printed media, ideas that transformed an understanding of reality, textuality, and the observing/thinking subject. To many, the psychology of reading and perceptual processes appeared to offer a rich seam of information that could open up the doors to human consciousness and answer questions about human perception, meaning, and reality that had been central issues of philosophical inquiry for centuries. Thus Edmund Huey wrote of the significance that the psychological understanding of reading processes held for modernity:

> And so to completely analyze what we do when we read would be the acme of a psychologist's achievements, for it would be to describe very many of the most intricate workings of the human mind, as well as to unravel the tangled story of the most remarkable specific performance that civilization had learned in all its history.[15]

Experiments to crack open the mysteries of the reading process appeared to have advanced in 1879 when French ophthalmologist Louis Émile Javal suceeded in measuring eye movement with electronics. Prior to this, only rough estimates had been possible, with experiments made using an observer to count movements whose own eyes were far from accurate as measuring instruments. Javal's experiment was carried out by connecting the eyelid to an electric circuit and then counting the series of sounds produced in a microphone by each eye movement. Javal noted that the movements were discontinuous and fragmented and didn't flow along the line of words in the correct order as would be expected, indicating for the first time that the way the human mind responded to language was less lineal than structural.[16] This new understanding of perception not only created new art forms but altered ideas about the structure of language and communication—opening up the possibilities for new structural and relativistic knowledge that has dominated theories of language since Ferdinand de Saussure taught his course in general linguistics in the early twentieth century.

Around the same time that Javal was experimenting with the motion of the eye, French physiologist Etienne-Jules Marey was performing experiments to record movement that would combine medical with cinematic history. As an experimental physiologist, Marey invented many instruments used by experimental psychologists, including the chronograph in 1888, out of which the cinematograph was eventually developed.[17] Marey's experiments with motion and speed paved the way for both the science of experimental psychology and cinema. The promise of Javal's experiment was not only to reveal the secret of the reading mind, but had wider implications for understanding how society functioned as a whole, with psychologists starting to apply their theories to form an understanding of the group and social psychology of the masses. How the mind of the mass viewing public functioned was a central concern at this time, a concern only enhanced by these parallel developments in nascent visual technologies that were to eventuate in a new mass media. New psychometric measurements and machinery, it was hoped, would provide a scientific picture of the functioning of the mind—less a snapshot of consciousness than a motion picture of it.

It was, however, the statistical accumulation of data that made it possible to map the limits and capabilities of the modern mind. Charles Darwin's cousin, Sir Francis Galton, had been studying individual and racial differences since the

1860s, storing measurements and test results to show that human characteristics and behaviors were inherited. In 1884, Galton opened his "Anthropometric Laboratory" at the International Health Exhibition in London, later moving to South Kensington Museum, where nine thousand people came to be tested and measured at three pence a time.[18] Prior to this, most tests had been undertaken by scientists or their students in laboratories well away from the public gaze. Galton aimed to prove his theories of racial inheritance through the analysis of huge quantities of data, making statistics an important aspect of the new science of the mind and body. The data gathered would provide a scientific X-ray of modernity, making it possible to discuss such things as "averages," as well as to predict the possibilities and the limits of the human mind—and by extension, civilization itself. Galton's tests were mainly physiological rather than psychological, measuring height, weight, arm span, head size, and so on, but they also included reaction time tests that were to feature prominently in later experiments undertaken in the United States. Mostly, Galton had introduced psychologists to the benefits and possibilities of psychometric testing on a mass scale. These tests and the method of gathering statistical information provided models for the development of mental testing by American experimental psychologists James McKeen Cattell and Joseph Jastrow, who were both followers and correspondents of Galton.[19]

With Galton's tests in mind, Cattell aimed to discover the possible limits of reading speed and comprehension in 1885, when he created further tests using letters pasted on a revolving drum, or kymograph.[20] The machine rotated nonsense words, real words, or random letters or numbers using a clockwork mechanism that displayed them through a small aperture for split-second moments. These machines timed the rate at which it was possible to read and make sense of words and sentences, enabling psychologists to get a greater understanding of the functioning of the mind and its response to structures of language. To judge the potential speed of reading, the scientists gave the utmost importance to determining the exact conditions for optimum transmission of information, such as the best typeface and letter size. Cattell concluded that ornamental type was "injurious" and that "the simplest geometrical forms seem the easiest to see."[21] The same tests also aimed to discover the "threshold of consciousness" or how much information the brain could record and retain. Numbers, letters, words, and sentences were exposed to view for .01 of a second to the discovery that whole words and sentences were retained over and above random letters and numbers, revealing that lack of ornamentation in type and lack of abstraction in meaning were optimal conditions for the transmission of information. The way that the brain reacted to these tests and responses to a variety of stimuli and conditions prompted the development of tests on vision and reading that would eventually enable scientists to establish "average" or "normal" categories, by which individuals could assess their personal performance. Experiments in design and type that aimed to attract the maximum response in the quickest period possible were complemented by changes in such discoveries about reading. Although scientifically problematic, these experiments ushered in an era of mental streamlining, as well as textual and literary experimentation.

In 1893, after consultation with Galton, Joseph Jastrow established the Section of Psychology at the Columbian Exposition in Chicago, where he tested thousands of individuals with the help of graduate student volunteers amassed for the occasion.[22] Using Galton's Anthropometric Laboratory as a model, this was the first public stage for experimental psychology in America, heralding psychometrics as a major new science as well as a new forum for transmitting knowledge about human subjectivity. In a series of psycho-physical tests lasting two hours, fairgoers could be tested with the latest equipment and methods to ascertain how "average" or normal they were, in relation to the accumulated data that the scientists had already acquired. Tests included judging weights and lengths, sensory acuteness and motor abilities, and included reading tests where:

> The subject looks at a vertical screen where, filing past an opening, are many cards displaying words and numbers: he must afterward write down all the figures and words that he can remember having seen, and in the same order as they were presented, if possible. This test indicates the extent of memory and the time of presentation that is sufficient for exact perception.[23]

Other tests measured the time taken to transcribe words and write another, suggested by association, or to identify previously seen words within a list of random words or images mixed in together. An experiment such as this, Henry de Varigny wrote, "gives some indication of the speed of perception and the vivacity of memory."[24] Experimental psychology combined with the new science of statistics thereby played a crucial role in beliefs about improvement in reading and enabled the reading subject and reading speeds to be measured in a comparative way on a national scale. While personal improvements in the reading process remained important to educators, the desire to apply these improvements more widely to improve national educational standards emerged out of such research. As entertainment and performance, the public nature of these tests also offered a way of measuring oneself and others in relation to the crowd, a feature of mass culture that reemerged in twentieth-century self-help popular psychology.

In 1898, persisting with more accurate measurement of eye movements, American psychologist Eugene Delabarre developed a plaster of Paris "contact lens" attached to a photographic recording device on a rotating kymograph (Figure 18.1).[25] Using this, Edmund Huey carried out experiments to compare reading horizontal with vertical lists of words.[26] These experiments indicated what technical adjustments could be made to words, text, and layout to facilitate the productivity of reading and "skimming." Experiments with narrow columns and page layout showed how certain styles forced the eye to make more movements, thus causing it to tire quickly. Words were "cut up" on the page, or "disarranged" in a variety of ways, to see how quickly sense could be made of partial words (or partial word images). Huey's tests showed that anticipation and association made reading faster and that words (even sentences) were not read in full or in the exact order they appeared on the page. What all this

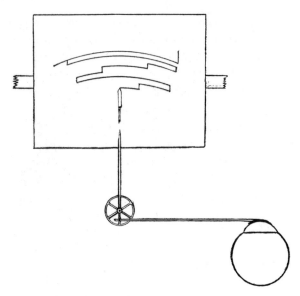

FIGURE 18.1 *Diagram of Delabarre's 1898 "contact lens" attached to a photographic recording device on a rotating kymograph.*

indicated was that rationalized reading was not in fact a speeding up the reading of individual words but training the mind to anticipate the structures of language and make associative connections more quickly.[27]

While changing the understanding of language reception it was quickly perceived that the reading subject could also be transformed by new processes of learning. Educational psychologists saw the pedagogical benefits of such knowledge and aimed to use it to "speed up" and improve learning abilities. In 1894, Adelaide Abell conducted experiments with students in a psychology class at Wellesley College that showed it "is evidently possible to apprehend almost simultaneously the words in a line, just as one reads by words instead of letters." Experiments confirmed Javal's research that the mind worked with structures of language rather than with units, and that simultaneity of perception was not only possible, it was now seen as preferable, and could even be an aid, rather than a hindrance, to comprehension. "The fundamental condition of swift reading" she concluded, "is undoubtedly quickness of association; so that any attempt to improve reading ability must include the effort to increase the rapidity of association by repeating and multiplying associations and by intensifying interest and attention." Attention, a problem that was central to late nineteenth-century fears over cultural decline, was thereby shown not to decrease with acceleration but to sharpen or improve—a significant discovery for the transition to modern reading.[28] Slow readers and readers who read one word at a time, the study found, could comprehend less, rather than more, and often hindered their speed of understanding by articulating words.

In the intervening years since Fowler had counseled against rapid reading, experiments in psychology and psychometrics had thus shown that to increase reading speed was "an evident pedagogical requirement."[29] Abell summarized that "though every individual has probably his maximum rate of reading, determined by his natural quickness of apprehension and association, it is yet possible and desirable to some extent to increase the ordinary rate."[30] This conclusion was to usher in a new era of speed reading training and educational psychometrics that could differentiate "slow" learners from the rest. As a new sign of success, reading rapidly would signify less a "novelistic" superficiality than a streamlined mind, trimmed and trained for modern times.

Despite this, problems encountered by scientists revealed how difficult they found it to make exact measurements of the mysterious process of reading. In reaction-time tests, the time lapse between seeing and reporting remained problematic. And while experimenters strove to standardize machinery, experimental subjects were rarely so accommodating. Likewise, most experiments were undertaken on graduate students, where the a priori standard of reading and education naturally had great effect on comprehension and reading speed—and experiments merely reflected these social and educational differences.[31] Although the limited number of subjects used in experiments also prevented experimenters from establishing overall averages or norms of movements, more problematic were the discoveries of psychologists, especially Sigmund Freud, that the mind could "misread" and misperceive. With the publication of *The Psychopathology of Everyday Life* in 1901, Freud illustrated that the mind could misread what the eye saw, showing how it creatively edited and transcribed the meaning of the perceived world or the text on a page. To Freud, such malfunctions could reveal the operation of the subconscious, but to others they illustrated the kind of personal weaknesses that made self-control, personality testing, and training even more essential for the industrial order.

The conditions of the experimental lab, Wesleyan professor Raymond Dodge noted, also altered the way the body would respond in a "natural, or unobserved environment," so that vision in darkened rooms thus "modified the conditions of attention."[32] Even Delabarre's contact lens, heralded as a major breakthrough, was problematic in that it was not known how the equipment might affect the eye movements (not to mention the effect of the anesthetizing cocaine that was used to numb the pain of the procedure). Dodge claimed to improve on previous experiments with the use of photographic film rotated on brass rails where reflections from the eye acted as inscriptors on a moving photographic film—literally making the eye write a record of its movement onto film.[33]

Despite the problems they encountered, and their own work that established the instability of the perceiving subject, experimental psychologists persisted with the goal of accurate reading measurements. By measuring eye movements, psychology aimed to become an exact science of the mind. Such experiments would unfold the mystery of the age—the modern response to stimuli—and give mathematical exactitude to how effectively, for example, a billboard advertisement might be perceived while passing at a certain distance and velocity from an electric tram. Experiments on eye movements grew alongside technologies developed by Marey, and then Edweard Muybridge, to record motion in animals, so that by 1905, researchers acknowledged the generosity of the Edison company in providing assistance with kinetoscope cameras to make more accurate records of eye movements. Educational psychologists Charles Judd, Henry McAllister, and William Steele undertook experiments with a hand-cranked camera, developing a new paraffin contact lens that improved on Delabarre's. The observing subject's head was held in a fixed position and wire glasses were used as points of reference—all in the hope that this would eliminate some of the inaccuracies of previous experiments (Figure 18.2).[34] Such experiments at the forefront of the silent film era may also have been of some benefit to the Edison company, offering a sense of how viewers "read" images and on-screen intertitles, which had begun to appear in moving picture "narratives."

Thus it was shown that reading could be measured and improvements in the individual could be recorded or even transformed by technology. The Taylorization of the workplace and the introduction of time and motion studies using photography and motion pictures by Lillian and Frank Gilbreth a few years later

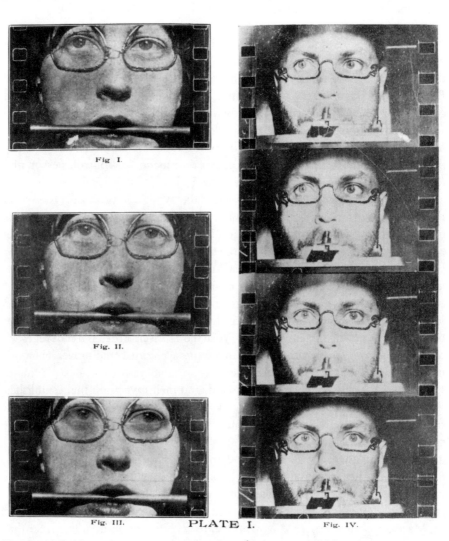

FIGURE 18.2 *Measuring eye movements using a kinetoscope, 1905.*

was thus anticipated and explored in studies of reading and perception around the turn of the century. As in the factory or workroom, improving reading appeared to involve the preservation of energy through increased efficiency of the machine and the elimination of unnecessary, or wasteful, movement. Because reading was not a "natural" function of the eye, such training to improve techniques would prevent certain "forms of degeneration" such as the "very evident inheritance of . . . myopia and nerve exhaustion . . . [that] warns us of the danger of race degeneration from this source."[35] Psychometric tests indicated to educators not just the level of the reader's abilities, but the extent to which they needed to be "improved."[36] So despite the belief that reading capacity was widely inherited, Walter Dearborn also claimed that reading rates and mental acuity could be improved by retraining the mind to function in new, and more modern, ways. "Bad form in reading is doubtless as distressingly common as bad form in swimming, skating or tennis," he claimed, and is established through poor educational techniques that were based on outmoded ideas about the reading process.[37]

Theories of degeneration and fatigue showed that the body and mind was as prone to mental inefficiency and loss of energy as any other machine.[38] Fears that muscular eye fatigue also represented atavism or racial deterioration were common enough by 1901 for physical culture guru Bernarr Macfadden, mostly famous for his bodybuilding and dietary advice, to publish the self-help book *Strong Eyes*. Macfadden recommended eye-training exercises that were similar to the exercises that he recommended for other muscular development, claiming that no one needed to wear glasses if they followed his exercise and dietary regimen. With many reprints over the next few decades, and a new title in 1926 of *Strengthening the Eyes: A Course in Scientific Eye Training*, the book was fundamental to notions that the "eye grasp" could be physically improved at will. Notably, his regime offered the chance to prevent mental flabbiness and weakness through exercising the eye muscles. Reading need not result in nerve exhaustion, he noted; neither did watching movies, which could even be of great benefit in improving vision if frequent attendance was used for practicing his exercises. "If properly used," he claimed, the eye "is fully able to withstand all the strains of modern life."[39] Macfadden thus offered a techno-scientific way of strengthening the eye, using the fast moving film images to increase the agility and speed of the eye, counteracting the common fear that new technologies weakened reading ability.[40]

The wider application of such exercises and the improvement of reading to keep up with modern transformations not only claimed to prevent mental flabbiness but were seen as fundamental to improving civilization itself. In his 1908 book *The Psychology and Pedagogy of Reading*, Edmund Huey considered the wider social implications of improvement in reading in a final chapter titled "The Future of Reading and Printing: The Elimination of Waste." "Improvement in the page or the method of reading means the rendering of a great service to the human race," he wrote, and further that:

> Human thought has been busy rationalizing. It has rationalized the traditional methods of transportation and locomotion until we have the steam

and electric locomotive and the economy and comfort of modern travel. Means of communication at a distance . . . have had the keenest and most persistent efforts of inventive genius, and the modern marvels in telegraphy and telephony are the results. Even printing and the making of books has had attentive study and continuous improvement. . . . Yet with it all the essential characteristics of the printed page itself, and of the reading process by which we gather its meaning for so many hours of the working day have never been rationalized in the reader's time or energy or comfort.[41]

Huey claimed that the scientific tests on reading and the eye had shown that the printed letter could be streamlined and rationalized to maximize reading efficiency and to prevent the waste of excessive eye movements and energy. The future of reading, he claimed, would involve, possibly, a new alphabet, type, methods of reading, phonetic spelling, pictographs, and abandoning archaisms that did nothing but impede the modern reader. "We are likely," he wrote, "soon to consider the possibilities of a total rearrangement of our printed symbols, in the interest of economy of time, energy, and effectiveness in getting thought from the page." Reading in vertical columns may also add to the economy of reading, "eliminating our own very evident wastes." In fact, new "telegraphic" methods of communicating may eventually override the need for reading, he predicted, because technologies would enable authors to "talk his thought directly into some sort of graphaphone-film book which will render it again to listeners."[42]

The future of the book, however, relied on a rationalization of the reader as well as the creation of new forms of texts. Speeding up brain function was seen as essential in the prevention of racial degeneration, and by the 1920s, educationalists had widely adopted the precepts of experimental psychology in the form of mass tests to measure intellectual ability, or IQ.[43] Mental training was increasingly perceived as a way to "Taylorize" white-collar work and the key to business success.[44] On a national scale, the testing of mental agility became fundamental to the implementation of immigration laws that would sort the "weak-minded" from the strong, and by 1925, colleges were beginning to introduce speed reading training as part of their general educational program.

The discoveries of psychologists reached an even wider mass market with the publication of professor of journalism Walter B. Pitkin's *The Art of Rapid Reading: A Book for People Who Want to Read Faster and More Accurately*. Written in 1929, for the man who is "dissatisfied with the amount of reading he does in the course of a year," the importance of keeping up to speed with the rapid developments in modern America was, he claimed, immense. By reading his book and undertaking regular speed drills, the "average" reader could, he proposed, easily improve his reading speed by 50 percent. Industrial developments simply required this speeding up of mental processing, he argued, "For man's mind is the most marvelous of all machines. To improve it is to improve all the things it moves and creates."[45]

Streamlining the reading process entailed the correct selection of matter, correct positioning, training the eye to select the important facts on the page, and committing the facts to memory. Building vocabulary was a necessary adjunct to this process, as was training the "eye grasp." Unlike advice in the late-nineteenth century that had condemned "skimming" as lazy and vulgar, Pitkin now counseled that skimming was perfectly acceptable and in many cases beneficial, if done according to the advice of experts. Mental training would unburden the reader from the difficulties associated with long and difficult reading, and speed was now most definitely akin to success. Through a series of exercises—some of which required an assistant with a stopwatch—the book enabled a rough measurement of the speed of reading and comprehension, as well as a measurement of the facts retained. As a layperson's version of psychometric testing, Pitkin's book took experimental psychology out of the laboratory and into the home.

Such homespun versions of psychometric testing proliferated over the 1930s. James Mursell's *Streamline Your Mind*, published in 1936, advised the reader on how to train for an efficient mind in an everyday environment. Quick and easy tests of mental efficiency could be carried out in the home or the street, and mail order reading tests could now be cheaply purchased on the market.[46] Mursell advised making simple estimates of ability level by comparisons with the speed at which others read in public, such as the movie theater crowd, for example. "Or time yourself on an article or story in *Liberty* magazine, where the 'reading time' is indicated," he advised."[47] Like Pitkin, his home-based experiments offered cheap and easy remedies to the machine age in a domestic and familiar environment.

Not only did Mursell compare the reader to a machine (one that could be efficient or inefficient according to training) but one that operated cinematically:

> Behind the lens of a motion-picture projector there is a shutter. This shutter synchronizes with the run of the film so that as each of the tiny pictures comes into position it is released onto the screen for a fraction of a second and then cut off as it slides away to make room for the next. If the shutter stopped working you would not see a series of clean-cut images, but only a moving blur. This is how you must learn to use your mind if you wish to memorize well. You must concentrate on the job, so that you have a series of definite, sharply defined pictures, and not a vague blur shading off into all sorts of irrelevancies.[48]

Just as Macfadden advised using the everyday experience of modern technology to improve strength, Mursell saw the mind as something that could be trained using "cinematic" techniques. More than a metaphor, however, by the 1930s, methods of training to speed up reading now involved training "by means of a motion picture technique."[49] Educational psychologist Walter Dearborn described the training as "photographing reading material on motion picture film in such a way that when the film is projected, successive units of the separate lines are seen exposed tachistoscopically across the screen . . . the reader's task . . . is to keep pace with the rate at which he is being directed through the material."[50] As shown in Figure 18.3, this cinematic method offered training that was "intended to improve reading ability by a type of practice which controls the eye movements of the unskilled reader in accordance with the pattern of the eyes of the skilled reader."[51] From the mechanics of measuring using new technologies, experimental psychologists had evolved "eye training" that would increase and standardize reading speeds at a mechanical rate, controlled by the machine itself.[52]

Avant-garde literary experiments also played with the possibility of filmed "books," or "logocinema," something that literature scholar Michael North describes as "a revolution of the word accomplished quite literally by bringing to language the physical dynamism and energy associated with film."[53] In the early 1930s, American journalist Bob Brown wrote that "modern word-conveyors are needed now, reading will have to be done by machine; microscopic type on a moveable tape running beneath a slot equipped with a magnifying glass and brought up to life size before the reader's birdlike eye."[54] Brown argued for the necessity of his "Readies" (books projected on moving film) as a way to keep up with "today's speed."[55]

The speeding up of reading may have seemed even more important with the introduction of microfilm technology in 1937, which appeared to turn books into films. *The Literary Digest* explained how a "full-sized newspaper can be reduced to a foot of movie film" and that the machinery for reading these new texts resembled movie cameras that projected words onto a backlit screen and into the eye.[56] Thus by the 1930s, it may certainly have seemed that changes in cultural production had indeed turned books into "machines." With this process, there seemed no limit to the potential information revolution, to the democratic possibilities of universal knowledge and an ever-increasing need to speed up reading.

FIGURE 18.3 *Eye training using the motion picture technique, 1938.*

Nevertheless, the utopian possibilities of these new technologies also harbored a dark pessimism about the future of language and civilization. The rise of the mental hygiene movement along with fears over personal and economic breakdown made self-improvement and education a panacea for what many saw as an ailing nation. Transforming the way people read was accorded so much importance that educator Irving Anderson wrote that while "This failure to acquire a language background is due partly to low intelligence . . . it is due as much, if not more, to a system of living and education which not only permits but sometimes fosters either verbalistic or imaginary, escape-like adjustments to the realities of everyday life."[57] The link between mental decline and the decline of the state was highlighted to many by the 1929 Wall Street crash and the onset of massive economic depression. If "cultural lag" was responsible for the collapse—as some asserted—the need to streamline the mind in line with technology appeared central to recovery.[58] By the end of 1929, Pitkin's comments, that "The fate of nations hangs on what he reads. So does the stock markets. So does the march of industry and business. So does the progress of education and every larger aspect of social welfare," may have seemed prophetic indeed.[59]

The centrality of self-improvement to the wider goals of social harmony and progress was embedded in the goal of rapid reading. The need to understand how the mind could cope with the influx and confusion of perceptions that characterized modernity led to new ways of understanding, and training, the modern mind. To avoid becoming lost in the welter of possible mental fragmentations, streamlined reading became a talisman to ward off the ills of postmodern subjectivity that was signified mostly by the loss of control over language.

Like skyscrapers and Fordism, speed reading is something with a particularly "American" flavor. After the 1930s, speed reading training lost its association with the avant-garde and utopian ideas about the liberatory potential of a "revolution of words." In the 1950s, speed reading became big business, with reading teacher Evelyn Wood dominating a now highly competitive market with her "Reading Dynamics," a course that became an international phenomenon.

Methods for training have continued to develop with the unwritten understanding that (modern) speed is better than (Victorian) slowness. Despite the increase of formal education and widening access to learning throughout the nineteenth and twentieth centuries, anxieties over educational standards continue to create a market for mass-market quick-fix solutions to the perception of mental "degeneracy" and educational decline. Self-help books devoted to enhancing the brain and mental efficiency continue to emerge, with titles that evoke the mechanistic imagery that has characterized modern perceptions of the mind: *Users Guide to the Brain; Owner's Manual for the Brain; Brain Building in Just 12 Weeks; Speed Reading; Photoreading; Quantum Learning; Triple Your Reading Speed; Twenty-First Century Guide to Increasing Your Reading Speed;* and *Mega Speed Reading*, to name a few. Further than this, the appearance of new styles of "machines to think with," along with the mass production of information via computers, have increased anxieties over reading and intelligence/culture while offering solutions to it. Advertisements for speed reading software offer faster reactions to faster machines, some now claiming to enable photo reading at 25,000 words a minute. Looking back at the history of such claims, it becomes ever more surprising that such old ideas about efficiency, intelligence, and speed are still sold as modern solutions to the digital age.

Notes

1. I. A. Richards, *Principles of Literary Criticism* (London: Routledge, 1924), preface. See also I. A Richards, *How to Read a Page: A Course in Effective Reading with an Introduction to a Hundred Great Words* (London: Routledge and Kegan Paul, 1936/1961), 1.

2. The search for a universal form of communication is beyond the subject of this chapter, but along with Richards, writers who also addressed this issue include C. K Ogden, Ezra Pound, Alfred Korzybski, Rudolf Carnap, Otto Neurath, and Stuart Chase.

3. Edmund Huey, *The Psychology and Pedagogy of Reading* (New York: Macmillan, 1908/1924), 248. For a discussion of debates over spelling and language reform, see David Baron, *Grammar and Good Taste* (New Haven: Yale University Press, 1982). For a social history of reading and books in America, see Cathy Davidson, ed., *Reading in America: Literature and Social History* (Baltimore: Johns Hopkins University Press, 1989). For other useful histories of reading, textuality, and literacy, see David R. Olsen, *The World on Paper: The Conceptual and Cognitive Implications of Writing and Reading* (Cambridge: Cambridge University Press, 1994), 143; Asa Briggs and Peter Burke, *A Social History of the Media: From Gutenberg to the Internet* (Oxford: Polity Press, 2002); Geoffrey Nunberg, *The Future of the Book* (Berkeley: University of California Press, 1984).

4. Henry James, "The Art of Fiction," in *America In Literature*, vol. 2, eds. Alan Trachtenberg and Benjamin De Mott (New York: John Wiley and Sons, 1978), 552.

5. Thomas de Quincey, *De Quincey's Writings* (Boston: Ticknor, Reed, and Fields, 1856), 38, 44; online at *Making of America* archive, http://www.hti.umich.edu/m/moagrp/. Walter Ong, in fact, dates concerns that an abundance of books would make men less studious to 1477 and the introduction of the printing press; see his *Orality and Literacy: The Technologizing of the Word* (London: Methuen, 1982), 79.

6. William Chauncey Fowler, *Essays: Historical, Literary, Educational* (Hartford: The Case, Lockwood and Brainard Company, 1876), 239; online at *Making of America* archive, http://www.hti.umich.edu/m/moagrp/.

7. Ibid., 238. Tim Armstrong examines this late nineteenth-century metaphor of eating and literature as an articulation of concerns over efficiency and mass production. See his *Modernism, Technology and the Body: A Cultural Study* (Cambridge: Cambridge University Press, 1998), 42–74.

8. By the 1890s, writers responded to the idea of an overabundance of culture as a pathological decadence that was undermining civilization. See, for example, Patrick Bratlinger, *The Reading Lesson: The Threat of Mass Literacy in Nineteenth-Century British Fiction* (Bloomington: Indiana University Press, 1998), 192–96. See also chapter 9, "The Overbooked Versus Bookless Futures in Late-Victorian Fiction."

9. For a discussion of the way that ideas about reading affected modernist productions, see Todd Avery and Patrick Bratlinger, "Reading and Modernism: 'Mind Hungers Common and Uncommon,'" in *Concise Companion to Modernism,* ed. David Bradshaw (Oxford: Blackwell, 2002), 243–65.

10. Jonathan Crary, "Modernizing Vision," in *Vision and Visuality,* ed. Hal Foster (Seattle: Bay Press, 1988), 43.

11. For a study of the fundamental role of perception and attention within modern discourses of art and culture, see Jonathan Crary, *Suspensions of Perception: Attention, Spectacle, and Modern Culture* (Cambridge, Mass.: MIT Press, 1999).

12. Deborah Coon, "Standardizing the Subject: Experimental Psychologists, Introspection, and the Quest for a Technoscientific Ideal," *Technology and Culture* 34, no. 4 (1993): 761. For a history of experimental psychology, see Edwin G. Boring, *A History of Experimental Psychology,* 2nd ed. (New York: Appleton, Century, Crofts, 1957).

13. Coon, "Standardizing the Subject," 759. On cultural manifestations and responses to Taylorization and industrial efficiency, see Martha Banta, *Taylored Lives: Narrative Productions in the Age of Taylor, Veblen and Ford* (Chicago: University of Chicago Press, 1993). For a history of scientific

management movement, see Samuel Haber, *Efficiency and Uplift: Scientific Management in the Progressive Era, 1890–1920* (Chicago: University of Chicago Press, 1964).

14. Coon, "Standardizing the Subject," 769. See also Crary, *Suspensions,* 302–22. For a list of typical laboratory equipment, see entry under L in James Mark Baldwin's *Dictionary of Philosophy and Psychology* (New York: Macmillan, 1902); also available online at http://psychclassics.yorku.ca/Baldwin/Dictionary/defs/L1defs.htm.

15. Huey, *The Psychology and Pedagogy of Reading,* 6.

16. Ibid., 16.

17. See Marta Braun, *Picturing Time: The Work of Etienne-Jules Marey* (Chicago: University of Chicago Press, 1992).

18. Jeffrey M. Blum, *Pseudoscience and Mental Ability: The Origins and Fallacies of the IQ Controversy* (New York and London: Monthly Review Press, 1978), 44. See also Stephen Jay Gould, *The Mismeasure of Man* (London: Penguin, 1981). For a study of Galton and the development of statistics, see Donald A. Mackenzie, *Statistics in Great Britain* (Edinburgh: Edinburgh University Press, 1981).

19. Michael M. Sokal, "James McKeen Cattell and Mental Anthropometry: Nineteenth-Century Science and Reform and the Origins of Psychological Testing," in *Psychological Testing and American Society 1890–1930,* ed. Michael M. Sokal (New Brunswick, N.J.: Rutgers University Press, 1990), 21–45. Cattell was a pupil of Wilhelm Wundt, the pioneer of reaction-time experiments.

20. James McKeen Cattell, "The Time It Takes to See and Name Objects," *Mind* 2 (1886): 63–65.

21. James McKeen Cattell, "The Inertia of the Eye and Brain," *Brain* 8, no. 3 (1885): 307.

22. Sokal, "James McKeen Cattell," 31.

23. M. Henry de Varigny, *The Experimental Psychology Laboratory of the University of Madison* (1894); originally published in *Revue Scientifique* 1, tome 1 (1894): 624–29; online at *Classics in the History of Psychology* Web site: http://psychclassics.yorku.ca/DeVarigny/madison-e.htm.

24. Ibid.

25. E. B. Delabarre, "A Method of Recording Eye Movements," *American Journal of Psychology* 9 (1898): 572–74.

26. Edmund B. Huey, "Preliminary Experiments in the Physiology and Psychology of Reading," *American Journal of Psychology* 9 (1898): 575–86.

27. Although the literary impact of these studies are beyond the scope of this chapter, it is notable that later modernist visual poetry—from Apollinaire's *calligrammes* and Marinetti's images of onomatopoeic language—bear a striking resemblance to some of the images used in word experiments at this time. See, for example, Joanna Drucker, "Visual Performance of the Poetic Text," in *Close Listening: Poetry and the Performed Word,* ed. Charles Bernstein (Chicago: Chicago University Press, 1998), 135–37.

28. Jonathan Crary discusses the modern concern with attention and how failure of attention had become associated with sociopathic behavior; see *Suspensions,* 1–26.

29. Adelaide Abell, "Rapid Reading: Advantages and Methods," *Educational Review* 8 (1894): 283–86.

30. Ibid., 286.

31. For example, see W. B Secor, "Visual Reading: A Study in Mental Imagery," *American Journal of Psychology* 11 (1900): 225–36; H. S. Curtis, "Automatic Movements of the Larynx," *American Journal of Psychology* 11 (1900): 237–38.

32. Raymond Dodge and Thomas Cline, "The Angle Velocity of Eye Movements," *Psychological Review* 8 (1901): 146.

33. For a history of inscription technologies, see Lisa Gitelman, *Scripts, Grooves, and Writing Machines: Representing Technology in the Edison Era* (Stanford: Stanford University Press, 1999). See also Friedrich A. Kittler, *Discourse Networks, 1800/1900,* trans. Michael Metteer (Stanford: Stanford University Press, 1990).

34. Charles H. Judd, C. N. McAllister, and W. M. Steele, "Introduction to a Series of Studies of Eye Movements by Means of Kinetoscopic Photographs," *Psychological Review,* Monograph Supplement 7, no. 1 (1905): 1–16.

35. Edmund Huey, *The Psychology and Pedagogy of Reading,* 8.

36. At the same time, the constitution of the individual appeared to be a deciding factor on the amount of improvement that was possible. Thus differences in reading were in some way naturalized as inherited rather than learned, despite the fact the "improvement" literature presupposes individual betterment. Secor, for example, claimed that "the mental constitution or type of the individual has

much to do in deciding the place that each sense shall have in a factor in reading"; see Secor, "Visual Reading," 225–36.

37. Walter Fenno Dearborn, *The Psychology of Reading: An Experimental Study of the Reading Pauses and Movements of the Eye* (New York: Science Press, 1906), 179.

38. On discourses of energy and bodily efficiency, see Armstrong, *Modernism, Technology and the Body*, 42–74, and Anson Rabinbach, *The Human Motor: Energy, Fatigue and the Origins of Modernity* (Berkeley: University of California Press, 1990). On the relationship among entropy, cinema, and statistics, see Mary Ann Doane, *The Emergence of Cinematic Time: Modernity, Contingency, the Archive* (Cambridge, Mass.: Harvard University Press, 2002), 114–27.

39. Bernarr Macfadden, *Strengthening the Eyes: A Course in Scientific Eye Training* (New York: Macfadden Publications, 1924), 191.

40. See Bratlinger, *The Reading Lesson*, 209–10.

41. Huey, *The Psychology and Pedagogy of Reading*, 421.

42. Ibid., 424, 427, 429.

43. See Jeffrey M. Blum, *Pseudoscience and Mental Ability: The Origins and Fallacies of the IQ Controversy* (New York: Monthly Review Press, 1978); Sokal, *Psychological Testing and American Society;* and Stephen Jay Gould, *The Mismeasure of Man* (New York: Norton, 1981).

44. See Richard T. Von Mayrhauser, "The Manager, The Medic, and the Mediator: The Clash of Professional Psychological Styles and the Wartime Origins of Group Mental Testing," in Sokal, *Psychological Testing and American Society*, 146.

45. Walter B. Pitkin, *The Art of Rapid Reading: A Book for People Who Want to Read Faster and More Accurately* (New York: McGraw-Hill, 1929), 12. My biographical research on Pitkin shows that as a graduate student in Berlin he performed experimental tests on reading speeds similar to those of Cattell.

46. James Mursell, *Streamline Your Mind* (Philadelphia: J. B. Lippincott, 1936), 212.

47. Ibid., 31. It is also pertinent to note that magazines now indicated the reading time for their features at the top of the page to enable readers to know how long it should take them to read the feature.

48. Ibid., 156.

49. Walter F. Dearborn, Irving H. Anderson, and James R. Brewster, "Controlled Reading by Means of a Motion Picture Technique," *Psychological Record* 2 (1938): 219–27.

50. Ibid., 219.

51. Irving H. Anderson, "A Motion Picture Technique for the Improvement of Reading," *Bulletin of the University of Michigan School of Education* 11 (November 1939): 27–30.

52. Students "quickly learn to read the film material with exceptional skill" at a rate of five hundred words per minute, Anderson claimed, and the "film technique" method trained them faster than all (*Bulletin of the University of Michigan*, 29). Speed in reading, and consequently in the abilities of the human mind, would thereby enable the brain to keep pace with technological developments, for "new habits of perceiving and comprehending" at faster speeds were shown to be possible.

53. Michael North, "Words in Motion: The Movies, the Readies, and 'the Revolution of the Word,'" *Modernism/Modernity* 9, no. 2 (2002): 206. Although North argues convincingly for the influence of cinema on literary invention, the experiments with language and rotating words in experimental psychology illustrate that the revolution of words—in a literal sense—emerged out of the same experiments with time and motion that led to cinema. The possibility of dynamic language was thus already present in technologies of "logocinema" within the laboratories of the experimental scientist, whose experiments with human responses to language had emerged contiguously with cinema.

54. Quoted in North, "Words in Motion," 215.

55. Ibid.

56. "Photolibraries: Microphotography Offers Possibility of Volume Reduction by Camera," *The Literary Digest*, January 13, 1937, 19.

57. Irving H. Anderson, "An Evaluation of Some Recent Research in the Psychology of Reading," *Harvard Educational Review* 7 (1937): 33–39.

58. See my "Life Begins at Forty: Self-Improvement and Eugenics during the Great Depression," in *Making it Modern: Popular Culture and Eugenics in 1930s America*, eds. Sue Currell and Christina Cogdell (Athens: Ohio University Press, 2006).

59. Pitkin, *The Art of Rapid Reading*, 2.

19 The Swift View: Tachistoscopes and the Residual Modern

Charles R. Acland

Speculative, cautionary, and comic renditions of the future have left us with a collection of immediately recognizable relics, among which we find moving sidewalks, hovering vehicles, ray guns, and soaring needlepoint buildings. Joining these identifiable motifs are machines for the acceleration of reading and learning. In the 1961 proto-hippie classic *Stranger in a Strange Land*, one character reads the morning newspaper on an automatically scrolling screen at his "optimum reading speed."[1] The sad and doomed guinea pig in the 1966 novel *Flowers for Algernon* narrates that part of his hyper-learning program involves an array of flashing television images.[2] Popular film has regularly reiterated this situation of force-feeding screen images, often as a mode of mind control. Recall Alex in *A Clockwork Orange* (Stanley Kubrick, 1971), Joseph in *The Parallax View* (Alan J. Pakula, 1974), and the brainwashing scene from the teen horror flick *Disturbing Behavior* (David Nutter, 1998). These film scenes—and there are many other comparable examples—are enactments of some dominant ideas about subject positioning in film spectatorship, with the viewer's attention planted unwaveringly on the screen and with consciousness written up by the stream of images. All of these illustrations depict scenes of extreme and intense education. The conditions created by these accelerated learning machines vary from the streamlining of a quotidian activity (reading the morning paper) to the outrageous anxieties about idea implantation. Reflecting this powerful strain, as brainwasher Dr. Yen Lo in Richard Condon's 1959 novel *The Manchurian Candidate* puts it, we have imagined the present or near-future application of a "radical technology for descent into the unconscious mind with the speed of a mine-shaft elevator."[3] More generally, these renditions of "machine instruction" are representations of the uses and effects of representation.

For commentary and research suggestions that helped the development of this chapter, I thank Ruth Benschop, Bill Buxton, Ron Greene, Keir Keightley, Jonathan Sterne, Will Straw, and Haidee Wasson. Expert research assistance came from Peter Lester, JoAnne Stober, Nikki Porter, and Michael Schulz.

FIGURE 19.1 *Gravity tachistoscope. From Rudolf Schulze,* Experimental Psychology and Pedagogy: For Teachers, Normal Colleges, and Universities *(London: George Allen, 1909), 205.*

As outrageous as these fantastic stories of screen instruction may be, we can pinpoint historical corollaries. Their authors have built on dominant sensibilities, experimental programs, and actual technologies. One key source for these imaginings, versions of which can be identified in the fictional illustrations just cited, is an underacknowledged and highly influential apparatus called a *tachistoscope*. Literally meaning "a swift view," a tachistoscope can be put to multiple uses, although it originally measured the speed with which visual stimulus is recognized. According to science historian Ruth Benschop, A. W. Volkmann first proposed a horizontally sliding instrument in the late 1850s as a way to reduce contaminating stimuli from a visual perception test, that is, to assure reactions could be traced to a specific exposure.[4] Wilhelm Wundt in Leipzig, Germany, in the 1870s and James M. Cattell in Germany in the 1880s and subsequently at Columbia University in the 1890s both figure in the formation of the young field of psychology and in the development of the tachistoscope. Cattell's fall tachistoscope was prototypical for future devices; he redesigned it as a miniature guillotine with a dropping mechanism that used gravity to move a screen downward. The screen had a small opening that passed by a stationary field on which icons could appear. The contents of the field would be visible for an extremely short, but calculable, amount of time, which could be lengthened by increasing the size of the passing window or shortened by decreasing the size (Figure 19.1). To assure that one was measuring rapid visual perception, everything else—the head, body, eyelids, eye movement, pupil dilation—had to be held in place as surely as possible. Thus the tachistoscope's brief stimulus appeared in an otherwise uninterrupted field of vision. Devices incorporated measures to capture the subject's speed of response. Cattell, for instance, had subjects place a telegraph key between their lips, which, when dropped, would break an electrical current and provide a record of split-second timing. During the six decades from the 1880s through the 1930s, psychologists used "swift-viewing" instruments to measure and test an assortment of tasks—reaction time, attention, recognition, retention, among others. Cattell adapted his device to include studies of reading.[5]

The little cultural analysis that has been done on this technology has dealt exclusively with its place in the field of psychology. Especially noteworthy is Benschop's research. She uses Ludwig Wittgenstein's "family resemblances," in which differences among members are as important as similarities in defining families, to unsettle "the idea that similarities among instruments should be taken as proof of a shared fundamental or transcendental quality."[6] Benschop largely limits her research to the late nineteenth and early twentieth centuries, charting the tachistoscope's movement from physiology to experimental psychology, with discussion of the current longing some contemporary psychological researchers have for precomputer instruments.[7] What impresses me, however, and what I focus on here, is that as we move into the twentieth century, the tachistoscopic function—the swift view—extends to unpredictable locations, away from the psychology lab to schools, reading courses, military training, and advertising firms. One of the tachistoscope's contributions here was to reorient reading from the page to the screen. With this movement, the technological interface of the screen connoted an efficient and optimizing function. The tachistoscope's initial substitution of the screen for the page, an exchange that would come to define a contemporary information era, signaled the new, improved, and up to date. Put differently, hyper-reading as promoted by screen projections and illuminations, whether of prose or advertisements, was a pursuit well tailored to the conditions of modern cultural acceleration.

I find the device's malleable nature intriguing, for it is not a single laboratory instrument but one that designates a set of ideas, impulses, and metaphors, or what Mark Seltzer calls "cultural logistics."[8] These logistics are not rigid and essential but establish an empirical and durable material dimension for a loose band of priorities, organizing the resources of sundry individuals, disciplines, and institutions and subsequently offering us a point of access to what is valued. With each iteration of the tachistoscopic idea, we have yet another instance of what I think of as a *residual modern*, that is, the persistence of certain particularly modern priorities, sensibilities, and materials. Many technologies and media have assisted in our grasping of foundational qualities of modernity; Benjamin's "Work of Art" essay is easily, and deservedly, a most ubiquitous point of reference for this claim.[9] My intention is not to throw one more "forgotten" technology to a collective critical potlatch, proffering one more object lesson for what went wrong with advanced capitalism. Although much work has treated motion pictures as having a special relationship with modernity, I contend that attention to the tachistoscope draws out two elemental aspects of contemporary visual media: a fascination with the rapid arrival and departure of texts and an epistemological tension between perceptual fragmentation and synthesis. The tachistoscopic idea carries with it a curiosity about the effects of the often imperceptible replacement of images. And as it moves out of the realm of psychological experimentation, it is rather like an everyday science of exploring the liminal zones of consciousness, becoming a quotidian perception test, one that may no longer be associated with that device specifically.

The tachistoscope was, in fact, several different technologies employing sundry mechanical principles for a widening array of experiments and purposes. Through most of its early years, no commercial manufacturer supplied tachistoscopes, and many of the modifications and improvements pertained to the demands and interests of individual experimental contexts. Many variations of these ostensibly makeshift gadgets intended to make the assembly of a device easier, using materials available in typical laboratories and classrooms of the time. Distortion and change in brightness had to be minimized because it would cause the eye to move and the pupil to change size. The use of camera-like shutter systems, with apertures that revealed fields evenly and rapidly, was one eventual option, although even this was not as fast as needed for some experiments. Raymond Dodge's early 1900s tachistoscope employed a mirror and gelatin film to eliminate the noisiness and distraction of the dropping screen, appropriating techniques used by stage illusionists.[10] Whipple's and Schumann's tachistoscopes both used a rotating disk and an adjustable window. Some were designed to expose actual objects rather than representations or characters.[11] Continuing the demand for more accurate, soundless, and motionless laboratory instruments, the modified Gulliksen tachistoscope borrowed from the mechanics of one built by Samuel Renshaw and I. L. Hampton.[12] Other new devices appeared in the 1930s,[13] some tailored to measure reading reactions[14] and some, takings cues from Dodge, drawing on stage magic for large exposed fields—and actual objects—at a distance from subjects.[15] During these first decades, tachistoscopes were largely apparatuses for individual testing, sometimes boxes into which one peered. Dozens of versions continued to appear through the 1930s, with flash projection beginning to be an equal competitor. Projection tachistoscopes initially aided classroom demonstrations for psychology students, and tried to replicate the experimental devices with reduced sound and motion perceptible as exposed fields appeared and disappeared. E. O. Lewis and C. W. Valentine used a projection tachistoscope at the turn of the twentieth century to explore spatial illusions.[16] In 1931, Harold Schlosberg and Sidney Newhall both introduce such devices, designed specifically for classroom use, the latter an ingenious way to convert a magic lantern or opaque projector into a tachistoscope with a rotating disk in front of the lens.[17]

Still a staple of psychology labs today, although mostly a computer software program that produces flashing screens, the tachistoscope captured a lasting investment in the quantification of essential human functions, as well as in the inspection of the tiniest, most fleeting, manifestations of behavior. In so doing, the study of reaction time, attention, reflex, recognition, and so on, involved the operationalization of procedures such that each would be observable on its own, essentially holding all other actions constant. As per the standards of experimental methods, along with measurability, generalizability, and reproducibility, the isolation of function, and observation, was the key. One studied vision as separate from hearing as separate from touch. One observed reflex as separate from identification, icons as separate from words, depth as separate from color, source as separate from pitch. The range of human sense perception consisted of component parts. The subject's body underwent a radical partitioning, one that matched a division of empirical experimental labor and one that presumably corresponded to a general human form. The brain may work to coordinate the incoming information into composite portraits, but the point of access, and hence of quantitative measurement, is the pickup device (e.g., the eye, the ear, the skin, the tongue, the nose).

Given its chronometric purposes, ameliorating the accumulation of precise data, the tachistoscope required an immobile subject. Friedrich Kittler says that the body was "chained" in place,[18] but this is misleading because the subject was more likely to have a cushioned headrest into which one was to lean.[19] (See Figure 19.2.) Still, it is correct to say that the research subject or, more accurately, the research eye waited expectantly for the infinitesimal flash that was to be reported on. Jonathan Crary investigates tachistoscopic experiments for the way they produced and studied "a fragmentation of vision perhaps even more thorough than anything in early forms of cinema and high-speed photography."[20] Crary has written about "attention" as a special problem of late nineteenth-century modernity, especially by investigating the role of the new field

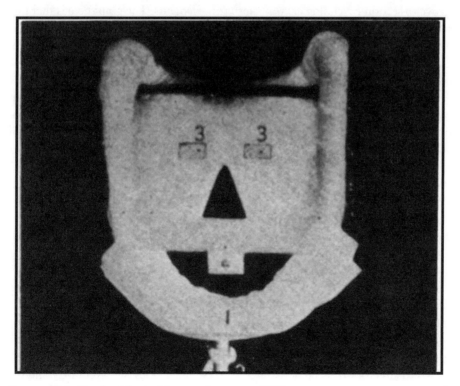

FIGURE 19.2 *Headrest for tachistoscopic test, ca. 1930.*

of psychology in the settlement of an epistemology of perception. He writes, "Attention as an object of knowledge involved the recognition that perception was essentially temporal and unstable but was also, if studied resolutely enough, capable of management and relative stabilization (as the example of the tachistoscope demonstrated)."[21] The work to which quantitative research is put explains the managing and stabilizing impulse of which Crary writes. His historical narrative focuses on the nascent stimulus-response, behaviorist strain of experimental psychology. He then elaborates on the paradoxes of modern sensory synthesis, that is, the impossible experience of perpetual reorientation of perception and its consequences for a fragmented and reconstituted self.

Crary's wide-ranging and subtle work notwithstanding, it seems to me that fragmentation and rationalization have been placed at the front of our understandings of modernity; one does not have to read too deeply in postmodernist critiques to stumble on such an inflection. This, however, has been weighted at the expense of synthesis. To the emphasis on the functionalist, behaviorist strain of experimental psychology, we need to add that there were other contemporaneous and competing frameworks. At precisely the same time that the first psychology labs were being built, William James proposed the experiential continuity of a stream of consciousness; the paths between physical and metaphysical worlds proposed by psychical research had not yet entirely waned in legitimacy; and Pierre Janet documented the "tension" existing in the synthesis of what some call the dynamic production of the self.[22] True enough, despite these contemporaneous and competing frameworks, the rising dominance of experimental psychology strove to section psychological traits into root elements. The point here is that this rise does not describe the entirety of the field. Moreover, even tachistoscopes were put to uses that revealed findings other than the fragmented human form. They were just as influential in the exploration of the perceptual linkages and continuities across sequences of exposed fields.

Particularly pioneering was Max Wertheimer, who in the early twentieth century advanced ideas about the interrelationship of sensation and the perceptual formation of wholeness while remaining inside a controlled experimental context. Considered the founder of Gestalt psychology, Wertheimer's first instrument of experimentation leading him to this new approach was the tachistoscope, used to show the perception of movement between two actually still exposed fields. In this instance in 1910, the tachistoscope was replicating an effect Wertheimer noticed in a stroboscope, a then century-old spinning protocinematic toy.[23] Other debates raged, for example that of Carr Harvey whose 1912 survey of tachistoscopy challenged Wundt, who concluded that there is no judgment in the briefest of perceptual events, which the latter described as reflexive or automatic.[24] Harvey pointed out that focus on isolated exposures fails to account for what happens among exposures. Wundt's blind spot is replicated today when Kittler mistakenly reads the tachistoscope for its reduction of perception to reflex, as in "Tachistoscopes measure automatic responses, not synthetic judgments."[25] Well, a century ago, psychologists debated the extent to which they measured both! As demonstrated by their role in Gestalt psychology and the range of debates that transpired, it is more accurate to say that tachistoscopes assisted in an investigation of *a tension between fragmentation and assemblage*, a tension that defines the modern sensory and perceptual subject. In short, assessments about fragmented and automated humans tell but half the tale of the epistemological battles engulfing the modern subject, not to mention half the tale of that particular instrument, given its equally central role in understanding the synthetic, combinatory function of human senses.

The apogee of this synthetic quality is evident in the analogy between tachistoscopes and motion pictures, another late nineteenth-century medium. This parallel is not lost on Crary, who notes perceptively that the tachistoscope's quick flashes of fields "certainly prefigure the effects of high-speed montage in cinema, where perceptual thresholds are approached and the question of subliminal images becomes important."[26] Kittler too is direct in pronouncing this relation: "The tachistoscope of the physiologists of reading was the twin of the movie projector, with the side effect of typographically optimizing the typewriter."[27] However, he soon flips this assertion to this:

> The film projector's twin thus functions in an opposite manner. The projector, in the unconscious of the movie house, presents a continuum of the imaginary, generated through a sequence of single images so precisely chopped up by and then fed through the projector's mechanism that the illusion of seamless unity is produced. With the tachistoscope, in the darkened laboratory of the alphabetical elite, a cut-up image assaults as a cut in order to establish out of the torment and mistaken readings of victims the physiologically optimal forms of letters and script.[28]

This comparison leads Kittler to claim that the program of reading promoted by the tachistoscope became noticeable differentials of material type, rather than reading per se.

Motion pictures and the tachistoscope both initially facilitated individual viewing and only later incorporated projection technologies for communal viewing. Crary notes elements cognate with Eadweard Muybridge's sequential photographs, seeing a similar operation in the isolation of visual fields and the sequential arrangement of those fields, although the tachistoscope's sequences were not necessarily consecutive.[29] I wish to draw special attention to this sequential aspect, that is, one image or text being replaced by another. In its predigital form, film is an arrangement of a series of still images that move at a constant rate, separated by imperceptible black fields. The tachistoscope is an arrangement of a still black field interrupted by nearly imperceptible images exposed at a variable rate. One instrument from the 1920s, the bradyscope, illustrates further this mechanical affinity between film and tachistoscopy. Used to assess memory, the bradyscope's innovation was to be able to flash its images or text at a constant rate, an idea that would develop into reading pacers.[30] Where the traditional tachistoscope was premised on a certain element of surprise, with flashes of light appearing randomly and suddenly, the bradyscope offered a way to measure the changes over time with ongoing exposures appearing at a steady rate, just as a metronome counts out beats. Although not designed to do this, if one could set the bradyscope to twenty-four flashes a second, what would you have but a standard film projector? Thus, at a slow frequency, this is a reading pacer; a faster rate produces tachistoscopic flicker; and even faster, reaching what is called a fusion frequency, we have motion pictures. *The only thing separating these devices is the speed of the arrival and departure of frames/fields.* I take time to establish these cognate elements because the ample, and convincing, claims about film's relationship to the modern fragmentation and elasticity of time and space (e.g., variable point of view and editing) neglect what examining the tachistoscope illuminates, that is, the speed of screen sequencing. In common is the high-speed mechanical replacement of images and text occupying a visual field. Even the relatively slow reading pacers were designed to increase, not slow down, reading speed.

As Benschop puts it, Cattell specifically used visual cues to measure other brain functions or "how fast we think," as he said.[31] As one might imagine, the speeds of perceptual and "thinking" functions vary among people, as well as with practice, and they appear to be a range rather than some absolute. The mounting number of trials in reading produced statistical evidence for standards of performance, that is, what constituted the normal, the exceptional, and the substandard. Charting how the brain learns, along with the other interests of experimental psychology, fashioned new ways to define, calculate, predict, and depict the sensory world of humans, knowledge that could be, and was, deployed in the service of maximum mechanical and cognitive efficiency (e.g., of labor). Where uncovering expected or normal rates of knowledge acquisition was a preoccupation, curiosity about improvements in reaction time followed. An encroaching attention to differentials then asked: How could the performance of the slower subjects be accelerated? Moreover what, if any, was the role of the tachistoscope not only in measuring but facilitating the improvement in response time? With this query we witness an extraordinary transposition: a measurement apparatus is understood as the agent of change. Intriguingly, and neglected in the existing histories of tachistoscopy, it became both a diagnostic and a therapeutic instrument in the service of improving rates. As Crary reminds us, the turn toward behaviorist experiments as offering base measures of individual performance, with a special relation between human perception and mechanical devices, is already evident in Cattell's work of the 1890s,[32] although this was not systematized into training regimes until much later.[33]

After decades of laboratory use, the tachistoscope was reconstructed, manufactured, and marketed as a device of instructional technology. Its tabulation of times was refashioned as that of scores and the very process of being hit repeatedly with flashing type became a form of lesson. Not all agreed about the tachistoscope's educational effectiveness, reasoning that mere repetition of actions, focus of attention, or innate perceptual capacity explained variations in results. Nevertheless, the move from lab to classroom is significant as an indication of the transportability of the "swift view" idea and because this gradually emerging linkage transpired during the initial rumblings of a discourse regarding educational technology. This classroom connection would never have developed without the budding field of visual, and later audiovisual, education. Just as machine-assisted reading instruction was beginning to establish its pedagogical paradigms, psychological research understood silent reading to be among the speediest of all human processes.[34] The relationship between behaviorist inquiry and modern mass education framed the question of learning as one of technological mediation.

The widespread turn to such instrumental training can be traced to an unlikely and innovative source. Pennsylvania teacher Catherine Aiken, in the 1890s, found a way to accelerate learning and to heighten concentration through processes she called "exercises in mind-training."[35] Her popular books proposed a separation of content from the "muscles" of perception and understanding, and she addressed the particularly modern threat of boredom. Students were put through a routine of brain exercise that echoed other increasingly popular forms of physical culture at the time. "Muscular Christianity" for Aiken could also be a brand of mental calisthenics. Her first improvised prop was a rotating blackboard, which she would spin so the girls in her class could see the material for but a brief instant. Arguably, this is the font from which later permutations of flashcards and tachistoscopic training emerge. Indeed, where psychologists endeavored to test Aiken's claims, a tachistoscope was used rather than the spinning blackboard.[36]

The classroom tachistoscope had the design advantage of repetitive, sequential presentation of material, such that one was not only learning about a subject, one was drilled on it. Teaching via a mechanized screen, as opposed to the page, was something an entire class did together and at the same time. These first educational devices would have been seen as souped-up magic lanterns, essentially a technology adaptable to any content. Explorations of possible uses in industrial training are also evident by the end of the 1920s. W. Hische determined that the speed of sequential presentation of material greatly affects how people retain and understand lessons, with too slow a pace leading to boredom. He indicated that this has implications for repetitive factory work.[37] Industrial applications tested, for instance, the reading rates of Johnson & Johnson executives and the eye–mind coordination of telephone operators.[38]

Tachistoscopes were especially useful in teaching and improving the core skill of reading, and they were regularly employed to study reading and reading disabilities in the first decade of the twentieth century.[39] One did not just read, but one read efficiently, swiftly. Kittler puts the first attempt to measure reading speed at 1803,[40] but most histories begin with Emile Javal in 1878 and his attention to the movement of the eye and the discovery of its saccadic, or jumpy, motion. The agreement among some psychologists was that reading was a perceptual activity that hence could be developed by altering perceptual skills.[41] The tachistoscope helped by disciplining the eye, training it to move in a regimented fashion, to take more in with each fixation of the eye's journey across the page, not to rest too long with each fixation, and not to wander without purpose. Moreover, these activities would have a measure for their change. Improvements could be documented, averages and expectations could be set, and "problems" could be identified. E. Grund, in Germany, conducted an experiment using tachistoscopes to assess reading skills, charting how children learn to see whole words rather than parts,[42] extending research by W. Stein from the previous year.[43] Tachistoscopic training was literally a form of discipline in some tests. For instance, one study divided a group of African American students into "troublemakers" and "non-troublemakers" and measured the effects of teaching with tachistoscopes on reading skills.[44] Other research employed tachistoscopes to show that speeding up reading first required correcting perceptual difficulties[45] and to argue that this laboratory instrument might be used to diagnose reading problems.[46] Not coincidentally, the official recognition of reading troubles like dyslexia transpired in the 1930s as well, part of the consolidation of new understandings of the operations of the brain and mass education.

Tachistoscopes had competition early on from other eye-training devices. Motion picture series were developed that sought to capitalize on these findings about perceptual skills and reading. Harvard, Iowa, and Purdue all had programs of 16mm film that taught controlled reading by projecting a filmed even-paced scroll of text, on which a class would be tested afterward. This approach, however, suffered in that the films did not offer flexibility in teaching and required the class to move at the pace set by the projector, with some teachers preferring 35mm filmstrips instead. The tachistoscope's variability in presentational speed made it the more felicitous technology, although it was best used to teach reduced eye movement and rapid recognition, rather than assimilation of material. Other basic instructional devices to teach reading in the early 1950s included controlled reading devices, which promoted left to right continuity, and accelerators. These latter were pacers, some of which had light bars moving smoothly down the page of a book and some of which were marketed to homes as well as schools. Many aids could be either handheld for individual use or projected for group instruction, with some instruments combining these functions.

Earl Taylor set a path-breaking course as a reading technology innovator and entrepreneur. His 1937 book, *Controlled Reading*, elaborated experiments in eye movement and training exercises for their improvement.[47] He had developed, in conjunction with his brothers James Y. and Carl C. Taylor, a three-field tachistoscope called a Metronoscope in 1931 and a device that could record eye movement called an Ophthalmograph. With the Metronoscope, each field would appear in sequence from left to right, working to condition the eye to move efficiently when reading. This was the first such device introduced to U.S. schools, distributed by American Optical Company (which later became Educational Development Laboratories), marking an initial reworking of the tachistoscope as a manufactured teaching tool.

In this reworking, few were as influential as the Ohio State University experimental psychology professor Samuel Renshaw. This enigmatic figure has appeared tangentially at best in the history of several fields, including film studies and education theory, although his influence has often gone unacknowledged. He was the main author of perhaps the most notorious of the Payne Fund studies in the 1930s in which his team documented the effects of motion pictures on children's sleep. One aspect of this research, conducted by students under his direction, used a tachistoscopic device to study the effect of visual flicker on a child's eyes.[48] For twenty-three years, he wrote the entirety of a monthly bulletin, called *Psychology of Vision,* then *Demonstrations in Psychological Optics,* and finally *Visual Psychology,* from 1939 to 1962.[49] Renshaw's program of training navy pilots to recognize aircraft and ships began in 1942. Using a tachistoscope, he documented precisely the rate at which he could speed up pilots' abilities to see and identify planes and ships. When Renshaw reported his findings, he noted that one class continued to improve its speed and accuracy through seventy sessions, and he claimed that the speed for some was at a rate ordinarily reserved for geniuses.[50] After his 120-hour course, men distinguished planes in 1/75th of a second and ships in 1 second.[51] Renshaw inferred that what the tachistoscope did was enlarge the subject's field of vision, training the eye to take in more at once.

Pilots identifying and reacting automatically, and attaining this machine-like reflex through a controlled stroboscopic device, captured the public imagination. These were the newest, most technologically advanced methods of instruction, ones that outshone the quaint, humanistic educational models of the past. Here, technologized humans produced through technologized programs fit with a sensibility about how modern mass education might be pursued. The Renshaw Recognition System received as much individual fame as any of the other significant educational programs of World War II. For his training regime, Renshaw received a Distinguished Public Service Award from the U.S. Navy in 1955, and the Smithsonian's National Air and Space Museum exhibited material from the program.[52] *Science Digest* drew attention to Renshaw's innovation in speed learning.[53] *Time* reported, on the occasion of his commendation, Renshaw's intention to speed up "sluggish readers" among the civilian population.[54] To this end, he helped develop and market his own device, the individually operated Stereo-Optical Tachitron, for which he wrote the manual.[55]

Buoyed by Renshaw's program, these instruments appeared to be the freshest offerings of a technologized age, ones that would address the post–World War II dilemma of an abundance of students to teach and an overload of information to impart.[56] Just as Renshaw was receiving attention for his military training work, Melcer and Brown documented improvements in 1945 on reading and intelligence tests for children after tachistoscopic training.[57] The U.S. air force sponsored tachistoscopic research in the area of target recognition,[58] and the navy funded hand–eye coordination tests that used a version of the device.[59] In 1953, Henry Smith and Theodore Tate found that students trained with a tachistoscope and an accelerator improved their reading speed, although with some loss in comprehension. Their test subjects appeared to increase speed more significantly with projected, rather than printed, material. They also had an inflated impression of their own improvement.[60] Even into the 1960s, research tested the effectiveness of tachistoscopes for training, generally finding that reading could be speeded up and fixations could be reduced, although errors would also increase.[61]

The commercial uses went beyond the educational market. Tachistoscopes have long been used to test advertisements, and they continue to be used in this way.[62] Early uses in advertising studies, in the 1930s, charted the best length of advertisements for people to recall accurately what they've seen.⁶³ One of the first such tachistoscopic tests assessed how rapidly men and women could distinguish a magazine page as either copy or ad.⁶⁴ Experiments paired tachistoscopes with other instruments. For example, the psychogalvanometer lie detector reduced the wily element of subject reporting of tachistoscopically flashed word association.⁶⁵ In the 1960s, Seymour Smith used a tachistoscope and survey questions for a cross-media comparison of advertising, studying film, print, brochures, publicity, phone, mail, trade shows, and internal business communication.⁶⁶ So widespread was the interest in qualities of swift-viewing training and testing that in 1964, a psychologist declared, "More than any other psychological apparatus, the tachistoscope ... has earned its place in commercial research."⁶⁷ Speed is an objective for marketers, who continue to test product packaging using tachistoscopes, flashing the design, color, and copy at subjects to chart the most easily recognizable.⁶⁸

Tachistoscopes for groups became more varied in function and were often adaptable to other projection uses,⁶⁹ and low-cost projection tachistoscopes were developed specifically for classroom instruction at the same time (Figure 19.3).⁷⁰ Even as late as the 1970s, new versions were being introduced, for instance, one that resembled a slide projector.⁷¹ Studies examined the varying applicability for different learners, from college students to clerical workers.⁷² In the 1950s, flash film for driving education combined filmstrips and a tachistoscope to surprise students with a driving decision.⁷³ During that decade, *Educational Screen* published several articles on tachistoscopic teaching.⁷⁴ Postwar mass-marketed devices included the Perceptoscope from Perceptual Development Laboratories, which developed and marketed its projection tachistoscope along with reading

FIGURE 19.3 *Keystone Flashmeter in use, ca. 1960.*

FIGURE 19.4 *Speed-I-O-Scope from the Society for Visual Education, 1957.*

instruction guides in 1957. The Speed-I-O-Scope, from the Society for Visual Education (SVE), appeared in the early 1950s (Figure 19.4). The Taylor family's firm, Educational Development Laboratories (EDL), introduced the Tach-X Tachistoscope in 1953, following a trademark battle blocking its initial use of Time-X as the name for this product (Figure 19.5).

The Keystone View Company, whose name had been made with stereoscopes and stereoscopic views, experimented with school uses in 1938, and it began to market its Flashmeter in 1944, a device with a top speed of 1/100th of a second.[75] Keystone actively produced manuals of instruction by educational psychologists with clear lesson plans for teachers to follow for tachistoscopic training, beginning in 1949.[76] These were whole teaching programs, entire course curricula, for subjects ranging from vocabulary to math. In effect, the Keystone View Company was selling what we now call both hardware and software to schools. One instructional set provided a way to use the tachistoscope as a timing device for other reading activities, such that students could assess how many words per minute they read during a particular trial.[77] Other Tachistoslides tested math skills, drilled object and geometric identification, improved vocabulary, presented picture–word pairs, assessed reading proficiency, and offered motivational slogans for efficient reading (e.g., "You Can Read Well," "Keep Your Mind Alert," and "Don't Move Your Lips") (Figure 19.6).[78] A reading program that began with the identification of shapes by young children was to train the controlled movement of the eye, a muscular preparation for the development of literacy.

Tachistoscopes for Group Use

EDL Tach-X, Educational Development Laboratories

Keystone Standard Tachistoscope, Keystone View Company

Rheem-Califone Percepta-matic, Carleton Films

T-ap All Purpose Tachistoscope Attachment (changes projectors into tachistoscopes), Lafayette Instrument, Co.

SVE Speed-I-O-Scope, Society for Visual Education, Inc.

Tachistoscopes for Individual Use

Electro-Tach, Lafayette Instrument, Co.

Tachist-O-Viewer and Tachist-O-Flasher, Learning Through Seeing, Inc.

AVR Eye-Span Trainer, Audio-Visual Research

Phrase Flasher, Reading Laboratories, Inc.

Reading Pacers

Shadowscope, Psychotechnics, Inc.

Prep-Pacer, Reading Laboratories, Inc.

AVR Reading Rateometer, Audio-Visual Research

SRA Reading Accelerator, Science Research Associates

Reader Pacer, Genco Educational Aids

Films or Filmstrips

Controlled Reader, Controlled Reader Jr., Educational Development Laboratories

Tachomatic 500 Reading Projector, Psychotechnics, Inc.

Craig Reader, Craig Research, Inc.

FIGURE 19.5 *Instructional technology for reading teachers and manufacturers. From Lou E. Burmeister,* Foundations and Strategies for Teaching Children to Read *(Reading, Mass.: Addison-Wesley, 1983), 466.*

Among other things, these companies successfully made the tachistoscope an indispensable tool for the 1950s fad of speed-reading, until the Evelyn Wood hegemony supplanted it the 1960s. Earlier reports about speed-reading in popular and trade publications did not focus on technique or instructional aids. For example, a *New York Times* report in 1923 made only brief reference to the possibility that movies might exercise the eyes and hence improve reading speeds,

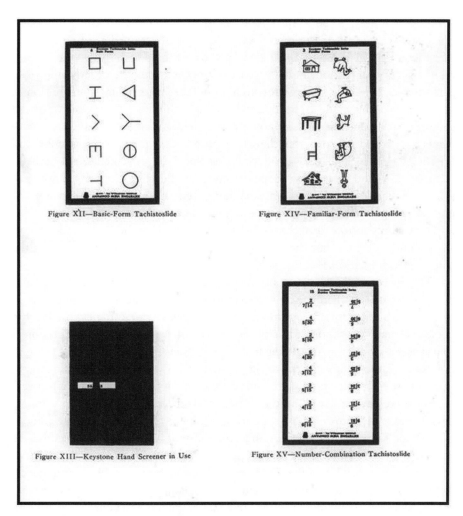

FIGURE 19.6 *Keystone Tachistoslides, 1954.*

concentrating instead on a long list of word counts for the Bible, newspapers, and novels and average word-per-minute scores for different levels of schooling.[79] This contrasts with post–World War II articles that reported on the technological aids to faster reading, with tachistoscopes highlighted.[80] In 1953, *Fortune* called it a full-blown "speed-reading craze," documenting the competing approaches: tachistoscopic training, reading films, and nonmechanical skimming.[81] At the time, speed-reading had become a question of career success; the captains of industry and leaders of government, so it was assumed, possess a valuable skill in their ability to scan and assimilate an abundance of material. If you will, they were seen as better armed to manage the abundance and clutter of changing information that characterized the modern world. Reports on the craze presented idolizing word-per-minute scores of famous fast readers, among them John F. Kennedy and Theodore Roosevelt, and corporations set up special seminars for their aspiring executives.

Competing experiments questioned the actual contribution of such devices. For instance, one study compared tachistoscopically trained reading students who were verbally directed and encouraged by teachers with those who did not receive such intervention, finding that the teacher's presence greatly improved performance.[82] EDL sponsored research to measure the singular contribution to improved reading made by different learning technologies. Previous research showed that a controlled reader could reduce fixation and regressive eye movements and could do so better than two brands of pacers (Shadowscope and Rateometer).[83] Even if they produced observable improvements, research on tachistoscopy had not shown conclusively that the technology itself was the *source* of changes in reading rates, including Renshaw's work that consisted of unusually highly educated subjects and had no control group.[84] EDL's study of different combinations of technology demonstrated that each does indeed contribute uniquely, and that the tachistoscope was most effective in lowering the length of each fixation.[85]

This speed-reading mania continued into the 1960s. As Lawrence Galton put it, "Already, indeed, there are reports that, in some circles, no longer is the content of a book a subject for conversation; all anybody wants to know is how fast the book was read."[86] Although this particular report reasserted the tachistoscope's role in speed-reading, again tracing its origins to Renshaw's World War II recognition training, the 1960s saw the rising popularity of Evelyn Wood's reading methods, whose self-help tone and simple hand-skimming technique effectively overtook the tachistoscope as the dominant speed-reading program. Stanford Taylor speaks for many of those supporting reading technologies when he admonishes Wood's students for not reading everything on the page. That method, he argues, was actually teaching skimming and scanning, not controlled reading, which involves disciplining the eye. Even as Wood's program rose to prominence, Keystone's 1962 client list of regular purchasers and users of its tachistoscope programs, slides, and equipment included over 2,300 schools and boards of education in the United States and 50 abroad.[87] By the end of the 1970s, Stanford Taylor, president of EDL, estimates 240,000 of their devices were in U.S. schools.[88] The extent to which they were being put to use, and not just cluttering up audiovisual closets, is another matter altogether. At the very least, the fact that his company continues to operate today after seventy years indicates that a market for these and comparable reading technologies persists.

Most infamously, some advertising firms in the late 1950s experimented with using tachistoscopes to flash messages at movies, sparking a momentary panic about the subconscious influences of imperceptible text, and incidentally moving the term "subliminal" from psychology into popular lexicon.[89] The twentieth century was awash with concerns about mass media and propaganda. Related, and springing from discourses of a modern technological character, were extreme visions of the production of a passive and irrational contemporary subject, as depicted in the fictional accelerated learning scenarios with which this chapter began. Some have surmised that media can create a special entrée to individual and collective dream life. Here, tachistoscopes have been thought to have "the power to activate psychodynamic processes," and with them "one

FIGURE 19.7 *Keystone catalog, 1966.*

could then subject to laboratory study the fascinating kinds of relationships that psychoanalysis has posited between behavior and unconscious mental processes."[90] In this way, they can replicate and produce the conditions of somnambulism. Since the 1970s, dubious business ventures using the device abound, developing in tandem with the industry of subliminal self-help. With quasi-evangelical zeal, founder of Subliminal Dynamics/Brain Management Richard Welch took credit for inventing in 1975 a new method of speed-reading based on what he called, alternately, subliminal and mental photography. As advertised, it offers a "Whole-Life Enhancement Course, not a photographic memory course."[91] This "new method" uses a tachistoscope flashing words at the impossible speed of 2 million per minute. Welch presents this as a missionary project to get the message out, "to help an ever overburdened world to find the answers for tomorrow."[92]

One should not be perplexed by this last direction. Instead, it is the realization of an underlying dream involving the necessity of meeting the demands of a speedy era. After all, Walter Benjamin long ago alerted us to our modern universe of milliseconds, calling it the "optical unconscious" of photography.[93] In a way, we are part of a Leibnitzian age, obsessed with the most infinitesimal *petit différence* that constitutes the barest minimum of an event. The tachistoscope's functional operations provided concrete figures for what had been imprecise tabulations of how fast, how much, and how well people learn, not to mention how much faster and how much more. In so doing, the device participated in moving part of the laboratory and the classroom into a fleeting realm of split seconds. The preoccupation with micro-manifestations of human behavior, and adjoining aspirations to accelerate those manifestations, eventually pushed responses to previously improbable rates (or at least the recordability of those responses).

We might think of the subliminal as the vanishing point of speed in human recognition, decision, and action, where a type of automatic reflex in thought is said to occur. Experimental psychologists pursued this for decades before the popular subliminal panic of the late 1950s. In 1933, O. A. Simley used a chronoscope with the intention of testing what sort of learning, if any, transpires above and below consciousness, that is, between supra- and sub-liminal learning, seeking to identify the bond that might build among memory, association, and a flickering stimulus.[94] As demonstrated in this chapter, the resonant concern represented by the tachistoscope's history had an initial incarnation as an empirical project to chart the range of normal human perception. The sure draw toward improvement saw questions about testing the limits of that threshold arise, and it saw the tachistoscope become an instrument for controlled experiments and a training aid. For all of the instrument's legitimate uses in reading instruction, underlying the movement from lab to classroom is a valorization of the efficient and the swift in perceptual faculties, as well as an exploration of automatic responsiveness over critical reasoning. The tachistoscope, as a medium of nearly automatic communication, helped us imagine a short-circuiting of rhetorical ploys. The extreme incarnation of its educational objective is the mind control trope. The wide acceptance of the *possible* manipulation of minds through subliminal means was a perfectly reasonable extension of the *actual* urgency to manage mass populations and the *actual* desire to do so efficiently, rapidly, and to special advantage. The tachistoscopic idea, then, circulated with it a language of media manipulation even as it offered remedies for media and textual abundance.

Friedrich Kittler argues that the tachistoscope was not only "the twin of the movie projector" but that it embodied elements of the typewriter, with flashing text striking eyes and brain. He writes, quite plainly, "The tachistoscope is a typewriter whose type hits the retina rather than paper,"[95] by which he ostensibly links it with the technology that is the primary focus of his study of modern notation and storage systems. The tachistoscope–typewriter association prompts Kittler to see the processes observed and measured as "mindless," that is, as an image of the direct and crisp imprinting in our skulls. When Kittler writes, "The automatism of tachistoscopic word exposition is not designed to transport thoughts,"[96] he neglects to appraise the technology's transformation into an instructional aid. And it is this idea of "mind writing" that carries through to the employment of the tachistoscope for subliminal influence. Such media structure the reigning understandings of the relative immobility, concentration, and suggestibility of subjects. And it is by no means the only one, once we begin to take note that these claims are reminiscent of those made of the purportedly manipulative properties of television, popular music, and video games. Pushing further, whether in imagining the forced learning or the overpowering nature of some media, critics contemplated how the same strategies and instruments might become vehicles for the injection of ideas and the overt control of action.

My research shows the sluggish historical trajectory of the tachistoscopic idea as it moved into a series of arenas, dragging along a set of logistics about knowing and influencing the workings of the mind. As a diagnostic and therapeutic instrument, it helped identify and rectify extraneous action in the movement of eyes. As an experimental instrument, it helped discover and measure optical and mental processes. As a teaching instrument, it helped accelerate recognition and memory and provided a quantifiable performance indicator. As a speed-reading training technique, it disciplined eyes to take in entire fields of information rapidly. As a military instrument, it pushed target recognition into a reflex response. As a marketing instrument, it helped streamline decisions about copy and layout. The tachistoscope began as an instrument for the measurement of physiological functions and in subsequent generations became an instrument for the measurement of student, teacher, and curriculum achievements (Figure 19.7). The logic of the performance indicator is evident in this technologization of instructional procedure, the guiding principles being more students and content, fewer teachers and digressions, and faster acquisition and course completion.

In sum, first, the tachistoscope reorients reading from pages to screens, highlighting that the ground for screen reading was well tilled long before computers. Second, the tachistoscope is perfectly continuous with the consequences of the accelerated appearance and disappearance—arrival and departure—of images and text that constitute what we came to think of as information overload. And third, it is continuous with the priorities of modernity and the markup placed on efficiencies of production, including the construction of a synthetic subject who could handle this abundance.

In light of these applications, the tachistoscopic idea is an epistemology of the orientation of attention, one that carries forward a set of ideas about the mass mind and the culture of speed. Crary calls the tachistoscope "paradigmatic of an ideal of experimental control."[97] The claim could apply to any number of equally prevalent psychological instruments. Nevertheless, he is right to see it as emblematic of a trajectory of research and as instrumental in a shifting understanding of the perceptual apparatus. I would add that *it is especially paradigmatic of media control, of technologized instruction, and of rapid mind manipulation*. It is evidence of the view that perception is a muscle to be trained and disciplined through a careful arrangement of a technological apparatus. The tachistoscope, in its many incarnations, is a material manifestation of what we take to be quintessential modern qualities: mechanized sight, Taylorist instruction, and contained and focused attention. It helped institutionalize subject performance into perceptual and cognitive processes, becoming a procedure to chart speeds and a program to improve them. Working on eye muscles and on mental faculties, efficiency and speed reigned as a goal for teachers and students and as a measure of success of that mine-shaft elevator to the brain. Speed, as investigated by commentators as varied as Paul Virilio and Stephen Kern, was especially salient. The tachistoscope designated a mental sound barrier to be smashed, this in the context of the fantastic warp speed in the universe of signs. Speed, it seems, is a modern environmental condition; the swift arrival and departure of words and images, the apparent fleeting presence of texts, necessitated acclimatization and instantaneous judgment. Moreover, the identification of these conditions catalyzed speculation about our relative inadaptability to such "unnatural" shocks.

Here we have a significant point, for once we consider the entirety of the tachistoscopic idea, we must acknowledge that the story of modernity is not only one of fragmentation, shock, and disorientation, as we are continuously reminded, but of the language and procedures of acclimatization, orientation, and synthesis, as well as mass instruction. What is drawn out is a lasting, residual array of concerns about adaptability to the pace of textual flow or the assembly of a survival kit for modern screen visuality.

Notes

1. Robert A. Heinlein, *Stranger in a Strange Land* (New York: Ace, 1961/1987), 72.
2. Daniel Keyes, *Flowers for Algernon* (New York: Harcourt, Brace & World, 1966).
3. Richard Condon, *The Manchurian Candidate* (New York: Quality Paperback, 1959/1988), 30.
4. Ruth Benschop, "What Is a Tachistoscope? Historical Explorations of an Instrument," *Science in Context* 11, no. 1 (1998): 30.
5. Robert S. Woodworth, *Experimental Psychology* (New York: Henry Holt and Company, 1938), 688; J. M. Cattell, "The Inertia of the Eye and Brain," *Brain* 8 (October 1885): 295–312.
6. Benschop, "What Is a Tachistoscope?" 27.
7. Ruth Benschop, *Unassuming Instruments: Tracing the Tachistoscope in Experimental Psychology* (Amsterdam: Proefschrift, Psychologische, Pedagogische en Sociologische Wetenschappen, Rijksuniversiteit Gröningen, 2001).
8. Mark Seltzer, *Bodies and Machines* (New York: Routledge, 1992).
9. Walter Benjamin, "The Work of Art in the Age of its Technological Reproducibility," in *Selected Writings, Volume 3, 1935–1938*, eds. Howard Eiland and Michael W. Jennings (Cambridge, Mass.: Belknap Press of Harvard University Press, 1936/2002), 101–33.
10. Raymond Dodge, "An Improved Exposure Apparatus," *Psychological Bulletin* 4, no. 1 (January 1907): 10–13; Raymond Dodge, "An Experimental Study of Visual Fixation," *Psychological Monograph* 8, no. 4 (1907), 1–95.
11. A. Netschajeff, "Zur Frage ueber die Qualitative Wahrnehmungsform," *Psychologische Studien* (1929): 114–18.

12. Daele L. Wolfle, "The Improved Form of the Gulliksen Tachistoscope," *Journal of General Psychology* 8 (1933): 479–84; Samuel Renshaw and I. L. Hampton, "A Combined Chronoscope and Interval Timer," *American Journal of Psychology* 43 (1931): 637–38.

13. G. C. Grindley, "A New Form of Tachistoscope," *British Journal of Psychology* 23 (1933): 405–7.

14. M. A. Tinker, "A Flexible Apparatus for Recording Reading Reactions," *Journal of Experimental Psychology* 15 (1932): 777–78.

15. John G. Jenkins, "A Simple Tachistoscope of Many Uses," *American Journal of Psychology* 45 (1933): 150; J. E. Evans, "A Tachistoscope for Exposing Large Areas," *American Journal of Psychology* 43 (1931): 285–86.

16. Carr Harvey, "Space Illusions," *Psychological Bulletin* 9, no. 7 (July 1912): 257–60.

17. Sidney M. Newhall, "Projection Tachistoscopy," *American Journal of Psychology* 48 (1936): 501–4; Harold Schlosberg, "A Projection Tachistoscope," *American Journal of Psychology* 43 (1931): 499–501.

18. Friedrich A. Kittler, *Discourse Networks, 1800/1900*, trans. Michael Metteer with Chris Cullens (Stanford, Calif.: Stanford University Press, 1985/1990), 222.

19. I have not found examples of bodies being secured in any violent fashion. After all, the tests were generally performed on colleagues, students, relatives, and acquaintances, whom (for the most part) one would not want to subject to petty tortures, even in the name of scientific investigation. Much of this early research was not the more ethically suspect work in asylums or work that exploited subjugated persons. For a good image of the relatively comfortable, if immobile, bodily situation, see Guy Montrose Whipple, "The Effect of Practice upon the Range of Visual Attention and of Visual Apprehension," *Journal of Educational Psychology* 1, no. 5 (May 1910): 252.

20. Jonathan Crary, *Suspensions of Perception: Attention, Spectacle and Modern Culture* (Cambridge, Mass.: MIT Press), 306.

21. Ibid., 311. Building on Crary's claims, Martin Thomasson offers a perceptive commentary linking Hugo Münsterberg and Wundt to argue that tachistoscopes and motion pictures together helped produce an idea of machine management of minds in his essay "Machines at the Scene: The Cutting-Up and Redistribution of the Sensorium in Nietzsche, Wundt and Münsterberg," *Site Magazine* (www.sitemagazine.net) 3-4 (2002): 10–11.

22. See Henri Ellenberger, *The Discovery of the Unconscious: The History and Evolution of Dynamic Psychiatry* (New York: Basic Books, 1970).

23. R. I. Waston Sr., *The Great Psychologists*, 4th ed. (New York: Lippincott, 1978).

24. Harvey, "Space Illusions."

25. Kittler, *Discourse Networks*, 223.

26. Crary, *Suspensions of Perception*, 307.

27. Kittler, *Discourse Networks*, 251.

28. Ibid., 252. Actually, unless it is not operating as it should, projectors do not "chop up" the flow of images. Kittler is most likely referring to the technologies of film editing, not projection.

29. Crary, *Suspensions of Perception*, 306.

30. E. A. Esper, "The Bradyscope: An Apparatus for the Automatic Presentation of Visual Stimuli at a Constant Slow Rate," *Journal of Experimental Psychology* 9 (1926): 56–59.

31. Benschop, "What Is a Tachistoscope?" 31.

32. Crary, *Suspensions of Perception*, 307.

33. Benschop, *Unassuming Instruments*, 90, cites a 1923 experimental classroom use of a tachistoscope.

34. Woodworth, *Experimental Psychology*, 696.

35. Catherine Aiken, *Method of Mind-Training: Concentrated Methods in Attention and Memory* (New York: Harper and Brothers, 1896); Catherine Aiken, *Exercises in Mind-Training* (New York: Harper and Brothers, 1899).

36. Guy Montrose Whipple, "The Effect of Practice upon the Range of Visual Attention and of Visual Apprehension," *Journal of Educational Psychology* 1, no. 5 (May 1910): 249–62.

37. W. Hische, "Identifikation und Psychotechnik," *Psychotechnisches Zeitschrift* 5 (1930): 95–97.

38. W. V. Machaver and W. A. Borrie, "A Reading Improvement Program for Industry," *Personnel* 28 (1951): 123–30; Mary Jane Hurt, "Perception Training for Telephone Information Operators," *American Journal of Optometry* 32 (1955): 546–52.

39. Viktor Sarris and Michael Wertheimer, "Max Wertheimer's Research on Aphasia and Brain Disorders: A Brief Account," *Gestalt Theory* 23, no. 2 (2001): 267–77.

40. Kittler, *Discourse Networks*, 223.

41. Louis Leon Thurstone, *Primary Mental Abilities* (Chicago: University of Chicago Press, 1938).

42. E. Grund, "Das Lessen des Wortanfanges bei Volkschulkindern Verschiedener Altersstufen," *Neue Psychologische Studien* 6 (1930): 311–16.

43. W. Stein, "Tachistoskopische Untersuchungen Ueber das Lesen mit Sukzessiver Darbietung," *Archiv für die Gesamte Psychologie* 64 (1928): 301–46.

44. C. O. Weber, "The Use of Tachistoscopic Exercises in the Improvement of Reading Speed," *Psychological Bulletin* 34 (1937): 533–34.

45. D. E. Swanson, "Common Elements in Silent and Oral Reading," *Psychological Monographs* 48 (1937): 36–60.

46. W. F. Dearborn, "The Tachistoscope in Diagnostic Reading," *Psychological Monographs* 47, no. 212 (1936): 1–19.

47. E. A. Taylor, *Controlled Reading* (Chicago: University of Chicago Press, 1937).

48. Samuel Renshaw, Vernon L. Miller, and Dorothy P. Marquis, *Children's Sleep* (New York: Macmillan, 1933).

49. John M. Larsen Jr., "Obituary: Samuel Renshaw (1892–1981)," *American Psychologist* 38, no. 2 (February 1983): 226.

50. Samuel Renshaw, "The Visual Perception and Reproduction of Forms by Tachistoscopic Methods," *Journal of Psychology* 20 (1945): 217–32.

51. "Recognizing Planes," *New York Times*, November 21, 1943, section 4, 13.

52. Larsen, "Obituary: Samuel Renshaw," 226.

53. Wesley Price, "New Ways to Learn Faster," *Science Digest* (March 1944): 12–16.

54. "Fast Looks," *Time*, September 24, 1945, 76.

55. Samuel Renshaw, *How to Use the Renshaw Tachistoscopic Trainer* (Chicago: Stereo Optical Company, 1950). In a curious parallel development, Michael Lobel examines how Roy Lichtenstein's paintings explore "the interrelation between machines and visual perception." Lobel argues that this interest grew from one of his early teachers, Hoyt Sherman, who used tachistoscopes at Ohio State University to flash images in a darkened room and had students draw what they had barely seen. Lichtenstein apparently taught in this flash lab between 1946 and 1949, and he tried to establish similar labs elsewhere. The artist himself implied Renshaw snatched the idea for tachistoscopic training from Sherman, which is unlikely given how long Renshaw had been doing such studies. Michael Lobel, "Technology Envisioned: Lichtenstein's Monocularity," *Oxford Art Journal* 24, no. 1 (2001): 131–54; and Michael Lobel, *Image Duplicator* (New Haven: Yale University Press, 2002). Where Lobel associates tachistoscopic training with the flatness of pop art, I would point to what we could reasonably call an emergent aesthetic of flicker in the early 1960s. Here, artists, and especially avant-garde filmmakers, take the raw physiological disorientation of flashing light as formal experiment and as reference to the psychological shock of modern life. The apogee here is perhaps Tony Conrad's film *Flicker* (1966).

56. Alan Snyder, "The Flashreader in the Reading Laboratory," *The English Journal*, May 1952, 269.

57. F. H. Melcer and B. G. Brown, "Tachistoscopic Training in the First Grade," *Optometric Weekly* 36 (1945): 1217–19.

58. Malcolm D. Arnoult, "Accuracy of Shape Discrimination as a Function of the Range of Exposure Intervals," *United States Air Force Human Resources Research Center Research Bulletin* 51, no. 32 (1951): v, 12.

59. D. K. Wilson, "The Manual-Verbal Response Tachistoscope: Distracting Device for Intelligibility Testing," Technology Report SDC104-2-20 (Port Washington, N.Y.: U.S. Navy Special Devices Center, 1950), ii, 17.

60. Henry P. Smith and Theodore R. Tate, "Improvements in Reading Rate and Comprehension of Subjects Training with the Tachistoscope," *Journal of Educational Psychology* 44 (1953): 176–84.

61. Amy Schaffer and John D. Gould, "Eye Movement Patterns as a Function of Previous Tachistoscopic Practice," *Perceptual and Motor Skills* 19, no. 3 (1964): 701–2.

62. Darvin Winick, "An Investigation of the Group Tachistoscopic Method of Evaluating Magazine Advertisements," *Dissertation Abstracts* 15 (1955): 2291; David Taylor, "A Study of the Tachistoscope," *British Journal of Marketing* 4 (Spring 1970): 22–28.

63. D. B. Lucas, "The Optimum Length of Advertising," *Journal of Applied Psychology* 18 (1934): 665–74.

64. H. K. Nixon, "A Study of Perception of Advertisements," *Journal of Applied Psychology* 11 (1927): 135–42.

65. John Caffyn, "Psychological Laboratory Techniques in Copy Research," *Journal of Advertising Research* 4, no. 4 (December 1964): 45–50.

66. Seymour Smith, "Total Measurements, Evaluating All Promotional Factors in a Campaign," *Proceedings of the 11th Annual Conference of the Advertising Research Foundation*, October 5, 1965, 76–80.

67. Caffyn, "Psychological Laboratory Techniques," 48.

68. "Tachistoscope," *Dictionary of Business* (Oxford: Oxford University Press, 1996); http://www.xrefer.com/entry/166745.

69. John L. Newson, "A Projection Tachistoscope," *Quarterly Journal of Experimental Psychology* 6 (1954): 93–94.

70. J. A. Deutsch, "The Reflecting Shutter Principle and Mechanical Tachistoscopes," *Quarterly Journal of Experimental Psychology* 12 (1960): 54–56.

71. E. G. Boyd, "Portable Tachistoscope Based on a Commercial Slide-Viewer," *Bulletin of the British Psychological Society* 25, no. 87 (April 1972): 113.

72. Marvin J. Dumler, "A Study of Factors Related to Gains in the Reading Rate of College Students Trained with the Tachistoscope and Accelerator," *Journal of Educational Research* 52 (1958): 27–30; Arthur C. Mackinney, "A Validation of Tachistoscopic Training for Clerical Workers," *Personnel Psychology* 11 (1958): 13–23.

73. Arnold E. Luce, "Flashfilm—Minnesota's Contribution to Better Driver Education," *Educational Screen and AV Guide* 37 (February 1958): 70, 71, 75.

74. James I. Brown, "Vocabulary via Tachistoscope," *Educational Screen* 30 (September 1951): 274, 287; Frances M. Benson, "We Improved Spelling with the Tachistoscope," *Educational Screen and AV Guide* 35 (November 1956): 408; Rolland A. Alterman, "Tachistoscopic Teaching," *Educational Screen and AV Guide* 37 (June 1958): 282, 283, 293.

75. *General Manual of Instructions: The Keystone Tachistoscope, Advanced Teaching Techniques* (Meadville, Penn.: Keystone View Company, 1954).

76. G. C. Barnette, *Learning through Seeing* (Meadville, Penn.: Keystone View Company, 1949).

77. James I. Brown and Eugene S. Wright, *Manual of Instructions for Use with the Minnesota Efficient Reading Series—Timing Group* (Meadville, Penn.: Keystone View Company, 1955).

78. James I. Brown and Eugene S. Wright, *Manual of Instructions for Use with the Minnesota Efficient Reading Series of Tachistoslides* (Meadville, Penn.: Keystone View Company, ca. 1955).

79. Thomas L. Masson, "Read with Speed," *New York Times*, February 18, 1923, section 4, 5, 13.

80. Alan Snyder, "The Flashreader in the Reading Laboratory," *The English Journal*, May 1952, 269; Katherine and Henry F. Pringle, "Time Yourself Reading This," *Nation's Business*, March 1953, 86–89.

81. "Fast Reading vs. Fast Thinking," *Fortune*, May 1953, 47.

82. John A. Marvel, "Acquisition and Retention of Reading Performance on Two Response Dimensions as Related to 'Set' and Tachistoscopic Training," *Journal of Educational Research* 52 (1959): 232–37.

83. William Mathis and Donald R. Senter, "Quantification of Contributions Made by Various Reading Instrument Combinations to the Reading Process," *Research and Information Report no. 7* (New York: Educational Developmental Laboratories, 1973), iii–v.

84. Ibid., v–vi.

85. Ibid., 13.

86. Lawrence Galton, "2,000 WPM—But Is it Reading?" *New York Times*, August 27, 1961, section 6, 60.

87. Keystone View Company, *Users of the Keystone Tachistoscopic Services* (Meadville, Penn.: Keystone View Company, 1962), Crawford County Historical Society, Keystone View Company Collection.

88. Stanford Taylor, personal interview, November 13, 2002.

89. I am currently working on a monograph on this topic, titled *Hidden Messages*.

90. Lloyd Silverman, quoted in Virginia Adams, "'Mommy and I are One': Beaming Messages to Inner Space," *Psychology Today*, May 1982, 24, 27, 28, 30–32, 34, 36.

91. Subliminal Dynamics, Centennial, Colorado, http://www.subdyn.com/subwel.html.

92. Ibid.

93. Walter Benjamin, "Little History of Photography," in *Selected Writings Volume 2, 1927–1934*, eds. Michael W. Jennings, Howard Eiland, and Gary Smith, trans. Rodney Livingstone and others (Cambridge, Mass.: Belknap Press of Harvard University Press, 1931/1999), 510.

94. O. A. Simley, "The Relation of Subliminal to Supraliminal Learning," *Archives of Psychology* 146 (January 1933): 5–40.

95. Kittler, *Discourse Networks*, 223.

96. Ibid., 223–24.

97. Crary, *Suspensions of Perception,* 305.

Contributors

Charles R. Acland is professor and Concordia University Research Chair in Communication Studies at Concordia University, Montreal. He is author of *Youth, Murder, Spectacle: The Cultural Politics of "Youth in Crisis"* and *Screen Traffic: Movies, Multiplexes, and Global Culture*, which won the 2004 Robinson Book Prize for best book in communication studies by a Canadian scholar. He is co-editor of *Harold Innis in the New Century: Reflections and Refractions* and is on the editorial boards of *Cultural Studies, The Velvet Light Trap,* and *Topia: Canadian Journal of Cultural Studies*.

Jennifer Adams is assistant professor in the Department of Communication and Theatre at DePauw University in Greencastle, Indiana. Her research in rhetorical theory and criticism, letter writing as a communicative genre, and dialogue has appeared in the *American Journal of Semiotics* and the *Southern Journal of Communication*. She is currently conducting archival research on women suffrage workers in Indiana.

Jody Berland is associate professor of humanities at York University, of Toronto. She has published widely on cultural studies, Canadian communication theory, music and media, cultural studies of nature, and the cultural technologies of space. She is co-editor of *Cultural Capital: A Reader on Modernist Legacies, State Institutions and the Value(s) of Art*, and editor of *Topia: A Canadian Journal of Cultural Studies* (www.yorku.ca/topia). Her book *North of Empire* is forthcoming with Duke University Press.

Sue Currell is lecturer in American literature at the University of Sussex, England. She is working on a history of self-improvement in the United States. She is co-editor of *Popular Eugenics: National Efficiency and Mass Culture in the 1930s* and author of *The March of Spare Time: The Problem and Promise of Leisure during the 1930s*.

John Davis is interested in the relationships between new media technologies, audiences/consumers/users, and commercial and noncommercial media producers. He teaches communication at several colleges in the greater Chicago area.

Maria DiCenzo is associate professor of English at Wilfrid Laurier University. Her work on the Edwardian suffrage press includes articles in *Media History*, *Women's History Review*, and the *Victorian Review*. She is also the co-editor, with Lucy Delap and Leila Ryan, of *Feminism and the Periodical Press, 1900-1918*.

Kate Egan is a lecturer in film and television studies at the University of Wales, Aberystwyth, UK. She has published essays on fan Web site discussions of the video nasties, and the marketing of and audience response to *The Lord of the Rings*. Her book *Trash or Treasure: Censorship and the Changing Meanings of the Video Nasties* is forthcoming from Manchester University Press.

Lisa Gitelman is associate professor and chair of the Department of Media Studies at Catholic University in Washington, D.C. She is author of *Scripts, Grooves, and Writing Machines: Representing Technology in the Edison Era* and *Always Already New: Media, History, and the Data of Culture*, as well as the coeditor of *New Media, 1740–1915*.

Alison Griffiths is associate professor of communication studies at Baruch College, City University of New York, where she teaches documentary film and media studies. She is author of *Wondrous Difference: Cinema, Anthropology, and Turn-of-the-Century Visual Culture*. Her work on precinema and early cinema has appeared in *Wide Angle*, *Visual Anthropology Review*, *Film History*, *Journal of Popular Film and Television*, and numerous anthologies.

James Hamilton is assistant professor of advertising and public relations at Grady College of Journalism and Mass Communication, University of Georgia, where he teaches courses on advertising and society, media history, and critical cultural history and theory. He has published essays on the history, theory, and practice of alternative media and is working on a critical history of alternative media.

James Hay is associate professor of media and cultural studies at the University of Illinois, Champaign-Urbana. His books include *Popular Film Culture in Fascist Italy*, *The Audience and Its Landscape*, and *Better Living Through Television* (forthcoming). He has written numerous essays about media and communication, technology, governmentality, and social space.

Michelle Henning is senior lecturer in media and cultural studies at the University of the West of England, Bristol, where she teaches photography practice and theory, cultural theory, and exhibition studies. She is author of *Museums, Media and Cultural Theory*, and her essays on photography and the digital image, and on the work of Walter Benjamin, are included in several anthologies. She is also a visual artist and photographer.

Lisa Parks is associate professor of film and media studies at the University of California at Santa Barbara. She is the author of *Cultures in Orbit: Satellites and the Televisual* and coeditor of *Planet TV: A Global Television Reader*. She serves on the editorial boards of *Film Quarterly, Mediascape*, and *The Velvet Light Trap* and is working on a new book titled *Mixed Signals: Media Technologies and Cultural Geographies*.

Hillegonda C. Rietveld is reader in cultural studies at London South Bank University, UK, where she teaches sonic-media-related subjects. Her publications address the development and experience of postdisco and postrave dance music cultures, and she is the author of *This Is Our House: House Music, Cultural Spaces and Technologies*. She has been involved professionally in club and DJ culture since 1982.

Leila Ryan is assistant clinical professor in the School of Nursing at McMaster University. Her research interests extend from literary studies to the history of nursing. She is currently working on a history of the School of Nursing at McMaster University. She is co-editor, with Lucy Delap and Maria DiCenzo, of *Feminism and the Periodical Press, 1900–1918*.

Collette Snowden has worked as a journalist, in public relations, and as a consultant and researcher in mobile communications and information technology. She was the inaugural Donald Dyer Research Scholar in the School of Communication at the University of South Australia where she is director of the Communication and Media Management Program.

Jonathan Sterne is associate professor in the Department of Art History and Communication Studies at McGill University. He is author of *The Audible Past: Cultural Origins of Sound Reproduction* and has written widely on media, technology, and the politics of culture. His next book is tentatively titled *MP3: The Meaning of a Format*. He is an editor of *Bad Subjects: Political Education for Everyday Life* (http://bad.eserver.org), one of the longest continuously running publications on the Internet.

JoAnne Stober is an archivist of photography at Library and Archives Canada, as well as a Ph.D. candidate in communication studies at Concordia University, Montreal. Her thesis examines Canadian cinemagoing during the adoption and integration of synchronized sound technology from 1926 to 1931.

Will Straw is associate professor of communications at McGill University in Montreal. He is the author of *Cyanide and Sin: Visualizing Crime in 50s America* and of more than fifty articles in journals and anthologies. He is on the editorial boards of *Cultural Studies, Space and Culture, Social Semiotics, Screen, Conjunctures, Journal of Canadian Studies, Canadian Journal of Communications, Semiotic Inquiry,* and *Culture, Theory, and Critique*.

Haidee Wasson is assistant professor of cinema studies at Concordia University, Montreal. She is author of *Museum Movies: MoMA and the Birth of Art Cinema* and co-editor of a collection of essays on the history of film studies, *Inventing Film Studies*. Her scholarship has appeared in *The Moving Image, The Canadian Journal of Film Studies, Continuum, Convergences,* and *Cineaction*. She is currently working on a history of 16mm projectors and mobile film screens.

Index

Abell, Adelaide, 350–51
abolitionism, broadcasting and, 286, 287–90
accumulation, xv–xvii; mobility and, 4–5; modes of, 5; salvaging from waste streams and, 33–47; via new storage media, 12; videocassettes and, 8
Ackery, Ivan, 145
Acland, Charles R., xiii–xxvii, 6–7, 361–84
activism, broadcasting and, 287–90
Adams, Jennifer, xvi, xxiv, 185–99
Ader, Clément, 294
advertising: museums and, 163, 172–78, 180; novelty emphasized in, 40; obsolescence and, 22; for pianos, 312, 317, 321; tachistoscopes and, 371–72
aesthetics: collecting and, 209–10; dance music and, 103; of letters, 196–97; pianos and, 312–13
affective attachments, 3
"African Voices" exhibit, 84–93 agency: broadcasting and, 288; museum exhibits and, 89–90; telephone use and, 117
The Age of Television (Bogart), 18
agriculture, broadcasting in, 283–86
Aiken, Catherine, 369
Albee, Edward F., 137
Al-Hurra, 280
Allen, Robert C., 133, 134, 135, 137, 139
ambivalence, 58–62
The Americanization of Edward Bok (Twain), 339
American Museum of Natural History (AMNH), 70–95, 166; film use by, 79–81; Milstein Hall of Ocean Life, 91; tuberculosis exhibit in, 72, 73
American Museum of the Moving Image, 84
American Philatelic Society, 192
American Premium Record Guide (Docks), 230
American Productivity and Quality Center, 276–77

analog technologies, 100–101; dance records, 97–114; photography, 48–65; vinyl records, 97–114, 222–36
Anderson, Irving, 357
Anderson, Maxwell L., 84
Angelus Novus (Klee), 60
annotation, 11–12, 13
Anytime (Mennan, Kutukcuoglu, & Yazgan), 5
Appadurai, Arjun, xv, 50, 51
Appelbaum, Ralph, 90
Apple Computer, 20
appropriation: resistant, 55–56
The Arcades Project (Benjamin), xvi–xvii, 59; on museums, 159; on time and progress, 60
archaic, the, xxi
architectonics, 79–80, 87–88
architecture, 5; museums and, 91–92, 164–65
archives, vinyl records as, 105–6
Arendt, Hannah, 195
Argento, Dario, 203
Arluke, Arnold, 120
art: commodification of, 159–84, 171–78; interwar period and, 179; letters as, 191–92, 196–97; photography and, 58–62; pianos as, 312; reproductions of, 74–75, 171–78, 197
Art and Experience (Dewey), 191
artifactuality: acquisition rituals and, 229–30; Internet and, 3–4; of vinyl records, 105–6
Art Institute of Chicago, 78
The Art of Rapid Reading (Pitkin), 354–55
attention, 365–66, 379–80
Atton, Chris, 242, 246–47
audiences: for museums, 167; social shaping theory and, 224; for vaudeville, 133–34, 136, 137, 141–42
auditioning rituals, 231–32
Augé, Marc, xv
Auslander, Peter, 328n50
Austin Powers: Goldmember, 9–11
authenticity: pianos and, 314–17; reproductions and, 197; vinyl recordings and, 225, 226–27

389

authority, museums and, 165, 178–79
authorship: DJs and, 98, 109; film and, 339–41; literary production and, 331; music and, 318 Twain and, 338
autobiography, collecting as, 210–12
automatic cinema, 93
avant-garde, 179

backward compatibility, 23
Baker, Arthur, 110
Baker, Houston A., 110
Baker, Josephine, 106
Bakhtin, M. M., 195–97
Bakshy, Alexander, 148
Baldwin, Craig, xvii
Balzac, Honoré de, xiii
Bambaataa, Afrika, 108, 110
bandwidth, 11
Barker, Robert, 335
Barnum, P. T., 78, 140
Barthes, Roland, 54, 62, 187, 188
Basel Action Network, 39
Bataille, Georges, 29
battery hen model of news, 127–29
Baudrillard, Jean, xv, 209–10, 212–13
Bava, Mario, 203
Beardsworth, Richard, 326
beat spinning, 107
Beck, Martin, 139
becoming, xviii
Beetham, Margaret, 244, 248
Belasco, David, 148
Belk, Russell W., 225, 228, 229, 231
Bell, Alexander Graham, 293
Bellotte, Pete, 110
Beninger, James R., 69
Benjamin, Walter: on auras, 197; on childhood, xvi–xvii; on collecting, 192–95; on film, 58–62; influence of, xvi–xvii; on memory, 194–95; modernity and, 363–64; on museums, 159; on the optical unconscious, 378; on personalization in collecting, 217; on progress, 59; video collecting criteria compared with, 202
Bennett, Tony, 165
Benschop, Ruth, 362, 363, 368
Bentham, Jeremy, 335, 342n17
Benton, Megan, 330
Berland, Jody, xxii, xxv, 303–28
bias: in communication technologies, 130–31; in telephone communication, 129; of time in the media, xvi
Bjarkman, Kim, 220n40
Blair, Jayson, 128
Blair, Tony, 258
Blauvelt, Andrew, 84
blaxploitation films, 9, 10
Blood Feast, 203
Bogart, Leo, 18
Bok, Edward, 338–39

Bolter, Jay, xxi–xxii, 49
Bond, James, 9–10
Bonnell, Victoria, xxii
Book of the Month Club (BMC), 159–62
Boston Museum of Fine Arts, 166
Bourdieu, Pierre, xv, 16, 162
Boyle, Danny, xv
Bragg, Rick, 128
branding: in "African Voices" exhibit, 86; of interactive media, 92–93; museums and, 178
break beat DJs, 104–5, 107–8
Brecht, Bertolt, 296–97, 322
Breslin, Jimmy, 129
Briggs, Asa, 294
British Board of Film Classification, 203
British Institute of Adult Education, 81
British Museum Association, 79, 80–81
British Science Museum, 80
broadcasting, 283–300; agriculture and, 283–86; evangelical Christianity and, 285, 286–91; human involvement in, 292–93; organization of, 291–96; as residual medium, 284–85; as social networks, 283; telephony in, 124–25
Broadcast over Britain, 75
Brokaw, Tom, 190
Brown, Bob, 356
Brown, James, 104, 108
Brown, John Seeley, 35, 185–86
Brown, Michelle, 128
Brügger, Neils, 241
Bryant, William Cullen, 284, 288
Buckhouse, James, 84
Buck-Morss, Susan, xvi–xvii
Bumpus, Herman Carey, 72
Burkett, Carey, 189
burlesque, 133, 134, 135–36. *See also* vaudeville
The Burning, 203, 207
Bush, George W., 127, 258, 259, 260, 277–79

Caine, Michael, 9
Camera Lucida (Barthes), 62
Campbell-Swinton, A. A., 45
Cannibal Holocaust, 203, 209
Cantwell, Robert, 11
capitalism: agricultural broadcasting in, 285–86; broadcasting and, 295–96; museums and, 163; obsolescence and waste in, 25–29; retro and, 14; waste streams created in, 33–47
car boot sales, 214, 215
Carey, James, 122, 123
cargo cults, 57
Carnegie Mellon, 35
Caruso, Enrico, 316
Casey, Bob, 103
Casey, Dawn, 87
Cattell, James McKeen, 340–41, 348, 362–63, 368

CD-Rs, 100, 112
celebrity, 8
centralization, broadcasting and, 291–96
Cerplex, 37
Certeau, Michel de, xiii, 141
Chanan, Michael, 102, 315
Chaplin, Charlie, 148
charge coupled devices (CCDs), 51
Charney, Leo, xxii, 330
Charter of the New Urbanism, 275–76
Chicago Daily News, 74
Chicago Indicator, 318–19
childhood, xvi–xvii
Christianity, evangelical, and broadcasting, 285, 286–91
chronotopes, 196
Cicero, 186–87
Cimarron, 147
cinema. *See* films
Cinema and the Invention of Modern Life (Charney & Schwartz), xxii, 330
City Lights, 148
City of God, 10
civic goods, 258
civic life: community and, 267–68; informatics and, 271–77; loss of social capital and, 268–70; New Urbanism and, 273–77
class systems: broadcasting and, 286, 287–91, 290; dance records and, 100; marginalization and, 54–55, 54–58, 70, 102–4; materialism and, xxi; pianos and, 304, 307; technological potential and, 101; vaudeville and, 133–34, 136, 138–39
clavecin, 309
Clinton, Bill, 259, 268, 274, 277–79
A Clockwork Orange, 361
coercive participation, 24–25
cognitive mapping, 93
Cohen, Joe, 149
collectivizing rituals, 232
collectors and collecting: autobiography and, 210–12, 228; comparative objects and, 209; as creators, 216–17; curatorial, 228; definition of, 225; history and, 202, 204, 208–10; homosocial nature of, 223; knowledge base in, 105–6, 207–10, 212; letters and, 192–95; market economies and, 213–16; memory and, 194–95; nostalgia and, 210–12; organization in, 202, 204, 211–12; as participation in culture, 192–95; public discourses in shaping of, 200–201; rarity and, 212–13; rituals in, 205–6, 229–234; sacralizing by, 231; technical orientation in, 201–202; terminal vs. instrumental, 229–30; value and, 209–10; of video nasties, 200–21; of vinyl records, 105–6, 223
Collier, John, 141
Colonel Chabert (Balzac), xiii
Columbian Exposition (1893), 349

commentary, 11–12, 13
commodification: of art by museums, 159–84; of entertainment, 303–28; of intimacy, 130; in museums, 92–93; of music making, 303–28; of video nasties, 212–13
commonality, 266
The Common Cause, 247, 251
communication: community and, 270–82; computer-mediated, 266–68; dyadic, 117, 118–119, 125, 129; effect of telegraphy on, 122–23; letters as, 185–99; public vs. private, 130; technologization of, 120–21; telephonic, 116–18, 120–21
communitarianism, 257–82; broadcasting and, 291–92; commonality and, 266; communities of practice and, 261–63; cyber-communities and, 266–68; gated communities and, 274–75; government rationalities and, 264–79; informatics and, 271–77; Internet, 268–82; new information technologies and, 273; power-geography of, 279–80; smart growth and, 273–76; as Western, 259, 260
communities of practice, 261–63, 276–79
Community Informatics (Schuler), 271
compact discs (CDs), 97, 113, 222–23
  CD-Rs, 100, 112
computers: disembodiment and, 17; disposability of, 16–31; Moore's law and, 19–20; as new media, 19–21; planned obsolescence of, 20–26; salvaging, 32–47; structured obsolescence of, 35; as trash, 24–28
Conboy, Martin, 241
concert saloons, 136
Conciliation Bill of 1912, 250–54
Condon, Richard, 361
Congress of the New Urbanism, 275
Conservative & Unionist Women's Franchise Review, 249
consumerism, xv–xvii; broadcasting and, 290–91; car boot sales and, 214, 215; democracy and, 315; gendered, 306–7; mass entertainment and, 138; museums and, 171–78; periodicals in, 248; planned obsolescence and, 21–25; progress and, 319–20; video collecting as, 200–221, 204–5; women in, 174–78, 306–7
context, 229
continuity, xix–xx, 134–35
Controlled Reading (Taylor), 370
convenience vs. duration, 304
convergence, 118–19, 128, 165
conviviality, 28
Coppola, Roman, 10
CQ, 10
Crafton, Donald, 142
Crary, Jonathan, 330, 346, 365–66, 379–80
Craw, George, 140
Crawford, Isabella Crawford, 307
Crewe, Louise, 215, 216

Crosby, Bing, 52
cultural knowledge, 7, 9–11
cultural logistics, 363
culture: collecting and, 192–95, 229; corporate, 29; development of forms of, 240–43; dominant and residual/emergent, xx–xxi, 53, 55–58; high vs. low, 138–39; marginalization and, 55–58; media history and, xxii; models for change in, xx–xxii; movement and, 4–5; recombinant, 11, 40; technologies as forms of, 223–24; video stores in reshaping of time in, 5–11
Curran, James, 240, 242–43
Currell, Sue, xxiii, xxv, 344–60
Curtis, David, 69
cyber-communities, 266–68

Dahir, Abirahman, 88
Dahl, Curtis, 332
Dahl, Hans Fredrik, 242
Dahl, Steve, 111
dance records, 97–114; development of, 101–2; disco, 105–7; hip hop, 107–8; as musical instrument, 108–9; public dancing and, 102–4; underground, 106–7
Darabont, Frank, 8
Dark Angel's Realm of Horror, 207
The Dark Side, 204–5, 207, 211
Davis, Fred, xiv
Davis, John, xxiv, 222–36
Dawn of the Dead, 208
28 Days Later, xv
dead media, xx, 224
Dearborn, William, 353, 355–56
Decasia, xvii
Decker, William, 188–89
decontextualization, 62
de Forest, Lee, 291, 294
Dekovic, Ivo, 32–33
Delabarre, Eugene, 349, 350, 351
delays, 5, 10–11
Dell, 36
democracy: broadcasting and, 290–91, 296–97; communitarianism and, 257–82; consumerism and, 315
Democracy in America (Tocqueville), 21
democratization, 71, 140–41
Demonstrations in Psychological Optics, 371
de Quincey, Thomas, 345
Derrida, Jacques, 187, 326
Descartes, Rene, 17
design, museums and, 174–78
Deutsches Museum, 80
Dewey, John, 191
dialogic utterance, 195–96
Diana cameras, 59
Dibbets, Kerel, 143
DiCenzo, Maria, xxii, xxiv, 239–56
Dickens, Charles, 336
Dickinson, Emily, 188

didactic, the ideal vs. the, 164–66
The Different Drum: Community-Making and Peace (Peck), 267–68
digital audiotape (DAT), 98, 100
Digital Equipment PDP-1, 19
digital media: analog vs., 98; competencies for, 93; as halfway technologies, 23; morphing in, xviii; in museums, 70; photography, 48–65, 51–55
dime museums, 136
Dirty Harry, 210, 211
Discipline and Punish (Foucault), 268
disco, 102, 105–8, 109; collapse of, 222; Disco Sucks campaign and, 111; electronic, 110–11; shuffle, 109–110
Disco (Goldman), 107
disembodiment, 17
distinction, 162
Disturbing Behavior, 361
DJs. *See also* dance records; authorship and, 98, 109; break beat, 104–5, 107–8
Doane, Mary Ann, 334
docents, 84, 165–66
Docks, Les R., 230
Dodge, Raymond, 351, 364
domesticity: museums and, 162–63, 174–77, 179; pianos and, 306–310, 312–13
Double Exposure, 109
Doughan, David, 244–45
Douglas, Gilbert, 136
Downing, Al, 109, 243, 246–47
Downing, John, 242
Dramagraph, 77, 78–79, 81, 82, 92
drum boxes, 111
dub plates, 104–5
du Gay, Paul, 224
Duguid, Paul, 35
Dunbar, Robin, 120
durability, 22–23
duration vs. convenience, 304

economics, salvaging and, 32–47
Edison, Thomas, 294
Edison Company, 224
eDocent, 84
education: in letter writing, 190, 192; museums and, 164–66, 165–66, 174–78; in speed reading, 350–51; subliminal, 376–78; tachistoscopes in, 368–77; women and, 174–78
Educational Development Laboratories (EDL), 373, 375
Edwards, Ernest, 336–39, 337
Egan, Kate, xvi, xxiv, 200-21
"Electrical Questionnaire for Young Visitors," 83
Electronics Industries Alliance (EIA), 36–37
Electronic Waste Recycling Act (2003), 36
elites. *See also* class systems: museums and, 164–66; pianos and, 312–13; Elliot, Huger, 168–69

emergent forms: culture and, 53, 55–58, 240; vaudeville and, 138–42
Empire Marketing Board, 81
Empire Theatre, Edmonton, Alberta, 133, 134
The Englishwoman, 239, 249, 250
The English Woman's Journal, 244–45
entelechy, 193
Environmental Protection Agency (EPA), 25, 35, 36
The Epic of Everest, 82
Epson, 36
equivalence, 48–51; fields of, 51, 59–60; obsolescence and, 54–55; photography and, 59–60
ethnicity: dance music and, 102–4, 106; disco and, 109–11; vaudeville and, 139
The Evil Dead, 203
e-waste. *See* waste.
exchangeability, 54
experimental psychology, 364–66; speed reading and, 346–54; the subliminal and, 378
"Exporting Harm: The High Tech Trashing of Asia," 39
Expositions des Arts Décoratifs et Industriels Modernes (1925), 174
extended product responsibility (EPR), 26–27

Faith-based and Community Initiative (FBCI), 277–79
Farnsworth, Philo, 45
feminism: letter writing and, 190–91; strategies and functions of feminist press and, 245–48; women's political periodicals and, 239–56
Fessenden, Reginald, 294
fetishism, 11–12, 13; branding and, 86; methodological, 50; museums and, 92–93
Field, Kate, 124
Fikentscher, Kai, 102
film industry: infrastructure of, 140; sound films and, 142–51; studio system in, 142–43
films: audience age and, 5–6, 7; authorship and, 339–41; automatic cinema, 93; as collectibles, 200–21; cultural knowledge and, 6–7; exhibition practices and, 143–44; fragmentation vs. assemblage in, 366–68; Imax, 82; lines of association among, 8–9; literary invention and, 360–53; logic of circulation and, 334; marginalization and, 70; matinees and, 145, 147; 16mm, 168–69, 182–24, 370; museum use of, 69–96, 166–69; panoramas and, 334–35; as private property, 217; reading and, 363, 374–75; sound, 139–40, 141–51; sound, as passing fad, 146–48; sound, cost of conversion to, 150; sound, shift to, 144–46; speed reading and, 355–356; subliminal text in, 376–78; tachistoscopes and, 367–68; vaudeville and, 135, 138–51; video nasties, 200–21; video stores' reshaping of cultural time and, 5–11; visions of the future in, 361
Final Scratch software, 105

Firearms of Our Forefathers, 167
"The First Writing-Machines" (Twain), 329, 338–39
"Five O'Clock Tea" (Crawford), 307
Flash, Grandmaster, 108
Flashmeter, 372, 373
Flint, Kate, 248
Flowers for Algernon, 361
format wars, 222, 225
Foucault, Michel, 268
Found Magazine, xiv, 13
found objects, xiv, 61
Fowler, William Chauncey, 345–46
fragmentation, 364–66
Franco, Jesus, 203
Fraser, Nancy, 247, 253
Frederick, Christine, 21–22
Freed, Alan, 103
Freedman-Harvey, Georgia, 69
The Freewoman, 249, 250–51, 252
Freud, Sigmund, 351
"From the Renaissance to the Present" series, 170
Frow, John, 29
Fuller, Wayne, 186
Furse, William T., 81
Fussell, G. E., 285, 286

Galton, Francis, 347–48
Galton, Lawrence, 376
gated communities, 274–75
Gates, Bill, 35
Gateway, 36
GE (General Electric), 101–2
gemeinschaft, 265, 270
General Motors, 20–24
genetic criticism, 329
gesellschaft, 265
Gestalt psychology, 366–67
ghettoization, 248
Gibbons, Walter, 109
Gibbs, Patricia, 253
Gilbreth, Lillian and Frank, 352
Gitelman, Lisa, xxiii, xxv, 34, 329–43; on hardware vs. software, 315; on the self, 318
Gitlin, Todd, 40
globalization, 12
Godino Siamese twins, 147
Godlovitch, Stan, 308–9
Goldberger, Paul, 83
Goldman, Albert, 107
Goldmine magazine, 230–31
Goldmine Record Album Price Guide (Neely), 230
Gomery, Douglas, 144
Goodwill Computer Works, 37
Gore, Al, 35, 268, 275, 277–79
Gossip: The Inside Scoop (Leven & Arluke), 120
government: broadcasting and, 286, 291–96; communitarianism and, 257–82; communities

of practice and, 262–63; in computer disposal, 26–28; economic, 260, 263–64; liberal conceptions of, 268–69; moral economy and, 263–64; museums and, 164; natural regulation and, 268; neoliberal, 277–79; New Urbanism and, 273–77; power-geography and, 279–80; rationalities and, 259–61, 264–79, 271–77; reading speed and, 357; smart growth and, 273–76
Graham, Margaret Baker, 191
Graham, Rodney, xvii
gramophones, 72, 73, 75–76
Grasso, Walter, 107
The Greatest Generation Speaks: Letter and Reflections (Brokaw), 190
The Great Good Place (Oldenburg), 267
Green, Dave, 205
Gregson, Nicky, 215, 216
Grierson, John, 81
Griffiths, Alison, xxii, xxiv, 69–96
Grund, E., 369–70
Grusin, Richard, xxi–xxii, 49
Gurstein, Michael, 272

halfway technologies, 23, 28
Hall, Peter, 87
Hall, Stuart, 224
Hamilton, James, xxiii, xxv, 253, 283–300
Hampton, I. L., 364
handwriting, 188–89, 192. *See also* letters, personal
Hanson, Miriam, 138, 143
Harken Energy, 127
Hartley, John, 46
Harvey, Carr, 366
Hawkes, Ellison, 292, 293
Hawkins, Gay, 27
Hay, James, xxiv–xxv, 257–82
Hayles, N. Katherine, 98–99
hazardous waste: computers as, 25–28, 35–40; exporting of, 27–28, 38–39; government regulation of, 26–28; TVs as, 36
Hebdige, Dick, 226
heliotype process, 336–339
The Heliotype Process (Edwards), 337–38
Henning, Michelle, xiv, xvi, 48–65
Henry VIII, 309
Herc, Kool, 108
Herrold, Charles, 294
Hewlett-Packard, xvii–xviii, xix, 36
Hidden Cameras, xiv
Hilliard, Robert L., 294
Hilmes, Michele, 296
hip hop, 99–100, 104, 107–8
Hische, W., 369
history: acceleration of, xv; acquaintance with the dead through, xiii; in "African Voices" exhibit, 87–88; sequence in, 11–12; television and, 45–46; transformation in, xvii–xviii; video collecting and, 202, 204, 208–10; vinyl records and, 228

Hitchhike, 211, 212
HMR (Harrington Metal Recyclers), 38–39
Holga cameras, 59
home. *See* domesticity
Hooper-Greenhill, Eilean, 181n11
Hope, Bob, 149–50
Hope VI program, 274, 278
house music, 110, 111, 112
Howells, William Dean, 329, 332
Huey, Edmund, 347, 349–50, 353–54
Hunt, Lynn, xxii
Hussein, Saddam, 260, 261, 280
Hutchby, Ian, 121
Huyssen, Andreas, 179
hygiene discourses, 60–61
hypermediacy, xxi

ICAP. *See* Iraq Community Action Program (ICAP)
Illich, Ivan, 24, 28
Illuminations (Arendt, introduction to), 195
illustrated radio, 72, 74–76
Imax, 82
immediacy, xxi, 130
Imperial War Museum (London), 78
Incredible Bongo Band, 108
indie rock music, 226–27
industrialization: broadcasting and, 287; museums and, 163–64; rationalization and, 265; reading and, 345–46, 354–55; speed reading; and, 344–45, 352–53; tachistoscopes and, 369; time economy and, 319–20
inertia, videocassettes and, 9–11
informatics, 265, 271–77
Innis, Harold, xvi, 8; on bias in communication, 129, 130; on destabilization among media, 151; on perception patterns, 135; on the telegraph, 122; on the telephone, 119; on time, 325
innovation, xv–xvii. *See also* novelty: computer industry, 20–26; in *Junkyard Wars*, 43–44; McLuhan on, 304; planned obsolescence and, 21–25; profit potential of, 55
institution, rites of, 16
interactive media: branding of, 92–93; in museums, 71, 80, 82–93
International Consortium of Investigative Journalists, 127
International Spy Museum, 83
Internet: collection/coherence via, 4; community and, 268–82; cultural past and, 3–4, 12–14; disembodiment and, 17; market creation via, 3; in museums, 91–92; vinyl records and, 231; visualization of artifacts via, 12–13
interviews, telephone vs. face-to-face, 116
In the Year 2889 (Verne), 124
intimacy: collecting and, 193; commodification of, 130; false, in telephony communication, 129
Iraq, communitarianism and, 257–62, 279–80

Iraq Community Action Program (ICAP), 259, 278–79
I Spit on Your Grave, 207, 217–18

Jamaica, urbanization in, 103–4
James, Henry, 345
James, William, 366
Janes, Linda, 224
Janet, Pierre, 366
Jastrow, Joseph, 348, 349
Javal, Louis Émile, 347, 369
The Jazz Singer, 144, 149
Jim Will Fix It, 103
Jolson, Al, 149
journalism: influence of telephone use in, 115–32; pack, 127–29; popular culture representations of, 125
"The Joy of Letter Writing" (Burkett), 189
Judd, Charles, 352
jukeboxes, 102
Junkyard Wars, 34, 40–46

Kafka, Franz, 188
kaolotype process, 336–37
Katz, James E., 117
Keene, Suzanne, 91–92
Keitel, Harvey, 8
Keith, B. F., 136–38, 140
Keith, Michael, 294
Kelly, Wayne, 305
Kennedy, John F., 375
Kent, H. W., 172, 175
Kenyon, F. G., 72
Kerekes, David, 204, 210
Kern, Stephen, 380
Keyes, Cheryl L., 103–104
Keystone View Company, 372, 373, 375, 376
Kittler, Friedrich, xvi, 322, 329, 365, 369; on media as storage, 51; on the tachistoscope, 378–79
Klee, Paul, 60
Klinger, Barbara, 200–202, 206, 217
Kluge, Alexander, 247
Knightley, Phillip, 127
knowledge: collecting and "insider," 205–6, 207–10, 212; cultural, 7, 9–11; democratization of, 71; exclusive, collecting and, 105; letters as repositories of, 189–91; television and production of, 45; vinyl records and, 230
Knuckles, Frankie, 106, 110
Kolstrup, Søren, 241
Koszarski, Richard, 142
Kracauer, Siegfried, xvi
Kraftwerk, 110
Kruger, Paul, 127
kymographs, 348

Ladies Home Journal, 170, 339
Lane, Nathan, 9
Langdon, Philip, 273

language, 123; effects of telephony on, 123; letters and, 195–96; memory and, 195–96; reading and, 346–54, 357
The Language of New Media (Manovich), 34
Laporte, Dominique, 26–28
The Last House on the Left, 203
Laurie, Joe, Jr., 149
Lear, Norman, 41
The Learning Channel, 40–46
Leary, Timothy, 106
Le Baron, William, 147
leisure: museums and, 179–80; music making and, 304, 314–17; skill vs. ease and, 318; vaudeville and, 138–39; Lensbaby, 59
letters, personal, 185–99, 339; absence and presence in, 186–88; as art, 191–92, 196–97; as collectibles, 192–95; handwriting and, 188–89; knowledge and memory in, 189–91; love letters, 189; as objects, 188–89
Levan, Larry, 106
Leven, Jack, 120
Levine, Ian, 110
Levine, Lawrence, 164
Lewis, E. O., 364
Lewis, Jacob, 142–43
Library of Congress, xvii
Lichtenstein, Roy, 382n55
lie detectors, 372
Life on the Mississippi (Twain), 329–43
Lindgren, Ernest H., 81–82
listening circles, 75
literary production, 329–43; visual culture and, 330–31, 339–41
Liveness: Performance in a Mediatized Culture (Auslander), 328n50
Lobel, Michael, 382n55
Loew, Marcus, 142
logic of circulation, 334
Lomo camera, 56–57
London, Barry, 5, 6
London School of Economics exhibition, 81
London Science Museum, 93
longlostperfume.com, xiv, 3–4, 12–14
Lost and Found, xiv
Love Is the Message, 110
Love to Love You Baby, 110
Lowe, E. E., 79, 80, 81
LP records, 101–2
Lull, James, 226
Lund, Katia, 10
luxury, pianos and, 312–13
Lyons, Martyn, 187

MacDonald, Dwight, 265
Macfadden, Bernarr, 353
Machine Art show (1934), 177–78
Mackay, Hugh, 224
Macy's design exposition, 176–77
magazines: abolitionist, 287–90; museums and, 174–78; women and, 239–56
magnetic recording, 52

maintenance, 22–23, 233
Maleuvre, Didier, 70
The Manchurian Candidate (Condon), 361
Mancuso, David, 106
Manifest Destiny, 291
Mann, Horace, 288
Manovich, Lev, 34
Marey, Etienne-Jules, 347, 351
marginalization: dance records and, 102–104; of nontheatrical film, 70; obsolescence and, 54–58
marketing: commodification and, 92–93; of digital cameras, 52–53; disco and, 109; exclusivity/scarcity in, 201; Internet markets in, 3; speech over music as, 103; video collecting and, 201
market saturation, 28
Mark of the Devil, 212
Marsden, Dora, 249
Martin, Michele, 124
Marvin, Carolyn, xxiii, 33–34, 124
Marx, Karl, xx, 57
mass media, 265; broadcasting as, 287–91; speed reading and, 344–360; time bias in, xvi
materiality: of feminist media, 248; history and, xx–xxii; of letters, 188–89, 191–92; literary production and, 336–341; of television, 45; transformation and, xvii–xvix; of videocassettes, 210, 211
Mayer, Louis, 147
McAllister, Henry, 352
McCallum, Doug, 145
McCarthy, Anna, 86
McCoy, Patrick, 61
McCracken, Grant, 228, 233
McDermott, Richard, 257
McGregor, Judy, 116
McLuhan, Eric, 323–24
McLuhan, Marshall, 119, 303; on the body, 324–25; emergent media laws of, 323–24; on obsolescence, 304, 320–21; on the telephone, 120–21, 127
meaning, 90–91, 346–54
Medavoy, Mike, 6
media: hypermediacy of, xxi–xxii; old vs. new, 33; storage, accumulation via, 12–14, 51; transparency/immediacy of, xxi–xxii
media studies, 240–43
mediation, 257–282, 328n50
Meirelles, Fernando, 10
memory: collecting and, 194–95; collective, 190; cultural, videocassettes and, 8; embedded, 3–15; language and, 195–96; letters and, 185–99; nostalgia vs., xiv; souvenirs and, 195; video collecting and, 204–5
mental hygiene movement, 357
Merritt, Russell, 140
Metaculture (Urban), 4–5
Metronoscope, 370
Metropolitan Museum of Art, 159–84; film use by, 168–69; radio show, 176

"Metropolitan Museum of Art Miniatures," 159–62
microfilm, 356
Microsoft Encarta, 85
military discourses, in *Junkyard Wars*, 43–44
Millard, Andre, 224–25
Mingay, G. E., 284
miniature golf, 146
Minow, Newton, 40, 45
minstrel shows, 133, 135–36, 139
"Mississippi from the Mouth of the Missouri to New Orleans" (Banvard), 334
mobility, 4–5, 8–9
modernity: attention and, 365–66; community and, 264, 265–79; literary production and, 330–43; museums and, 163–64; pianos and, 312–13
Modern Times, 148
Mohammed, Faduma, 88
Monitors (Dekovic), 32–33
monopolies, 24–25
Moondog Coronation Ball, 103
Moore, Bobby, 109
Moore, Gordon, 19–20, 35
Moore's law, 19–20, 35
morality: government and, 263–64; leisure and, 318; New Urbanism and, 275–76; vaudeville and, 141
Moroder, Giorgio, 110
morphing, digital, xviii
Morrison, Bill, xvii
Moulton, Tom, 109
"Movies and Museums," 76, 80
"Moving Image as Art: Time-Based Media in the Art Gallery," 69
MP3 format, 97, 105, 112
Mullin, Jack, 52
Mumford, Lewis, 122
Münsterberg, Hugo, 346
Mursell, James, 355–56
Museum of Modern Art (MOMA), 80, 170, 177–78
museums, 69–96; advertising and, 163, 172–78, 180; architectonics and, 79–80, 87–88; architecture of, 91–92, 164–65; authority and, 165, 178–79; branding by, 86, 178; conceptualizations of, 76; film use by, 76–81, 166–69; home influenced by, 162–63, 179; illustrated radio in, 72, 74–76; merchandising by, 159–84; newspapers and, 169–70; private vs. public spheres and, 162–63; radio and, 169, 176; reproductions by, 74–75, 171–78; television and, 170–71; timelessness and, 165, 180; urban space and, 163–64, 184–77
Museums Journal, 76, 77
Musicking: The Meanings of Performing and Listening (Small), 304
music making, 303–28; authenticity in, 314–17; digital, 322–25; history of the keyboard in, 305–9; mechanization of, 316,

317–20; the physical in, 322–25; time in, 310–11, 313–15, 325
music programming, 109
Muybridge, Edweard, 351, 367
My Polar Adventures with Shakleton in the Antarctic, 82

narrative, 90–91, 196. *See also* literary production
Nash, Christopher, 83
National Association of Record Manufacturers (NARM), 223
National Broadcasting Company (NBC), 293
National Council for the Training of Journalists, 116
National Film Library, 81–82
nationalism, 285
National Museum of Australia, 87
National Museum of Natural History (NMNH), 70–71, 82–93, 88–90
National Public Radio, xiv
natural coding, 54
Neely, Tim, 230
Negt, Oskar, 247
Negus, Keith, 224
neoliberalism, 277–79
neotraditionalism, 273
Nerone, John, xxii
Nessheim, Ragnild, 249–50
Networks in the Global Village (Wellman), 270
Newhall, Sidney, 364
new media: content of, 322, 323; defining, 18, 22–25; mythologization of, xviii–xvix; as social transformation, 49–50
New Media, 1740–1915 (Gitelman & Pingree), xxiii, 34
news media: audience roles in, 119; battery hen model of news and, 127–29; content influences on, 115–16; falsification in, 128–29; gossip and, 119–20; importance of orality in, 129–30; interpretation of, 248–49; news cycle in, 122–23; shorthand in, 338; telegraphy and, 120–23; telephone reporting in, 115–32; via telephone, 115–32
newspapers: abolitionist, 287–90; museums and, 169–70; telephony and, 124–27; women and, 239–56
newswire services, 122–23
New York City Tract Society, 287
New York Times, 163, 374; fabricated news in, 128–29; museums and, 170
nickelodeons, 140–41
Noranda, 37
Nord, David Paul, 287
North, Michael, 356
Northern Soul, 105
The North Star, 287–90
nostalgia, xiv–xv, 56–57, 62, 210–12
novelty, xv–xvii, 40, 139, 224
nuclear waste, 40

Obscene Publications Act (OPA) of 1959, 203
obsolescence: of computers, 20–26; definition of, 223–24; devaluation in, 24–25; durability and, 22–23; equivalences and, 54–55; of films, video stores and, 5–11; forced/engineered, 22–23; hygiene discourses and, 60–61; marginalization and, 55–58; photography and, 48–65; pianos and, 320–24; production of, 51–55; progress and, 21–22; remediation and, 53; salvaging and, 33–47; social, 51; as spatial problem, 25–26; structured, 34, 35–40; stylistic, 20–21; technological, 20, 21–22; utopianism and, 57–58; value and, 53–54; videocassettes in slowing of, 11; vinyl records and, 223–30; waste and, 16–31, 60–61
Oettermann, Stephan, 330
Oklahoma City National Memorial, 83
Oldenburg, Ray, 267, 274
"Old Times on the Mississippi" (Howells), 329
"Old Times on the Mississippi" (Twain), 331,332
Olins, Wally, 86
O'Malley, Tom, 242
Ophthalmograph, 370
organicity, 265–66
organizational categories, in collecting, 202, 204, 211–12
organ (musical instrument), 309
Osgood, James Ripley, 336–38
other, the, 106
Outposts of Empire, 82

Pacey, Arnold, 23, 28
Palmer, Bill, 52
Pankhurst, Christabel, 250
Pankhurst, Emmeline, 250
panopticon, 342–17
panoramas, 333–335, 342n17, 343–23; Twain and, 332–33, 334–35
The Parallax View, 361
Parks, Lisa, xviii, xxiv, 32–47
participation, coercive, 24–25
Pastor, Tony, 136
Payne Fund Studies, 76–78
Peck, M. Scott, 267–68
perception, 135; errors in, 351; fragmentation vs. synthesis in, 363–64; frames for, 3; psychology of, 364–66; reading and, 346–54
performance: mediation of, 328–50; pianos and, 316, 317–20; recording technologies and, 315–17; vaudeville and individual, 137–38
Perskyi, Constantin, 44
Peters, John Durham, 283
Phelps, Roswell, 338
photography, 48–65; digital, 48; found, 61–62; lo-fi, 56–57; museums and, 163, 171; obsolescence and, 51–55; panoramas and, 343–23; pinhole, 59–60; as social practice, 50; transformative power of, 58–62

physicality, 17
pianos, 303–28; development of, 305–9; manufacturing of, 305–6; mechanical, 308–16; obsolescence and, 320–24; performance and, 316, 317–20; player, 314–16, 318–19; sheet music and, 311; time investment and, 313–15; women and, 307–10
Pingree, Geoffrey B., xxiii, 34
Pippin, Steven, 60
Pitkin, Walter B., 354–55, 357
political organization, broadcasting in, 287–90
political periodicals, women's, 239–56
Pool, Ithiel de Sola, 269
Pope John Paul II Cultural Center, 91
Porter, Charles A., 188
possession rituals, 232
The Postcard: From Socrates to Freud and Beyond (Derrida), 187
postmodernism, 71
Powell, Morgan, 147
power: broadcasting and, 284, 296; communitarianism and, 276–79; -geography, 279–80
Prelinger, Rick, xvii
The Prince and the Pauper (Twain), 336
private vs. public spheres: counterpublics and, 253; museums and, 162–63; music making and, 308–9; social movements and, 247; telephonic communication and, 130; video collecting and, 200
progress, 60; agricultural broadcasting as, 285–86; Benjamin on, 59; mechanization as, 319–20; obsolescence and, 21–22
psychology: Gestalt, 366–67; reading and, 346–54, 364–66; subliminal text and, 376–78
Psychology and Industrial Efficiency (Münsterberg), 346
The Psychology and Pedagogy of Reading (Huey), 353–54
Psychology of Vision, 371
psychometric testing, 347–49, 353
The Psychopathology of Everyday Life (Freud), 351
public sphere. *See* private vs. public spheres
Putnam, Robert, 268–70

Quinnell, Justin, 60

Rachmaninoff, Sergei, 316
radical monopolies, 24–25
radio, 19; illustrated, 72, 74–76; museums and, 72, 74–76, 169, 176; piano production and, 320; reverse adaptation and, 322; use of the telephone in, 127–28
Radio Photologues, 74, 169
Rare Groove, 105
rarity, in collecting, 212–13
rationalization, 259–61, 264–79; broadcasting and, 291–92; of community, 271–77; fragmentation and, 364–66; of government, 271–77; of readers, 354–55; reading and, 364–66
RCA-Victor, 102, 293, 296
readability theory, videocasettes and, 9–11
reading: attention and, 365–66; communities, 248; comprehension and, 350–51; film and, 355–356, 363; heredity and, 359–60–36; hyper-, 363; mechanization of, 345–60; physiology of, 340–41; quality of fast, 345–46; social implications of, 353–54; speed, 344–60; subliminal text, 376–78; tachistoscopes and, 361–84; vision studies and, 346–54; as visual process, 340–41
reality television, 41–42
À la Recherches du temps perdu (Proust), 13
recombinant culture, 11, 40
recontextualization, 229
recording industry: analog vs. digital, 100–101, 222–36; development of, 224–25; globalization and, 12; obsolescence of vinyl by, 222–23
recording technologies, 52–53, 320. *See also* pianos; authenticity and, 315; gramophones, 72, 73, 75–76; magnetic sound, 52–53; performance and, 315–17; videocasettes, 5–11, 200–21
recycling, 36, 39–40
reform, broadcasting and, 286–91
remediation, xxi–xxii, 48–51; definition of, 3, 49; equivalences and, 51; obsolescence and, 53; social practice and, 50
Renshaw, Samuel, 364, 370–71
reporting, news. *See* news media
repurposing, 38–40
residual media, definition of, xix–xx
residuals, 33, 35. *See also* waste
retro, xiv, 14
reverse adaptation, 321–22
rewind technique, 104
Rheingold, Howard, 266–68, 269, 273, 275
"Rheinmetall/Victoria 8" (Graham), xvii
Richards, I. A., 344
Rietveld, Hillegonda C., xv, xxiv, 97–14
rites of institution, 16
Robinson, Edward, 170
Roell, Craig H., 314, 315
Rogoff, Iris, 165
The Romance and Reality of Radio (Hawkes), 292, 293
Romero, George, 208
Ronell, Avital, 117, 129
Roosevelt, Theodore, 375
Rose, Nikolas, 260
Rosenzweig, Roy, 138, 140
Rosing, Boris, 45
Ross, Andrew, 274
Rossetti, Dante, 330
Roto-Radio Talks, 74, 169
Rourke, Mickey, 8
Royal, John F., 149

rubbish theory, xvi
Rubinstein, Daniel, 97, 105
Ruggles, Wesley, 147
Rumarson Technologies, 37
Rumsfeld, Donald, 280n4
Ryan, Leila, xxii, xxiv, 239–56

sacred, collecting and, 231
salvage, 32–47. *See also* waste; definition of, 34 self-regulation in, 36–37; vinyl records and, 230, 231
sampling, digital, 111, 304–5, 322–24
Sanford and Son, 41
satellite telephones, 118–19, 123
Saturday Night Fever, 110
Saussure, Ferdinand de, 347
Savile, Jimmy, 103
Scannell, Paddy, 283
Schlosberg, Harold, 364
Schmid, Joachim, 61
Schuler, Doug, 271, 272
Schultz, Lucille M., 190
Schwartz, Vanessa R., xxii, 330
Scientific American, 78
Scrapheap Challenge, 41
Scully, Vincent, 273
Seattle Community Network, 271–72
Sedorkin, Gail, 116
self, the, 366
self-help, 353, 357
Selling Mrs. Consumer (Frederick), 21–22
Seltzer, Mark, 363
The Shawshank Redemption, 8
Sheen, Charlie, 8
sheet music, 311
Sherman, Daniel, 165
Sherman, Hoyt, 382–55
Sherman, Larry, 112
Sherry, John, 225, 231
shorthand, 338
Siano, Nicky, 106
Siepmann, C. A., 75
Silicon Salvage, 37
Silicon Valley Toxic Waste Coalition, 39
Simley, O. A., 378
Slater, David, 204, 210, 217
Slater, Jay, 210
Sloan, Alfred, 21
slow mixes, 107
Small, Christopher, 304
smart growth, 273–76
Smith, Henry, 371
Smith, Seymour, 372
Smith, Virginia, 91, 92
Snowden, Collette, xvi, xxiv, 115–32
Snyder, William M., 257
Sobchack, Vivian, xviii
social capital, 268–70
social class. *See* class systems
social computing, 84

A Social History of the English Countryside (Mingay), 284
social necessity, 102–3
social networks. *See also* communitarianism: broadcasting as, 283; communities as, 269–73; news reporting and, 119–20
social shaping theory, 224
Soloman, Martha, 248
The Song of the Sower (Bryant), 284, 288
Sony Walkman, 224
Soul Sonic Force, 110
sound clashes, 104
sound recordings: CDs, 97, 113, 222–23; DAT, 98, 100; 12-inch dance singles, 97–114; vinyl records, 222–36
Sousa, John Philip, 318, 319, 320
souvenirs, xiv; collections vs., 193–94; memory and, 195
space: community and, 264, 270–82; dissolution of in communication, 122–23; museums and, 172–78; music making and, 310–11; narrative and, 196; urban, 163–64, 184–77
Sparke, Penny, 313
Spiegel, Stacey, 84
Spigel, Lynn, 170
SS Experiment Camp, 203
statistics, 348
Stead, W. T., 253
Steele, William, 352
Stein, W., 370
Sterling, Bruce, xx, 224
Sterne, Jonathan, xiv, xviii, xxiii–xxiv, 16–31
Stewart, Jon, xvi
Stewart, Susan: on aesthetics and collecting, 209–10; on organization and collecting, 202; on remembering, xiv; on souvenirs, 193–94, 195, 197
Stober, JoAnne, xxii, xxiv, 133–55
Stone, Sandy, 17
"Stories of Everyday Living: The Life and Letters of Margaret Bruin Machette" (Graham), 191
Stranger in a Strange Land, 361
Strasser, Susan, 20
strategies, 141, 245–48
Straw, Will, xiv, xxiii, 3–15, 26, 223
Streamline Your Mind (Mursell), 355–56
Strong Eyes (Macfadden), 353
style, museums and, 174–78
subliminal text, 376–78
suffrage press, 239–56; as alternative media, 246–47; comparison of oppositional periodicals in, 250–54; on Conciliation Bill, 250–54; positioning and interpreting, 248–49; strategies and functions of, 245–48
Summer, Donna, 110
supermodernity, xv

surrealism, 61
Swanwick, Helena, 247

tachistoscopes, 361–84; advertising and, 371–72; fragmentation vs. assemblage and, 364–68
tactics, 141
tactility, of vinyl recordings, 98, 100–101
Talbott, Steve, 326
Tannock, Stuart, xiv
"Tap" (Buckhouse), 84
Tapsall, Suellen, 127
Tashiro, Charles, 201–2, 209, 212
Tate, Theodore, 371
Tate International Council Conference (2001), 69
Taylor, Carl C., 370
Taylor, Earl, 370
Taylor, Frederick, 319
Taylor, James Y., 370
Taylor, Stanford, 376
techno-fatique, 93
technoideological coding, 54
Telefon Hirmondo, 124, 125, 295
telegraphy, 18, 121–23
The Telephone Herald, 124
telephony, 18, 19; broadcasting and, 294; as broadcast medium, 124–25; content influenced by, 115–116; convergence of with other media, 118–19, 128; effects of on communication, 120–21; fabricated news and, 128–29; false intimacy of, 129; fears in early, 293; as machine, 116; newspapers and, 124–27; in news reporting, 115–32; as overlooked medium, 115–16; in radio, 127–28; as residual media, 116–17; satellite, 118–19, 123; studies of, 117–18; in Watergate reporting, 126
television, 18; development of, 44–45; hardware as trash, 36; *Junkyard Wars*, 34, 40–46; LP records and, 101–2; museums and, 170–71; New Urbanism and, 275–76; significance of, 34; variety shows, 149–50; vaudeville's influence on, 139, 148–50; as wasteland, 40–46
"Theses on the Philosophy of History" (Benjamin), 60
Thompson, Michael, xvi, 22–23
Thorne, Christian, 14
Thornton, Sarah, 223, 226
time: cultural change and, 4–5; duration vs. convenience and, 304; historical change and, xvii–xviii; industrialization and, 319–20; lags in cultural change, xx–xxi, 254; mass media bias toward, xvi; museums and timelessness, 165, 180; music making and, 310–11, 313–15, 325; narrative and, 196; subliminal learning and, 378; videocassettes and reordering of, 9; video stores in reshaping of cultural, 5–11
Titanic, sinking of the, 253

Tonnies, Ferdinand, 265
Toop, David, 108
Toqueville, Alexis de, 21
Trachtenburg Family Slideshow Players, xiii–xiv
tract societies, broadcasting and, 287
Trans-Europe Express, 110
transformation, xvii–xvix, 49–50, 58–62
trash. *See* waste
Tribulation 99; Alien Anomalies under America, xvii
Tubby, King, 104
Tuchman, Gaye, 248–49
Tunstall, Jeremy, 127
turntables, twin, 103, 108–9
Twain, Mark, 329–343
typing and typewriters, 329–43, 378–79

underground dance music (UDM), 106–7
University of Florida, 36
Urban, Greg, xvi, 4–5, 9
urbanism: delay and, 5; museums and, 163–64, 184–77; new, citizenship and, 273–77;
U-Roy, 104;
Urrichio, William, 44;
U.S. Agency for International Development (USAID), 258, 259, 261
U.S. Holocaust Memorial Museum, 90
USAID. *See* U.S. Agency for International Development (USAID)
"Useful Household Objects under Five Dollars" show, 178
utopianism, 57–58

Valentine, C. W., 364
value: circuits of, xv; collectibles and, 201, 204, 209–10, 212–16; computers and, 23–24; obsolescence and, 24–25, 53–54; of old vs. new media, 50–51; reproductions and, 233–34; video collecting and, 204; vinyl records and, 233–34; waste and the creation of, xvi, xvii
Varigny, Henry, 349
Varley, Carolyn, 127
vaudeville, 133–55; circuits, 145; dominant, residual, and emergent forms in, 138–42; films and, 138–51; history of, 135–38; individual performance in, 137–38; morality and, 141; program organization in, 139; revival of after sound films, 144–45, 148–50; sound films and, 139–40, 142–51; target audiences of, 133–34, 136, 137, 141–42; television and, 148–50; variety entertainment and, 134–35, 136–37
Verne, Jules, 124
Victoria and Albert Museum, 181n10
Victorian studies, 243–45
Victor Talking Maching Company, 224
videocassettes: cultural memory and, 8; cultural time and, 5–11; length of actors' careers and, 8; mobility of, 8–9; reordering of time via, 9

video nasties, 200–21; classification of, 211–12; definition of, 203; DVD availability of, 217–18; knowledge base in, 205–206, 207–10, 212; market economies and, 213–16; nostalgia and, 210–212, 220–40; personal rituals in collecting, 205–6; renovation of, 216–17; versions of, 207–10
Video Recordings Act (VRA) of 1984, 203
video stores, cultural time and, 5–11
vinyl records, 222–36; acquisition of, 229–32; auditioning rituals for, 231–32; authenticity and, 225, 226–27; collectivizing rituals with, 232; dance records, 48–65; format wars and, 222, 225; history and collecting, 228; novelty vs. fidelity and, 224; obsolescence process of, 222–23; past orientation and, 228; personal rituals and collecting, 229–34; personal significance of, 228; playing, 226; possession of, 232–33; preservation of, 233; reproductions of, 233–34; use rituals with, 233–34
Virilio, Paul, 380
The Virtual Community: Homesteading on the Electric Frontier (Rheingold), 266–68, 273, 275
virtuosity, 315
A Visit to the Armor Galleries of the Metropolitan Museum of Art, 167
visual culture: literary production and, 339–41; of museum artifacts in mass media, 74–75; storage media and, 12–14
"Vital Space" exhibit, 84
Volkmann, A. W., 362
Volosinov, V. N., 50
volunteerism, 272
The Vote, 251
Votes for Women, 239, 250, 251

Walker Art Center, 84
Wallendorf, Melanie, 225, 231
Walsh, Nonee, 128
Ward, Brian, 107
War of Desire and Technology at the Close of the Mechanical Age (Stone), 17
Wasson, Haidee, xxii, xxiv, 159–84
waste, xxiii–xxiv; computers as, 23–28, 35–40; economic growth and, 20–26; exporting of, 27–28, 38–39; government intervention in, 26–28; hazardous, computers as, 25–28; hygiene discourses and, 60–61; nuclear, 40; obsolescence and, 16–31; salvaging and, 32–47; on television, 40–46; transformation to, xvii–xviii; value attributions and, xvi, xvii
Watergate story, 126
Watkins, Evan, 51, 54, 56–57
Watson, Janelle, 5
Weber, Max, 309
Welch, Richard, 377–78
Wellman, Barry, 270
WELL (Whole Earth 'Lectronic Link), 266–67
Wenger, Etienne, 257, 262
Wertheimer, Max, 366
When Old Technologies Were New (Marvin), xxiii, 33–34
Whitney Museum of American Art, 84
Williams, Raymond: on continuity, 134; on cultural forms, 223–24; on dominant, emergent, and residual culture, xx–xxi, 55–56, 59, 240; on media as concept, 49–50; on participation, 296–97; on political and economic description, 283
Wilson, Frank, 105
Winner, Langdon, 321–22
Winston, Brian, 101, 103, 125
With Captain Scott in the Antarctic, 82
Wittgenstein, Ludwig, 363
Wolfe, Cary, 326
Woman's Dreadnought, 249
Woman's Home Companion, 172, 175
women: domesticity discourses and, 174–77; letter-writing by, 190–191; media's influence on roles of, 241–42; museums and, 174–78; pianos and, 303–28; print media and, 239–56; studies on feminist/suffrage media of, 243–45
Women's Franchise, 249
Women's Freedom League, 252
Women's Social and Political Union, 248, 249, 251, 252
Women's Suffrage Journal, 249
Wood, Evelyn, 357, 376
work ethic, 314
"The Work of Art in the Age of Its Technical Reproducibility" (Benjamin), 58
Wright, Almroth, 250, 252
Wundt, Wilhelm, 340, 362, 366

Yearning for Yesterday (Davis), xiv
Yorkin, Bud, 41
Young, Earl, 110

Zworykin, Vladimir, 45